Chemical and Functional Properties of Food Components Series

Food Flavors

Chemical, Sensory and Technological Properties

Chemical and Functional Properties of Food Components Series

SERIES EDITOR
Zdzisław E. Sikorski

Food Flavors: Chemical, Sensory and Technological Properties
Edited by Henryk Jelen

Environmental Effects on Seafood Availability, Safety, and Quality
Edited by E. Grażyna Daczkowska-Kozon and Bonnie Sun Pan

Chemical and Biological Properties of Food Allergens
Edited by Lucjan Jędrychowski and Harry J. Wichers

Food Colorants: Chemical and Functional Properties
Edited by Carmen Socaciu

Mineral Components in Foods
Edited by Piotr Szefer and Jerome O. Nriagu

Chemical and Functional Properties of Food Components, Third Edition
Edited by Zdzisław E. Sikorski

Carcinogenic and Anticarcinogenic Food Components
Edited by Wanda Baer-Dubowska, Agnieszka Bartoszek and Danuta Malejka-Giganti

Methods of Analysis of Food Components and Additives
Edited by Semih Ötleş

Toxins in Food
Edited by Waldemar M. Dąbrowski and Zdzisław E. Sikorski

Chemical and Functional Properties of Food Saccharides
Edited by Piotr Tomasik

Chemical and Functional Properties of Food Lipids
Edited by Zdzisław E. Sikorski and Anna Kolakowska

Chemical and Functional Properties of Food Proteins
Edited by Zdzisław E. Sikorski

Chemical and Functional Properties of Food Components Series

Food Flavors

Chemical, Sensory and Technological Properties

EDITED BY

Henryk Jeleń

CRC Press is an imprint of the
Taylor & Francis Group, an **informa** business

CRC Press
Taylor & Francis Group
6000 Broken Sound Parkway NW, Suite 300
Boca Raton, FL 33487-2742

First issued in paperback 2016

Version Date: 2011912

ISBN 13: 978-1-138-03497-6 (pbk)
ISBN 13: 978-1-4398-1491-8 (hbk)

Library of Congress Cataloging-in-Publication Data

Food flavors : chemical, sensory and technological properties / [edited by] Henryk Jelen.
p. cm. -- (Chemical & functional properties of food components)
Includes bibliographical references and index.
ISBN 978-1-4398-1491-8 (hardback)
1. Flavor. 2. Food--Analysis. 3. Food--Composition. 4. Food--Biotechnology. I. Jelen, Henryk.

TX531.F66 2011
664--dc23 2011022870

Visit the Taylor & Francis Web site at
http://www.taylorandfrancis.com

and the CRC Press Web site at
http://www.crcpress.com

Contents

Preface

Flavor is one of the main food sensory attributes of crucial importance for consumers acceptance of food. Therefore, it attracts the attention of not only food technologists, but also psychologists and neurophysiologists. Flavor compounds are challenging ones to investigate for chemists and biochemists.

This book was intended to provide a concise one volume selection of flavor topics especially important for food technologists and students in food technology/chemistry, who are main target reader groups, and all those who want to have a starting point in a more in-depth exploration of the field of food flavors.

Having this in mind, the book chapters can be grouped into five areas generally focused on the following aspects: *i)* introductory information on flavor compounds and odor and taste perception; *ii)* basics of aroma compounds formation; *iii)* flavor compounds specificity in food technology; *iv)* examples of flavors of selected foods; and *v)* analytical approaches to characterize food flavor compounds.

Chapters 1 and 2 provide an introduction into the chemistry of food odorants and food tastants, whereas Chapter 3 covers the area of flavor perception and provides fundamentals, as well as recent accomplishments in this field. Chapters 4 through 6 are organized based on flavor precursors (lipids, carbohydrates, and aminoacids), presenting universal mechanisms and pathways in aroma compounds biogenesis or formation of process flavors. This gives readers a broad outlook of the common points in the formation of flavors and should help to understand the process of flavor formation in technological processes. Chapter 7 is related to interaction of food matrix with aroma compounds in the process of their binding and release, whereas Chapter 8 describes an significant issue of flavor suppression and enhancement, important especially in functional food production and flavor perception. To guide reader through the legislative meanders of food flavors and flavorings, Chapter 9 provides important data in this respect. Chapters 10 and 11 are also helpful for food technologists, providing information on spices and essential oils, and functional properties of flavor compounds. Chapters 12 through 15 provide insight into various food products and their characteristic aroma. Because of the ample variety of food products with a distinct flavor and, simultaneously, because of the limited space in the book, a choice of products must be a compromise. Cheese flavors, flavor characterization of meat, odorants in wines, and formation of flavor in bread and bakery products represent diversified character of flavor and aroma compounds, their formation, sensory implication, and the roles of microorganisms and technological processes in their formation. Chapter 16 describes the problems of food taints and off-flavors, their origin in foods, and the strategy for their identification. The last part of the book, comprising four chapters, is devoted to analytical aspects. Chapter 17 describes the use of volatile compounds in food authenticity and traceability testing, whereas Chapter 18 presents analytical approach to the determination of key aroma and taste compounds that play a crucial role in formation of food flavor. This chapter links results of instrumental analysis with sensory impressions. Chapter 19 is focused on the techniques used in the sensory characterization of food, whereas Chapter 20 provides the idea, theory, and instrumentation of machine olfaction.

The chapters in this book have been written by specialists from academia and industry based on their teaching and research experience and contain both fundamentals, required to understand basic processes in flavor chemistry/biochemistry and flavor perception and the evaluation of literature to present recent trends in flavor research.

I hope that information provided in the book shall give an outlook of the various aspects of flavor chemistry to the novices in the field, as well as useful information for more experienced readers, and can be a concise starting book for all interested in a role of flavors in food industry.

Editor

Henryk Jeleń received his MS, PhD, and DSc from the Faculty of Food Science and Nutrition, Poznań University of Life Sciences, Poznań, Poland, where he holds a position of professor. He spent his postdoctoral fellowship at the University of Minnesota, worked also as a visiting professor at North Dakota State University, and completed several short term trainings and research assignments at various European universities and institutes. His scientific interests and teaching are focused on food chemistry, mainly flavor chemistry, sample preparation, chromatography and mass spectrometry in food analysis, especially of volatile/flavor compounds. He is a member of the Committee on Food Sciences of the Polish Academy of Sciences, the Chromatography and Related Techniques Commission at the Committee of Analytical Chemistry of the Polish Academy of Sciences, and ACS. He has published nearly 80 journal papers and 8 book chapters.

Editor

Contributors

Javaid Aziz Awan
National Institute of Food Science and Technology
University of Agriculture
Faisalabad, Pakistan

Tomas Cajka
Department of Food Chemistry and Analysis
Institute of Chemical Technology
Prague, Czech Republic

Chris R. Calkins
Animal Science Department
University of Nebraska
Lincoln, Nebraska

Corrado Di Natale
Department of Electronic Engineering
University of Rome Tor Vergata
Rome, Italy

Karl-Heinz Engel
Lehrstuhl für Allgemeine Lebensmitteltechnologie
Technische Universität München
Freising, Germany

Vicente Ferreira
Department of Analytical Chemistry
University of Zaragoza
Zaragoza, Spain

Elisabeth Guichard
UMR CSGA (Centre des Sciences du Goût et de l'Alimentation)
INRA (Institut National de Recherche Agronomique)
Dijon, France

Jana Hajslova
Department of Food Chemistry and Analysis
Institute of Chemical Technology
Prague, Czech Republic

Jennie M. Hodgen
Ruminant Technical Services
Intervet/Schering Plough Animal Health
DeSoto, Kansas

Thomas Hofmann
Lehrstuhl für Lebensmittelchemie und molekulare Sensorik
Technische Universität München
Garching, Germany

Thomas Hummel
Department of Otorhinolaryngology
University of Dresden Medical School
Dresden, Germany

Henryk Jeleń
Faculty of Food Science and Nutrition
Poznań University of Life Sciences
Poznań, Poland

Danuta Kalemba
Faculty of Biotechnology and Food Sciences
Institute of General Food Chemistry
Łódź, Poland

Gerhard Krammer
Symrise AG
Global Innovation Flavor & Nutrition
Holzminden, Germany

S. P. D. Lalljie
Unilever Safety and Environment Assurance Center
Bedfordshire, United Kingdom

Jakob Ley
Symrise AG
Global Innovation Flavor & Nutrition
Holzminden, Germany

Manfred Lützow
saqual GmbH
Wettingen, Switzerland

Małgorzata Majcher
Faculty of Food Science and Nutrition
Poznań University of Life Sciences
Poznań, Poland

Katja Obst
Lehrstuhl für Allgemeine Lebensmitteltechnologie
Technische Universität München
Freising, Germany

Salim-ur-Rehman
National Institute of Food Science and Technology
University of Agriculture
Faisalabad, Pakistan

Katharina Reichelt
Symrise AG
Global Innovation Flavor & Nutrition
Holzminden, Germany

Kathy Ridgway
Reading Scientific Services Limited
Berkshire, United Kingdom

Felipe San Juan
Department of Analytical Chemistry
University of Zaragoza
Zaragoza, Spain

Han-Seok Seo
Department of Otorhinolaryngology
University of Dresden Medical School
Dresden, Germany

Peter Schieberle
Lehrstuhl für Lebensmittelchemie
Technische Universität München
Garching, Germany

Takayuki Shibamoto
Department of Environmental Toxicology
University of California–Davis
Davis, California

Henry-Eric Spinnler
UMR de Génie et Microbiologie des Procédés Alimentaires
AgroParisTech/INRA
Thiverval-Grignon, France

Anna Wajs
Faculty of Biotechnology and Food Sciences
Institute of General Food Chemistry
Łódź, Poland

Erwin Wąsowicz
Faculty of Food Science and Nutrition
Poznań University of Life Sciences
Poznań, Poland

Alfreda Wei
Department of Molecular Biosciences
University of California–Davis
Davis, California

Renata Zawirska-Wojtasiak
Faculty of Food Science and Nutrition
Poznań University of Life Sciences
Poznań, Poland

1 Specificity of Food Odorants

Henryk Jeleń

CONTENTS

1.1 INTRODUCTION

Food is one of the main stimuli to our senses in everyday life. Apart from providing nutritious constituents during consumption, food engages the human senses: not only taste and smell, but also sight, hearing, and touch. Sensory properties resulting from the involvement of all senses provide a wholesome picture of food, which is either accepted or rejected by a consumer. Food appearance, texture, and mainly food flavor are the sensory properties that influence food acceptance. Among the sensory properties, flavor is usually the decisive factor for the choice of a particular product.

According to the *New Oxford American Dictionary*, flavor (Brit. *flavour*) is the distinctive quality of a particular food and drink as perceived by the taste buds and the sense of smell. The origin in late Middle English (in the sense fragrance, aroma) is from Old French *flaor*, perhaps based on a blend of Latin flatus "blowing" and foetor "stench." It can be assumed that flavor is the sensation produced by material taken in the mouth perceived principally by the senses of taste and smell, and also by the general pain, tactile, and temperature receptors in the mouth.

Food flavor is of cardinal importance not only for consumers at the moment of choosing a particular product, but also an important feature for breeders of fruit and vegetable varieties and in selection of raw materials used for food production. Flavor is an issue for food technologists, when developing new products, meeting consumers' requirements, and controlling it during processing and storage. Finally, flavor is one of the main factors that determine the shelf life of a particular food product. Development of off-flavors as a result of enzymatic, chemical, or microbial changes can make food products unpalatable. Therefore, maintaining the proper flavor of food products is in the interest of both consumers and producers.

1.2 FOOD VOLATILES AND FOOD ODORANTS

Odorants have to be volatile to reach the human olfactory system; therefore, it is accepted that, generally, odorants are molecules characterized by relatively high vapor pressures of molecular weight lower than 300 Da, although there are odorants that are relatively nonvolatile [5α-androst-16-en-3-one—mammalian pheromone having sweaty, ruinous, unpleasant woody odor; odor threshold (OT) = 0.00062 mg/kg; vapor pressure = 4.22×10^{-3}]. The vapor pressure of odorants can vary over several orders of magnitude. Majority of volatiles are also relatively nonpolar (hydrophobic) compounds, which favors their partition in aqueous media.

A differentiation should be made between volatile and odoriferous compounds. More than 11000 volatile compounds have been identified in food. They have been compiled as a database accessible on the Internet (VCF Volatile Compounds in Foods, 2010, www.vcf-online.nl) and fill 18 different chemical classes (hydrocarbons, aldehydes, ketones, esters, acids, lactones, halogens, sulfur compounds, etc.). However, it is estimated that only 5%–10% of them play a significant role in the formation of specific aromas of food products. The importance of particular compounds in flavor formation is related to their concentration and their odor thresholds. Volatile compounds influence the odor of a particular food when present in concentrations higher than their odor threshold, or they can also influence the flavor when present in mixtures that exceed these odor thresholds as a result of additive or synergistic effects.

Increasing the potential of separation techniques, and developments in gas chromatography especially comprehensive gas chromatography (GC × GC) allow the detection of hundreds or even thousands of peaks. Therefore, among the bulk of volatile compounds, key odorants of a product are of special importance in flavor analysis. As a consequence, although profiles of volatile compounds are useful in metabolomic or authenticity/traceability testing (Chapter 17), the analysis of food aroma compounds should be sensory guided. In analysis of food odorants, gas chromatography-olfactometry (GC-O) allows selection of aroma important compounds from numerous volatiles (van Ruth 2001). This approach to the analysis of food odorants and tastants is discussed in detail in Chapter 18.

1.3 ODOR THRESHOLDS AND AROMA DESCRIPTION

Odoriferous compounds are present in food in very low concentrations, usually in milligram per kilogram amounts, but very often in much lower concentrations—microgram per kilogram or even nanogram per kilogram of the product. Our olfactory system is able to detect some odorants present in extremely low concentrations. Odoriferous molecules are sensed by the olfactory epithelium located in the nasal cavity, which can be reached entering a nasal passage via the nose (orthonasal) or via the mouth (retronasal path). In humans, introduction of an odorant above a certain threshold into the nasal cavity triggers a response to the stimulus (see Chapter 3). Odor threshold can be defined as the lowest concentration of a compound in a specified medium that is sufficient for the recognition of a particular odor. In flavor description, two thresholds are used: detection threshold defined as the lowest physical intensity at which a stimulus is perceptible, and the recognition threshold (odor threshold), which is the lowest intensity in which the stimulus can be correctly defined or identified.

To characterize aroma compounds and their contribution to food flavor, odor thresholds need to be determined. Traditionally, this has been carried out in air using olfactometers; however, for food products, a more reasonable solution is to determine odor thresholds in water (Buttery 1999). This is based on the assumption that in determining a water threshold, the odor threshold of a compound in air, where the compound is at equilibrium between the water solution and air, is determined (Buttery et al. 1973). Odor threshold in air (T_a) can be determined from the following equation:

$$T_a = T_w \times K_{aw}$$

where T_w is a threshold concentration in water and K_{aw} is the air-to-water partition of the compound at the testing temperature. Similarly, threshold in oil can be calculated as follows:

$$T_{ol} = T_w \times (K_{aw}/K_{aol})$$

where K_{aw}/K_{aol} is equal to the oil-to-water partition.

To quantify the influence of a particular odor compound on the aroma of a product, aroma values (AV) are calculated by integrating the concentration of a particular compound and its odor threshold—AV is the ratio of a compound's concentration to its odor threshold (Rothe and Thomas 1963). An example of such approach is shown in Table 1.1. Furthermore, the log of concentration/threshold ratios is used to express the contribution of a compound to a product's overall aroma.

Basic tastes can be relatively easily described (sweet, salty, sour, bitter, and with umami being classified as the fifth basic taste). Contrary to this, description of odors is often extremely difficult to do in unequivocal terms. Complex mixtures are difficult to describe unless there is one dominant compound that influences flavor. Odors are described using adjectives comparing odors with known products/impressions (e.g., hay-like, fruity). Many terms describing odors include animal (musk, civet), camphoraceous, citrus, earthy, fatty, floral, green, herbaceous, medicinal, resinous, spicy, waxy, or woody. For complicated (from a flavor point of view) products, such as whisky, wine, or beer, flavor wheels have been constructed to help in describing main and additional odors and notes associated with a product (Figure 1.1).

1.4 CHEMICAL PROPERTIES AND PERCEPTION OF ODORANTS

Perception of food odorants is related to the nature of food product from which an aroma compound is released (food matrix) and the chemical/spatial nature of

TABLE 1.1
Aroma Compounds in Rye Bread Crust

Compound	Concentration (ppm)	Aroma Value[a]
Ethanol	1100	110
Acetaldehyde	23	25
Acetoin	1	1
Diacetyl	1.3	330
2/3-Methylbutanal	15	1900
Pyruvic aldehyde	9	20
2-Methylpropanal	6	6000
Furfural	19	300

Source: Rothe, M., and Kruse, H.-P., in *Flavor Chemistry. Thirty Years of Progress*, ed. R. Teranishi, E.L. Wick, and I. Horstein, 367–375, Kluwer Academic/Plenum Publishers, New York, 1999. With permission.

[a] Concentration (ppm)/threshold in water (ppm).

the odorant molecule. Aroma compounds interact with food macro constituents—proteins, lipids, and carbohydrates. The interactions influence release of aroma compounds from the matrix, resulting in partition coefficients that in consequence influence the levels of aroma compounds in the headspace. Odor threshold values of 2,4,6-trichloroanisol vary substantially in matrices of different viscosity and composition: in water, it is estimated at 7.6×10^{-8}; in beer, 7×10^{-6}; in edible oil, 7×10^{-3}; and in egg yolk, 2.4×10^{-3} (Maarse et al. 1987).

The food macroconstituents can govern aroma binding and release in various ways: the presence of polysaccharides can alter partition of volatiles due to modified viscosity or formation of inclusion complexes. Binding of flavor compounds to proteins is dependent on protein type. The nature of aroma binding to protein molecules can be reversible, based on hydrogen bonding and hydrophobic interactions, or can be irreversible, such as in the case of sulfur compounds reacting with proteins. Lipids influence partition of volatile compounds in foods to a great extent. In emulsions, even a low level of fat can substantially influence absorption of volatiles (Roberts and Pollien 2000). The process of aroma release was discussed in detail in several review papers (Druaux and Voilley 1997; Guichard 2002). The problem is of high importance for food manufacturers especially considering the trends for production of fat-free or low-fat products. Binding and release of aroma compounds are discussed in detail in Chapter 7.

The chemical nature of odorants influences the way they are perceived. Functional groups position influences odor threshold. As an example, isomers of trichloroanisole—2,4,6-trichloroanisole and 2,3,6-trichloroanisole can be cited having odor thresholds of 0.03 and 0.0003 ppb, respectively (Maarse et al. 1987). Size and character of the functional groups may also influence odor thresholds via the influence of

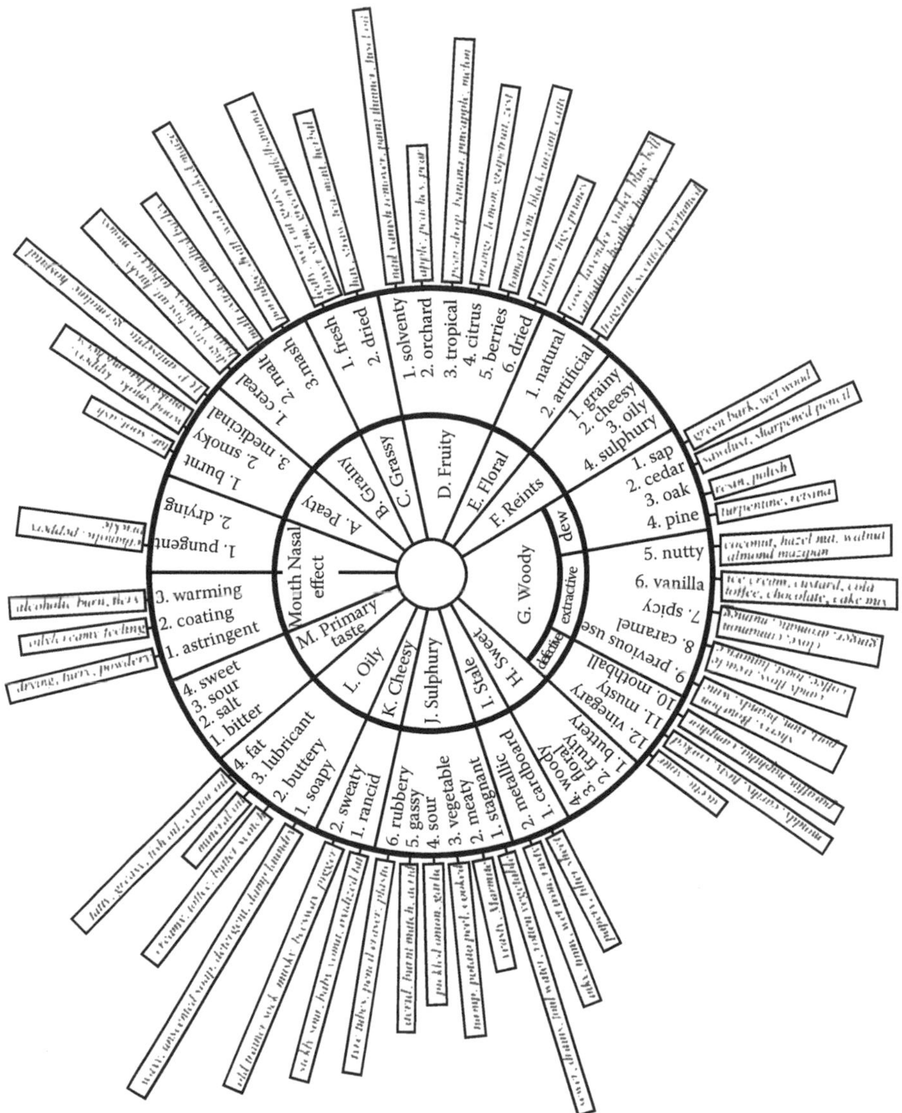

FIGURE 1.1 Whisky flavor wheel. (From Lee, K.-Y.M., Paterson, A., Piggot, J.R., *J. Inst. Brewing* 107, 287–313, 2001. With permission.)

odorant receptor spatial interaction. A good example showing the influence of functional groups on odor thresholds are alkylpyrazines. Werner and coworkers studied the structure–odor activity relationships of 80 alkylpyrazines. Tetramethylpyrazine had an odor threshold of >2000 ng/L, whereas trimethylpyrazine had a substantially lower odor threshold of 50 ng/L (in air). When one of the methyl groups was

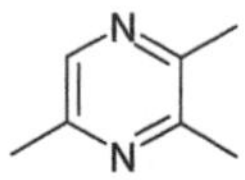

Trimethylpyrazine
OT = 50 ng/L

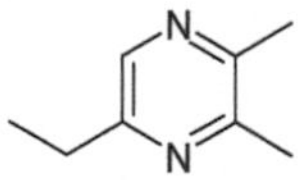

5-Ethyl-2,3-dimethylpyrazine
OT = 200 ng/L

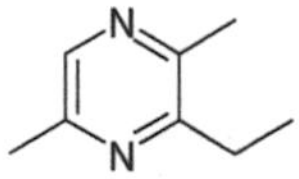

3-Ethyl-2,5-dimethylpyrazine
OT = 3.6 ng/L

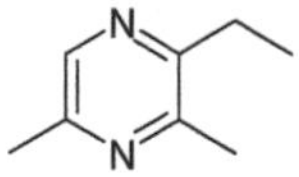

2-Ethyl-3,5-dimethylpyrazine
OT = 0.011 ng/L

FIGURE 1.2 Structures of selected alkylpyrazines and their odor thresholds.

TABLE 1.2
Odor Properties of Selected Chiral Compounds

Compound	Odor Description
Linalool	(+): Sweet, petigrain
	(–): Woody, lavender
Carvone	(+): Caraway
	(–): Spearmint
Nootkatone	(+): Grapefruit
	(–): Woody, spicy
Nerol oxide	(+): Green, floral
	(–): Green, spicy, geranium
Menthol	(+): Dusty, vegetable, less minty, less cooling than (–)
	(–): Sweet, fresh minty, strong cooling effect
Limonene	(+): Orange
	(–): Turpentine

Source: Brenna, E., Fuganti, C., Serra, S., *Tetrahedron: Asymmetry* 14, 1–42, 2003. With permission.

replaced with the ethyl group in positions 3 and 2, a decrease in odor thresholds was observed; in the case of 2-ethyl-3,5-dimethylpyrazine, a decrease to 0.01 ng/L was observed (Figure 1.2). However, 5-ethyl-2,3-dimethylpyrazine showed a higher odor threshold of 200 ng/L. When propyl or butyl groups replaced the ethyl group in 2-ethyl-3,5-dimethylpyrazine, the volume of the functional group became too bulky, resulting in an increase of odor thresholds to 23 and 180 ng/L, respectively. Moreover, based on the investigated compounds and their sensory properties, it is evident that a specific spatial configuration is required to induce earthy/roasty odor, and a model receptor for it could be proposed (Werner et al. 1999).

Enantiomers exhibit different behaviors when interacting with organisms, which is of immense importance in pharmacology, physiology, and also in odorants perception. Biogenesis of odoriferous compounds in nature yields specific enantiomers (for compounds possessing chiral centers). Enantiomers of the same compound can exhibit different odor notes (Table 1.2). Odor thresholds of enantiomers of the same compound may vary: they may have the same odor, or some of the enantiomers can be nonodoriferous: (1*R*,2*R*)-(–)-methyl-(*Z*)-jasmonate and its diastereoisomer (1*R*,2*S*)-(+)-methyl-(*Z*)-jasmonate are the key odorants in jasmine flower oil (*Jasminum grandiflorum* L.) and occur in the proportion of 97:3. They both have floral, jasmine, slightly fruity odor with odor thresholds of 70 and 3 μg/ml, respectively, whereas their enantiomers (Figure 1.3) are almost odorless (Acree et al. 1985).

1.5 FORMATION OF AROMA COMPOUNDS IN FOOD

Food is a complex matrix rich in flavor precursors. Food flavors (aromas) can be divided based on the origin/nature of compounds into three main categories: primary aromas, secondary aromas, and off-flavors (Figure 1.4). Such classification reflects to a certain extent the pathways of formation of odorants. Primary aromas are produced mainly in enzyme-catalyzed reactions in raw foods, whereas secondary aromas are those comprising products of thermal reactions in food processing and also compounds obtained as a result of microbial activity in fermentation processes. "Off-flavors" is a term related to compounds of various origins and nature, which impairs the natural flavor of food.

Characteristic pathways of formation of particular flavor compounds are unique for a given food, although the pool of precursors is very often similar.

1.5.1 Primary Aromas

Primary aromas are usually associated with enzymatic reactions in raw materials. They are the most characteristic for plants, fruits, and vegetables.

In the formation of aroma compounds in fruits, the main precursors are lipids, carbohydrates, and amino acids. Moreover, mevalonic acid (terpene metabolism) and cinnamic acid metabolism is involved in the formation of fruit flavors. In fruits and vegetables, many flavor compounds are released from their nonvolatile precursors in the process of enzymatic hydrolysis. The main groups of compounds forming primary aromas are esters, alcohols, aldehydes, and terpenes. The formation of

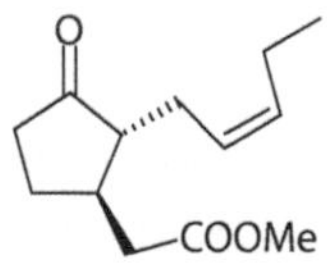

COOMe

(1*R*,2*R*)-(−)-Methyl-(*Z*)-jasmonate [OT > 70 μg/L]

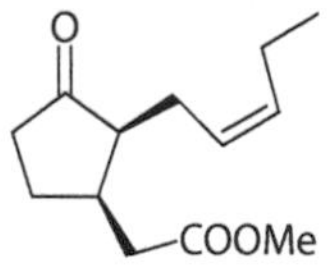

COOMe

(1*R*,2*S*)-(+)-Methyl-(*Z*)-jasmonate [OT = 3 μg/L]

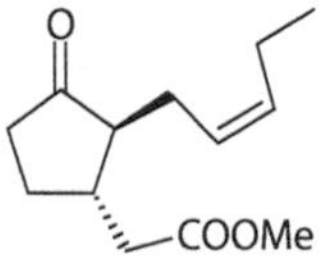

COOMe

(1*S*,2*S*)-(−)-Methyl-(*Z*)-jasmonate [odorless]

COOMe

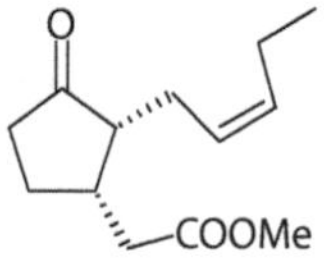

(1*S*,2*R*)-(+)-Methyl-(*Z*)-jasmonate [odorless]

FIGURE 1.3 Enantiomers of methyl-(*Z*)-jasmonate: (1*S*,2*S*) and (1*S*,2*R*) isomers are odorless; (1*R*,2*R*) and (1*R*,2*S*) isomers have jasmine odor.

aroma compounds from fatty acid precursors as a result of lipoxygenase activity is presented for tomatoes, olives, and cucumbers in Chapter 4.

An important group of compounds in essential oils, spices, and many fruits are of terpenes. Monoterpenes (C10), sesquiterpenes (C15), diterpenes (C20), triterpenes (C30), and tetraterpenes (C40) are formed in plants; however, only monoterpenes, and to a lesser extent, sesquiterpenes, play an important role in the formation of

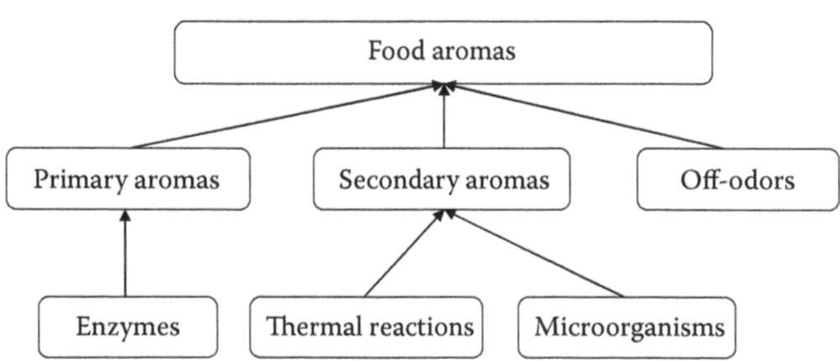

FIGURE 1.4 Classification of food flavors (aromas) and their main categories of origin.

flavor. Terpenes are formed in plants from an isoprene C5 unit via the head-to-tail addition (Ruzicka et al. 1953). The biosynthesis of terpenes commences from acetyl-CoA via acetate and mevalonate pathway (Bach et al. 1999). The independent pathway for formation of isoprenoids in plants is the deoxyxylulose phosphate/methylerythritol phosphate pathway (Lichtenhalter et al. 1997, 1998). Plants form volatile monoterpenes and sesquiterpenes as main constituents of their essential oils in specialized cells of *Lamiaceae*, *Myrtaceae*, *Pinaceae*, *Rosaceae*, *Umbelliferae* and others. Terpenes form a vast group of compounds—mainly hydrocarbons, alcohols, and aldehydes that are relatively labile and can undergo oxidations and transformations. Chapter 10 describes in detail terpenes present in essential oils and spices.

1.5.2 Secondary Aromas

Secondary aromas according to the classification provided in this chapter are those formed as a result of microbial activity, mainly in fermentation processes (i.e., alcoholic beverages), controlled enzymatic reactions, and in thermal reactions (mainly Maillard reaction). There are a number of fermented foods in which microorganisms cause the formation of characteristic odoriferous compounds. Pure cultures of microorganisms are used (starter cultures) or microorganisms native to raw material with their specific enzyme profile. In cheese, starter cultures contribute to their characteristic smell and produce flavor compounds from lactose, citrate, proteins, and lipids (Marilley and Casey 2004). Depending on a cheese type, different pathways prevail and microorganisms form distinct compounds (methyl ketones in blue-veined cheese, volatile fatty acids, and sulfur compounds in Cheddar). Chapter 12 describes the aroma of cheese in detail. In traditional yeast fermentations used in the production of bread and alcoholic beverages, a number of volatile compounds, mainly alcohols, esters, and aldehydes are formed. *Saccharomyces cerevisiae* form compounds characteristic for yeast leavened bread: lower alcohols, acetaldehyde, propanal, pentanal, furfural, and ethyl esters lactic and acetic acid.

Heat-generated aroma compounds form the biggest group of odorants obtained in technological processes. Thermal processes that lead to heat-generated aromas not only include coffee and cocoa roasting, boiling, frying, grilling of meat, baking, but also milder processes, such as pasteurization of milk. The number of compounds obtained as a result of thermal processing is estimated at 3500–4500. Heating of foods that contain proteins, peptides, and amino acids with reducing sugars forms a vast array of volatile compounds as a result of Maillard reaction. It is one of the most important sets of reactions in food chemistry and technology and yields, apart from flavor compounds, nonvolatiles and characteristic brown pigments (melanoidins) (Hodge 1953). Maillard reaction is a source of many compounds produced from a limited number of precursors. Although it is a source of numerous volatiles, the number of odoriferous compounds is much lower. Flavor compounds produced in thermal reactions include Strecker aldehydes, oxidation products formed in free radical lipids auto-oxidation, and heterocyclic compounds being secondary products of Maillard reaction/carbohydrate degradation products. In heated foods, flavor

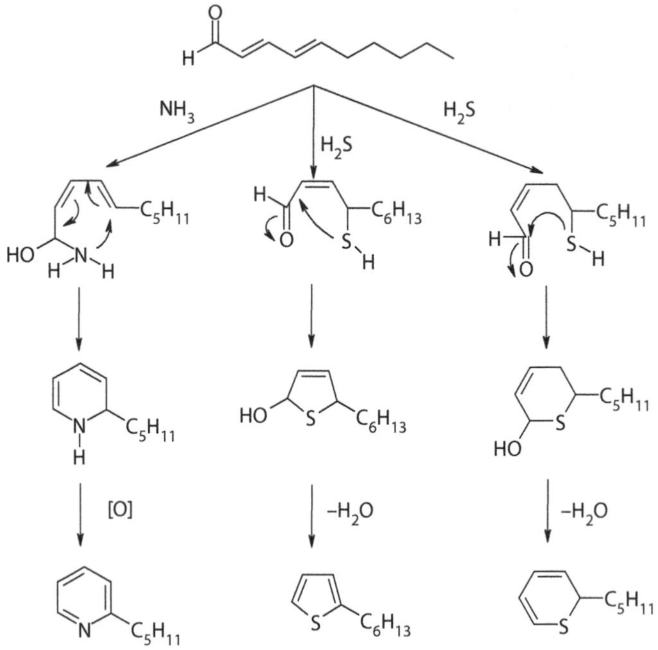

FIGURE 1.5 Formation of 2-pentylpyridine, 2-hexylthiophene, and 2-pentyl-2*H*-thiapyran from decadienal, ammonia, and hydrogen sulfide. (From Mottram, D.S., *Food Chemistry* 62, 415–424, 1998. With permission.)

compounds may interact with each other. As an example, interactions of volatile compounds formed in Maillard reactions with fatty acids oxidation products can be shown (Figure 1.5). This example shows an array of heterocyclic compounds obtained in the reaction of reactive cysteine degradation products (ammonia and hydrogen sulfide) with decadienal.

Heterocyclic compounds formed in thermal reactions include furanones, pyranones, thiazoles, pyridines, pyroles, and pyrazines. Examples of heterocyclic compounds found in foods are shown in Figure 1.6. Furanones are compounds that result from carbohydrates degradation and usually have a caramel odor note, besides various other characteristic notes. Heteroatomic compounds containing nitrogen(s) atoms are usually formed in heated foods; however, some can be formed in enzymatic reactions or in biological systems in a combination of enzyme-catalyzed and nonenzymatic reactions. Pyridines can be formed in a reaction of unsaturated aldehydes with ammonia or primary amines. Pyrazines are formed in heat-treated foods by nonenzymatic reactions, or in a combination of nonenzymatic and enzymatic reactions. Alkylpyrazines are usually associated with heat-treated foods, formed in

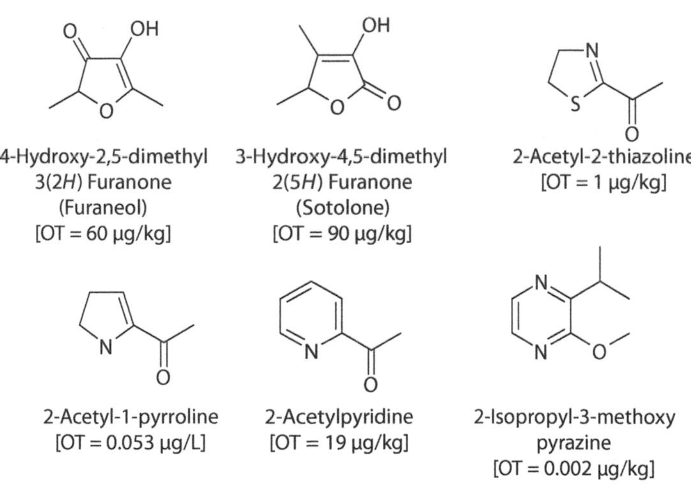

FIGURE 1.6 Selected heterocyclic odor compounds in foods.

a chemical reaction from hydroxyacetone or related hydroxycarbonyls, ammonia/aminoacetone (Rizzi 1988). On the contrary, methoxy pyrazines are usually associated with living organisms and fermented foods and their number in food products is limited to the main odorants: 2-methoxy-3-isopropyl, 2-methoxy-3-isobutyl, and 2-methoxy-3-secbutylpyrazine. Their biosynthesis was studied in *Bacillus subtilis*, *Aspergillus oryzae*, and *Pseudomonas* spp., indicating amino acids as precursors (Rizzi 2002).

2-Acetyl-1-pyrroline has been associated with the aroma of white bread crust and contributes to the popcorn flavor compound and identified in wheat bread crust, cooked beef, and corn tortillas. It is formed in Maillard reaction using 2-oxopropanal (formed in the degradation of reducing sugars) or 1-pyrroline (which can be formed from ornithine and/or proline) (Hoffmann and Schieberle 1998). However, 2-acetyl-1-pyroline can be also microbially formed in cheeses and air-dried sausages.

Sulfur-containing compounds are an important group of compounds found in heated foods. They are aliphatic thiols, sulfides, and mainly heterocyclic compounds. Heterocyclic aroma compounds often contain sulfur in conjunction with nitrogen. Sulfur-containing compounds comprise more than 250 volatiles and are abundant in meat. The main precursors of sulfur-containing aroma compounds are sulfur-containing amino acids cysteine and methionine. Strecker degradation of cysteine leading to formation of hydrogen sulfide, ammonia, and acetaldehyde provides intermediates for many interactions especially with lipids (Figure 1.5). The important pathway in the formation of sulfur-containing compounds is thermal degradation of thiamine-yielding thiazoles, furanthiols, and thiophenes. Excellent reviews of sulfur odor compounds in foods have been provided in the literature (Mottram and Mottram 2002; Vermeulen et al. 2005).

1.5.3 Off-Odors

Off-odors in food are associated with the presence of compounds that either migrate into food products from external sources (taints), or are formed from food constituents as a result of chemical, biological, or microbial deterioration (off-flavors). They generate problems in food processing that affect the whole food chain. Chapter 16 is devoted to off-flavors and taints in food. There are also entire books on taints and off-odors in foods (Saxby 1993; Baigrie 2003).

Off-odorants are represented by haloanisoles, halophenols, phenolics, sulfur compounds, carbonyls, esters, amines, and fatty acids. They can be classified according to their origin—microbially derived; compounds originating from packagings, cleaning agents, and disinfectants; and compounds formed in auto-oxidation or other reactions taking place in stored foods.

One of the most important groups of off-odorants comprises compounds that contribute to musty and earthy off-odors in foods. They include geosmin, methylisoborneol, chloro-, bromoanisoles, octa-1,3-diene, 4,4,6-trimethyl-1,3-dioxan, 2,6-dimethyl-3-methoxypyrazine, and 2-methoxy-3-isopropylpyrazine. Chloroanisoles were detected as compounds responsible for musty taint first time in eggs and broilers (Engel et al. 1966); 2,4,6-trichloroanisol is responsible for a Rio off-flavor in coffee (Spadone et al. 1990) and corky taint in wine (Buser et al. 1982). Geosmin and/or 2-methylisoborneol mainly produced by *Strepromyces*, *Nocardia*, *Oscillatoria*, and various *Penicillium* species (Bőrjesson et al. 1993; Mattheis and Roberts 1992) are compounds related to the occurrence of muddy–musty off-flavor in water (Watson et al. 2000), fish (Conte et al. 1996), stored grain (Jeleń et al. 2003), and in red beet (Lu et al. 2003). Phenolic compounds form an important group of off-odorants in foods contributing to smoky odors (guaiacol, 4-ethylphenol, 4-ethyl guaiacol) or pharmaceutical odors (4-vinylphenol and 4-vinylguaiacol) (Saxby 1993). They contribute to Brett defect in wines described as barnyard or band aid odor (Licker et al. 1998). Sulfur compounds, besides contributing to the characteristic flavor of many vegetables, fruit, meat, and cheeses, can be a potent group of off-flavors: 3-methyl-2-butene-thiol (MBT) in beer, possessing a skunky odor, is derived from photochemical degradation of *iso*-α-acids in beer exposed to light. Volatile lipid oxidation products, which are also an important group of off-odorants, are responsible for the rancid off-flavor (1-octene-3-one is responsible for the off-flavor of rancid butter), whereas 1-pentene-3-one has a characteristic fishy off-flavor (Badings 1970; Belitz et al. 2008). Off-flavors in fat-containing foods are also related to the presence of methyl ketones, which are responsible for the ketonic rancidity–musty, stale off-flavor of desiccated coconut (Kellard et al. 1985), caused by oxidation of medium-chain fatty acids by molds (Kochar 1993).

1.6 BOUND FLAVOR COMPOUNDS

Aroma compounds exist in plants in their free form; however, in many plants, they also exist in the form of glycoconjugated precursors. There are about 50 plant families in which glycoconjugated aroma compounds have been detected. Moreover, the pool of glycosidically bound compounds exceeds in some cases the free flavor compounds fraction in a ratio range of 2:1 to 10:1 (Winterhalter et al. 1999). Bound aroma compounds are characteristic to fruits and vegetables and do not contribute to food aroma until released. Odoriferous aglycons are bound as glycosides with mainly monosaccharides and disaccharides. Aglycons may be released from carbohydrates as a result of tissue disruption, during maturation, storage, processing, and aging. The aglycon release is usually catalyzed by enzymes, acids, or heat. Examples of bound aroma compounds are glucosinolates in *Cruciferae*, bound terpenes, norisoprenoids, and phenolic compounds in grapes and other fruits.

Glucosinolates are metabolites in *Cruciferae* family that include among others radish, horseradish, cabbage, cauliflower, kale, and wasabi. The enzyme that catalyzes the hydrolysis of thioglucosidic bond in glucosinolates is myrosinase (Figure 1.7). Myrosinase gains access to glycosinolates in tissue damage, as both the enzyme and a substrate are located within vacuoles, although in different cell types and subcellular structures. Of the presented hydrolysis compounds, isothiocyanates are the most important in forming the sharp and pungent flavor of *Cruciferae* plants. However, epitiospecific proteins that influence the formation of glucosinolates hydrolization products favor the formation of nitriles and thiocyanates, which are less toxic than isothiocyanates.

Since the first studies on bound aroma compounds—geraniol in rose flowers (Francis and Allock 1969)—investigations have focused mainly on grapes, and also on other fruits—apricot, peach (Krammer et al. 1991), passiflora (Chassagne et al. 1996), nectarines (Aubert et al. 2003), mango (Ollé et al. 1998), strawberries (Roscher

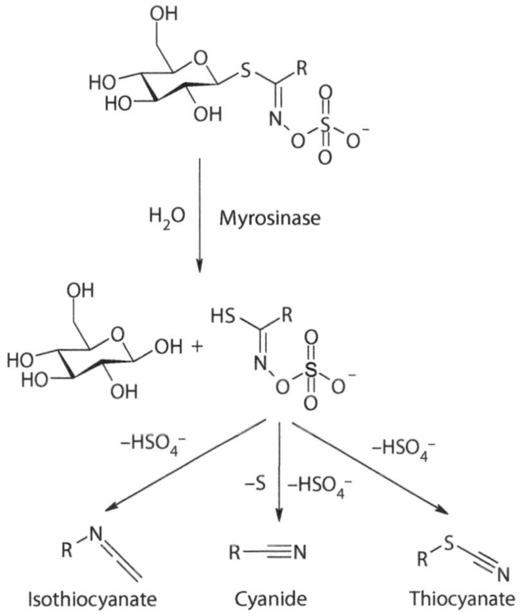

H_2O Myrosinase

$-HSO_4^-$ $-S$ $-HSO_4^-$ $-HSO_4^-$

Isothiocyanate   Cyanide   Thiocyanate

FIGURE 1.7 Formation of aroma compounds released by myrosinase in *Cruciferae*.

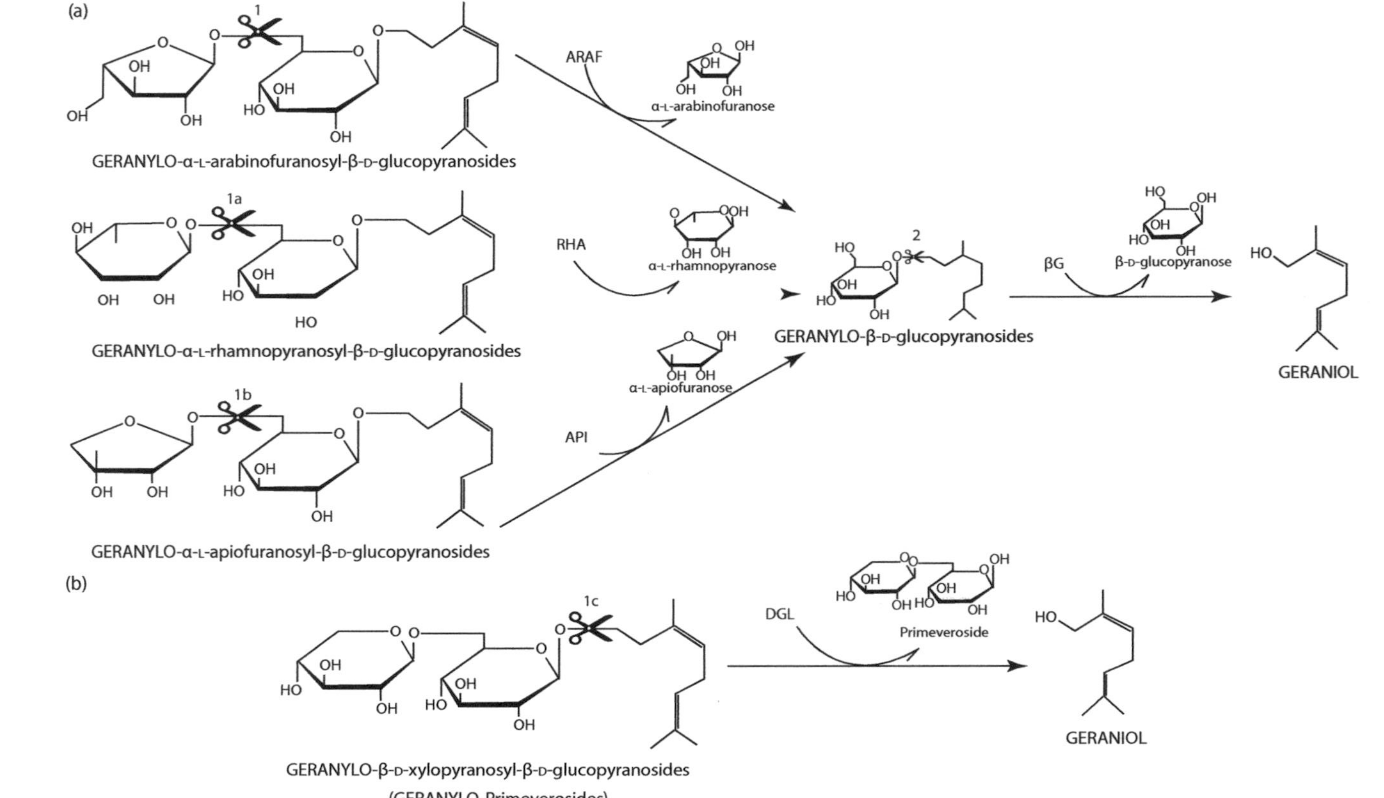

FIGURE 1.8 Structures of glycosidic monoterpene precursors from grape berries and their hydrolysis by specific enzymes (geraniol shown as aglycone). (From Dziadas, M., and Jeleń H., Acta Sci. Pol. Technol. Aliment., 10, 7–17, 2011. With permission.)

et al. 1997), cape gooseberry (Mayorga et al. 2001) and saffron (Straubinger et al. 1998).

The odoriferous aglycons of glycosides in plants include monoterpenes, C13-norisoprenoids, benzene derivatives, and alcohols. The sugar moieties are usually glucose or disaccharides in 6-*O*-α-L-arabino-pyranosyl-β-D-glucopyranosides (vicianosides), 6-*O*-α-L-arabinofuranosyl-β-D-glucopyranosides, 6-*O*-α-L-rhamnopyranosyl-β-D-glucopyranosides (rutinosides), 6-*O*-β-D-apiofuranosyl-β-D-glucopyranosides, 6-*O*-β-D-xylopyranosyl-β-D-glucopyranosides (primeverosides), and 6-*O*-β-D-glucopyranosyl-β-D-glucopyranosides (gentibiosides) (Stahl-Biskup et al. 1993; Vasserot et al. 1995; Cabaroglu et al. 2003).

Grape varieties, rich in monoterpenes, are used for white wine production; in particular, Muscat and aromatic varieties such as Gewürtztraminer, Riesling, and Sylvaner provide rich floral aromas because of the presence of monoterpene odoriferous alcohols, which exist in free and bound forms (Güth 1997; Mateo and Jimenez 2000). In the winemaking process, volatile compounds from glycosides can be released by enzyme or acid hydrolyses (Williams et al. 1982; Ibarz et al. 2006; Tamborra et al. 2004; Günata and Sarry 2004; Mateo and Jimenez 2000; Mateo and Maicas 2005). The acid hydrolysis progresses relatively slowly in winemaking conditions, being dependent on the pH and temperature of the medium and on the structure of the aglycone moiety. Sequential reactions take place in the enzymatic hydrolysis of diglycosides involving several glycosidases specific for the sugar moiety of the substrates (Cabaroglu et al. 2003) (Figure 1.8).

1.7 CONCLUSION

Food volatiles are thousands of compounds that represent different chemical classes. Among them, odoriferous compounds form a smaller (5%–10% of compounds) group, which is responsible for food aroma. Food odorants are present in concentrations ranging from milligram per kilogram down to nanogram per kilogram. Key odorants also represent diverse chemical classes, are very often unstable, and react with various other compounds. Formation of aroma compounds is a very dynamic process in heated foods as well as in enzyme-catalyzed reactions. Understanding the formation of aroma compounds and their chemical and sensory properties allows better understanding of the technological processes and reactions involved during food manufacturing and storage.

REFERENCES

Acree, T.E., R. Nishida, and H. Fukami. 1985. Odor thresholds of the stereoisomers of methyl jasmonate. *J. Agric. Food Chem.* 33: 425–427.

Aubert, C., C. Ambid, R. Baumes, and Z. Günata. 2003. Investigation of bound aroma constituents of yellow-fleshed nectarines (*Prunus persica* L. Cv. Springbright). Changes in bound aroma profile during maturation. *J. Agric. Food Chem.* 51: 6280–6286.

Bach, T.J., A. Boronat, N.J. Campos, A. Ferrer, and K.-U. Vollack. 1999. Mevalonate biosynthesis in plants. *Crit. Rev. Biochem. Mol. Biol.* 34: 107–122.

Badings, H.T. 1970. Cold storage defects in butter and their relation to auto-oxidation of unsaturated fatty acids. *Neth. Milk Dairy J.* 24: 147–256.

Baigrie, B., ed. 2003. *Taints and Off-Flavours in Food.* Cambridge: Woodhead Publishing Limited.

Belitz, H.D., W. Grosch, and P. Schieberle. 2008. *Food Chemistry.* Berlin: Springer-Verlag.

Bőrjesson, T., U. Stőllman, and J. Schnűrer. 1993. Off odours compounds produced by molds on oatmeal agar; Identification and relation to other growth characteristics. *J. Agric. Food Chem.* 41: 2104–2111.

Brenna, E., C. Fuganti, and S. Serra. 2003. Enantioselective perception of chiral odorants. *Tetrahedron: Asymmetry* 14: 1–42.

Buser, H.-R., C. Zanier, and H. Tanner. 1982. Identification of 2,4,6-trichloroanisole as a potent compound causing cork taint in wine. *J. Agric. Food Chem.* 30: 359–362.

Buttery, R. 1999. Flavor chemistry and odor thresholds. In *Flavor Chemistry. Thirty Years of Progress*, ed. R. Teranishi, E.L. Wick, and I. Hornstein, 353–365. New York: Kluwer Academic.

Buttery, R.G., D.G. Gaudagni, and L.C. Ling. 1973. Flavor compounds: Volatiles in vegetable oil and oil/water mixtures. Estimation of odor thresholds. *J. Agric. Food Chem.* 21: 198–201.

Cabaroglu, T., S. Selli, A. Canbas, J.P. Leputre, and Z. Günata. 2003. Wine flavor enhancement through the use of exogenous fungal glycosidases. *Enzyme Microbial Technol.* 33: 581–587.

Chassagne, D., J. Crouzet, C.L. Bayonove, J.-M. Brillouet, and R.L. Baumes. 1996. 6-*O*-α-L-arabinopyranosyl-β-D-glucopyranosides as aroma precursors from passion fruit. *Phytochemistry* 41: 1497–1500.

Conte, E.D., C.-Y. Shen, P.W. Perschbacher, and D.W. Miller. 1996. Determination of geosmin and methylisoborneol in catfish tissue (*Ictalurus punctatus*) by microwave-assisted distillation-solid phase adsorbent trapping. *J. Agric. Food Chem.* 44: 829–835.

Druaux, C., and A. Voilley. 1997. Effect of food composition and microstructure on volatile flavour release. *Trends in Food Sci. Technol.* 8: 364–368.

Dziadas, M., and H. Jeleń. 2011. Influence of glycosidases addition on selected monoterpenes contents in musts and white wines from two grape varieties grown in Poland. *Acta Sci. Pol. Technol. Aliment.* 10: 7–17.

Engel, C., A.P. de Groot, and C. Wuermann. 1966. Tetrachloroanisole: A source of musty taint in eggs and broilers. *Science* 154: 270–271.

Francis, M.J.O., and C. Allock. 1969. Gerianiol β-D-glucoside: Occurrence and synthesis in rose flowers. *Phytochemistry* 8: 1339–1347.

Guichard, E. 2002. Interactions between flavor compounds and food ingredients and their influence on flavor perception. *Food Rev. Int.* 18: 49–70.

Güth, H. 1997. Quantitation and sensory studies of character impact odorants of different white wine varieties. *J. Agric. Food Chem.* 45: 3027–3035.

Hodge, J.E. 1953. Dehydrated foods: Chemistry of browning reactions in model systems. *J. Agric. Food Chem.* 1: 928–943.

Hoffmann, T., and P. Schieberle. 1998. 2-Oxopropanal, hydroxy-2-propanone and 1-pyrroline—important intermediates in the generation of the roast smelling food flavor compounds 2-acetyl-1-pyrroline and 2-acetyltetrahydropyridine. *J. Agric. Food Chem.* 46: 2270–2277.

Ibarz, M.J., V. Ferreira, P. Hernandez-Orte, N. Loscos, and J. Cacho. 2006. Optimization and evaluation of a procedure for the gas chromatographic–mass spectrometric analysis of the aromas generated by fast acid hydrolysis of flavor precursors extracted from grape. *J. Chromatogr. A* 1116: 217–229.

Jeleń, H.H., M. Majcher, R. Zawirska-Wojtasiak, M. Wiewiórowska, and E. Wąsowicz. 2003. Determination of geosmin, 2-methylisoborneol, and a musty–earthy odor in wheat grain by SPME-GC-MS, profiling volatiles and sensory analysis. *J. Agric. Food Chem.* 51: 7079–7085.

Kellard, B., D.M. Busfield, and J.L. Kinderlerer. 1985. Volatile off-flavour compounds in desiccated coconut. *J. Sci. Food Agric.* 36: 415–420.

Kochar, S.P. 1993. Oxidative pathways to the formation of off-flavours. In *Food Taints and Off-Flavours*, ed. M.J. Saxby. Glasgow, UK: Blackie Academic & Professional.

Krammer, G., P. Winterhalter, M. Schwab, and P. Schreier. 1991. Glycosidally aroma bound compounds in the fruits of *Prunus* species: apricot (*P. armeniaca* L.), peach (*P. persica* L.), yellow plum (*P. domestica* L. ssp. *syriaca*). *J. Agric. Food Chem.* 39: 778–781.

Lee, K.-Y.M., A. Paterson, and J.R. Piggot. 2001. Origins of flavor in whiskies and a revised flavor wheel: A review. *J. Inst. Brewing* 107: 287–313.

Lichtenhalter, H.K. 1998. The plants' 1-deoxy-D-xylulose-5-phosphate pathway for biosynthesis of isoprenoids. *Fett/Lipid* 100: 128–138.

Lichtenhalter, H.K., M. Rohmer, and J. Schwender. 1997. Two independent biochemical pathways for isopentenyl diphosphate (IPP) and isoprenoid biosynthesis in higher plants. *Physiol. Plant* 101: 643–652.

Licker, J.L., T.E. Acree, and T. Henick-Kling. 1998. What is "Bret" (*Bretanomyces*) flavor. A preliminary investigation. In *Chemistry of Wine Flavor*, ed. A.L. Waterhouse, and S.E. Ebeler, 96–115. ACS Symposium Series, Vol. 714. Washington, DC: American Chemical Society.

Lu, G., C.G. Edwards, J.K. Fellman, D.S. Mattinson, and J. Navazio. 2003. Biosynthetic origin of geosmin in red beets (*Beta vulgaris* L.). *J. Agric. Food Chem.* 51: 1026–1029.

Maarse, H., L.M. Nijsen, and S.A.G.F. Angelino. 1987. Halogenated phenols and chloroanisoles: Occurrence, formation and prevention. In *Proceedings of the Second Wartburg Aroma Symposium*, Wartburg, Nov. 16–19, ed. M. Rothe, 43–61. Berlin: Akademie Verlag.

Marilley, L., and M.G. Casey. 2004. Flavours of cheese products: Metabolic pathways, analytical tools and identification of producing strains. *International Journal of Food Microbiology* 90: 139–159.

Mateo, J.J., and M. Jimenez. 2000. Monoterpenes in grape juice and wines. *J. Chromatogr. A.* 881: 557–567.

Mateo, J.J., and S.J. Maicas. 2005. Hydrolysis of terpenyl glycosides in grape juice and other fruit juices: A review. *Appl. Microb. Biotechnol.* 67: 322–335.

Mattheis, J.P., and R.G. Roberts. 1992. Identification of geosmin as a volatile of *Penicillium expansum*. *Appl. Environ. Microbiol.* 58: 3170–3172.

Mayorga, H., H. Knapp, P. Winterhalter, and C. Duque. 2001. Glycosidically bound flavor compounds of cape gooseberry (*Physalis peruviana* L.). *J. Agric. Food Chem.* 49: 1904–1908.

Mottram, D.S. 1998. Flavor formation in meat and meat products: a review. *Food Chem.* 62: 415–424.

Ollé, D., R.L. Baumes, C.L. Bayonove, Y.E. Lozano, C. Szpaner, and J.-M. Brillouet. 1998. Comparison of free and glycosidally linked volatile components from polyembrionic and monoembrionic mango. *J. Agric. Food Chem.* 46: 1094–1100.

Rizzi, G.P. 1988. Formation of pyrazines from acyloin precursors under mild conditions. *J. Agric. Food Chem.* 36: 349–352.

Rizzi, G.P. 2002. Biosynthesis of aroma compounds containing nitrogen. In *Heteroatomic Aroma Compounds*, ACS Symposium Series 826, ed. G.A. Reineccius, and T.A. Reineccius, 132–147. Washington, DC: American Chemical Society.

Roberts, D.D., and P. Pollien. 2000. Relative influence of milk components on flavor compound volatility. In *Flavor Release*, ACS Symposium Series 763, ed. D.D. Roberts, and A.J. Taylor, 321–332. Washington, DC.

Roscher, R., M. Herderich, J.P. Steffen, P. Schreier, and W. Schwab. 1997. 2,5-Dimethyl-4-hydroxy-3(2*H*)-furanone 6′-*O*-malonyl β-D-glucopyranosidase in strawberry fruits. *Phytochemistry* 43: 155–159.

Rothe, M., and H.-P. Kruse. 1999. Solving flavor problems by sensory methods: A retrospective view. In *Flavor Chemistry. Thirty Years of Progress*, ed. R. Teranishi, E.L. Wick, and I. Horstein, 367–375. New York: Kluwer Academic/Plenum Publishers.

Rothe, M., and B. Thomas. 1963. Aromastoffe des Brotes. Versuch einer Auswertung chemischer geschmackanalysen mit hilfe des Schwellenwertes. *Z. Lebensm. Unters. U. Forsch.* 119: 302–310.

Ruzicka, L., A. Eschenmoser, and H. Heusser. 1953. The isoprene rule and the biogenesis of isoprenoid compounds. *Experientia* 9: 357–396.

Sarry, J.E., and Z. Gunata. 2004. Plant and microbial glycoside hydrolases: Volatile release from glycosidic aroma precursors. *Food Chem.* 87: 509–524.

Saxby, M.J., ed. 1993. *Food Taints and Off-Flavours*. Glasgow: Blackie Academic & Professional.

Spadone, J.C., G. Takeoka, and R. Liardon. 1990. Analytical investigation of Rio off-flavour in green coffee. *J. Agric. Food Chem.* 38: 226–231.

Stahl-Biskup, E., F. Intert, J. Holthuijzen, M. Stengele, and G. Schulz. 1993. Glycosidically bound volatiles—A review 1986–1991. *Flavor Fragrance J.* 8: 61–80.

Straubinger, M., B. Bau, S. Eckstein, M. Fink, and P. Winterhalter. 1998. Identification of novel glycosidic aroma precursors in saffron (*Crocus sativus* L.). *J. Agric. Food Chem.* 46: 3238–3243.

Tamborra, P., N. Martino, and M. Esti. 2004. Laboratory tests on glycoside preparations in wine. *Anal. Chim. Acta* 513: 299–303.

van Ruth, S.M. 2001. Methods for gas chromatography–olfactometry: A review. *Biomol. Eng.* 17: 121–128.

Vasserot, V., A. Arnaud, and P. Galzy. 1995. Monoterpenol glycosides in plants and their biotechnological transformation. *Acta Biotechnol.* 15: 77–95.

VCF Volatile Compounds in Food: Database/Nijssen, L.M., C.A. Ingen-Visscher, and J.J.H. van Donders, ed. Version 12.3. Zeist (The Netherlands): TNO Quality of Life, 1963–2010.

Vermeulen, C., L. Gijs, and S. Colin. 2005. Sensorial contribution and formation pathways of thiols in foods. *Food Rev. Int.* 21: 69–137.

Watson, S.B., B. Brownlee, T. Satchwill, and E.E. Hargesheimer. 2000. Quantitative analysis of trace levels of geosmin and MIB in source and drinking water using headspace SPME. *Wat. Res.* 34: 2818–2828.

Werner, R., M. Czerny, J. Bielohradsky, and W. Grosch. 1999. Structure–odour–activity relationships of alkylpyrazines. *Z. Lebensm. Unters. Forsch. A* 208: 308–316.

Williams, P.J., C.R. Strauss, B. Wilson, and R.A. Massy-Westropp. 1982. Studies on the hydrolysis of *Vitis vinifera* monoterpene precursor compounds and model monoterpene beta-D-glucosides rationalizing the monoterpene composition of grapes. *J. Agric. Food Chem.* 30: 1219–1223.

Winterhalter, P., H. Knapp, and M. Straubinger. 1999. Water soluble aroma precursors. In *Flavor Chemistry. Thirty Years of Progress*, ed. R. Teranishi, E.L. Wick, and I. Hornstein, 255–264. New York: Kluwer Academic.

2 Important Tastants and New Developments

Jakob Ley, Katharina Reichelt, Katja Obst, Gerhard Krammer, and Karl-Heinz Engel

CONTENTS

2.1 INTRODUCTION

Taste and odor are among the most important factors influencing the selection of food by humans. Tastants are perceived mainly in the oral cavity, especially on the tongue and the soft palate (Smith and Margolskee 2001). Here, a food is "analyzed" for the last time before being swallowed to avoid the intake of tainted food or toxic compounds. Five basic tastes are distinguished: sweet, sour, salty, bitter, and umami. Salty taste is involved in regulating ion and water homeostasis in the body, whereas sweet and umami tastes are responsible for estimating the energy content of food (especially carbohydrates and amino acids). Both sour and bitter tastes may act as warning mechanisms to avoid the consumption of unripe fruits or the intake of toxins (Glendinning 1994; Lindemann 1996). In addition, sour taste is considered to be a protection against immoderate acid intake to avoid mismatches in acid/base balance in the body (Roper 2007).

Apart from the five basic tastes, a number of further taste qualities are under discussion. Among them are fat taste (Laugerette et al. 2007), calcium taste (Tordoff and Sandell 2009), and kokumi taste, standing for mouthfulness, mouth feel, and richness of taste (Dunkel et al. 2007; Schlichtherle-Cerny et al. 2003; Toelstede and Hofmann 2009; Ueda et al. 1997).

Another important taste-related impression is the perception of pain caused by a number of compounds, responsible for cold, hot, and spicy taste of a food. Many traditionally used spices, such as chilli, pepper, paracress, mustard, or ginger, contain pungent, tingling, or "heating" compounds. Cooling sensations are caused by compounds such as (–)-menthol. Astringency is a very important (off-) taste quality. It is heavily discussed whether it may be mainly a physical effect, which is caused by precipitation of saliva proteins by astringent compounds, or a real taste or trigeminal quality, or even a combination of both (Bajec and Pickering 2008).

In this short review, it is not possible to cover all known tastants, and therefore we will focus on a selection of the most important compounds of each category and some highlights found just recently. Flavor- and/or taste-modifying compounds are reviewed in Chapter 8 of this book.

2.2 BITTER TASTANTS

In contrast to most other taste qualities, the number of bitter-tasting molecules is very high and shows a wide structural variance. For food bitterness, some common polyphenols such as the widely distributed catechins (e.g., epigallocatechin gallate from green tea) and the lower molecular weight proanthocyanidins (e.g., procyanidine B2 from roasted cocoa nibs) (Stark et al. 2006); phenolic glycosides such as naringin and neohesperidin from citrus pericarps; β-amygdalin from bitter almonds; terpenoids such as thujone from sage oils; limonin from citrus fruits; curcubitacins occurring in the family of *Cucurbitaceae*; *iso*-α-acids such as *cis*-isohumolone from hop (Fritsch and Shellhammer 2008); iridoids, for example, oleuropein from olives; alkaloids such as caffeine from coffee or tea; and theobromine from cocoa are important examples (Figure 2.1). Over the centuries, one of the intentions of plant selection, cultivation, breeding, and processing has been to reduce the content of these bitter compounds. In contrast to these undesired compounds, certain bitter principles are added to food or beverages for taste reasons, for example, quinine to bitter lemon soft drinks, and hop extracts to beer.

Another class of bitter components are the small hydrophobic peptides. These peptides, especially the ones with hydrophobic amino acids, such as leucine, isoleucine, valine, or proline at the N or C terminus, frequently exhibit a bitter taste and are often generated during aging or fermentation processes starting from larger proteins, for example, during cheese maturing (Maehashi and Huang 2009; Solms 1969; Toelstede and Hofmann 2008). As a recent example, the pentapeptide Asn-Ala-Leu-Pro-Glu isolated from soybean glycinine shows a bitter threshold of 74 μmol/L (Kim et al. 2008).

Unlike other taste-active compounds, bitter-tasting molecules often show low thresholds to avoid poisoning by high concentrations of toxins (Roper 2007). Examples for extremely bitter-tasting compounds are denatonium benzoate with a taste threshold of 0.01 μmol/L (Meyerhof 2005), quinine (1.6 μmol/L), limonin (2.1 μmol/L) (Glendinning 1994), and *cis*-isohumolone (10 μmol/L) (Intelmann et al. 2009), whereas caffeine (133 μmol/L) (Stevens et al. 2001) and epigallocatechin gallate (190 μmol/L) (Stark et al. 2006) are moderate bitter taste-eliciting substances.

Identification of new bitter components in raw or processed foods is also very important to support quality assurance and to prevent taste defects. The bitter off-taste observed in some batches of carrots that can deteriorate carrot-based baby food

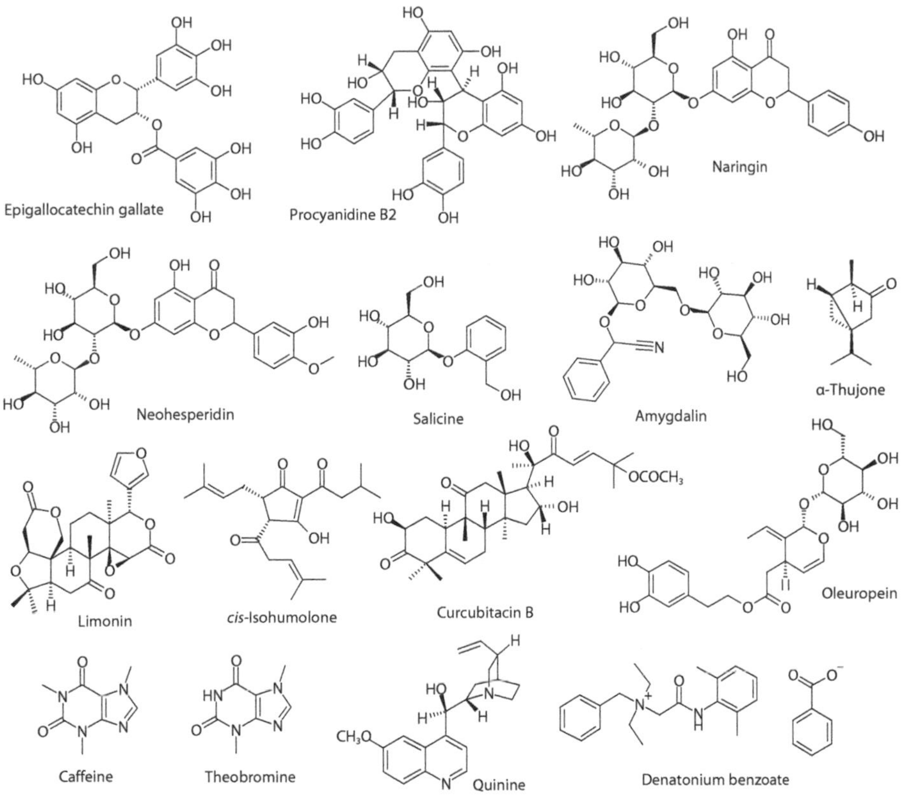

FIGURE 2.1 Typical bitter-tasting molecules that occur naturally or are added to foods.

was identified by using taste dilution analysis (TDA) to be falcarindiol (Czepa and Hofmann 2003, 2004) (Figure 2.2). Some chlorogenic acid-derived lactones (e.g., 5-*O*-caffeoyl-muco-γ-quinide, detection threshold 29 µmol/L) could be detected to contribute significantly to the bitter taste of coffee brew (Frank et al. 2008). The cyclic octapeptide cyclolinopeptide E was found to contribute to the bitter off-taste of some linseed oil qualities after storage (Brühl et al. 2008). Some 1-oxo-2,3-dihydro-1*H*-indolizinium-6-olates could be identified via TDA as very potent bitter (threshold 0.25 µmol/kg water) compounds of Maillard reactions based on xylose, rhamnose, and alanine (Frank et al. 2003).

2.3 SWEET TASTANTS

In food preparations, carbohydrates, especially sucrose, glucose, and fructose, are widely used because of their sweet taste and mouth feel. However, not only carbohydrates evoke sweet taste in humans but also a number of structurally diverse compounds, showing high sweet intensities at low levels of use. These high intensity sweeteners (HIS) can be both artificial and of natural origin, their structures ranging from small molecules such as saccharin to highly complex proteins, such as

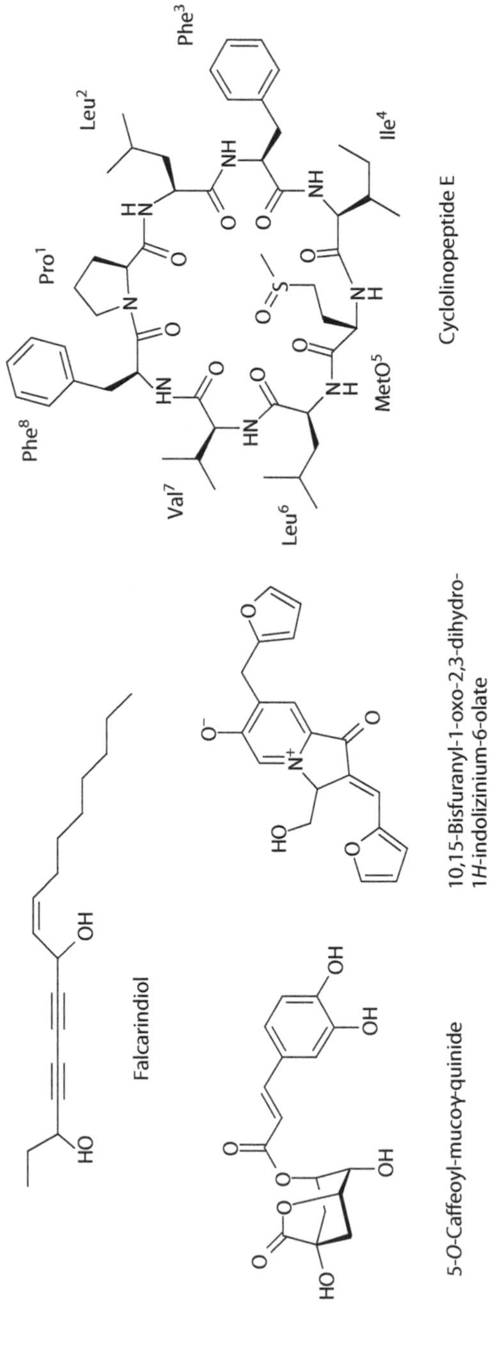

FIGURE 2.2 New bitter compounds found in foods by modern approaches of taste analysis.

thaumatin or monellin (Duffy et al. 2004; Kinghorm and Soejarto 1986). Commonly used artificial HIS (Figure 2.3) include saccharin (approved as food additive in United States and the European Union [EU]), which was the first commercially used artificial HIS; cyclamate (EU only); aspartame (United States, EU); acesulfame K (United States, EU); and sucralose (United States, EU).

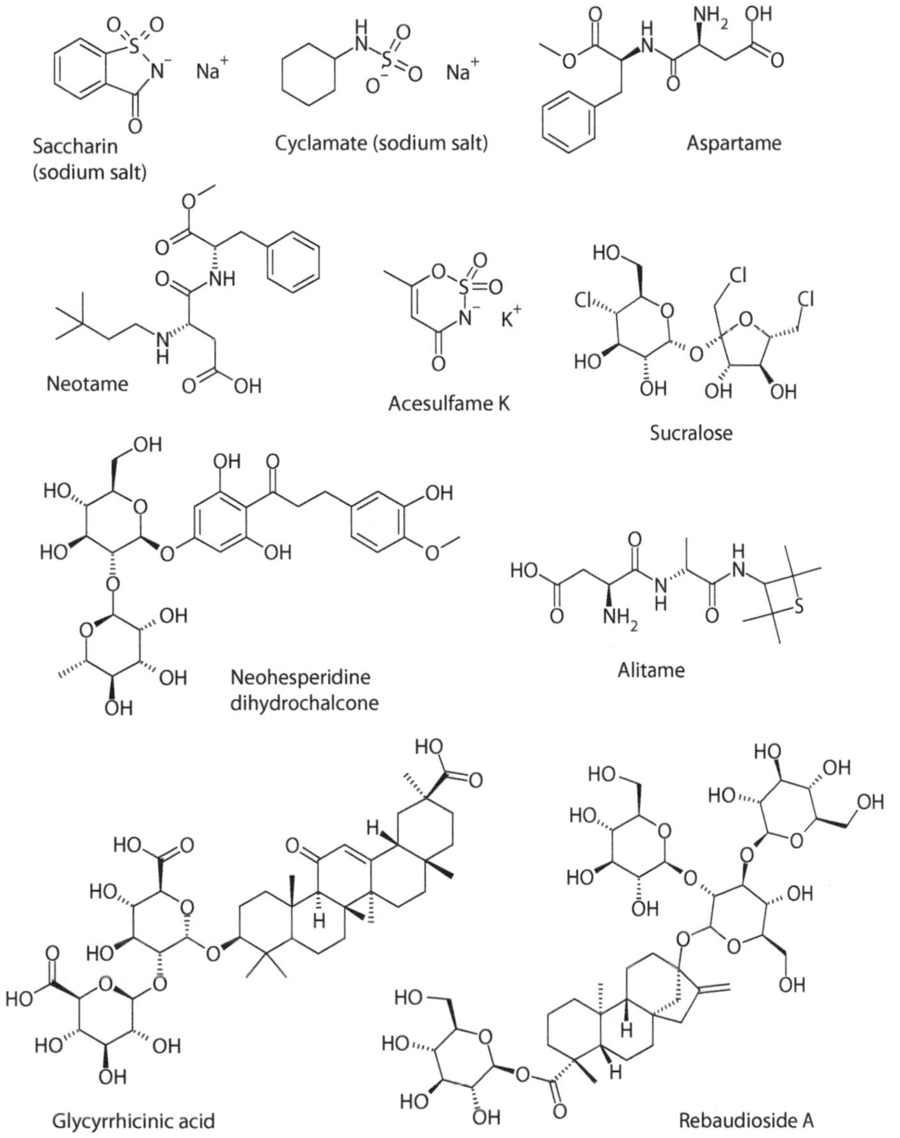

Saccharin (sodium salt)

Cyclamate (sodium salt)

Aspartame

Neotame

Acesulfame K

Sucralose

Neohesperidine dihydrochalcone

Alitame

Glycyrrhicinic acid

Rebaudioside A

FIGURE 2.3 Important HIS approved as food additives by FDA and/or European Commission.

Saccharin is considered to be 450 times sweeter than a 10% sucrose solution, but a typically bitter taste occurs at higher saccharin concentrations, which can be detected only by about 25% of the European population (Helgren et al. 1955). Cyclamate is 40 times sweeter than a 2% sucrose solution. Aspartame shows a 340 times higher sweet intensity in comparison to a 0.34% iso-sweet sucrose solution (Belitz et al. 2007) and is described to have a clean sweet taste without any bitter or metallic off-notes. The new high-potency sweetener neotame (United States only), a derivative of aspartame, is considered to be 30–60 times sweeter than aspartame. Because of its lower use levels, it can also, unlike aspartame, be consumed by patients with phenylketonuria (Stargel et al. 2001). Sucralose is approximately 750 times sweeter compared to a 2% sucrose solution, while the taste is perceived similar to that of sucrose: sweet without any off-tastes (Goldsmith and Merkel 2001). Other HIS include neohesperidine dihydrochalcone and alitame, a representative of the series of L-aspartyl-D-alanine amides.

HIS, however, are not only obtained as artificial compounds, but occur also naturally (Kim and Kinghorn 2002; Kinghorn and Compadre 2001). Currently, glycyrrhicinic acid from the rhizomes and roots of licorice (*Glycyrrhiza glabra* L. Fabaceae), thaumatin from the fruit of *Thaumatococcus daniellii* (Bennett) Benth (Marantaceae), and rebaudioside A from *Stevia rebaudiana* (Carakostas et al. 2008) are the most important naturally occurring HIS of commercial interest.

2.4 UMAMI AND KOKUMI TASTANTS

Until recently, only the amino acids glutamic acid and to a lesser extent aspartic acid and their corresponding salts, especially the monosodium glutamate (MSG), and some peptides were described as umami tastants. Some nucleotides such as guanosine 5′-monophosphate or inosine 5′-monophosphate also show a weak intrinsic umami taste, but more important is their ability to enhance synergistically the umami taste of MSG. In the past 5 years, several new high intensity umami-tasting compounds were found; recently, a review regarding the newest developments was given (Winkel et al. 2008), but meanwhile some new structures were published (Looft et al. 2008a, 2008b). In contrast to MSG (umami threshold about 0.2 mmol/L; Kaneko et al. 2006) and all other known umami tastants, these particular molecules are much stronger in activity (100–1000 times compared to MSG) and exhibit umami taste without any additional sodium ions or other peptides. Examples are shown in Figure 2.4.

Some new peptide-like kokumi-tasting compounds were described recently (Figure 2.4). Starting from earlier reports that glutathione-rich foods (Ueda et al. 1997) as well as some seasonings containing fermented peptides (Yamanaka 2006) exhibit a "kokumi" taste quality, some γ-glutamyl peptides such as γ-glutamylcysteine-β-alanine occurring in beans were found to contribute to this taste direction (Dunkel et al. 2007). Furthermore, γ-glutamyl dipeptides such as γ-Glu–Met in mature Gouda cheese (Toelstede et al. 2009) and other cheeses produced or ripened by *Penicillium roqueforti* (Toelstede and Hofmann 2009) imparting kokumi taste could be identified. From dry scallop, (*R*)-strombine was isolated as a component contributing to the dry mouth feel of umami taste (Starkenmann 2009).

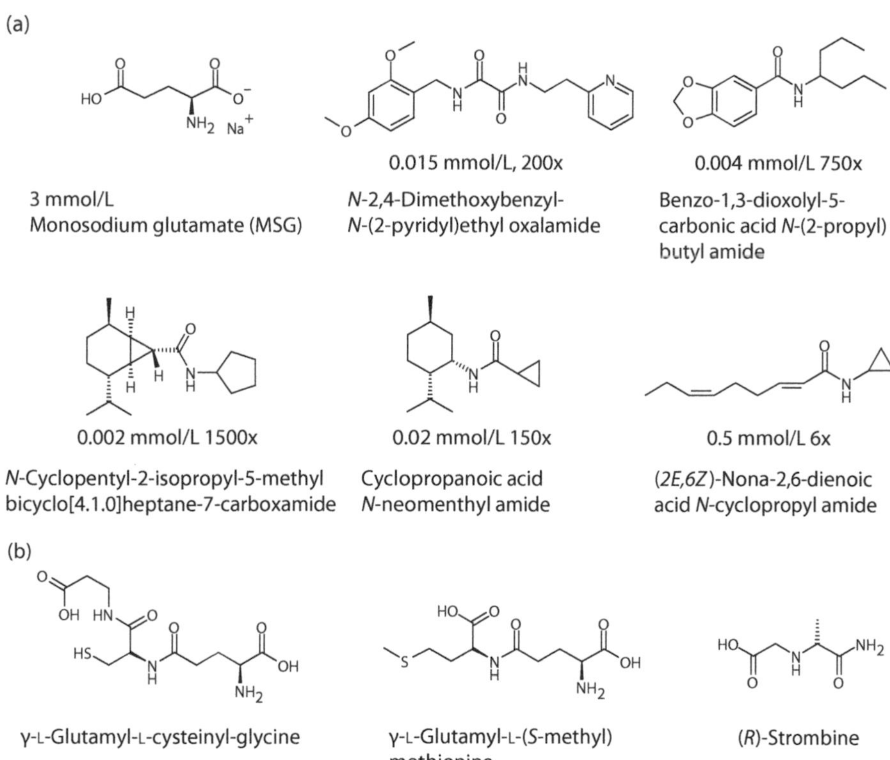

FIGURE 2.4 Umami tastants. (a) Concentrations given are those exhibiting nearly the same umami impression as 3 mmol/L MSG. (b) New kokumi tastants.

2.5 SALTY AND SOUR COMPOUNDS

A clean salt taste can only be found for sodium chloride, the most intense salty-tasting salt. LiCl and NH_4Cl are also salty but cannot be used for human consumption because of safety concerns or off-tastes, respectively. Therefore, KCl as the next relative is often used as a NaCl replacer, but shows strong bitter and other off-tastes. In the literature, reports can be found about salty-tasting compounds such as glycine ethyl ester (Kawai et al. 2008) or choline chloride (Locke and Fielding 1994), but in general, the compounds do not intrinsically taste salty but in fact enhance saltiness of sodium salts.

There are no reliable reports about molecules exhibiting a sour taste without being protic acids. Various studies, however, showed that naturally occurring weak organic acids, such as citric, succinic, malic, or lactic acid, are perceived as more sour in the oral cavity than hydrochloric acid at the same pH (Ganzevles and Kroeze 1987; Makhlouf and Blum 1972). This leads to the conclusion that undissociated acids are also involved in the sour taste perception of organic acids. Unlike pH value, titratable acidity involves both free and bound hydrogen ions in a solution (Da Conceicao Neta et al. 2007). It was shown that sourness intensity is positively correlated with increasing titratable acidity at a given pH (Makhlouf and Blum 1972). The perceived sourness of different organic acids is positively correlated with their dissociation constant. Titration experiments with several organic acids showed correlations with the rank order of the perceived sourness (Ganzevles and Kroeze 1987).

2.6 HEATING, PUNGENT, AND TINGLING COMPOUNDS

Warming, heating, or pungent sensations are caused mostly by substances influencing the vanilloid receptor TRPV1 or the pain receptor TRPA1 expressed on free nerve endings of the trigeminal nerve system (Patapoutina et al. 2003). These effects are not elicited by real increase in physical temperature but by *lowering* the temperature threshold of the free nerve endings. As a result, the nerves are firing at body temperature and not only at moderate or noxious heat as in the nonactivated state. The exact description of the so-called chemestethic effects depends on the compound and its concentration; whereas low amounts of, for example, capsicum extract elicits only a mild and pleasant warming effect, the sensation caused by higher concentrations is described as "pungent," "hot," or even "burning like fire." The most important flavor compounds showing these effects (Figure 2.5) are capsaicin and nonivamide from *Capsicum* extracts (*Capsicum* ssp.); piperine from pepper (*Piper nigrum*) (Szallasi 2005); [8]-gingerole and related compounds from ginger (*Zingiber* ssp.) and related species (Banno and Mukaiyama 1976; Kikuzaki 2000); [6]-paradol from "Grains of Paradise" (*Aframomum melegueta*) (Fernandez et al. 2006); some alkamides such as *cis*-pellitorine from tarragon (*Artemisia dracunculus*) (Ley et

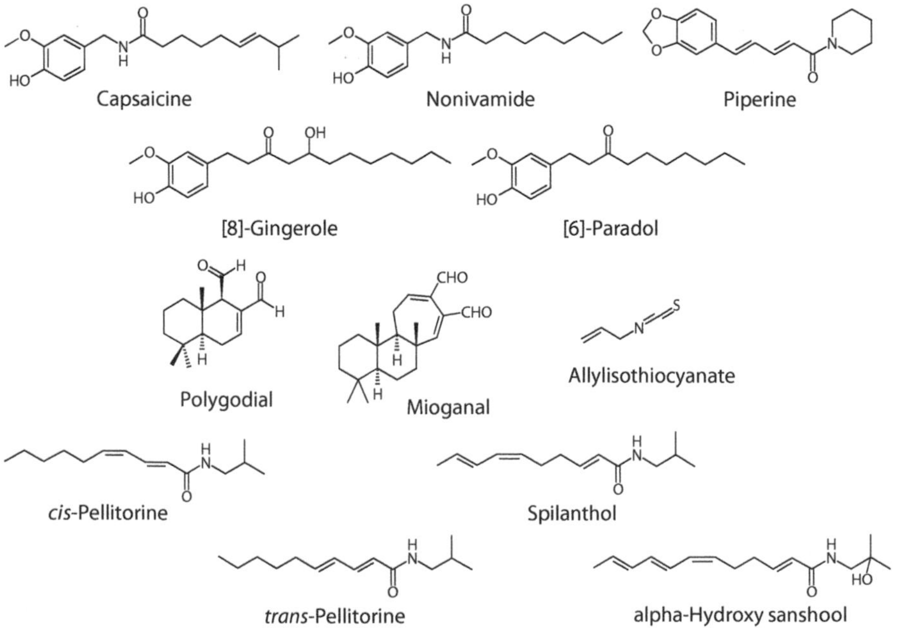

Capsaicine

Nonivamide

Piperine

[8]-Gingerole

[6]-Paradol

Polygodial

Mioganal

Allylisothiocyanate

cis-Pellitorine

Spilanthol

trans-Pellitorine

alpha-Hydroxy sanshool

FIGURE 2.5 Typical heating, pungent, and tingling compounds found in nature.

al. 2004); terpenoidal dialdehydes such as polygodial from hydropiper (*Polygonum hydropiper*) (Starkenmann et al. 2006); mioganal from *Zingiber mioga* (Abe et al. 2008); and isothiocyanates such as the lachrymator allyl isothiocyanate from Brassicacea (Jordt et al. 2004).

The alkamides *trans*-pellitorine (found in, e.g., *P. nigrum*; Subehan et al. 2006), spilanthol (found in, e.g., paracress, *Spilanthes acmella*; Ramsewak et al. 1999), and alpha-hydroxy sanshool (from *Zanthoxylum schinifolium*; Iseli et al. 2007) show a so-called "tingling" effect on the tongue. It is more an irritating sensory experience and accompanied by a numbing feeling. In addition, they can induce salivation (Ley and Simchen 2007).

2.7 COOLING COMPOUNDS

The most important flavor compound exhibiting a physiological cooling effect is (–)-menthol. Similar to heating, the cooling effect is not caused by a real physical decrease in temperature but by increasing the temperature threshold of the cool-sensitive free nerve endings of the trigeminal nerve (or the corresponding nerves of the dorsal root ganglion for the lower parts of the body) (Zanotto et al. 2007) resulting in activation of the nerves at body temperature. This pseudo-cooling effect can be caused by binding of (–)-menthol to the ion channel TRPM8 (Patapoutina et al. 2003).

(–)-Menthol is broadly used for mint-flavored products such as toothpaste, chewing gum, and mouthwash. Unfortunately, it is difficult to apply to non-mint-flavored products because the threshold for its aroma value (0.1–0.2 mg/kg water) is roughly 10 times lower compared to the threshold for its cooling effect (0.9–1.9 mg/kg water) (Ottinger et al. 2001). Consequently, cooling compounds with a much weaker aroma value but with a comparable cooling effect were developed during the past decades. Several comprehensive reviews about the most important cooling compounds were given in recent years (Eccles 1994; Erman 2004; Leffingwell 2009; Pringle and Brassington 2007; Watson et al. 1978). Therefore, only the most important cooling compounds as well as the established natural alternatives to menthol and some new developments are shown in Figure 2.6.

The menthol derivatives menthyl lactate and menthyl glycerol ether (Cooling Agent 10), and menthyl carbonates such as the propylene glycol carbonate (Frescolat MGC) are established cooling compounds as well as the menthone ketal of glycerol (Frescolat MGA). The menthane carbonic acid amides and some structural related hindered carbonic acid amides (WS compounds) were developed in the 1970s and show a very low aroma value and an increased cooling strength compared to menthol (Watson et al. 1978). Icilin was found to elicit a strong cooling effect in vitro and in vivo (Chuang et al. 2004) but was never commercially used. By using TRPM8 assays and structure–activity concepts, the cooling strength of the menthane carbonic acid amides could be potentiated, and as a result, the menthane carbonic acid 4-cyano-methylphenyl amide (Evercool 180) was found (Furrer et al. 2008).

Apart from (–)-menthol, only a small number of cooling compounds were found in nature: menthyl lactate isomers were described to occur in fermented *Mentha piperita* leaves (Gassenmeier 2006), menthyl succinate in *M. piperita* and *Lycium*

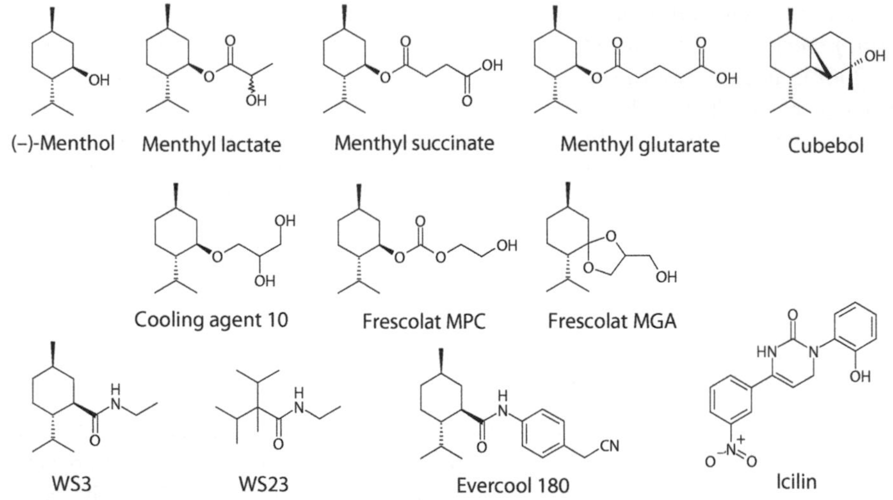

FIGURE 2.6 Important cooling compounds including natural alternatives to (–)-menthol (first row).

barbarum (Hiserodt et al. 2004; Marin and Schippa 2006), and menthyl glutarate in *Litchi chinesis* (Hiserodt et al. 2004). Cubebol is known as a constituent of fruits of false cubeb (*Piper lowong*) (De Rosa 1994; Velazco et al. 2000).

2.8 ASTRINGENCY

The perception of astringency is still not well understood: it may be caused by a more macroscopic event via precipitation of saliva proteins and subsequent alteration of the lubrication property of the saliva, binding and cross-linking of proteins by astringent compounds to the surface of the mucous membrane, or even binding of the compounds to still unknown receptors or ion channels (Bajec and Pickering 2008; Rossetti et al. 2009). Typical astringents are alumina, ferrous and zinc salts; catechol- and especially gallate-type polyphenols such as epigallocatechin gallate; gallic acid esters of carbohydrates (ester type tannins such as tannic acid, hamameli tannins); condensed catechins such as proanthocyanidine A2 (Haslam 2007); and flavonol glycosides such as rutin (Scharbert and Hofmann 2005) (Figure 2.7).

The detection thresholds differ widely for several well-known astringent compounds. Rutin shows a very low threshold of 1 μg/kg (Scharbert et al. 2004), a concentration at which the protein precipitation cannot play any significant role. In contrast, epigallocatechin gallate has a threshold of 87 mg/kg (Scharbert et al. 2004) and is also able to cause protein aggregation (Nayak and Carpenter 2008). Further astringent substances with low thresholds and uncommon structures are shown in Figure 2.8. As examples, 3-carboxymethyl-indole-1*N*-β-D-glucopyranoside (1 μmol/L) and 3-methylcarboxymethyl-indole-1*N*-β-D-glucopyranoside (0.3 μmol/L), and the

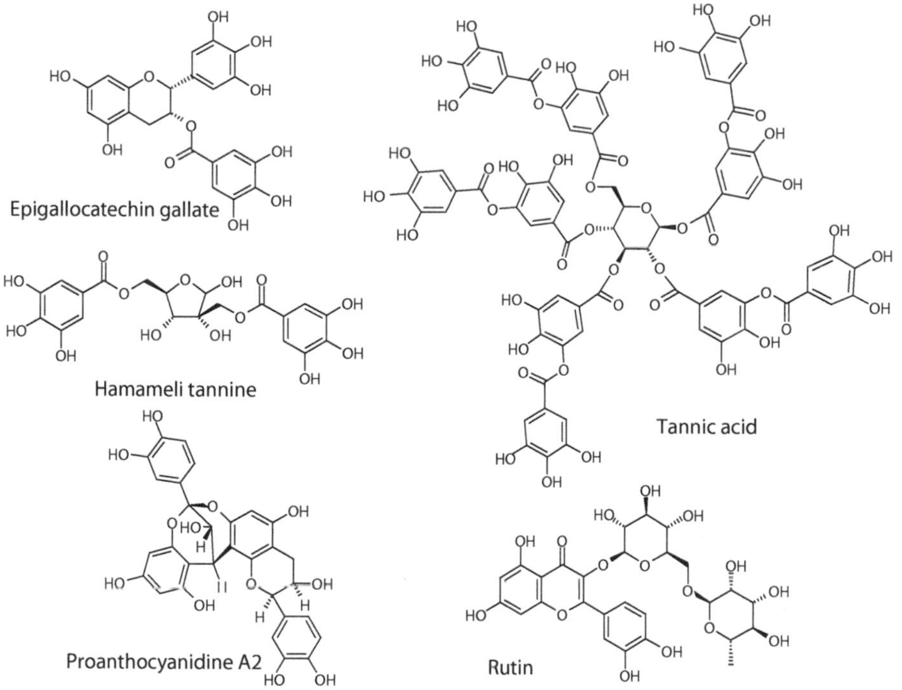

FIGURE 2.7 Polyphenolic astringent compounds.

non-cyanogenic nitriles 2-(4-hydroxybenzoyloxymethyl)-4-β-D-glucopyranosyloxy-2(*E*)-butenenitrile (5.9 mmol/L) and 2-(4-hydroxy-3-methoxybenzoyloxymethyl)-4-β-D-glucopyranosyloxy-2(*E*)-butenenitrile (1.2 mmol/L) were isolated from red currant (Schwarz and Hofmann 2007). Cinnamic acid amides of amino acids such as clovamide (9.3 mg/kg) could be identified in cocoa nibs (Stark et al. 2006; Stark and Hofmann 2005).

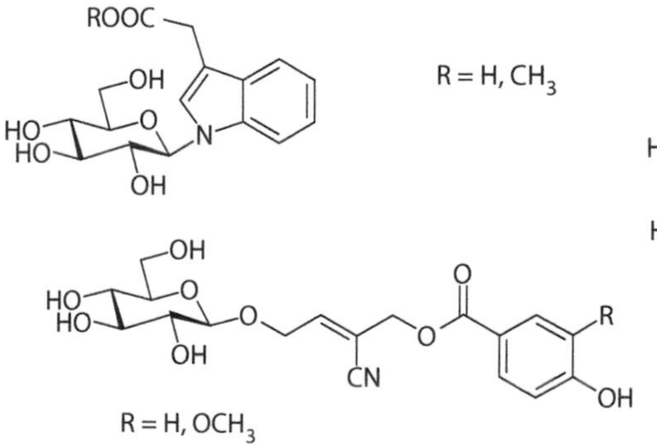

FIGURE 2.8 Newly found astringent compounds with considerably low taste thresholds.

2.9 CONCLUSION AND OUTLOOK

Most tastants do not occur as neat compounds in nature, but in mixtures and even as single molecules, they can exhibit more than one taste (or even aroma) quality. Consequently, the detection and structure elucidation of newly found tastants is still a challenge for analytical and flavor chemists. Nevertheless, there are still surprising new compounds as exemplified for the class of high intense umami-tasting molecules that were not known until several years ago. Whereas detection of bitter molecules is crucial for the identification of off-flavors and especially off-tastes, the finding of new high intensity sweet, umami, or other taste qualities is of high importance for the development of new flavorings. Additionally, the discussion is ongoing whether further taste qualities and their specific tastants can be found and unequivocally be described.

REFERENCES

Abe, M., Y. Ozawa, Y. Morimitsu, and K. Kubota. 2008. Mioganal, a novel pungent principle in myoga (*Zingiber mioga* Roscoe) and a quantitative evaluation of its pungency. *Biosci. Biotechnol. Biochem.* 72: 2681–2686.

Bajec, M.R., and G.J. Pickering. 2008. Astringency: Mechanisms and perception. *Crit. Rev. Food Sci. Nutr.* 48: 858–875.

Banno, K., and T. Mukaiyama. 1976. A new synthesis of the pungent principle of ginger—Zingerone, gingerol and shogaol. *Bull. Chem. Soc. Jpn.* 49: 1453.

Belitz, H.D., W. Grosch, and P. Schieberle. 2007. *Lehrbuch der Lebensmittelchemie*, 6th edn. Berlin: Springer.

Brühl, L., B. Matthäus, A. Scheipers, and T. Hofmann. 2008. Bitter off-taste in stored cold-pressed linseed oil obtained from different varieties. *Eur. J. Lipid Sci. Technol.* 110: 625–631.

Carakostas, M.C., L.L. Curry, A.C. Boileau, and D.J. Brusick. 2008. The history, technical function and safety of rebaudioside A, a naturally occurring steviol glycoside, for use in food and beverages. *Food Chem. Toxicol.* 46: S1–S10.

Chuang, H.H., W.M. Neuhausser, and D. Julius. 2004. The super-cooling agent icilin reveals a mechanism of coincidence detection by a temperature-sensitive TRP channe. *Neuron* 43: 859–869.

Czepa, A., and T. Hofmann. 2003. Structural and sensory characterization of compounds contributing to the bitter off-taste of carrots (*Daucus carota* L.) and carrot puree. *J. Agric. Food Chem.* 51: 3865–3873.

Czepa, A., and T. Hofmann. 2004. Quantitative studies and sensory analyses on the influence of cultivar, spatial tissue distribution, and industrial processing on the bitter off-taste of carrots (*Daucus carota* L.) and carrot products. *J. Agric. Food Chem.* 52: 4508–4514.

Da Conceicao Neta, E.R., S.D. Johanningsmeier, and R.F. McFeeters. 2007. The chemistry and physiology of sour taste—A review. *J. Food Sci.* 72: R33–R38.

De Rosa, S., S. De Guilio, C. Iodice, and N. Zavodink. 1994. Sesquiterpenes from the brown alga *Taonia atomaria*. *Phytochemistry* 37: 1327–1330.

Duffy, V.B., M. Sigman-Grant, M.A. Powers, et al. 2004. Position of the American dietetic association: Use of nutritive and nonnutritive sweeteners. *J. Am. Diet. Assoc.* 104: 255–275.

Dunkel, A., J. Koester, and T. Hofmann. 2007. Molecular and sensory characterization of gamma-glutamyl peptides as key contributors to the kokumi taste of edible beans (*Phaseolus vulgaris* L.). *J. Agric. Food Chem.* 55: 6712–6719.

Eccles, R. 1994. Menthol and related cooling compounds. *J. Pharm. Pharmacol.* 46: 618–630.

Erman, M. 2004. Progress in physiological cooling agents. *Perfumer & Flavorist* 29: 34–50.
Fernandez, X., C. Pintaric, L. Lizzani-Cuvelier, A.-M. Loiseau, A. Morello, and P. Pellerin. 2006. Chemical composition of absolute and supercritical carbon dioxide extract of *Aframomum melegueta*. *Flavour Fragrance J.* 21: 162–165.
Frank, O., S. Blumberg, G. Krümpel, and T. Hofmann. 2008. Structure determination of 3-*O*-caffeoyl-epi-quinide, an orphan bitter lactone in roasted coffee. *J. Agric. Food Chem.* 56: 9581–9585.
Frank, O., M. Jezussek, and T. Hofmann. 2003. Sensory activity, chemical structure, and synthesis of maillard generated bitter-tasting 1-*Oxo*-2,3-dihydro-1*H*-indolizinium-6-olates. *J. Agric. Food Chem.* 51: 2693–2699.
Fritsch, A., and T.H. Shellhammer. 2008. Relative bitterness of reduced and nonreduced *iso*-alpha-acids in lager beer. *J. Am. Soc. Brew. Chem.* 66: 88–93.
Furrer, S.M., J.P. Slack, S.T. McCluskey, et al. 2008. New developments in the chemistry of cooling compounds. *Chemosens. Percept.* 1: 119–126.
Ganzevles, P.G.J., and J.H.A. Kroeze. 1987. The sour taste of acids. The hydrogen ion and the undissociated acid as sour agents. *Chem. Sens.* 12: 563–576.
Gassenmeier, K. 2006. Identification and quantification of L-menthyl lactate in essential oils from *Mentha arvensis* L. from India and model studies on the formation of L-menthyl lactate during essential oil production. *Flavour Fragrance J.* 21: 725–730.
Glendinning, J.I. 1994. Is bitter taste rejection always adaptive? *Physiol. Behav.* 56: 1217–1227.
Goldsmith, L.A., and C.M. Merkel. 2001. Sucralose. In *Alternative Sweeteners*, ed. L. O'Brien Nabors, 185–207. Basel, Switzerland: Marcel Dekker, Inc.
Haslam, E. 2007. Vegetable tannins—Lessons of a phytochemical lifetime. *Phytochemistry* 68: 2713–2721.
Helgren, F.J., M.J. Lynch, and F.J. Kirchmeyer. 1955. A taste panel study of saccharin 'off-taste'. *J. Am. Pharmaceut. Assoc.* 44: 353–355.
Hiserodt, R.D., J. Adedeji, T.V. John, and M.L. Dewis. 2004. Identification of monomenthyl succinate, monomenthyl glutarate, and dimenthyl glutarate in nature by high performance liquid chromatography–tandem mass spectrometry. *J. Agric. Food Chem.* 52: 3536–3541.
Intelmann, D., C. Batram, C. Kuhn, G. Haseleu, W. Meyerhof, and T. Hofmann. 2009. Three TAS2R bitter taste receptors mediate the psychophysical responses to bitter compounds of hops (*Humulus lupulus* L.) and beer. *Chemosens. Percept.* 2: 118–132.
Iseli, V., O. Potterat, L. Hagmann, J. Egli, and M. Hamburger. 2007. Characterization of the pungent principles and the essential oil of *Zanthoxylum schinifolium* pericarp. *Pharmazie* 62: 396–400.
Jordt, S.E., D.M. Bautista, H.H. Chuang, et al. 2004. Mustard oils and cannabinoids excite sensory nerve fibres through the TRP channel ANKTM1. *Nature* 427: 260–265.
Kaneko, S., K. Kumazawa, H. Masuda, A. Henze, and T. Hofmann. 2006. Molecular and sensory studies on the umami taste of Japanese green tea. *J. Agric. Food Chem.* 54: 2688–2694.
Kawai, T., Y. Kusakabe, and T. Ookura. 2008. Behavioral estimation of enhancing effects in taste enhance. *Chem. Sens.* 33: J6.
Kikuzaki, H. 2000. Ginger for drug and spice purposes. In: *Herbs, Botanicals and Teas*, ed. Mazza, G.O., and B. Dave, 75–105. Lancaster, USA: Technomic Publishing Ltd.
Kim, M.-R., Y. Kawamura, K.M. Kim, and C.-H. Lee. 2008. Tastes and structures of bitter peptide, asparagine–alanine–leucine–proline–glutamate, and its synthetic analogues. *J. Agric. Food Chem.* 56: 5852–5858.
Kim, N.C., and A.D. Kinghorn. 2002. *Sweet Tasting and Sweetness-Modifying Constituents of Plant*. New York: Elsevier Science B.V.
Kinghorm, D.A., and D.D. Soejarto. 1986. Sweetening agents of plant origin. *CRC Crit. Rev. Plant Sci.* 4: 79–120.
Kinghorn, A.D., and C.M. Compadre. 2001. Less common high-potency sweeteners. *Food Sci. Technol. (NY)* 112: 209–233.

Laugerette, F., D. Gaillard, P. Passilly-Degrace, I. Niot, and P. Besnard. 2007. Do we taste fat? *Biochimie* 89: 265–269.

Leffingwell, J.C. 2009. Cool without menthol and cooler than menthol and cooling compounds as insect repellents. Available at http://www.leffingwell.com/cooler_than_menthol.htm.

Ley, J.P., J.-M. Hilmer, B. Weber, G. Krammer, I.L. Gatfield, and H.-J. Bertram. 2004. Stereoselective enzymatic synthesis of *cis*-pellitorine, a taste active alkamide naturally occurring in tarragon. *Eur. J. Org. Chem.* (24):5135–5140.

Ley, J.P., and U. Simchen. 2007. Quantification of the saliva-inducing properties of pellitorine and spilanthol. In *Recent Highlights in Flavor Chemistry and Biology*, ed. T. Hofmann, W. Meyerhof, and P. Schieberle, 365–368. Proceedings of the Wartburg Symposium on Flavor Chemistry and Biology, 8th, Eisenach, Germany, Feb. 27–Mar. 2, 2007. Garching, Germany: Deutsche Forschungsanstalt für Lebensmittelchemie.

Lindemann, B. 1996. Taste reception. *Physiol. Rev.* 76: 719–766.

Locke, K.W., and S. Fielding. 1994. Enhancement of salt intake by choline chloride. *Physiol. Behav.* 55: 1039–1046.

Looft, J., T. Voessing, and M. Backes. 2008a. Substituted bicyclo[4.1.0]heptane-7-carboxylic acid amides and derivatives thereof as food flavor substances. WO 2008 046,895. Germany, Symrise GmbH & Co. KG.

Looft, J., T. Voessing, J. Ley, M. Backes, and M. Blings. 2008b. Substituted cyclopropanecarboxylic acid (3-methyl-cyclohexyl)amide as flavoring substance. EP 1,989,944. Germany, Symrise GmbH & Co. KG.

Maehashi, K., and L. Huang. 2009. Bitter peptides and bitter taste receptors. *Cell. Mol. Life Sci.* 66: 1661–1671.

Makhlouf, G.M., and A.L. Blum. 1972. Kinetics of the taste response to chemical stimulation: A theory of acid taste in man. *Gastroenterology* 63: 67–75.

Marin, C., and C. Schippa. 2006. Identification of monomenthyl succinate in natural mint extracts by LC-ESI-MS-MS and GC-MS. *J. Agric. Food Chem.* 54: 4814–4819.

Meyerhof, W. 2005. Elucidation of mamalian bitter taste. *Rev. Physiol. Biochem. Pharmacol.* 154: 37–72.

Nayak, A., and G.H. Carpenter. 2008. A physiological model of tea-induced astringency. *Physiol. Behav.* 95: 290–294.

Ottinger, H., T. Soldo, and T. Hoffmann. 2001. Systematic studies on structure and physiological activity of cyclic alpha-keto enamines, A novel class of "cooling compounds." *J. Agric. Food Chem.* 49: 5383–5390.

Patapoutina, A., A.M. Peier, G.M. Story, and V. Viswanath. 2003. ThermoTrp channels and beyond: Mechanism of temperature sensation. *Nat. Rev.* 4: 529–539.

Pringle, S., and D. Brassington. 2007. Physiological coolants—New materials and emerging applications. *Perfum. Flavor.* 32: 38–42.

Ramsewak, R.S., A.J. Erickson, and M.G. Nair. 1999. Bioactive *N*-isobutylamides from the flower buds of *Spilanthes acmella*. *Phytochemistry* 51: 729–732.

Roper, S.D. 2007. Signal transduction and information processing in mammalian taste buds. *Eur. J. Physiol.* 454: 759–776.

Rossetti, D., J.H.H. Bongaerts, E. Wantling, J.R. Stokes, and A.-M. Williamson. 2009. Astringency of tea catechins: More than an oral lubrication tactile percept. *Food Hydrocolloids* 23: 1984–1992.

Scharbert, S., and T. Hofmann. 2005. Molecular definition of black tea taste by means of quantitative studies, taste teconstitution, and omission experiments. *J. Agric. Food Chem.* 53: 5377–5384.

Scharbert, S., N. Holzmann, and T. Hofmann. 2004. Identification of the astringent taste compounds in black tea infusions by combining instrumental analysis and human bioresponse. *J. Agric. Food Chem.* 52: 3498–3508.

Schlichtherle-Cerny, H., M. Affolter, and C. Cerny. 2003. Taste-active glycoconjugates of glutamate: New umami compounds. In *Challenges in Taste Chemistry and Biology*, ed. T. Hofmann, C.H. Ho, and W. Pickenhagen, 210–222. Washington, DC: American Chemical Society.
Schwarz, B., and T. Hofmann. 2007. Isolation, structure determination, and sensory activity of mouth-drying and astringent nitrogen-containing phytochemicals isolated from red currants (*Ribes rubrum*). *J. Agric. Food Chem.* 55: 1405–1410.
Smith, D., and R. Margolskee. 2001. Making sense of taste. *Sci. Am.* 284: 32–39.
Solms, J. 1969. Taste of amino acids, peptides, and proteins. *J. Agric. Food Chem.* 17: 686–688.
Stargel, W.W., D.A. Mayhew, C.P. Comer, S.E. Andress, and H.H. Butchko. 2001. Neotame. In *Alternative Sweeteners*, ed. L. O'Brien Nabors, 129–145. Basel, Switzerland: Marcel Dekker, Inc.
Stark, T., S. Bareuther, and T. Hofmann. 2006. Molecular definition of the taste of roasted cocoa nibs (*Theobroma cacao*) by means of quantitative studies and sensory experiment. *J. Agric. Food Chem.* 54: 5530–5539.
Stark, T., and T. Hofmann. 2005. Isolation, structure determination, synthesis, and sensory activity of *N*-phenylpropenoyl-L-amino acids from cocoa (*Theobroma cacao*). *J. Agric. Food Chem.* 53: 5419–5428.
Starkenmann, C. 2009. Contribution of (*R*)-strombine to dry scallop mouthfeel. *J. Agric. Food Chem.* 57: 7938–7943.
Starkenmann, C., L. Luca, Y. Niclass, E. Praz, and D. Roguet. 2006. Comparison of volatile constituents of *Persicaria odorata* (Lour.) Sojak (*Polygonum odoratum* Lour.) and *Persicaria hydropiper* L. Spach (*Polygonum hydropiper* L.). *J. Agric. Food Chem.* 54: 3067–3071.
Stevens, D.R., R. Seifert, B. Bufe, et al. 2001. Hyperpolarization-activated channels HCN1 and HCN4 mediate response to sour stimuli. *Nature* 413: 631–635.
Subehan, T. U., S. Kadota, and Y. Tezuka. 2006. Alkamides from *Piper nigrum* L. and their inhibitory activity against human liver microsomal cytochrome P450 2D6 (CYP2D6). *Nat. Prod. Commun.* 1: 1–7.
Szallasi, A. 2005. Piperine: Researchers discover new flavor in an ancient spice. *Trends Pharmacol. Sci.* 26: 437–439.
Toelstede, S., A. Dunkel, and T. Hofmann. 2009. A series of kokumi peptides impart the long-lasting mouthfulness of matured Gouda cheese. *J. Agric. Food Chem.* 57: 1440–1448.
Toelstede, S., and T. Hofmann. 2008. Sensomics mapping and identification of the key bitter metabolites in Gouda cheese. *J. Agric. Food Chem.* 56: 2795–2804.
Toelstede, S., and T. Hofmann. 2009. Kokumi-active glutamyl peptides in cheeses and their biogeneration by *Penicillium roqueforti*. *J. Agric. Food Chem.* 57: 3738–3748.
Tordoff, M.G., and M.A. Sandell. 2009. Vegetable bitterness is related to calcium content. *Appetite (Amsterdam, Netherlands)* 52: 498–504.
Ueda, Y., M. Yonemitsu, T. Tsubuku, M. Sakaguchi, and R. Miyajima. 1997. Flavor characteristics of glutathione in raw and cooked foodstuffs. *Biosci. Biotechnol. Biochem.* 61: 1977–1980.
Velazco, M.I., L. Wuensche, and P. Deladoey. 2000. Use of cubebol as a flavoring ingredient. EP 1,040,765. Switzerland: Firmenich S.A.
Watson, H.R., R. Hems, D.G. Rowsell, and D.J. Spring. 1978. New compounds with menthol cooling effect. *J. Soc. Cosmet. Chem.* 29: 185–200.
Winkel, C., A. de Klerk, J. Visser, et al. 2008. New developments in umami (enhancing) molecules. *Chem. Biodivers.* 5: 1195–1203.
Yamanaka, T. 2006. Peptide with strong "kokumi" in fermented seasonings. *Bio Industry* 23: 35–41.
Zanotto, K.L., A.W. Merrill, M.I. Carstens, and E. Carstens. 2007. Neurons in superficial trigeminal subnucleus caudalis responsive to oral cooling, menthol, and other irritant stimuli. *J. Neurophysiol.* 97: 966–978.

3 Smell, Taste, and Flavor

Han-Seok Seo and Thomas Hummel

CONTENTS

3.1 INTRODUCTION

During food consumption, we experience dynamic stimulations through the gustatory, olfactory, and trigeminal sensory systems. Accordingly, the International Standards Organization (ISO 1992, 2008) has defined flavor as a "complex combination of the olfactory, gustatory and trigeminal sensations perceived during tasting.

The flavor may be influenced by tactile, thermal, painful and/or kinesthetic effects." In addition, many studies have demonstrated that visual or auditory cues also influence flavor perception (Verhagen and Engelen 2006; Zampini et al. 2007, 2008; Shankar et al. 2009, 2010). That is, flavor perception is a multisensorial and dynamic sensation.

In this chapter, we will try to explain how flavors are perceived. First, the basic anatomy and physiology of three main sensory systems involved in flavor sensation (i.e., gustatory, trigeminal, and olfactory system) will be explored. Next, the two olfactory pathways will be introduced: ortho- and retronasal olfaction. In fact, during food consumption, aromatic volatile compounds of foods are delivered via both ortho- and retronasal pathways, but odors from either nose or mouth are perceived differently. Finally, interactions between olfactory and gustatory stimuli will be addressed.

3.2 CHEMOSENSORY SYSTEM

3.2.1 Gustatory System

3.2.1.1 Human Taste

In many cases, people use the word "taste" to describe sensations during eating and drinking. However, from the biological point of view, taste is restricted to sensations mediated by an anatomically and physiologically defined gustatory system (Bachmanov and Beauchamp 2007). As one of the basic sensory modalities, the sense of taste is essential to evaluate whether foods or drinks are safe or poisonous. Broadly speaking, sweet taste reflecting calorie or energy is accepted, whereas bitter taste signaling toxic or hazardous substances is rejected. In particular, a sense of taste (gustation), in addition to a sense of smell (olfaction), is closely related to food consumption. For example, a large portion of people with taste loss or distortion report that their eating behavior and/or nutritional status is altered (Kokal 1985; Mattes-Kulig and Henkin 1985; Mattes et al. 1990; Langius et al. 1993; Ritchie et al. 2002; Aschenbrenner et al. 2008b). In addition, a line of studies have demonstrated the association between eating disorder and gustatory function (Casper et al. 1980; Nakai et al. 1987; Tóth et al. 2004; Frank et al. 2006; Aschenbrenner et al. 2008b; Wagner et al. 2008; Wöckel et al. 2008). Specifically, people with eating disorders (e.g., anorexia nervosa and bulimia nervosa) show lower gustatory sensitivity than healthy controls, and they prefer sweet taste, but not fat stimuli (Casper et al. 1980; Nakai et al. 1987; Aschenbrenner et al. 2008b; Wöckel et al. 2008). Furthermore, compared to healthy controls, people with eating disorders show a different pattern of neural processing for the taste stimulation in the electroencephalogram (Tóth et al. 2004) and brain imaging (Frank et al. 2006; Wagner et al. 2008). For example, using functional magnetic resonance imaging (fMRI), Wagner et al. (2008) reported differences of neural activation for sucrose or water stimulation between women recovered from restricting-type anorexia nervosa and healthy control women. Specifically, in the primary gustatory cortical region (e.g., insular), neural activation was significantly lower in women who recovered from restricting-type anorexia nervosa than in healthy controls.

The current consensus is that human taste sensation is divided into five taste qualities labeled as sweet, sour, salty, bitter, and umami [or savory; the prototypical substance is the sodium salt of the amino acid glutamate (monosodium glutamate or MSG) and 5′-ribonucleotides, e.g., inositol monophosphate or guanidine monophosphate] (Bachmanov and Beauchamp 2007; Ninomiya and Beauchamp 2009). Human neonates seem to discriminate those taste qualities (Steiner 1974; Rosenstein and Oster 1988; Bergamasco and Beraldo 1990). For example, neonates show differential facial expressions depending on the taste quality presented (Steiner 1974; Rosenstein and Oster 1988; Bergamasco and Beraldo 1990). However, taste qualities are still dependent on experience; for example, the term "umami" is still unfamiliar to most people in the Western Hemisphere, although they are able to perceive and discriminate umami-tasting substances (Singh et al. 2010).

Moreover, there is growing evidence that "fat taste" is another basic taste quality (Khan and Besnard 2009; Mattes 2009a). Previous studies have demonstrated that dietary fat and its metabolites activate somatosensory, olfactory, and gustatory systems (Mattes 2005; Bachmanov and Beauchamp 2007; Khan and Besnard 2009; Mattes 2009a). Esterified fatty acids (e.g., triacylglycerol fatty acids) are not effective as gustatory stimuli (Mattes 2009a). However, in the oral cavity, lingual lipase hydrolyzes triglycerides and releases free fatty acids, which can access taste receptor cells and inhibit the delayed rectifying potassium channels in rats (Gilbertson et al. 1997; Bachmanov and Beauchamp 2007). Additionally, in mice and rats, the fatty acid transporter CD36 that is localized along the apical side of taste buds of foliate and circumvallate papillae, has a high affinity for long-chain fatty acids and is expressed in taste receptor cells (Laugerette et al. 2005; Bachmanov and Beauchamp 2007; Gaillard et al. 2008). That is, free fatty acids appear to be effective as gustatory stimulus (Mattes 2009a). It has been shown that humans may detect free fatty acids varying in saturation and chain length (C6–C18) in the condition designed to minimize nongustatory sensory cues (Chale-Rush et al. 2007; Mattes 2009a, 2009b). However, there is individual variability in detection of free fatty acids and there is little evidence for the existence and function of fatty acid transduction systems on human taste receptor cells (Mattes 2009a). In addition, data verify that free fatty acids are effective stimuli only for the gustatory system, but not for other related sensory systems (Mattes 2009a). Taken together, it is still debatable whether "fat taste" constitutes a sixth primary taste—further study is needed to clarify this issue.

The taste qualities seem to interact at central nervous levels (e.g., cognition) as well as at peripheral levels (e.g., perception) in terms of enhancement/suppression or synergy/masking (Keast and Breslin 2002; Breslin and Huang 2006). Keast and Breslin (2002) suggested three levels of taste interaction that must be considered when examining interaction of taste mixture: (1) chemical interaction between tasting components in the mixture; (2) oral physiological interaction between one of mixture components and the taste receptor/transduction of the other component; and (3) cognitive interactions between different taste qualities being perceived together. Typically, interactions between same taste qualities are fitted to a sigmoidally shaped psychophysical function, with expansive, linear, and compressive phases (Keast and Breslin 2002). Rather, when two taste compounds eliciting the same quality are mixed at high concentration/intensity, suppressive interactions are more likely to occur (Breslin and Beauchamp 1995; Keast and Breslin 2002). In addition, the types of binary interactions between different taste qualities may be dependent on the concentration/intensity of taste compounds. That is, at low concentration/intensity, enhanced interaction is more common, whereas at high concentration/intensity, suppressive interaction is more often reported (Keast and Breslin 2002).

3.2.1.2 Anatomy and Physiology

Humans express "taste receptor cells" within the oral cavity including the tongue (with few or no receptors in the middle of its dorsal surface), soft palate, and pharyngeal and laryngeal regions of the throat (Witt et al. 2003; Breslin and Huang 2006). Interestingly, taste receptors are found not only in the oral cavity, but also in the stomach and the gut (Rozengurt 2006; Bezençon et al. 2007). Taste receptor cells are mostly found within multicellular rosette clusters termed "taste buds" (Northcutt 2004; Breslin and Huang 2006). The majority of taste buds on the dorsal surface and edges of the tongue are located in "taste papillae" (Smith and Margolskee 2001; Breslin and Huang 2006).

Once tasting substances (tastants) reach taste receptors in the oral cavity, taste signals are generated; taste signals are transmitted to the brain via branches of three "cranial nerves" (CN): facial (CN VII), glossopharyngeal (CN IX), and vagus (CN X) nerves, depending on the site of taste stimulation (Spector 2000; Bachmanov and Beauchamp 2007; Huart et al. 2009). The axons of gustatory neurons terminate in the rostral part of the nucleus of the solitary tract in the medulla (Beckstead and Norgren 1979) and the neurons project to the ventro-postero-medial nucleus (VPMn) of the thalamus through the central tegmental tract (Huart et al. 2009). From there, neurons project to the primary taste cortex located in the frontal operculum and adjoining insula (Huart et al. 2009). The primary taste cortex projects to the secondary taste cortex including caudolateral orbitofrontal cortex (OFC), cingulated gyrus, amygdala, hypothalamus, and basal ganglia (Rolls and Scott 2003; Sewards 2004; Breslin and Huang 2006; Huart et al. 2009).

3.2.1.2.1 Taste Receptors

Each taste quality has a specific transduction mechanism mediated by specific taste receptors (Bachmanov and Beauchamp 2007). It is known that T1R and T2R receptors [belonging to a superfamily of G protein-coupled receptors (GPCRs)] are involved in the perception of sweet, umami, or bitter tastes (Bachmanov and Beauchamp 2007). Specifically, sweet and umami tastes are mediated by a family of three class C GPCRs: T1R1, T1R2, and T1R3. These receptors have a seven-helix transmembrane domain and a large extracellular domain (termed a Venus flytrap domain) that contains the active site for taste ligands (Temussi 2009). Although T1R3 alone can serve as a taste receptor for sucrose and other sugars (Nelson et al. 2001; Zhao et al. 2003), only the heterodimers of T1R2 and T1R3 function as a receptor for all sweet-tasting substances including sugars, amino acids, sweet proteins, and synthetic sweeteners (Breslin and Huang 2006; Temussi 2009). Another heterodimeric complex of T1R3 with T1R1 serves as a taste receptor for umami-tasting stimuli such as L-amino acids, L-glutamate, and MSG (Zhao et al. 2003). Bitter-tasting compounds bind to a large family (about 30 members) of class A GPCRs, T2Rs (Chandrashekar et al. 2006), and each T2R can detect a wide range of bitter-tasting compounds (Temussi 2009).

It is generally agreed upon that salty (Na^+) and sour (H^+) tastes are mediated by ion channels (Bigiani et al. 2003; DeSimone and Lyall 2006). Permeation of Na^+ and other ions into taste cells depolarizes membrane potentials and triggers influx of calcium and then releases neurotransmitters (Schiffman et al. 1983; Breslin and Huang 2006). At least in rodents, Na^+ taste reception is related to the selective epithelial amiloride-sensitive sodium channel (ENaC) (Bachmanov and Beauchamp 2007), but the evidence for ENaC in humans is not entirely clear.

Several candidate receptors for sour taste have been proposed, for example, the neuronal amiloride-sensitive cation channel 1 (ACCN1), HCN1 and HCN4 of a family of hyperpolarization-activated cyclic nucleotide-gated (HCN) channels, TASK-1, and Na^+–H^+-exchanger isoform 1 (NHE-1) (Bachmanov and Beauchamp 2007), but further study is needed for a better understanding of the sour taste mechanism.

3.2.1.2.2 Taste Buds

The taste bud is mainly observed in the papillae of the tongue, the epithelium of the palate, oropharynx, larynx, and upper esophagus (Witt et al. 2003; Breslin and Huang 2006). The taste bud is a bulb- or rosebud-shaped structure containing between 50 and 120 bipolar epithelial cells (Witt et al. 2003).

Using confocal microscopy, it is observed that taste bud volume changes over periods of several weeks (Srur et al. 2010). Each taste bud usually contains a small-sized opening (e.g., between 135 and 225 mm^2 in fungiform papillae) called "taste pore" on the upper surface taste papillae (Just et al. 2005). Through this taste pore, the bodies of taste receptor cells are directly in contact with tastants (Smith and Margolskee 2001; Breslin and Huang 2006).

There are four types of cells in a taste bud: basal cells at the base of the taste bud, type I (dark) cells with long microvilli, type II (light) cells with shorter microvilli, and type III (intermediate) cells (Azzali 1997; Witt et al. 2003; Konstantindis 2009; Roper and Chaudhari 2009). The round-type basal cells are located at the base of the taste bud and the other three cells are elongated, stretching from the base to the apical end of the bud (Azzali 1997; Breslin and Huang 2006). The majority of taste receptors and enzymes involved in signal transduction are found only in type II (light) cells, whereas most synapses with afferent axons are observed in type III (dark) cells (Yang et al. 2000; Huang et al. 2005; Breslin and Huang 2006). The information from type II to type III cells can be passed on electrically via gap junctions; however, taste cells seem to communicate each other through another neurotransmitter-like serotonin (Yoshii 2005; Breslin and Huang 2006).

3.2.1.2.3 Taste Papillae

There are four types of taste papillae. Filiform ("thread-like") papillae without taste buds are present on the surface of tongue, and they have both soft (epithelia) and hard (trichocytic) keratins, which reflects the dual requirement of the tongue epithelium (Dhouailly et al. 1989; Manabe et al. 1999; Konstantindis 2009). It must be both hard and flexible for resisting the friction and expansion accompanying tongue movements during feeding and drinking (Dhouailly et al. 1989; Manabe et al. 1999; Konstantindis 2009). Thus, the filiform papillae seem to be involved in making the tongue surface mechanically rough for food manipulation, or in distributing saliva on the tongue and food bolus, or in somotosensory function (Smith and Margolskee 2001; Konstantindis 2009). Out of the other three papillae having taste buds, the fungiform ("mushroom-like") papillae are distributed on the anterior two-thirds of the tongue (Smith and Margolskee 2001; Konstantindis 2009) and are also more concentrated on the sides (Kullaa-Mikkonen et al. 1987). The area of taste stimulation or the number of fungiform papillae that are stimulated is related to taste intensity (Miller and Reedy 1990; Delwiche et al. 2001). Furthermore, patients with eating disorders have a significantly lower number of fungiform papillae than healthy people, and this reduced number is more apparent in patients with restrictive-type eating disorder (e.g., restrictive anorexia nervosa) (Wöckel et al. 2008). Moreover, women have, on average, more fungiform papillae and taste buds than men (Bartoshuk et al. 1994). Moreover, the circumvallate ("wall-like") papillae containing roughly 12 large taste buds are present in the shape of an inverted "V" at the posterior third of the tongue (Smith and Margolskee 2001; Konstantindis 2009). Finally, the foliate ("leaf-like") papillae are observed in the posterior lateral edges, adjacent to the lower molar teeth, of the tongue (Smith and Margolskee 2001; Witt et al. 2003; Konstantindis 2009).

3.2.1.2.4 Taste Nerves

As mentioned above, three cranial nerves are involved in the transmission of gustatory signals. The facial nerve (CN VII) has its cell bodies in the geniculate ganglion and its taste signal is carried via the chorda tympani nerve and the greater superficial petrosal nerve (Spector 2000; Breslin and Huang 2006). The chorda tympani nerve innervates the anterior two-thirds of the tongue including fungiform and anterior foliate papillae, and the greater superficial petrosal nerve innervates taste buds on the soft palate (Spector 2000; Witt et al. 2003; Breslin and Huang 2006; Konstantindis 2009). Also, both the chorda tympani and the greater superficial petrosal nerves carry parasympathetic fibers to their related salivary glands (Witt et al. 2003). The glossopharyngeal nerve (CN IX) has its cell bodies in the petrosal ganglion and innervates the majority of the circumvallate papillae of the posterior tongue via the lingual-tonsillar branch (Spector 2000; Witt et al. 2003; Breslin and Huang 2006; Konstantindis 2009). The vagus nerve (CN X) has its cell bodies in the inferior (nodose) vagal ganglion and innervates the taste buds located on the laryngeal surface of the epiglottis, larynx, and the proximal part of the esophagus via the superior laryngeal nerve (Spector 2000; Witt et al. 2003; Breslin and Huang 2006; Huart et al. 2009).

3.2.2 Trigeminal System

3.2.2.1 Human Trigeminal Sensation

In daily life, we often experience certain sensations characterized as irritation, burning, stinging, tickling, or cooling during eating or drinking (e.g., chilly foods or carbonated beverages). To discriminate this sensation mediated via free spinal nerve endings from the olfactory or gustatory sensations, Parker (1912) used the term "common chemical sense"; however, the word "common" has been under debate (Green et al. 1990). Afterward, the term "chemesthesis" was introduced to describe the distinct chemical sensibility more clearly (Green et al. 1990; Green and Lawless 1991; Doty and Cometto-Muñiz 2003). However, because "chemesthesis" is very broad, in this chapter we will use the term "trigeminal sensation" to focus on the idea that the chemosensory sensation is mediated by trigeminal nerve.

The primary function of the trigeminal system appears to detect and protect us from potentially life-threatening substances (Hummel and Livermore 2002). For example, activation of the intranasal trigeminal system reflexively stops inspiration to protect inhalation of harmful or toxic substances, and this preventive response to trigeminal stimulants is likely to be innate (Rieser et al. 1976; Doty and Cometto-Muñiz 2003). However, it is interesting to note that humans enjoy irritant or spicy foods eliciting trigeminal sensation, even though they appear to show negative or neutral response toward these foods at initial exposure (Rozin and Schiller 1980). Similarly, in an initial experience, animals show aversive responses to irritant or spicy foods (Hilker et al. 1967; Rozin et al. 1979), but their preference for spicy foods may develop after social interactions (Rozin and Kennel 1983). In humans, the preference for spicy foods seems to be acquired through repetitive exposure to the spicy food, depending on various psychological and physiological factors (Rozin and Schiller 1980; Stevenson and Yeomans 1995; Carstens et al. 2002). For example, Rozin and Schiller (1980) suggested that the enjoyment of spicy or irritant foods may result from the recognition that the sensation elicited by those foods is harmless. In addition, it may be an example showing enjoyment of "constrained risks" or thrill seeking (Rozin and Schiller 1980).

Moreover, it is known that the majority of chemosensory stimuli (e.g., odorants and tastants) produce, in addition to olfactory or gustatory sensations, trigeminally mediated sensations, typically in a higher concentration of the stimuli (Doty et al. 1978; Hummel and Livermore 2002). For example, Doty and colleagues (1978) reported that among 47 odorants tested, only two odorants, vanillin and decanoic acid, could not be detected by participants lacking olfactory, but not trigeminal, nerve function. In contrast, only one chemical, carbon dioxide (CO_2), has been found to relatively selectively stimulate the trigeminal nerve producing little or no olfactory sensation (Cain 1976; Kobal and Hummel 1989; Hummel and Livermore 2002). These results indicate that few chemosensory stimulants yield selectively olfactory or trigeminal sensations (Hummel and Livermore 2002). Indeed, many studies have demonstrated that olfactory and trigeminal systems interact by mutually enhancing or suppressing each other (Cain and Murphy 1980; Kobal and Hummel 1988; Livermore and Hummel 2004; Cashion et al. 2006; Rombaux et al. 2006; Frasnelli et al. 2007; Hummel et al. 2009); this interaction may occur at both peripheral and central nervous levels (Livermore et al. 1993; Frasnelli and Hummel 2007; Iannilli et al. 2008; Hummel et al. 2009). There are several possible sites where interactions between trigeminal and olfactory stimulation may take place (Frasnelli and Hummel 2007): at central nervous sites, for example, mediodorsal thalamus (Inokuchi et al. 1993); at the level of the olfactory bulb (Schaefer et al. 2002); at the level of olfactory epithelium (Schaefer et al. 2002); and indirectly via nasal trigeminal reflex (Finger et al. 1990). In addition, the mutual interaction between olfactory and trigeminal perception has been observed in patients with olfactory impairment. Specifically, loss of olfactory sensitivity leads to a decrease of intranasal trigeminal sensitivity (Hummel et al. 1996a). In addition, brain imaging studies demonstrated that trigeminally mediated information is differently processed in the absence and presence of an intact sense of smell (Iannilli et al. 2007; Hummel et al. 2009). However, anosmic patients with long-standing olfactory loss exhibit bigger electrophysiological responses to trigeminal stimulation (e.g., CO_2) than patients with recently acquired anosmia (Hummel 2000). That is, the trigeminal sensitivity may increase gradually in compensation for the loss of olfactory function (Hummel 2000).

3.2.2.2 Anatomy and Physiology

In humans, somatosensory stimuli are mediated via the fifth cranial nerve (CN V) (Doty and Cometto-Muñiz 2003). The sensory nerve endings from branches of the trigeminal nerve are found in the epithelia of the nose and sinuses, the oral cavity, the cornea, and the conjunctivae (Doty and Cometto-Muñiz 2003). The trigeminal nerve has three major branches: the ophthalmic nerve (V1), the maxillary nerve (V2), and the mandibular nerve (V3) (Doty and Cometto-Muñiz 2003; Huart et al. 2009). The ophthalmic nerve branch carries sensory message originating from the upper part of the head, the forehead, the upper eyelid, the tip of the nose, the nasal mucosa, entering the skull through the upper orbital fissure and running in the wall of the cavernous sinus (Huart et al. 2009). The maxillary nerve branch carries sensory information coming from the middle third of the head, the lower eyelid and cheek, the nares, and the upper lip, and the upper teeth, entering through the round foramen and running in the wall of the cavernous sinus (Huart et al. 2009). More specifically to the nasal cavity, the anterior and lateral areas of the nasal cavity are innervated by the lateral and medial nasal branches of the ethmoidal nerve, which comes from the nasociliary branch of the ophthalmic region of the fifth cranial nerve from the trigeminal ganglion (Doty and Cometto-Muñiz 2003), whereas the posterior areas of the nasal cavity are innervated by the nasopalatine nerve (Doty and Cometto-Muñiz 2003). The mandibular nerve transmits both sensory and motor information coming from the lower third of the head, the lower lip, the lower teeth, the chin, and the jaw, arriving at the skull through the oval foramen (Huart et al. 2009). To form the fifth cranial nerve, the three branches converge on the trigeminal ganglion (also called Gasserian Ganglion or Semilunar Ganglion) located within Meckel's cave (Huart et al. 2009). The axons of the first-order neurons from the trigeminal ganglion enter the brainstem at the level of the pons and descend to the bulb through the spinal trigeminal tract to arrive at the trigeminal spinal nucleus (Doty and Cometto-Muñiz 2003; Huart et al. 2009). The second-order neurons cross to the contralateral side and ascend through the trigemino-thalamic tract toward the VPMn of the thalamus (Huart et al. 2009). Next, the third-order neurons project to the primary somatosensory cortex (SI) via the posterior arm of the internal capsule (Huart et al. 2009). Other important processing areas include secondary somatosensory cortex (SII), amygdala, or the hippocampus (Huart et al. 2009).

3.2.2.2.1 Unmyelinated C Fibers and Myelinated A_δ Fibers

The afferent chemosensory innervation of the nasal respiratory epithelium is composed of two fiber systems: unmyelinated C fibers and myelinated A_δ fibers (Anton and Peppel 1991; Hummel 2000). Due to lack of myelination, C fibers have a slower conduction velocity than A_δ fibers (Treede et al. 1995). Typically, C fibers mediate burning painful sensations, whereas A_δ fibers are involved in the mediation of stinging sensations (Mackenzie et al. 1975; Hummel 2000).

Trigeminal sensations mediated by C fibers or A_δ fibers, respectively, are elicited depending on stimulus intensity or stimulus time course (Hummel 2000). Specifically, at low concentrations, nicotine evokes burning sensation more than stinging sensation (Hummel et al. 1992). While olfactory and stinging sensation elicited by nicotine appear immediately after stimulus onset, burning sensation starts after several seconds ("second pain") (Hummel et al. 1992). In addition, C and A_δ fibers respond differently to repeated trigeminal stimulation depending on the interstimulus interval (ISI). For example, Hummel et al. (1994) reported that at an ISI of 8 s, participants' ratings of overall intensity for gaseous CO_2 decreased and mostly stinging sensations were perceived. However, at shorter ISI (e.g., 2 s), the overall perceived intensity was increased being accompanied by the buildup of burning pain probably relating to the "wind-up" of spinal neurons (Hummel et al. 1996b). In contrast, no such summation for stinging painful sensations exhibited (Hummel 2000). A majority of afferent A_δ fibers adapt during sustained painful stimulation (Sumino and Dubner 1981; Hummel 2000). Both peripheral adaptation and/or central habituation may result in the decrease in stinging painful sensation after repetitive stimulation (Hummel 2000).

3.2.3 Olfactory System

3.2.3.1 Human Olfaction

Olfaction is the sensation elicited by volatile compounds reaching the olfactory epithelium via the nasal or oral cavity. Although the sense of smell is much less important for humans than other senses such as vision or audition, it plays a significant role in a wide range of functions in daily life. Specifically, Stevenson (2010) systematically classified olfactory function into three main categories: (1) function relating to ingestion behavior, (2) avoidance of environmental hazards, and (3) social communication. In particular, the sense of smell appears to play an important role in ingestion behavior. For example, the sense of smell can help to detect (Porter et al. 2007) and identify food sources suitable for eating (Fallon and Rozin 1983). In addition, the sense of smell can influence appetite, eating behavior, food preparation, or nutritional status (Duffy et al. 1995; Aschenbrenner et al. 2008a; Seo and Hummel 2009), although this effect is not always consistent (Mattes 2002). The crucial role of the olfactory sense in ingestive behavior is also found in reports on patients suffering from olfactory dysfunction. For example, among hazardous events reported by patients with olfactory dysfunction, cooking-related incidents (45%) were most common, followed by reports of ingestion of spoiled foods (25%) (Santos et al. 2004). Similarly, Miwa et al. (2001) reported that eating- or safety-related activities (e.g., ability to detect spoiled foods, gas leaks, or smoke; eating; cooking; and buying fresh foods) were the most frequently cited daily activities impaired by olfactory dysfunction, based on survey results for patients with olfactory impairments.

3.2.3.2 Anatomy and Physiology

When we breathe or sniff via our noses, airborne chemicals pass through the nasal valve area and contact the olfactory epithelium that contains mucus-coated olfactory receptors (Hornung 2006; Huart et al. 2009). After the interactions between odorant molecules and olfactory receptors on the cilia located on the end of olfactory receptor cells, the structure of the membrane-bound proteins changes to allow extracellular calcium ions to enter the cell (Hornung 2006). Consequently, a generator potential is created, which finally may lead to discharge of the neuron. Olfactory signals are projected ipsilaterally via the olfactory nerve to the olfactory bulb through the cribriform plate of the ethmoid bone (Bensafi et al. 2004; Hornung 2006). From the olfactory bulb, the olfactory signals are further transmitted ipsilaterally via the lateral olfactory tract to the primary olfactory cortex (Bensafi et al. 2004). Primary olfactory cortex consists of all brain areas receiving directly sensory signals from the axons of mitral and tufted cells in the olfactory bulb (Bensafi et al. 2004). Mostly based on animal studies, the primary olfactory cortex includes anterior olfactory cortex (nucleus), piriform cortex, periamygdaloid cortex, anterior cortical nucleus of the amygdala, entorhinal cortex, ventral tenia tecta, and olfactory tubercle (Carmichael et al. 1994; Bensafi et al. 2004; Gottfried 2006). Of these, the piriform cortex is sometimes referred to as primary olfactory cortex because it receives the majority of signal input from the olfactory bulb (Gottfried 2006). Projections between the olfactory bulb and most areas of the primary olfactory cortex, with the exception of the olfactory tubercle, are typically reciprocal (Gottfried 2006). Also, connections between areas of the primary olfactory cortex but the olfactory tubercle are bidirectional (Gottfried 2006). In addition, unlike other senses such as vision and audition, where peripheral signals project to the secondary cortex initially via a thalamic relay, olfactory signals project ipsilaterally, without an obligatory thalamic delay, to multiple brain regions such as the secondary olfactory cortex (Bensafi et al. 2004; Huart et al. 2009).

3.2.3.2.1 Olfactory Epithelium

The olfactory epithelium is located bilaterally about 7 cm up from the nasal floor, covering the cribriform plate medially to the insertion of the middle turbinate and extending to the superior turbinate (Moran et al. 1982; Leopold et al. 2000; Bensafi et al. 2004; Huart et al. 2009). The size of the epithelium is related to the olfactory acuity of the animal (Watelet et al. 2009). For example, a dog's sense of smell is more sensitive than the human sense of smell, which can be explained by a difference in the surface area of the epithelium (Watelet et al. 2009). Specifically, the surface area of a dog's epithelium is estimated to be about 17 times as large (170 cm^2) as that of humans (10 cm^2); in addition, it has been estimated that dogs have more than 100 times more receptors per area than humans (Stoddart 1979).

The olfactory epithelium consists of at least three distinct types of cells: the basal cells, the supporting cells, and the olfactory receptor neurons (Watelet et al. 2009). Basal cells are capable of division and differentiation into olfactory receptor neurons or other cell types throughout adulthood (Stoddart 1979). Supporting cells are analogous to the neural glial cells and help to produce mucus and to balance the mucous environment of olfactory cilia (Menco 1994; Watelet et al. 2009). Olfactory receptor neurons are bipolar cells; importantly, they are frequently replaced (Schultz 1960). The olfactory receptor neurons involved in signal transduction (Menco 1997) and their dendritic extensions are directed toward the olfactory cleft (Huart et al. 2009) and their cilia project into the mucus, allowing contact with the odorant molecules (Huart et al. 2009). As noted before, the odorant molecules reaching the olfactory cleft probably bind to olfactory binding proteins and are carried to the olfactory receptor neurons (Huart et al. 2009). Subsequently, when odorants make contact with the olfactory receptors, the olfactory transduction starts and axons of the olfactory receptor neurons converge into the olfactory nerve (CN 1), passing though the cribriform plate of the ethmoid to the olfactory bulb (Huart et al. 2009).

3.2.3.2.2 Olfactory Bulb

The olfactory bulb is located in the anterior cranial fossa, above the cribriform plate of the ethmoid (Huart et al. 2009). The volume of olfactory bulb shows relatively large individual variations, ranging from 37 to 98 mm^3 in healthy adults (Buschhüter et al. 2008). On average, the olfactory bulb volumes of men are larger than those of women (Buschhüter et al. 2008). Interestingly, this volume of the olfactory bulb is correlated with olfactory function in healthy people (Buschhüter et al. 2008) as well as in people with olfactory impairments (Haehner et al. 2008; Rombaux et al. 2010). The olfactory bulb contains glomeruli, which are the first synaptic relay during olfactory information processing (Huart et al. 2009). Each glomerulus collects axons of olfactory receptor neurons that express the same type of odorant receptor; the sphere-shaped glomeruli contain synapses between olfactory receptor neurons and the dendrites of the mitral cells (Huart et al. 2009). The axons of mitral cells and other cells form the olfactory tract (Huart et al. 2009).

3.3 ORTHONASAL AND RETRONASAL OLFACTION

3.3.1 Conceptual Definition

Rozin (1982) proposed that "olfaction is the only dual sensory modality, in that it senses both objects in the external world and objects in the body." In fact, aromatic volatile compounds reach the olfactory epithelium by two different pathways: via the nose ("external" nares) during sniffing or nasal inhalation (referred to as "orthonasal olfaction"), and via the mouth (oral cavity or nasopharynx), during eating, drinking, or exhalation (referred to as "retronasal olfaction") (Rozin 1982; Shepherd 2006). A large number of aromatic volatile compounds perceived via the orthonasal route provide information about the external world related to food, toxicity/danger, prey/predator, or social interaction (Rozin 1982; Doty 1986; Shepherd 2006). Retronasal olfaction routinely occurs during food ingestion. Aromatic volatile compounds are released, by mastication, from the food matrix, and are "pumped," by mouth movements, exhalation, or swallowing, from the back of the oral cavity up through the nasopharynx to the olfactory epithelium (Shepherd 2006; Negoias et al. 2008). As retronasal sensations are in many cases generated by food-related stimuli and are localized to the mouth, people frequently refer to retronasal olfaction as "taste," that is, they commit the common "smell–taste confusion" (Murphy et al. 1977; Murphy and Cain 1980; Rozin 1982). Furthermore, it is also often observed that people who lose their sense of smell complain of taste loss, even though their sense of taste is normal (Fujii et al. 2004).

3.3.2 Localization of Ortho- and Retronasal Stimulation

Humans seem not to be able to localize purely olfactory stimuli when they are presented to the left or right nostril (Schneider and Schmidt 1967; Prah and Benignus

1984; Kobal et al. 1989; Frasnelli et al. 2008, 2009), which is different from stimuli activating the trigeminal nerve. However, several studies have shown that participants perceive the orthonasal odor as coming from the front of the nasal cavity and the retronasal odor as arising from the back of the nasal/oral cavity (Small et al. 2005; Hummel 2008).

3.3.3 Methodological Issues for Retronasal Stimulation

In line with Rozin's suggestion: "The same olfactory stimulation may be perceived and evaluated in two qualitatively different ways, depending on whether it is referred to the mouth or the external world," many studies have demonstrated differential perception and judgments between ortho- and retronasal olfaction (see below).

Before demonstrating the difference in ortho- and retronasal sensation, it is worthwhile to consider how to present odorous stimuli toward the olfactory mucosa through the two distinct routes, without generating additional gustatory or mechanical stimuli (Halpern 2004). So far, in many studies investigating retronasal olfaction, researchers have presented liquid-, semisolid-, or solid-type stimuli directly to the oral cavity. However, this oral administration of odors may induce unexpected thermal or mechanical sensations elicited by interactions with olfactory stimuli (Welge-Lüssen et al. 2005). To avoid the potential interactions produced by contacting the oral mucosa, Halpern and his colleagues used odorant delivery containers (Pierce and Halpern 1996; Sun and Halpern 2005; Chen and Halpern 2008; Bolton and Halpern 2010). Others (Voirol and Daget 1986) asked participants to sniff the headspace of odorous solutions (for orthonasal stimulation) or inhale the same headspace of solutions through the mouth followed by nasal exhalation (for retronasal stimulation). However, a major limitation of these methods is that the odor concentration in the oral cavity is unknown (Hummel 2008; Negoias et al. 2008). In addition, with these methods, the stimulation of intraoral surfaces, for example, tongue, palate, or teeth, is not controlled (Hummel 2008; Negoias et al. 2008). Indeed, retronasal olfaction is found to be a dynamic process influenced by many mechanical or chemical interactions in the oral cavity (Burdach and Doty 1987; Land 1996; Buettner et al. 2001, 2002; Hodgson et al. 2003, 2005; Haahr et al. 2004; Ruijschop et al. 2009).

To allow for a more defined retronasal odor stimulation, Heilmann and Hummel (2004) designed a new model that allows the release of odors directly to the epipharynx above the soft palate. More specifically, as shown in Figure 3.1, two plastic tubes of 3.3 mm outer diameter and 15 cm length are attached to each other so that the openings of the two tubes are 6 cm apart. The tubes are cut from sterile suction catheters made from soft polyvinyl chloride and their ends are bent at an angle of 45°. Two tubes are placed in the nasal cavity under endoscopic control such that the opening of one of the tubes is just beyond the nasal valve (1.5 cm from the naris) and the opening of the other tube is in the epipharynx (7.5 cm from the naris). For presentation of odor stimulus, the tubes are connected to outlets of a computer-controlled air-dilution olfactometer. Odor stimuli presented through either tube reach the olfactory epithelium at approximately the same concentration and with the same time course. Consequently, this new model allows avoiding concomitant oral gustatory,

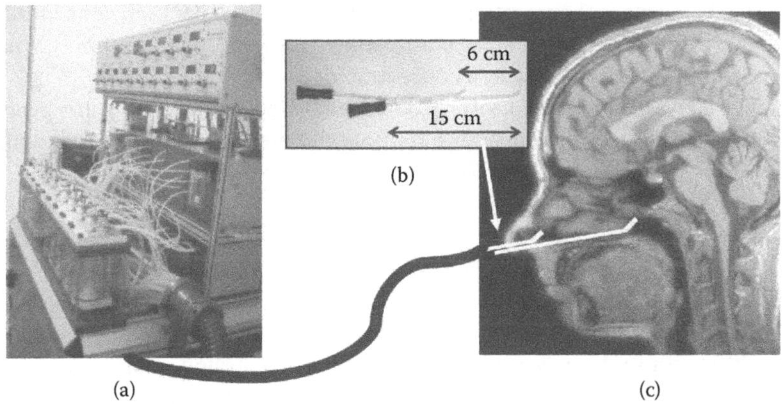

FIGURE 3.1 A new model for more defined retronasal odor stimulation, designed by Heilmann and Hummel (2004). For presentation of odor stimulus, two plastic tubes (b) are connected to outlets of a computer-controlled air dilution olfactometer (a). Two tubes of 3.3 mm outer diameter and 15 cm length are attached to each other so that their openings are 6 cm apart. These tubes are placed in nasal cavity under endoscopic control such that the opening of one of them is just beyond the nasal valve (1.5 cm from naris) and the opening of the other tube is in the epipharynx (7.5 cm from naris) (c). Odor stimuli presented through either tube reach the olfactory epithelium at approximately the same concentration and with the same time course. Consequently, this new model allows avoiding concomitant oral gustatory, thermal and mechanical stimulation that can be elicited by oral presentation of odors.

thermal, and mechanical stimulation that can be elicited by oral presentation of odors (Heilmann and Hummel 2004; Hummel 2008; Negoias et al. 2008); however, this model may be to some extent different from retronasal odor stimulation elicited during normal breathing or food consumption (Welge-Lüssen et al. 2009). Generally, during normal food consumption, retronasal olfaction occurs from not only variable mechanical or chemical interactions of olfactory stimuli with other senses' stimuli in the oral cavity (Burdach and Doty 1987; Land 1996; Buettner et al. 2001, 2002; Hodgson et al. 2003, 2005; Haahr et al. 2004; Ruijschop et al. 2009), but also oropharyngeal processes during mastication, deglutition, or swallowing (Burdach and Doty 1987). Thus, further studies are needed to fully understand the differences between orthonasal and retronasal olfaction.

3.3.4 Comparison between Ortho- and Retronasal Olfaction

3.3.4.1 Psychophysical Perception

3.3.4.1.1 Odor Threshold/Intensity

For both food and nonfood odors, orthonasal thresholds are lower than retronasal thresholds (Voirol and Daget 1986; Heilmann and Hummel 2004). That is, people are more sensitive when they smell the odors orthonasally than via the retronasal route. Heilmann and Hummel (2004) argued that the higher thresholds (i.e., lower sensitivity) for retronasal odors are compatible with our daily experience during eating. That is, retronasal perception is typically encountered at high concentrations, as a result of warming, salivation, or mastication (Burdach and Doty 1987), whereas orthonasal stimuli are typically less concentrated.

Moreover, at suprathreshold levels, odors delivered orthonasally are generally perceived as more intense compared to retronasally presented odors (Heilmann and Hummel 2004; Small et al. 2005; Visschers et al. 2006; Ishii et al. 2008).

3.3.4.1.2 Odor Identification

Under the condition where relatively low concentrated odorants are used or that only normal breathing is allowed, participants identify the odors from an orthonasal location more accurately than those from a retronasal location (Pierce and Halpern 1996; Halpern 2004, 2009). However, if relatively high concentrated odorants are used or if deep breathing is permitted, the differences between ortho- and retronasal identification accuracy are reduced or disappear (Halpern 2009).

3.3.4.2 Cortical Electrophysiological Responses

Using the aforementioned new model of retronasal stimulation, Heilmann and Hummel (2004) demonstrated that the peak amplitudes or latencies of olfactory electrophysiological responses (ERPs) to retronasal stimulation were smaller or longer, respectively, compared to orthonasal stimulation (Heilmann and Hummel 2004; Welge-Lüssen et al. 2009). However, as these results were not consistently obtained for all odorants, they may be best described by contextual explanation (Heilmann and Hummel 2004; Ishii et al. 2008). For example, when nonfood odor (e.g., woody odor) is delivered via the retronasal route, in comparison to food odor (fruity odor), larger amplitudes and longer P2 latencies are observed as a result of the novelty of the unusual stimulation (Ishii et al. 2008). In other words, this result suggests that context is one of the factors influencing olfactory perception.

3.3.4.3 Brain Imaging

Brain imaging techniques allow us to compare neural responses between ortho- and retronasal olfaction. Several studies have found that although many brain regions (e.g., piriform cortex, anterior ventral insula, OFC, hippocampus, and entorhinal cortex) are activated by either ortho- or retronasal stimulation, there is some overlap between the activated areas (Cerf-Ducastel and Murphy 2001; de Araujo et al. 2003; Hummel 2008); certain brain areas are activated depending on the odor delivery route (Small et al. 2005). For example, Small and colleagues (2005) presented four different odors that can be classified by food association (food odor: chocolate vs. nonfood odor: lavender) or physicochemical property (hydrophilic butanol vs. lipophilic farnesol) via either ortho- or retronasal route during fMRI scanning. Regardless of odorants, brain regions (e.g., Rolandic operculum at the base of the central sulcus) corresponding to primary representation of the oral cavity (Pardo et al. 1997; Yamashita et al. 1999; Boling et al. 2002) were more activated by retronasal than orthonasal stimulation, reflecting that retronasal, but not orthonasal, odors are perceived as originating from the oral cavity (Murphy et al. 1977; Small et al. 2005). For example, when analyzing the effects of chocolate odor (which was perceived as similarly intense and pleasant across both ortho- and retronasal stimulation), orthonasal stimulus presentation showed higher activation in the insular/operculum, thalamus, hippocampus, amygdala and caudolateral OFC, whereas retronasal stimulation with exactly the same odor preferentially activated multiple brain regions including perigenual cingulate, medial OFC, posterior cingulated, and superior temporal gyrus. This result indicates that the neural response elicited by an odor may be influenced by its delivery route and supports Rozin's (1982) hypothesis of olfaction as a dual sense modality (see above). However, this effect by odor administration route was pronounced in food odors, but not in nonfood odors, which reflects that different neural responses between ortho- and retronasal odor stimulations may be dependent on conditioned associations (experience), that is, whether the odors have been experienced in the retronasal pathway such as a food odor (Small et al. 2005). Moreover, although orthonasal olfaction activated brain regions related to the anticipatory phase in food rewards (i.e., food availability), retronasal olfaction produced higher activation in brain regions related to the consummatory phase, receipt of a reward (i.e., food is being eaten) (Small et al. 2005). Although this study did not show major differences in the processing of hydrophillic or lipophilic odors, this effect was clearly shown in animals, emphasizing the significance of sorption for the sense of smell (Scott et al. 2007). For example, Scott et al. (2007) demonstrated that hydrophilic odorants delivered via the orthonasal pathway activated the dorsal olfactory epithelium of rats, whereas they were less effective when delivered via the retronasal pathway.

3.4 INTERACTIONS BETWEEN SMELL AND TASTE

3.4.1 Effect of Congruency on Smell–Taste Interaction

3.4.1.1 Psychophysical Perception

Many studies have elucidated that "congruency," being defined as "the extent to which two stimuli are appropriate for combination in a food product" (Schifferstein and Verlegh 1996), plays a key role in modulating the interactions between olfactory and gustatory stimuli. It is assumed that congruent odor (or taste) magnifies the perceived intensity of taste (or odor), although the enhancement of perceived taste intensity is not always obtained in all studies (Bingham et al. 1990; Frank et al. 1993; Welge-Lüssen et al. 2005, 2009). For example, sweet-congruent odors (e.g., strawberry or vanilla) enhance the intensity of sweetness (Frank and Byram 1988; Frank et al. 1989, 1993; Bingham et al. 1990; Clark and Lawless 1994; Schifferstein and Verlegh 1996; Sakai et al. 2001; de Araujo et al. 2003; Djordjevic et al. 2004). Specifically, Frank and Byram (1988) showed that tasteless strawberry flavor, but not peanut butter flavor, enhanced sweetness in whipped cream; conversely, saltiness was not increased by the strawberry flavor. Similarly, Dalton et al. (2000) demonstrated that the orthonasal detection threshold for the odor of benzaldehyde (cherry/almond odor) was significantly decreased (i.e., more sensitive to the odor) in the presence of a subthreshold concentration of congruent taste (saccharin) in the mouth, but not in the presence of incongruent taste (monosodium glutamate) or of deionized water. Additionally, several psychophysical studies have addressed that salty-congruent odors (e.g., soy sauce or bacon) increased saltiness (Djordjevic et al. 2004;

Lawrence et al. 2009). For example, Djordjevic et al. (2004) demonstrated that soy sauce odor presented via orthonasal route could increase perceived saltiness but not sweetness. Of interest, although soy sauce odor was not presented, perceived saltiness was to some extent enhanced by imagining the soy sauce odor. Furthermore, people seem to be able to expect saltiness of food flavors by the labels attached to them (Lawrence et al. 2009).

3.4.1.2 Cortical Electrophysiological Responses

Only a few studies have demonstrated the smell–taste interaction on an electrophysiological level in the human brain (Welge-Lüssen et al. 2005, 2009). Using olfactory ERP, Welge-Lüssen et al. (2005) presented one of two stimulants (i.e., vanillin or gaseous CO_2) with one of three taste conditions (i.e., sweet or sour taste, or intraoral presentation of an empty taste dispenser). Congruent stimuli (i.e., vanillin odor and sweet taste) produced higher amplitudes and shorter latencies of N1 and P2 peaks compared to incongruent combination (i.e., vanillin odor and sour taste). As the interactions between odor and taste conditions were found in the early ERP peaks P1 and N1, Welge-Lüssen et al. (2005) suggested that the olfactory–gustatory interaction occurs at relatively early levels of processing.

3.4.1.3 Brain Imaging

Using single-cell recordings in the macaque, Rolls and Baylis (1994) demonstrated that convergence between gustatory, olfactory, and visual stimuli occurs in the OFC. Since then, many human brain imaging studies have found that independent presentation of an olfactory, a gustatory, or a somatosensory stimulus induces brain activations in overlapping regions such as the insula/frontal operculum (Small et al. 1997, 2005; Francis et al. 1999; Savic et al. 2000; Cerf-Ducastel and Murphy 2001; Cerf-Ducastel et al. 2001; O'Doherty et al. 2001; Poellinger et al. 2001; de Araujo et al. 2003; Haase et al. 2009), the OFC (Small et al. 1997, 2005; Francis et al. 1999; Savic et al. 2000; Cerf-Ducastel and Murphy 2001; O'Doherty et al. 2001; Poellinger et al. 2001; de Araujo et al. 2003; Haase et al. 2009), and the anterior cingulate cortex (Savic et al. 2000; O'Doherty et al. 2001; Poellinger et al. 2001; de Araujo et al. 2003; Small et al. 2005; Haase et al. 2009). Thus, these studies suggest that the insular/frontal operculum, the OFC, and the anterior cingulate cortex play an important role in integrating diverse sensory information involved in flavor perception (Small 2006).

Furthermore, several brain imaging studies compared brain regions responding to unimodal or bimodal stimuli of odor and taste (Small et al. 1997, 2004; de Araujo et al. 2003). Specifically, de Araujo et al. (2003) demonstrated that both olfactory and gustatory stimuli activated the anterior insular, the amygdala, the caudal OFC, the ventral forebrain, and the anterior cingulate cortex and its adjoining areas. Rather, the bimodal interaction of odor and taste stimuli was more involved in a lateral part of the anterior OFC, and the subjective ratings of congruency between odor and taste stimuli were correlated with brain activity in a medial part of the anterior OFC. In addition, Small et al. (2004) reported that multiple brain regions (frontal operculum, ventral insula, caudal OFC, and anterior cingulate cortex) were activated in the mixture of odor and taste stimuli. In particular, in those brain areas, the brain activity for a congruent mixture was higher (i.e., supraadditive effect) than the sum of brain activities for unimodal stimuli of odor and taste. Taken together, it seems that the insula/operculum, caudal OFC, and anterior cingulate cortex are significantly involved in flavor processing.

3.4.2 Effect of Odor Delivery Route on Smell–Taste Interaction

Many psychophysical studies have elucidated that the congruent odor-induced taste enhancement was obtained regardless of the odor delivery pathway (i.e., ortho-/retronasal route; Frank and Byram 1988; Frank et al. 1989, 1993; Bingham et al. 1990; Clark and Lawless 1994; Schifferstein and Verlegh 1996; Sakai et al. 2001; Djordjevic et al. 2004; Lawrence et al. 2009). For example, Sakai et al. (2001) demonstrated that participants rated an aspartame solution as being sweeter when a vanilla odor was presented either by the retronasal route or by the orthonasal route. When the odor stimulus was separately administered via the orthonasal route, there was little possibility of the interactions taking place in the mouth. In other words, this result suggests that the taste enhancement by congruent odor takes place in the central level, but not in the peripheral level (Djordjevic et al. 2004).

The effect of odor delivery route on integration between odor and taste stimuli was also observed in the electrophysiological setting (Welge-Lüssen et al. 2009) and brain imaging studies (Small et al. 1997, 2004; de Araujo et al. 2003). For example, using olfactory ERPs, Welge-Lüssen et al. (2009) examined whether orthonasal and retronasal odors (food: vanillin and nonfood: phenylethanol, rose-like) are differently processed in the simultaneous presentation of a related or unrelated taste stimulus. For orthonasal stimulation, vanillin odor produced longer latencies of P2 peak in the presence of the sweet taste in the mouth, compared to the sour taste condition. In contrast, retronasal stimulation with vanillin odor yielded shortened latencies of the P2 peak in the presence of the sweet taste in the mouth. In particular, as compared to retronasal stimulation, for orthonasal stimulation, both food (vanillin) and nonfood (phenylethanol) odors were processed more quickly even in the presence of the incongruent taste stimulus; this result is opposed to the other results reporting that congruent combination of odor and taste stimuli shorten the peak latencies in the olfactory ERPs (Welge-Lüssen et al. 2005). The authors suggested that arousal and attention were increased by incongruent stimuli resulting in faster sensory processing (i.e., "conflict priming").

Such an attention to spatial difference (i.e., orthonasal presentation of odor) induced neural deactivations in the insula, operculum, caudal OFC, and anterior cingulated cortex (Small et al. 1997). This result is in contrast to the supraadditive activities observed in these same regions when odors are administered via a retronasal pathway (Small 2006). That is, as the odor and taste stimuli are perceived as arising from a common source, the possibility that one stimulus influences the other stimulus is increased. Consequently, this increased possibility is likely to induce responses such as supraadditive neural responses (Small 2006).

3.4.3 Effect of Selective Attention on Smell–Taste Interaction

Selective attention (i.e., focusing on either odor or taste) and strategy (i.e., analytic vs. synthetic) are among the factors influencing smell–taste interactions (Small and

Prescott 2005). Specifically, Ashkenazi and Marks (2004) demonstrated that for a pair of congruent odor and taste (e.g., vanillin and sucrose), selective attention to the taste component (sucrose) improved the ability to discriminate the taste component from the pair, but this enhanced ability was not observed in the selective attention to the odor component (vanillin). Moreover, when this selective attention is attributed to different qualities within a flavor, the odor-induced taste enhancement is likely to be reduced (Frank et al. 1990, 1991, 1993; Clark and Lawless 1994; van der Klaauw and Frank 1996). For example, Frank et al. (1993) found that strawberry odor increased the sweetness of sucrose in the mixture only when participants were asked to judge sweetness alone, whereas the enhanced sweetness disappeared when participants rated other attributes such as sourness and fruitiness of the mixture as well. Furthermore, when participants were asked to first rate total intensity of the mixture and then this was broken down into other qualities such as sweetness, saltiness, sourness, bitterness, and fruitiness, the sweetness of the mixture was suppressed.

Moreover, the odor-induced taste enhancement is dependent on whether the combination of odor and taste is considered as an analytic perceptual strategy (i.e., as a set of discrete qualities) or as a synthetic perceptual strategy (i.e., as a flavor) (Prescott et al. 2004). In other words, when participants were required to rate a single sensory attribute (e.g., sweetness) in the mixture of odor and taste, the synthetic perceptual strategy for both sensory modalities is brought forward; when participants were asked to rate multiple sensory attributes in the mixture, the analytic perceptual strategy for one separate modality is enhanced (Prescott et al. 2004). Specifically, the odor-induced taste enhancement was present during the coexposure of odor and taste only when a synthetic strategy emphasizing the unitary nature of the components as a flavor was applied, but not when an analytic strategy focusing on separate components was used (Prescott et al. 2004). Similarly, Bingham et al. (1990) reported that untrained participants rated a sucrose–maltol solution as being sweeter than an equivalent concentration of sucrose solution; in contrast, trained participants using an analytical strategy (i.e., concentrating only on taste) showed no odor-induced taste enhancement.

In addition, as noted above, selective attention-induced suppression was observed in brain imaging studies (Small et al. 1997). Specifically, attention to spatial source of odor stimulus resulted in massive neural deactivations in the brain regions showing supraadditive activations in the synchronized spatial source (Small et al. 1997).

3.4.4 Effect of Other Sensory Cues on Smell–Taste Interaction

As noted earlier, flavor perception is not restricted to crossmodal interactions between olfactory and gustatory stimuli. That is, flavor perception is also influenced by other sensory cues.

In particular, many studies have demonstrated that visual cue such as color plays an important role in flavor perception (Delwiche 2004; Zampini et al. 2007, 2008; Shankar et al. 2010); however, a large portion of them focused on color effects on either odor or taste perception. In general, congruent color appears to facilitate flavor discrimination or identification (Moir 1936; DuBose et al. 1980; Stillman 1993; Philipsen et al. 1995; Zampini et al. 2007) compared to incongruent color. For example, DuBose et al. (1980) demonstrated that participants identified the flavor of fruit-flavored beverage less accurately when they were unable to perceive its color (i.e., color masking using red fluorescent lighting and red eye-goggles). Also, flavor identification was much better for appropriately colored flavored beverages than for inappropriately colored ones. Morrot et al. (2001) showed that white wines colored red with an odorless dye were described using odor terms more related to red wine than to white wine. In addition, it seems that color intensity influences odor/taste/flavor intensities (Delwiche 2004). That is, a higher level of color intensity produces stronger odor/taste/flavor intensities (DuBose et al. 1980; Johnson and Clydesdale 1982; Johnson et al. 1982); however, nonsignificant effects of color intensity have also been reported (Philipsen et al. 1995; Lavin and Lawless 1998; also see Frank et al. 1989). Taken together, colors appear to modulate odor/taste/flavor perceptions; effects of color on odors seem to be more reliable than effects of color on taste/flavor (Shankar et al. 2010). In addition, this influence of colors on odor/taste/flavor perceptions is affected by many other factors such as participants' age, expectation, experimental contexts, and cultural contexts (Philipsen et al. 1995; Lavin and Lawless 1998; Koza et al. 2005; Zampini et al. 2008; Shankar et al. 2009, 2010). For example, when presented with a brown-colored drink, 70% of British participants associate the brown color with a cola flavor, whereas 49% of Taiwanese participants associate the color with a grape flavor (Shankar et al. 2010).

Moreover, somatosensory tactile stimuli influence odor/taste/flavor perceptions (Delwiche 2004; Verhagen and Engelen 2006). Typically, as the viscosity of the flavor medium increases, odor/taste/flavor sensitivity (Mackey and Valassi 1956; Stone and Oliver 1966) and perceived intensity (Moskowitz and Arabie 1970; Pangborn et al. 1973, 1978; Pangborn and Szczesniak 1974; Christensen 1980; Cook et al. 2003; Bayarri et al. 2006; Bult et al. 2007) decrease. The mechanism for this suppressed flavor intensity by higher viscosity remains unclear. The viscosity of medium may trigger trigeminal sensation, which may influence flavor perception (Bayarri et al. 2006); or the viscosity of medium may affect the release and/or transport of flavor components to the receptors (Bayarri et al. 2006).

Furthermore, in addition to effects of colors and somatosensory cues, odor/taste/flavor perceptions are modulated by other sensory cues including irritation (see above) and temperature (Olson et al. 1980; Kähkönen et al. 1995; Cruz and Green 2000; Engelen et al. 2003; Ventanas et al. 2010), and sound (Vickers and Wasserman 1980; Christensen and Vickers 1981; Zampini and Spence 2004, 2005; Crisinel and Spence 2010; Simmer et al. 2010). Therefore, flavor perception should be considered as a Gestalt concept.

3.5 CONCLUSION

Taken together, flavor perception is a complex and dynamic process. Although gustatory, olfactory, and somatosensory systems are different at the peripheral level, they are highly integrated at a central nervous level. Furthermore, many factors, for example, congruency, spatial synchrony, and selective attention and other sensory cues, influence flavor perception, with the chemical senses being a unique example for integrative neuronal processing.

REFERENCES

Anton, F., and P. Peppel. 1991. Central projections of trigeminal primary afferents innervating the nasal mucosa: A horseradish peroxidase study in the rat. *Neuroscience* 41: 617–628.

Aschenbrenner, K., C. Hummel, K. Teszmer, F. Krone, T. Ishimaru, H.-S. Seo, and T. Hummel. 2008a. The influence of olfactory loss on dietary behaviors. *Laryngoscope* 118: 135–144.

Aschenbrenner, K., N. Scholze, P. Joraschky, and T. Hummel. 2008b. Gustatory and olfactory sensitivity in patients with anorexia and bulimia in the course of treatment. *J. Psychiatr. Res.* 43: 129–137.

Ashkenazi, A., and L.E. Marks. 2004. Effect of endogenous attention on detection of weak gustatory and olfactory flavors. *Percept. Psychophys.* 66: 596–608.

Azzali, G. 1997. Ultrastructure and immunocytochemistry of gustatory cells in man. *Ann. Anat.* 179: 37–44.

Bachmanov, A.A., and G.K. Beauchamp. 2007. Taste receptor genes. *Annu. Rev. Nutr.* 27: 389–414.

Bartoshuk, L.M., V.B. Duffy, and I.J. Miller. 1994. PTC/PROP tasting: Anatomy, psychophysics, and sex effects. *Physiol. Behav.* 56: 1165–1171.

Bayarri, S., A.J. Taylor, and J. Hort. 2006. The role of fat in flavor perception: Effect of partition and viscosity in model emulsions. *J. Agric. Food Chem.* 54: 8862–8868.

Beckstead, R.M., and R. Norgren. 1979. An autoradiographic examination of the central distribution of the trigeminal, facial, glosspharyngeal, and vagal nerves in the monkey. *J. Comp. Neurol.* 184: 455–472.

Bensafi, M., C. Zelano, B. Johnson, J. Mainland, R. Khan, and N. Sobel. 2004. Olfaction: From sniff to percept. In *The Cognitive Neurosciences*, ed. M.S. Gazzaniga, 259–280. Cambridge, MA: The MIT Press.

Bergamasco, N.H., and K.E. Beraldo. 1990. Facial expressions of neonate infants in response to gustatory stimuli. *Braz. J. Med. Biol. Res.* 23: 245–249.

Bezençon, C., J. le Coutre, and S. Damak. 2007. Taste-signaling proteins are coexpressed in solitary intestinal epithelial cells. *Chem. Senses* 32: 41–49.

Bigiani, A., V. Ghiaroni, and F. Fieni. 2003. Channels as taste receptors in vertebrates. *Prog. Biophys. Mol. Biol.* 83: 193–225.

Bingham, A.F., G.G. Birch, C. de Graaf, J.M. Behan, and K.D. Perring. 1990. Sensory studies with sucrose–maltol mixtures. *Chem. Senses* 15: 447–456.

Boling, W., D.C. Reutens, and A. Olivier. 2002. Functional topography of the low postcentral area. *J. Neurosurg.* 97: 388–395.

Bolton, B., and B.P. Halpern. 2010. Orthonasal and retronasal but not oral-cavity-only discrimination of vapor-phase fatty acids. *Chem. Senses* 35: 229–238.

Breslin, P.A., and G.K. Beauchamp. 1995. Suppression of bitterness by sodium: Variation among bitter taste stimuli. *Chem. Senses* 20: 609–623.

Breslin, P. A., and L. Huang. 2006. Human taste: Peripheral anatomy, taste transduction, and coding. In *Taste and Smell. An Update. Advances in Oto-Rhino-Laryngology*, eds. T. Hummel and A. Welge-Lüssen, 152–190. Basel: Karger.

Buettner, A., A. Beer, C. Hannig, and M. Settles. 2001. Observation of the swallowing process by application of videofluoroscopy and real-time magnetic resonance imaging—Consequences for retronasal aroma stimulation. *Chem. Senses* 26: 1211–1219.

Buettner, A., A. Beer, C. Hannig, M. Settles, and P. Schieberle. 2002. Physiological and analytical studies on flavor perception dynamics as induced by the eating and swallowing process. *Food Qual. Prefer.* 13: 497–504.

Bult, J.H., R.A. de Wijk, and T. Hummel. 2007. Investigations on multimodal sensory integration: texture, taste, and ortho- and retronasal olfactory stimuli in concert. *Neurosci. Lett.* 411: 6–10.

Burdach, K.J., and R.L. Doty. 1987. The effects of mouth movements, swallowing, and spitting on retronasal odor perception. *Physiol. Behav.* 41: 353–356.

Buschhüter, D., M. Smitka, S. Puschmann, J.C. Gerber, M. Witt, N.D. Abolmaali, and T. Hummel. 2008. Correlation between olfactory bulb volume and olfactory function. *Neuroimage* 42: 498–502.

Cain, W.S. 1976. Olfaction and the common chemical sense: Some psychophysical contrasts. *Sens. Processes* 1: 57–67.

Cain, W.S., and C.L. Murphy. 1980. Interaction between chemoreceptive modalities of odour and irritation. *Nature* 284: 255–257.

Carmichael, S.T., M.C. Clugnet, and J.L. Price. 1994. Central olfactory connections in the macaque monkey. *J. Comp. Neurol.* 346: 403–434.

Carstens, E., M.I. Carstens, J.-M. Dessirier, M. O'Mahony, C.T. Simons, M. Sudo, and S. Sudo. 2002. It hurts so good: Oral irritation by spices and carbonated drinks and the underlying neural mechanisms. *Food Qual. Prefer.* 13: 431–443.

Cashion, L., A. Livermore, and T. Hummel. 2006. Odour suppression in binary mixtures. *Biol. Psychol.* 73: 288–297.

Casper, R.C., B. Kirschner, H.H. Sandstead, R.A. Jacob, and J.M. Davis. 1980. An evaluation of trace metals, vitamins, and taste function in anorexia nervosa. *Am. J. Clin. Nutr.* 33: 1801–1808.

Cerf-Ducastel, B., and C. Murphy. 2001. fMRI activation in response to odorants orally delivered in aqueous solutions. *Chem. Senses* 26: 625–637.

Cerf-Ducastel, B., P.F. Van de Moortele, P. MacLeod, D. Le Bihan, and A. Faurion. 2001. Interaction of gustatory and lingual somatosensory perceptions at the cortical level in the human: A functional magnetic resonance imaging study. *Chem. Senses* 26: 371–383.

Chale-Rush, A., J.R. Burgess, and R.D. Mattes. 2007. Evidence for human orosensory (taste?) sensitivity to free fatty acids. *Chem. Senses* 32: 423–431.

Chandrashekar, J., M.A. Hoon, N.J. Ryba, and C.S. Zuker. 2006. The receptors and cells for mammalian taste. *Nature* 444: 288–294.

Chen, V., and B.P. Halpern. 2008. Retronasal but not oral-cavity-only identification of "purely olfactory" odorants. *Chem. Senses* 33: 107–118.

Christensen, C.M. 1980. Effects of solution viscosity on perceived saltiness and sweetness. *Percept. Psychophys.* 28: 347–353.

Christensen, C.M., and Z.M. Vickers. 1981. Relationships of chewing sounds to judgments of food crispness. *J. Food Sci.* 46: 574–578.

Clark, C.C., and H.T. Lawless. 1994. Limiting response alternatives in time-intensity scaling: An examination of the halo-dumping effect. *Chem. Senses* 19: 583–594.

Cook, D.J., T.A. Hollowood, R.S.T. Lindforth, and A.J. Taylor. 2003. Oral shear stress predicts flavour perception in viscous solutions. *Chem. Senses* 28: 11–23.

Crisinel, A.S., and C. Spence. 2010. A sweet sound? Food names reveal implicit associations between taste and pitch. *Perception* 39: 417–425.

Cruz, A., and B.G. Green. 2000. Thermal stimulation of taste. *Nature* 403: 889–892.

Dalton, P., N. Doolittle, H. Nagata, and P.A. Breslin. 2000. The merging of the senses: Integration of subthreshold taste and smell. *Nat. Neurosci.* 3: 431–432.

de Araujo, I.E., E.T. Rolls, M.L. Kringelbach, F. McGlone, and N. Phillips. 2003. Taste-olfactory convergence, and the representation of the pleasantness of flavour, in the human brain. *Eur. J. Neurosci.* 18: 2059–2068.

Delwiche, J. 2004. The impact of perceptual interactions on perceived flavor. *Food Qual. Prefer.* 15: 137–146.

Delwiche, J.F., Z. Buletic, and P.A. Breslin. 2001. Relationship of papillae number to bitter intensity of quinine and PROP within and between individuals. *Physiol. Behav.* 74: 329–337.

DeSimone, J.A., and V. Lyall. 2006. Taste receptors in the gastrointestinal tract: III. Salty and sour taste: Sensing of sodium and protons by the tongue. *Am. J. Physiol. Gastrointest. Liver Physiol.* 291: G1005–G1010.

Dhouailly, D., C. Xu, M. Manabe, A. Schermer, and T.-T. Sun. 1989. Expression of hair-related keratins in a soft epithelium: Subpopulations of human and mouse dorsal tongue keratinocytes express keratin markers for hair-, skin- and esophageal-types of differentiation. *Exp. Cell Res.* 181: 141–158.

Djordjevic, J., R.J. Zatorre, and M. Jones-Gotman. 2004. Odor-induced changes in taste perception. *Exp. Brain Res.* 159: 405–408.

Doty, R.L. 1986. Odor-guided behavior in mammals. *Experientia* 42: 257–271.

Doty, R.L., W.P.E. Brugger, P.C. Jurs, M.A. Orndorff, P.J. Snyder, and L.D. Lowry. 1978. Intranasal trigeminal stimulation from odorous volatiles: Psychometric responses from anosmic and normal humans. *Physiol. Behav.* 20: 175–185.

Doty, R.L., and I.E. Cometto-Muñiz. 2003. Trigeminal chemosensation. In *Handbook of Olfaction and Gustation*, 2nd edn., ed. R.D. Doty, 981–999. New York: Marcel Dekker.

DuBose, C.N., A.V. Cardello, and O. Maller. 1980. Effects of colorants and flavorants on identification, perceived flavor intensity, and hedonic quality of fruit-flavored beverages and cake. *J. Food Sci.* 45: 1393–1399.

Duffy, V.B., J.R. Backstrand, and A.M. Ferris. 1995. Olfactory dysfunction and related nutritional risk in free-living, elderly women. *J. Am. Diet. Assoc.* 95: 879–884.

Engelen, L., R.A. de Wijk, J.F. Prinz, A.M. Janssen, H. Weenen, and F. Bosman. 2003. The effect of oral and product temperature on the perception of flavor and texture attributes of semi-solids. *Appetite* 41: 273–281.

Fallon, A., and P. Rozin. 1983. The psychological bases of food rejection by humans. *Ecol. Food Nutr.* 13: 15–26.

Finger, T.E., M.L. Getchell, T.V. Getchell, and J.C. Kinnamon. 1990. Affector and effector functions of peptidergic innervation of the nasal cavity. In *Chemical Senses, Irritation*, eds. B.G. Green, J.R. Mason, and M.R. Kare, 1–20. New York: Marcel Dekker.

Francis, S., E.T. Rolls, R. Bowtell, F. McGlone, J. O'Doherty, A. Browning, S. Clare, and E. Smith. 1999. The representation of pleasant touch in the brain and its relationship with taste and olfactory areas. *NeuroReport* 10: 453–459.

Frank, R.A., and J. Byram. 1988. Taste–smell interactions are tastant and odorant dependent. *Chem. Senses* 13: 445–455.

Frank, R.A., K. Ducheny, and S.J.S. Mize. 1989. Strawberry odor, but not red color, enhances the sweetness of sucrose solutions. *Chem. Senses* 14: 371–377.

Frank, R.A., G. Shaffer, and D.V. Smith. 1991. Taste–odor similarities predict taste enhancement and suppression in taste–odor mixtures. *Chem. Senses* 16: 523.

Frank, R.A., N.J. van der Klaauw, and H.N.J. Schifferstein. 1993. Both perceptual and conceptual factors influence taste–odor and taste–taste interactions. *Percept. Psychophys.* 54: 343–354.

Frank, G.K., A. Wagner, S. Achenbach, C. McConaha, K. Skovira, H. Aizenstein, C.S. Carter, and W.H. Kaye. 2006. Altered brain activity in women recovered from bulimic-type eating disorders after a glucose challenge: A pilot study. *Int. J. Eat. Disord.* 39: 76–79.

Frank, R.A., N. Wessel, and G. Shaffer. 1990. The enhancement of sweetness by strawberry odor is instruction-dependent. *Chem. Senses* 15: 576–577.

Frasnelli, J., G. Charbonneau, O. Collignon, and F. Lepore. 2009. Odor localization and sniffing. *Chem. Senses* 34: 139–144.

Frasnelli, J., and T. Hummel. 2007. Interactions between the chemical senses: Trigeminal function in patients with olfactory loss. *Int. J. Psychophysiol.* 65: 177–181.

Frasnelli, J., B. Schuster, and T. Hummel. 2007. Interactions between olfaction and the trigeminal system: What can be learned from olfactory loss. *Cereb. Cortex* 17: 2268–2275.

Frasnelli, J., M. Ungermann, and T. Hummel. 2008. Ortho- and retronasal presentation of olfactory stimuli modulates odor percepts. *Chemosens. Percept.* 1: 9–15.

Fujii, M., K. Fukazawa, Y. Hashimoto, S. Takayasu, M. Umemoto, A. Negoro, and M. Sakagami. 2004. Clinical study of flavor disturbance. *Acta Otolaryngol.* 553(Suppl): 109–112.

Gaillard, D., F. Laugerette, N. Darcel, A. El-Yassimi, P. Passilly-Degrace, A. Hichami, N.A. Khan, J.P. Montmayeur, and P. Besnard. 2008. The gustatory pathway is involved in CD36-mediated orosensory perception of long-chain fatty acids in the mouse. *FASEB J.* 22: 1458–1468.

Gilbertson, T.A., D.T. Fontenot, L. Liu, H. Zhang, and W.T. Monroe. 1997. Fatty acid modulation of K^+ channels in taste receptor cells: Gustatory cues for dietary fat. *Am. J. Physiol.* 272: C1203–C1210.

Gottfried, J.A. 2006. Smell: Central nervous processing. In *Taste and Smell. An Update. Advances in Oto-Rhino-Laryngology*, eds. T. Hummel and A. Welge-Lüssen, 44–69. Basel: Karger.

Green, B.G., and H.T. Lawless. 1991. The psychophysics of somatosensory chemoreception in the nose and mouth. In *Smell and Taste in Health and Disease*, eds. T.V. Getchell, R.L. Doty, L.M. Bartoshuk, and J.B. Snow Jr. New York: Raven Press.

Green, B.G., J.R. Mason, and M.R. Kare. 1990. Preface. In *Chemical Senses, Irritation*, eds. B.G. Green, J.R. Mason, and M.R. Kare, v–vii. New York: Marcel Dekker.

Haahr, A.-M., A. Bardow, C.E. Thomsen, S.B. Jensen, B. Nauntofte, M. Bakke, J. Adler-Nissen, and W.L.P. Bredie. 2004. Release of peppermint flavor compounds from chewing gum: Effect of oral functions. *Physiol. Behav.* 82: 531–540.

Haase, L., B. Cerf-Ducastel, and C. Murphy. 2009. Cortical activation in response to pure taste stimuli during the physiological states of hunger and satiety. *Neuroimage* 44: 1008–1021.

Haehner, A., A. Rodewald, J.C. Gerber, and T. Hummel. 2008. Correlation of olfactory function with changes in the volume of the human olfactory bulb. *Arch. Otolaryngol. Head Neck Surg.* 134: 621–624.

Halpern, B.P. 2004. Retronasal and orthonasal smelling. *Chemosense* 6: 1–7.

Halpern, B.P. 2009. Retronasal olfaction. In *Encyclopedia of Neuroscience*, ed. L.R. Squire, 297–304. Oxford: Academic Press.

Heilmann, S., and T. Hummel. 2004. A new method for comparing orthonasal and retronasal olfaction. *Behav. Neurosci.* 118: 412–419.

Hilker, D.M., J. Hee, J. Higashi, S. Ikehara, and E. Paulsen. 1967. Free choice consumption of spiced diet by rats. *J. Nutr.* 91: 129–131.

Hodgson, M.D., J.P. Langridge, R.S.T. Linforth, and A.J. Taylor. 2003. Simultaneous real-time measurements of mastication, swallowing, nasal airflow, and aroma release. *J. Agric. Food Chem.* 51: 5052–5057.

Hodgson, M.D., J.P. Langridge, R.S.T. Linforth, and A.J. Taylor. 2005. Aroma release and delivery following the consumption of beverages. *J. Agric. Food Chem.* 53: 1700–1706.

Hornung, D.E. 2006. Nasal anatomy and the sense of smell. In *Taste and Smell. An Update. Advances in Oto-Rhino-Laryngology*, eds. T. Hummel and A. Welge-Lüssen, 1–22. Basel: Karger.

Huang, Y.J., Y. Maruyama, K.S. Lu, E. Pereira, I. Plonsky, J.E. Baur, D. Wu, and S.D. Roper. 2005. Mouse taste buds use serotonin as a neurotransmitter. *J. Neurosci.* 25: 843–847.

Huart, C., S. Collet, and P. Rombaux. 2009. Chemosensory pathways: From periphery to cortex. *B-ENT* 5(Suppl. 13): 3–9.

Hummel, T. 2000. Assessment of intranasal trigeminal function. *Int. J. Psychophysiol.* 36: 147–155.

Hummel, T. 2008. Retronasal perception of odors. *Chem. Biodivers.* 5: 853–861.

Hummel, T., S. Barz, J. Lötsch, S. Roscher, B. Kettenmann, and G. Kobal. 1996a. Loss of olfactory function leads to a decrease of trigeminal sensitivity. *Chem. Senses* 21: 75–79.

Hummel, T., M. Gruber, E. Pauli, and G. Kobal. 1994. Chemo-somatosensory event-related potentials in response to repetitive painful chemical stimulation of the nasal mucosa. *Electroencephalogr. Clin. Neurophysiol.* 92: 426–432.

Hummel, T., E. Iannilli, J. Frasnelli, J. Boyle, and J. Gerber. 2009. Central processing of trigeminal activation in humans. *Ann. N. Y. Acad. Sci.* 1170: 190–195.

Hummel, T., and A. Livermore. 2002. Intranasal chemosensory function of the trigeminal nerve and aspects of its relation to olfaction. *Int. Arch. Occup. Environ. Health* 75: 305–313.

Hummel, T., A. Livermore, C. Hummel, and G. Kobal. 1992. Chemosensory event-related potentials in man: Relation to olfactory and painful sensations elicited by nicotine. *Electroencephalogr. Clin. Neurophysiol.* 84: 192–195.

Hummel, T., C. Schiessl, J. Wendler, and G. Kobal. 1996b. Peripheral electrophysiological responses decrease in response to repetitive painful stimulation of the human nasal mucosa. *Neurosci. Lett.* 212: 37–40.

Iannilli, E., C. Del Gratta, J.C. Gerber, G.L. Romani, and T. Hummel. 2008. Trigeminal activation using chemical, electrical, and mechanical stimuli. *Pain* 139: 376–388.

Iannilli, E., J. Gerber, J. Frasnelli, and T. Hummel. 2007. Intranasal trigeminal function in subjects with and without an intact sense smell. *Brain Res.* 1139: 235–244.

Inokuchi, A., C.P. Kimmelman, and J.B. Snow Jr. 1993. Convergence of olfactory and nasotrigeminal inputs and possible trigeminal contributions to olfactory responses in the rat thalamus. *Eur. Arch. Otorhinolaryngol.* 249: 473–477.

Ishii, A., N. Roudnitzky, N. Béno, M. Bensafi, T. Hummel, C. Rouby, and T. Thomas-Danguin. 2008. Synergy and masking in odor mixtures: An electrophysiological study of orthonasal vs. retronasal perception. *Chem. Senses* 33: 553–561.

ISO. 1992. Standard 5492: *Terms Relating to Sensory Analysis*. International Organization for Standardization.

ISO. 2008. Standard 5492: *Terms Relating to Sensory Analysis*. International Organization for Standardization.

Johnson, J., and F.M. Clydesdale. 1982. Perceived sweetness and redness in colored sucrose solutions. *J. Food. Sci.* 47: 747–752.

Johnson, J.L., E. Dzendolet, R. Damon, M. Sawyer, and F.M. Clydesdale. 1982. Psychophysical relationships between perceived sweetness and color in cherry-flavored beverages. *J. Food Protect.* 45: 601–606.

Just, T., J. Stave, H.W. Pau, and R. Guthoff. 2005. In vivo observation of papillae of the human tongue using confocal laser scanning microscopy. *ORL J. Otorhinolaryngol. Relat. Spec.* 67: 207–212.

Kähkönen, P., H. Tuorila, and L. Hyvönen. 1995. Dairy fat content and serving temperature as determinants of sensory and hedonic characteristics in cheese soup. *Food Qual. Prefer.* 6: 127–133.

Keast, R.S.J., and P.A.S. Breslin. 2002. An overview of binary taste–taste interactions. *Food Qual. Prefer.* 14: 111–124.

Khan, N.A., and P. Besnard. 2009. Oro-sensory perception of dietary lipids: New insights into the fat taste transduction. *Biochem. Biophys. Acta* 1791: 149–155.

Kobal, G., and C. Hummel. 1988. Cerebral chemosensory evoked potentials elicited by chemical stimulation of the human olfactory and respiratory nasal mucosa. *Electroencephalogr. Clin. Neurophysiol.* 71: 241–250.

Kobal, G., and T. Hummel. 1989. Brain responses to chemical stimulation of the trigeminal nerve in man. In *Chemical Senses. Irritation*, eds. B.G. Green, J.R. Mason, and M.R. Kare, 123–139. New York: Marcel-Dekker, Inc.

Kobal, G., S. van Toller, and T. Hummel. 1989. Is there directional smelling? *Experientia* 45: 130–132.

Kokal, W.A. 1985. The impact of antitumor therapy on nutrition. *Cancer* 55(Suppl): 273–278.

Konstantindis, I. 2009. The taste peripheral system. *B-ENT* 5(Suppl. 13): 115–121.

Koza, B.J., A. Cilmi, M. Dolese, and D.A. Zellner. 2005. Color enhances orthonasal olfactory intensity and reduces retronasal olfactory intensity. *Chem. Senses* 30: 643–649.

Kullaa-Mikkonen, A., A. Koponen, and A. Seilonen. 1987. Quantitative study of human papillae and taste buds: Variation with aging and in different morphological forms of the tongue. *Gerodontics* 3: 131–135.

Land, D.G. 1996. Perspectives on the effects of interactions on flavor perception: An overview. Flavor–food interactions, *ACS Symp. Ser.* 633: 2–11.

Langius, A., H. Bjorvell, and M.G. Lind. 1993. Oral- and pharyngeal-cancer patients' perceived symptoms and health. *Cancer Nurs.* 16: 214–221.

Laugerette, F., P. Passilly-Degrace, B. Patris, I. Niot, M. Febbraio, J.P. Montmayeur, and P. Besnard. 2005. CD36 involvement in orosensory detection of dietary lipids, spontaneous fat preference, and digestive secretions. *J. Clin. Invest.* 115: 3177–3184.

Lavin, J.G., and H.T. Lawless. 1998. Effects of color and odor on judgments of sweetness among children and adults. *Food Qual. Prefer.* 9: 283–289.

Lawrence, G., C. Salles, C. Septier, J. Busch, and T. Thomas-Danguin. 2009. Odour–taste interactions: A way to enhance saltiness in low-salt content solutions. *Food Qual. Pref.* 20: 241–248.

Leopold, D.A., T. Hummel, J.E. Schwob, S.C. Hong, M. Knecht, and G. Kobal. 2000. Anterior distribution of human olfactory epithelium. *Laryngoscope* 110: 417–421.

Livermore, A., and T. Hummel. 2004. The influence of training on chemosensory event-related potentials and interactions between the olfactory and trigeminal systems. *Chem. Senses* 29: 41–51.

Livermore, A., T. Hummel, E. Pauli, and G. Kobal. 1993. Perception of olfactory and intranasal trigeminal stimuli following cutaneous electrical stimulation. *Experientia* 49: 840–842.

Mackenzie, R.A., D. Burke, N.F. Skuse, and A.K. Lethlean. 1975. Fibre function and perception during cutaneous nerve block. *J. Neurol. Neurosurg. Psychiatry* 38: 865–873.

Mackey, A.O., and K. Valassi. 1956. The discernment of primary tastes in the presence of different food textures. *Food Technol.* 10: 238–240.

Manabe, M., H.W. Lim, M. Winzer, and C.A. Loomis. 1999. Architectural organization of filiform papillae in normal and black hairy tongue epithelium: Dissection of differentiation pathways in a complex human epithelium according to their patterns of keratin expression. *Arch. Dermatol.* 135: 177–181.

Mattes, R.D. 2002. The chemical senses and nutrition in aging: Challenging old assumptions. *J. Am. Diet. Assoc.* 102: 192–196.

Mattes, R.D. 2005. Fat taste and lipid metabolism in humans. *Physiol. Behav.* 86: 691–697.

Mattes, R.D. 2009a. Is there a fatty acid taste? *Annu. Rev. Nutr.* 29: 305–327.

Mattes, R.D. 2009b. Oral detection of short-, medium-, and long-chain free fatty acids in humans. *Chem. Senses* 34: 145–150.

Mattes, R.D., B.J. Cowart, M.A. Schiavo, M.A., C. Arnold, B. Garrison, M.R. Kare, and L.D. Lowry. 1990. Dietary evaluation of patients with smell and/or taste disorders. *Am. J. Clin. Nutr.* 51: 233–240.

Mattes-Kulig, D.A., and R.I. Henkin. 1985. Energy and nutrient consumption of patients with dysgeusia. *J. Am. Diet. Assoc.* 85: 822–826.

Menco, B.P. 1994. Ultrastructural aspects of olfactory transduction and perireceptor events. *Semin. Cell Biol.* 5: 11–24.

Menco, B.P. 1997. Ultrastructural aspects of olfactory signaling. *Chem. Senses* 22: 295–311.

Miller, I.J., and F.E. Reedy. 1990. Variations in human taste bud density and taste intensity perception. *Physiol. Behav.* 47: 1213–1219.

Miwa, T., M. Furukawa, T. Tsukatani, R.M. Costanzo, L.J. DiNardo, and E.R. Reiter. 2001. Impact of olfactory impairment on quality of life and disability. *Arch. Otolaryngol. Head Neck Surg.* 127: 497–503.

Moir, H.C. 1936. Some observations on the appreciation of flavor in foodstuffs. *Chem. Ind.* 14: 145–148.

Moran, D.T., J.C. Rowley III, and B.W. Jafek. 1982. Electron microscopy of human olfactory epithelium reveals a new cell type: The microvillar cell. *Brain Res.* 253: 39–46.

Morrot, G., F. Brochet, and D. Dubourdieu. 2001. The color of odors. *Brain Lang.* 79: 309–320.

Moskowitz, H.R., and P. Arabie. 1970. Taste intensity as a function of stimulus concentration and solvent viscosity. *J. Texture Stud.* 1: 502–510.

Murphy, C., and W.S. Cain. 1980. Taste and olfaction: Independence vs interaction. *Physiol. Behav.* 24: 601–605.

Murphy, C., W.S. Cain, and L.M. Bartoshuk. 1977. Mutual action of taste and olfaction. *Sens. Processes* 1: 204–211.

Nakai, Y., F. Kinoshita, T. Koh, S. Tsujii, and T. Tsukada. 1987. Taste function in patients with anorexia nervosa and bulimia nervosa. *Int. J. Eat. Disord.* 6: 257–265.

Negoias, S., R. Visschers, A. Boelrijk, and T. Hummel. 2008. New ways to understand aroma perception. *Food Chem.* 108: 1247–1254.

Nelson, G., M.A. Hoon, J. Chandrashekar, Y. Zhang, N.J. Ryba, and C.S. Zuker. 2001. Mammalian sweet taste receptors. *Cell* 106: 381–390.

Ninomiya, Y., and G.K. Beauchamp. 2009. Symposium overview: Umami reception in the oral cavity: Receptors and transduction. *Ann. N. Y. Acad. Sci.* 1170: 39–40.

Northcutt, R.G. 2004. Taste buds: Development and evolution. *Brain Behav. Evol.* 64: 198–206.

O'Doherty, J., E.T. Rolls, S. Francis, R. Bowtell, and F. McGlone. 2001. Representation of pleasant and aversive taste in the human brain. *J. Neurophysiol.* 85: 1315–1321.

Olson, D.G., F. Caporaso, and R.W. Mandigo. 1980. Effects of serving temperature on sensory evaluation of beef steaks from different muscles and carcass maturities. *J. Food Sci.* 45: 627–631.

Pangborn, R.M., Z.M. Gibbs, and C. Tassan. 1978. Effect of hydrocolloids on apparent viscosity and sensory properties of selected beverages. *J. Texture Stud.* 9: 415–436.

Pangborn, R.M., and A.S. Szczensniak. 1974. Effect of hydrocolloids and viscosity on flavor and odor intensities of aromatic flavor compounds. *J. Texture Stud.* 4: 467–482.

Pangborn, R.M., I.M. Trabue, and A.S. Szczesniak. 1973. Effect of hydrocolloids on oral viscosity and basic taste intensities. *J. Texture Stud.* 4: 224–241.

Pardo, J.V., T.D. Wood, P.A. Costello, P.J. Pardo, and J.T. Lee. 1997. PET study of the localization and laterality of lingual somatosensory processing in humans. *Neurosci. Lett.* 234: 23–26.

Parker, G.H. 1912. The relation of smell, taste and the common chemical sense in vertebrates. *J. Acad. Nat. Sci. Phila.* 15: 221–234.

Philipsen, D.H., F.M. Clydesdale, R.W. Griffin, and P. Stern. 1995. Consumer age affects response to sensory characteristics of a cherry flavored beverage. *J. Food Sci.* 60: 364–368.

Pierce, J., and B.P. Halpern. 1996. Orthonasal and retronasal odorant identification based upon vapor phase input from common substances. *Chem. Senses* 21: 529–543.

Poellinger, A., R. Thomas., P. Lio, A. Lee, N. Makris, B.R. Rosen, and K.K. Kwong. 2001. Activation and habituation in olfaction—an fMRI study. *Neuroimage* 13: 547–560.

Porter, J., B. Craven, R. Khan, S.-J. Chang, I. Kang, B. Judkewitz, J. Volpe, G. Settles, and N. Sobel. 2007. Mechanisms of scent tracking in humans. *Nat. Neurosci.* 10: 27–29.

Prah, J.D., and V.A. Benignus. 1984. Trigeminal sensitivity to contact chemical stimulation: A new method and some results. *Percept. Psychophys.* 35: 65–68.

Prescott, J., V. Johnstone, and J. Francis. 2004. Odor–taste interactions: Effects of attentional strategies during exposure. *Chem. Senses* 29: 331–340.

Rieser, J., A. Yonas, and K. Wikner. 1976. Radial localization of odors by human newborns. *Child Dev.* 47: 856–859.

Ritchie, C.S., K. Joshipura, H.C. Hung, and C.W. Douglass. 2002. Nutrition as a mediator in the relation between oral and systemic disease: Associations between specific measures of adult oral health and nutrition outcomes. *Crit. Rev. Oral Biol. Med.* 13: 291–300.

Rolls, E.T., and L.L. Baylis. 1994. Gustatory, olfactory, and visual convergence within the primate orbitofrontal cortex. *J. Neurosci.* 14: 5437–5452.

Rolls, E.T., and T.R. Scott. 2003. Central taste anatomy and neurophysiology. In *Handbook of Olfaction and Gustation*, ed. R.L. Doty, 679–705. New York: Marcel Dekker, Inc.

Rombaux, P., A. Mouraux, B. Bertrand, J.M. Guerit, and T. Hummel. 2006. Assessment of olfactory and trigeminal function using chemosensory event-related potentials. *Neurophysiol. Clin.* 36: 53–62.

Rombaux, P., H. Potier, E. Markessis, T. Duprez, and T. Hummel. 2010. Olfactory bulb volume and depth of olfactory sulcus in patients with idiopathic olfactory loss. *Eur. Arch. Otorhinolaryngol.* 267: 1551–1556.

Roper, S.D., and N. Chaudhari. 2009. Processing umami and other tastes in mammalian taste buds. *Ann. N. Y. Acad. Sci.* 1170: 60–65.

Rosenstein, D., and H. Oster. 1988. Differential facial responses to four basic tastes in newborns. *Child Dev.* 59: 1555–1568.

Rozengurt, E. 2006. Taste receptors in the gastrointestinal tract. I. Bitter taste receptors and alpha-gustducin in the mammalian gut. *Am. J. Physiol. Gastrointest. Liver Physiol.* 291: G171–G177.

Rozin, P. 1982. "Taste–smell confusions" and the duality of the olfactory sense. *Percept. Psychophys.* 31: 397–401.

Rozin, P., L. Gruss, and G. Berk. 1979. Reversal of innate aversions: Attempts to induce a preference for chili peppers in rats. *J. Comp. Physiol. Psychol.* 93: 1001–1004.

Rozin, P., and K. Kennel. 1983. Acquired preferences for piquant foods by chimpanzees. *Appetite* 4: 69–77.

Rozin, P., and D. Schiller. 1980. The nature and acquisition of a preference for chili pepper by humans. *Motiv. Emotion* 4: 77–101.

Ruijschop, R.M.A.J., M.J.M. Burgering, M.A. Jacobs, and A.E.M. Boelrijk. 2009. Retro-nasal aroma release depends on both subject and product differences: A link to food intake regulations? *Chem. Senses* 34: 395–403.

Sakai, N., T. Kobayakawa, N. Gotow, S. Saito, and S. Imada. 2001. Enhancement of sweetness ratings of aspartame by a vanilla odor presented either by orthonasal or retronasal routes. *Percept. Mot. Skills* 92: 1002–1008.

Santos, D.V., E.R. Reiter, L.J. DiNardo, and R.M. Costanzo. 2004. Hazardous events associated with impaired olfactory function. *Arch. Otolaryngol. Head Neck Surg.* 130: 317–319.

Savic, I., B. Gulyas, M. Larsson, and P. Roland. 2000. Olfactory functions are mediated by parallel and hierarchical processing. *Neuron* 26: 735–745.

Schaefer, M.L., B. Böttger, W.L. Silver, and T.E. Finger. 2002. Trigeminal collaterals in the nasal epithelium and olfactory bulb: A potential route for direct modulation of olfactory information by trigeminal stimuli. *J. Comp. Neurol.* 444: 221–226.

Schifferstein, H.N.J., and P.W.J. Verlegh. 1996. The role of congruency and pleasantness in odor-induced taste enhancement. *Acta Psychol. (Amst.)* 94: 87–105.

Schiffman, S.S., E. Lockhead, and F.W. Maes. 1983. Amiloride reduces the taste intensity of Na^+ and Li^+ salts and sweeteners. *Proc. Natl. Acad. Sci. U. S. A.* 80: 6136–6140.

Schneider, R.A., and C.E. Schmidt. 1967. Dependency of olfactory localization on non-olfactory cues. *Physiol. Behav.* 2: 305–309.

Schultz, E.W. 1960. Repair of the olfactory mucosa with special reference to regeneration of olfactory cells (sensory neurons). *Am. J. Psychol.* 37: 1–19.

Scott, J.W., H.P. Acevedo, L. Sherrill, and M. Phan. 2007. Responses of the rat olfactory epithelium to retronasal air flow. *J. Neurophysiol.* 97: 1941–1950.

Seo, H.-S., and T. Hummel. 2009. Effects of olfactory dysfunction on sensory evaluation and preparation of foods. *Appetite* 53: 314–321.

Sewards, T.V. 2004. Dual separate pathways for sensory and hedonic aspects of taste. *Brain Res. Bull.* 62: 271–283.

Shankar, M.U., C.A. Levitan, J. Prescott, and C. Spence. 2009. The influence of color and label information on flavor perception. *Chemosens. Percept.* 2: 53–58.
Shankar, M.U., C.A. Levitan, and C. Spence. 2010. Grape expectations: The role of cognitive influence on color–flavor interactions. *Conscious Cogn.* 19: 380–390.
Shepherd, G.M. 2006. Smell images and the flavour system in the human brain. *Nature* 444: 316–321.
Simmer, J., C. Cuskley, and S. Kirby. 2010. What sounds does that taste? Cross-modal mappings across gustation and audition. *Perception* 39: 553–569.
Singh, P.B., H.-S. Seo, and B. Schuster. 2010. Variation in umami taste perception in the German and Norwegian population. *Eur. J. Clin. Nutr.* 64: 1248–1250.
Small, D.M. 2006. Central gustatory processing in humans. In *Taste and Smell. An Update. Advances in Oto-Rhino-Laryngology*, eds. T. Hummel and A. Welge-Lüssen, 152–190. Basel: Karger.
Small, D.M., J.C. Gerber, Y.E. Mak, and T. Hummel. 2005. Differential neural responses evoked by orthonasal versus retronasal odorant perception in humans. *Neuron* 48: 593–605.
Small, D.M., M. Jones-Gotman, R.J. Zatorre, M. Petrides, and A.C. Evans. 1997. Flavor processing: More than the sum of its parts. *NeuroReport* 8: 3913–3917.
Small, D.M., and J. Prescott. 2005. Odor/taste integration and the perception of flavor. *Exp. Brain Res.* 166: 345–357.
Small, D.M., J. Voss, Y.E. Mak, K.B. Simmons, T. Parrish, and D. Gitelman. 2004. Experience-dependent neural integration of taste and smell in the human brain. *J. Neurophysiol.* 92: 1892–1903.
Smith, D.V., and R.F. Margolskee. 2001. Making sense of taste. *Sci. Am.* 284: 32–39.
Spector, A.C. 2000. Linking gustatory neurobiology to behavior in vertebrates. *Neurosci. Biobehav. Rev.* 24: 391–416.
Srur, E., O. Stachs, R. Guthoff, M. Witt, H.W. Pau, and T. Just. 2010. Change of the human taste bud volume over time. *Auris Nasus Larynx* 37: 449–455.
Steiner, J.E. 1974. Discussion paper: Innate, discriminative human facial expressions to taste and smell stimulation. *Ann. N. Y. Acad. Sci.* 237: 229–233.
Stevenson, R.J. 2010. An initial evaluation of the function of human olfaction. *Chem. Senses* 35: 3–20.
Stevenson, R.J., and M.R. Yeomans. 1995. Does exposure enhance liking for the chilli burn? *Appetite* 24: 107–120.
Stillman, J.A. 1993. Color influences flavor identification in fruit-flavored beverages. *J. Food Sci.* 58: 810–812.
Stoddart, D.M. 1979. External nares and olfactory perception. *Experientia* 35: 1456–1457.
Stone, H., and S. Oliver. 1966. Effect of viscosity on the detection of relative sweetness intensity of sucrose solutions. *J. Food Sci.* 31: 129–134.
Sumino, R., and R. Dubner. 1981. Response characteristics of specific thermoreceptive afferents innervating monkey facial skin and their relationship to human thermal sensitivity. *Brain Res. Rev.* 3: 105–122.
Sun, B.C., and B.P. Halpern. 2005. Identification of air phase retronasal and orthonasal odorant pairs. *Chem. Senses* 30: 693–706.
Temussi, P.A. 2009. Sweet, bitter and umami receptors: A complex relationship. *Cell* 34: 296–302.
Tóth, E., F. Túry, Á. Gáti, J. Weisz, I. Kondákor, and M. Molnár. 2004. Effects of sweet and bitter gustatory stimuli in anorexia nervosa on EEG frequency spectra. *Int. J. Psychophysiol.* 52: 285–290.
Treede, R.D., R.A. Meyer, S.N. Raja, and J.N. Campbell. 1995. Evidence for two different heat transduction mechanisms in nociceptive primary afferents innervating monkey skin. *J. Physiol.* 483: 747–758.

van der Klaauw, N.J., and R.A. Frank. 1996. Scaling component intensities of complex stimuli: The influence of response alternatives. *Environ. Int.* 22: 21–31.

Ventanas, S., S. Mustonen, E. Puolanne, and H. Tuorila. 2010. Odour and flavour perception in flavoured model systems: Influence of sodium chloride, umami compounds and serving temperature. *Food Qual. Prefer.* 21: 453–462.

Verhagen, J.V., and L. Engelen. 2006. The neurocognitive bases of human multimodal food perception: Sensory integration. *Neurosci. Biobehav. Rev.* 30: 613–650.

Vickers, Z.M., and S.S. Wasserman. 1980. Sensory qualities of foods sounds based on individual perceptions. *J. Texture Stud.* 10: 319–332.

Visschers, R.W., M.A. Jacobs, J. Frasnelli, T. Hummel, M. Burgering, and A.E.M. Boelrijk. 2006. Cross-modality of texture and aroma perception is independent of orthonasal or retronasal stimulation. *J. Agric. Food Chem.* 54: 5509–5515.

Voirol, E., and N. Daget. 1986. Comparative study of nasal and retronasal olfactory perception. *Lebensm.-Wiss. Technol.* 19: 316–319.

Wagner, A., H. Aizenstein, L. Mazurkewicz, J. Fudge, G.K. Frank, K. Putnam, U.F. Bailer, L. Fischer, and W.H. Kaye. 2008. Altered insula response to taste stimuli in individuals recovered from restricting-type anorexia nervosa. *Neuropsychopharmacology* 33: 513–523.

Watelet, J.-B., M. Katotomichelakis, P. Eloy, and V. Danielidis. 2009. The physiological basics of the olfactory neuro-epithelium. *B-ENT* 5(Suppl. 13): 11–19.

Welge-Lüssen, A., J. Drago, M. Wolfensberger, and T. Hummel. 2005. Gustatory stimulation influences the processing of intranasal stimuli. *Brain Res.* 1038: 69–75.

Welge-Lüssen, A., A. Husner, M. Wolfensberger, and T. Hummel. 2009. Influence of simultaneous gustatory stimuli on orthonasal and retronasal olfaction. *Neurosci. Lett.* 454: 124–128.

Witt, M., K. Reutter, and I.J Miller. 2003. Morphology of the peripheral taste system. In *Handbook of Olfaction and Gustation*, 2nd edn., ed. R.D. Doty, 651–677. New York: Marcel Dekker, Inc.

Wöckel, L., A. Jacob, M. Holtmann, and F. Poustka. 2008. Reduced number of taste papillae in patients with eating disorders. *J. Neural Transm.* 115: 537–544.

Yamashita, H., Y. Kumamoto, T. Nakashima, T. Yamamoto, A. Inokuchi, and S. Komiyama. 1999. Magnetic sensory cortical responses evoked by tactile stimulations of the human face, oral cavity and flap reconstructions of the tongue. *Eur. Arch. Otorhinolaryngol.* 256(Suppl. 1): S42–S46.

Yang, R., S. Tabata, H.H. Crowley, R.F. Margolskee, and J.C. Kinnamon. 2000. Ultrastructural localization of gustducin immunoreactivity in microvilli of type II taste cells in the rat. *J. Comp. Neurol.* 425: 139–151.

Yoshii, K. 2005. Gap junctions among taste bud cells in mouse fungiform papillae. *Chem. Senses* 30(Suppl): i35–i36.

Zampini, M., D. Sanabria, N. Phillips, and C. Spence. 2007. The multisensory perception of flavor: Assessing the influence of color cues on flavor discrimination responses. *Food Qual. Prefer.* 18: 975–984.

Zampini, M., and C. Spence. 2004. The role of auditory cues in modulating the perceived crispness and staleness of potato-chips. *J. Sens. Stud.* 19: 347–63.

Zampini, M., and C. Spence. 2005. Modifying the multisensory perception of a carbonated beverage using auditory cues. *Food Qual. Prefer.* 16: 632–641.

Zampini, M., E. Wantling, N. Phillips, and C. Spence. 2008. Multisensory flavor perception: Assessing the influence of fruit acids and color cues on the perception of fruit-flavored beverages. *Food Qual. Prefer.* 19: 335–343.

Zhao, G.Q., Y. Zhang, M.A. Hoon, J. Chandrashekar, I. Erlenbach, N.J. Ryba, and C.S. Zuker. 2003. The receptors for mammalian sweet and umami taste. *Cell* 115: 255–266.

4 Lipid-Derived Flavor Compounds

Henryk Jeleń and Erwin Wąsowicz

CONTENTS

4.1 INTRODUCTION

Lipid-derived volatile compounds form one of main groups of food flavors. Lipid oxidation, which influences food acceptability by consumers, results in formation of volatile compounds responsible for rancid off-odors as a result of autooxidation reactions. Lipid oxidation mediated by enzymes that takes place in plant tissues during growth, maturation, or tissue disruption is responsible for the characteristic flavor of many fruits and vegetables.

Lipid oxidation, apart from formation of flavor compounds, causes nutritional losses, influences color, and can lead to formation of toxic compounds. As one of the most important reactions in food systems containing lipids, it is an important subject of numerous textbooks, books, book chapters, and review papers (Belitz et al. 2009; Damodaran et al. 2008; Ho and Hartman 1994b; Grosch 1982; St. Angelo 1996; Min and Smouse 1985). However, lipid oxidation is not the sole source of lipid-derived food flavors. Lipolysis resulting in the release of short chain free fatty acids has a substantial role in development of flavor of dairy products. Similarly, β-oxidation and decarboxylation reactions result in formation of odoriferous ketones, and lactones can be formed in the cyclization reaction of hydroxy fatty acids.

In Section 4.2, we discuss the nature of reactions and processes that lead to the formation of lipid-derived flavor compounds in autooxidation processes. Unsaturated fatty acid autooxidation reactions following a free radical mechanism and/or photosensitized oxidation are discussed in this section. Section 4.3 is devoted to oxidation mediated by lipoxygenase (LOX) in biological systems, focused on flavor formation in selected plants and animal sources as examples. Section 4.4 describes other mechanisms of formation of volatile compounds of sensory importance: methylketones, lactones, and free fatty acids.

4.2 FLAVOR COMPOUNDS RESULTING FROM AUTOOXIDATION OF FATTY ACIDS

The formation of secondary, volatile products of fatty acid oxidation is an important reaction not only for consumers because it impairs food quality, but also for food technologists because it influences the sensory, nutritional quality, and shelf life of a product. Nearly all food products are exposed to oxidation processes, as the reactive oxygen species (ROS) in foods and in the environment are abundant. Reactive oxygen species, which are formed as a result of chemical, enzymatic, and photochemical reactions, react with all main classes of food macroconstituents—proteins, sugars, and lipids producing among other constituents flavor compounds. The term ROS includes radicals and nonradical derivatives of oxygen. The radicals include superoxide anion (O_2•⁻), hydroxy (HO•), peroxy (ROO•), hydroperoxy (HOO•), and alkoxy (RO•) radicals of which hydroxy radical is the most reactive. Free radicals can vary widely in terms of energy, and energetic ones such as hydroxy radicals easily oxidize unsaturated fatty acids by abstracting hydrogen. The nonradical forms include the most important—singlet oxygen ($^1O_2$)—and also hydrogen peroxide (H_2O_2) and ozone (O_3) (Choe and Min 2006). The fatty acid oxidation process that results in the formation of flavor compounds follows two pathways: free radical autooxidation mechanism and oxidation with a singlet oxygen.

4.2.1 Autooxidation Mechanism

The kinetics of lipid oxidation in foods typically has a lag phase, followed by an exponential increase phase during which oxidation products are formed and rancidity off-flavor develops rapidly. Autooxidation is affected by fatty acids composition, mainly their degree of unsaturation, the partial pressure of oxygen, the presence of prooxidants and antioxidants, and oxygen contact surface and storage conditions of the product, mainly water activity (RVP), temperature, and light. Autooxidation is the prevailing mechanism for the oxidation of unsaturated acyl lipids in contact with air. Autooxidation of unsaturated fatty acids proceeds mainly by free radical mechanism. It consists of three main stages: initiation, propagation, and termination (Figure 4.1).

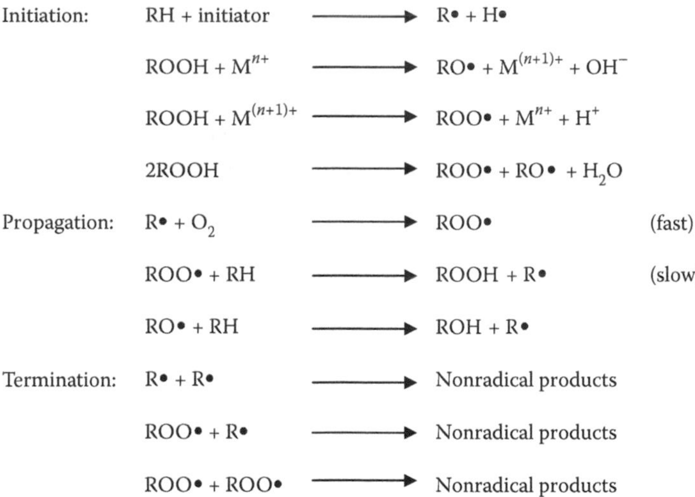

FIGURE 4.1 Schematic reactions in free radical autooxidation of unsaturated fatty acids. RH—fatty acid or ester; R•—alkyl radical; RO•—alkoxy radical; ROO•—peroxy radical; ROOH—hydroperoxide; M—transition metal.

In the initiation step, peroxy- (ROO•), alkoxy- (RO•), or alkyl- (R•) radicals, which are formed or are already present in a food system, start the reaction of abstraction of a hydrogen from a fatty acid to form alkyl (fatty acid) radical (R•). Direct reaction of the bimolecular oxygen attack on a fatty acid molecule (RH + O_2 → R• + HOO•) due to the high activation energy would have an insignificant initial role in autooxidation based on the thermodynamics of the reaction. Radical formation is induced by heat, light, presence of metals, and ROS. The initiation of radical formation in fatty acids occurs at the carbon atom that requires the least energy for hydrogen removal. The energies for the abstraction of hydrogen atom from the functional groups, which can occur in fatty acids, range from ≈65 kcal/mol for methylene group of a 1,4-pentadiene system to 101 kcal/mol for CH_3 bond. Various strengths of carbon–hydrogen bonds in fatty acids result in differences in oxidation rates of fatty acids of different degrees of unsaturation (Privett and Blank 1962). Oleic, linoleic, and linolenic acids' oxidation rate is 1:12:25 in the process of triplet oxygen (radical) oxidation on the basis of peroxide development (Akoh and Min 2008). Once the fatty acid free radical is formed, it is stabilized by delocalization of the double bond or bonds, which results in double bond shifting and in formation of conjugated double bonds in polyunsaturated fatty acids (PUFAs). Of *cis* and *trans* configurations, *trans* configuration dominates because of its better stability.

Propagation stage involves the formation of peroxy radicals from alkyl radicals and oxygen (very fast reaction in the presence of air), and reaction of peroxy radicals and alkoxy radicals with fatty acids that yields alkyl radicals and hydroperoxides and hydroxides, respectively. Peroxy radicals formed are highly energetic and can abstract hydrogen from another molecule. Formation of hydroperoxides from peroxy radicals (ROO• + RH → ROOH + R•) is the limiting step for the whole autooxidation process, being the slowest reaction in the sequence of autooxidation reactions. Hydroperoxides formed will correspond in position to the location of alkyl radicals. For oleate, hydrogen abstraction (on C-8 and C-11) yields two allylic radicals, which in the reaction with oxygen produce a mixture of 8-, 9-, 10-, and 11-allylic hydroperoxides. In the process of linoleate autooxidation, pentadienyl radical is formed as an intermediate, which yields a mixture of conjugated 9- and 13-diene hydroperoxides. For linolenate, which comprises two 1,4-diene systems, hydrogen abstraction will take place on methylene groups at C-11 and C-14. Intermediates that are formed produce conjugated dienes in which hydroperoxides are on C-9 and C-13, or C-12 and C-16. The third double bond remains unaffected. The formed hydroperoxides (ROOH) decompose into alkoxy (RO•) and hydroxy (HO•) radicals. This step provides radicals for the initiation step where, at the beginning, the number of hydroperoxides is low and is responsible for the branching reaction and of the increase in reaction rate in autocatalysis (Grosch 1982). Decomposition of hydroperoxides is influenced by the presence of transition metals, heme compounds, and is regarded as the initial point in the synthesis of flavor compounds. Transition metal ions in their lower valency states (M^{n+}) react rapidly with hydroperoxides, and the reduced state of metal can be regenerated by another hydroperoxide (Figure 4.1).

The termination step is usually described as the set of collision reactions in which radicals react with each other forming stable, nonradical, final products. In the presence of sufficient amount of air, all acyl radicals are converted into peroxy radicals and termination undergoes only via the collision of two peroxy radicals. In the presence of insufficient amount of available oxygen, termination can proceed via the (R• + R•) and (R• + ROO•) collisions.

4.2.2 Photooxidation vs. Autooxidation

Initiation step is critical in the start-up of an autooxidation process. The reaction of unsaturated fatty acid with oxygen in its standard (triplet) form has a very high activation energy (35–65 kcal/mol), because it requires a change in total electron spin, as the substrate and the product are in singlet state and oxygen in triplet. To overcome the energy barrier for the formation of free radicals in the beginning of an autooxidation process, photooxidation (photosensitized oxidation) plays an essential role. In the photooxidation process, singlet oxygen is responsible for the oxygen reactivity. It is suggested that singlet oxygen is involved in the initial step of lipid oxidation, as nonradical singlet oxygen can react directly with double bonds of fatty acids, not forming free radicals. The significance of singlet oxygen in oxidation process is due to the fact that singlet oxidation rate is much higher than triplet oxygen; singlet oxidation reaction can take place at very low temperatures and lead to formation of products partially different from those formed in triplet oxygen oxidation. Sensitized photooxidation in which allylic hydroperoxides are formed from unsaturated fatty acids by exposure to light in the presence of oxygen and sensitizers should be distinguished from UV-catalyzed decomposition of hydoperoxides that takes place according to the free radical mechanism. Differences in the chemical properties of triplet and singlet oxygen result from their differences in molecular orbitals. The molecular orbital of O_2 is $(\sigma 2s)^2$ $(\sigma^* 2s)^2$ $(\sigma 2p)^2$ $(\pi 2p)^4$ $(\pi^* 2p)^2$. In the ground state,

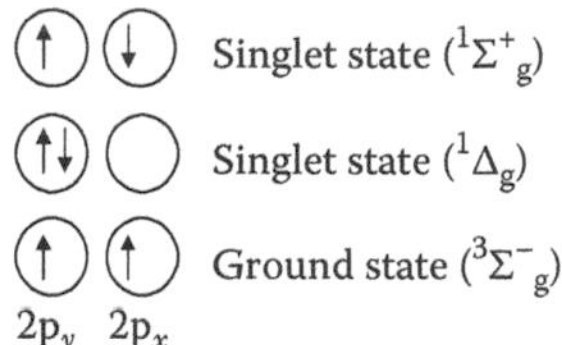

FIGURE 4.2 Comparison of triplet (ground state) and singlet oxygen electron configuration in an oxygen molecule on orbitals $2p_x$ and $2p_y$.

oxygen is a triplet: of two antibonding π orbitals available $\pi^*(2p_y$ and $2p_z)$, two electrons occupy these orbitals alone. Singlet oxygen molecular orbital differs from that of triplet oxygen, where electrons in the π antibonding orbital are paired. The total spin quantum number (S) for singlet oxygen is +½ – ½ = 0. In contrast, for triplet oxygen, it is ½ + ½ = 1. For singlet oxygen, Hund's rule is violated (Figure 4.2). As a consequence, singlet oxygen is a highly energetic molecule. The higher excited state $^1\Sigma^+_g$ (but of a very low stability) has an energy of 37 kcal/mol and the lower state $^1\Delta_g$ (responsible for most singlet oxidation processes in foods) has an energy of 22 kcal/mol above the ground state. It is also stable enough (up to 700 μs in CCl_4; Long and Kearns 1975) to react—as nonradical species with nonradical compounds with double bonds. The term singlet oxygen ($^1O_2$) used in describing oxidation reactions in fatty acids is usually synonymous with $^1\Delta_g$ state. Because of its low activation energy, temperature has little effect on the oxidation rate contrary to triplet oxygen.

Singlet oxygen can be formed chemically, enzymatically, or photochemically, the most important being photosensitization. Such compounds as methylene blue, curcumin, eosin, chlorophyll, riboflavin, myoglobin, and pheophytins serve as sensitizers. Sensitizers are divided into two types: I and II. Type I mechanism involves activation of the substrate by light and the reaction proceeds with hydrogen abstraction and in effect yields substrate radicals as in free radical autooxidation. Type II sensitizers activate the ground state of oxygen into the singlet ($^1O_2$) state ($^3Sens^* + {}^3O_2 \rightarrow {}^1O_2^* + {}^1Sens$). This singlet state oxygen reacts subsequently with fatty acids forming hydroperoxide ($^1O_2 + RH \rightarrow ROOH$) directly by cycloaddition into an unsaturated fatty acid. The insertion point is at either carbon atom of the double (C=C) bond and as a result, an allylic hydroperoxide is formed. The double bond is shifted to yield a hydroperoxide in *trans*-configuration (Figure 4.3). In this, there are no radicals involved as intermediates. It was found that $^1O_2$ reacts at least 10^3 to 10^4 times faster than $^3O_2$ with pure methyl linoleate (Rawls and Van Santen 1970). In contrast to autooxidation, the relative reactivities of mono-, di-, and tri-enoid fatty acids with $^1O_2$ do not differ markedly (Terao and Matsushita 1977). The consequence of the different mechanisms of fatty acids oxidation by $^3O_2$ and $^1O_2$ is the difference in the hydroperoxide profiles of unsaturated fatty acids. The hydroperoxides formed in the reaction of oleic, linoleic, and linolenic acids oxidation via free radical and photosensitized routes are shown in Table 4.1. The difference between free radical and photosensitized oxidation is that the latter route is not inhibited by antioxidants that operate as radical scavengers. However, photosensitized oxidation can be inhibited by compounds that quench singlet oxygen to the ground state, such as β-carotene and α-tocopherol (Yang and Min 1994).

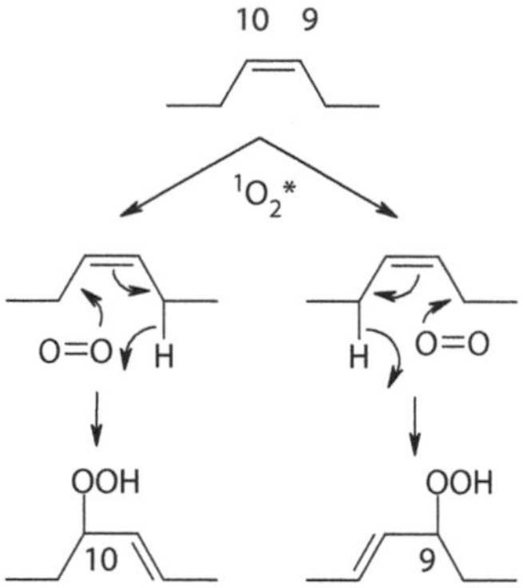

FIGURE 4.3 Reaction of oleate with a singlet oxygen ($^1O_2^*$) leading to formation of 9- and 10-hydroperoxides.

TABLE 4.1
Hydroperoxides Formed in Fatty Acids Esters Oxidized According to Free Radical and Photosensitized Oxidation Mechanisms

		Free Radical Autooxidation		Photosensitized Oxidation	
Substrate	**Temp* (°C)**	**Hydroperoxides**	**Distribution (%)**	**Hydroperoxides**	**Distribution (%)**
Methyl oleate	25–80	8-OOH	26–28		
		9-OOH	22–25	9-OOH	48–51
		10-OOH	22–24	10-OOH	49–52
		11-OOH	26–28		
Methyl linoleate	40–80	9-OOH	48–53	9-OOH	32
				10-OOH	16–17
				12-OOH	17
		13-OOH	48–53	13-OOH	34–35
Methyl linolenate	25–80	9-OOH	28–35	9-OOH	20–23
				10-OOH	13
		12-OOH	8–13	12-OOH	12–14
		13-OOH	10–13	13-OOH	14–15
				15-OOH	12–13
		16-OOH	41–52	16-OOH	25–26

Source: Frankel EN Chemistry of Autoxidation: Mechanism, products and flavor significance. In: Min DB and Smouse TH (Eds.) Flavor Chemistry of Fats and Oils. American Oil Chemists' Society, Champaign Ill. (1985)

Note: Temp*—temperatures provided are for free radical mechanism. Photosensitized oxidation was at 0°C.

4.2.3 Formation of Volatile Compounds from Hydroperoxides Decomposition

Hydroperoxides formed in the course of autooxidation are odorless, but break down easily into a broad range of volatile products with distinct flavors impairing food quality. Nonvolatile products are also formed, but they will not be discussed in this chapter.

Alkoxy radicals produced from hydroperoxides by loss of hydroxy radical undergo hemolytic scission, which is the most important (simplified) reaction in the formation of flavor compounds in the autooxidation process. As a result of this reaction, aldehyde and an alkyl radical are formed. Because of the complexity of fatty acids taking part in the autooxidation and the possibilities of carbon–carbon cleavages on either side of the carbon bearing the oxygen, different aldehydes and radicals can be formed. The reaction for various alkoxy radicals was summarized by Grosch (1982). The general reaction schemes presented in Figure 4.4 show a diversity of products. Structures of volatile compounds formed depend on the residue R′ type of scission

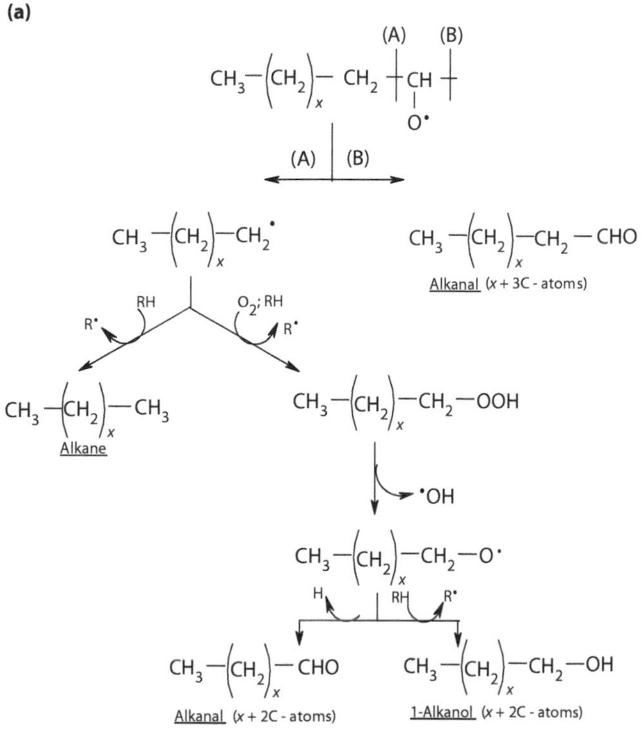

FIGURE 4.4 Pathways for formation of volatile compounds as a result of hydroperoxides decomposition. (a) R′ is saturated; (b) R′ contains an ene system; (c) R′ contains a diene system; (d) R′ contains an allylic system. (From Grosch, W., in *Food Flavours. Part A. Introduction*, ed. I.D. Morton and A.J. Macleod, 325–398, Elsevier Scientific Publishing Company, Amsterdam, 1982. With permission.)

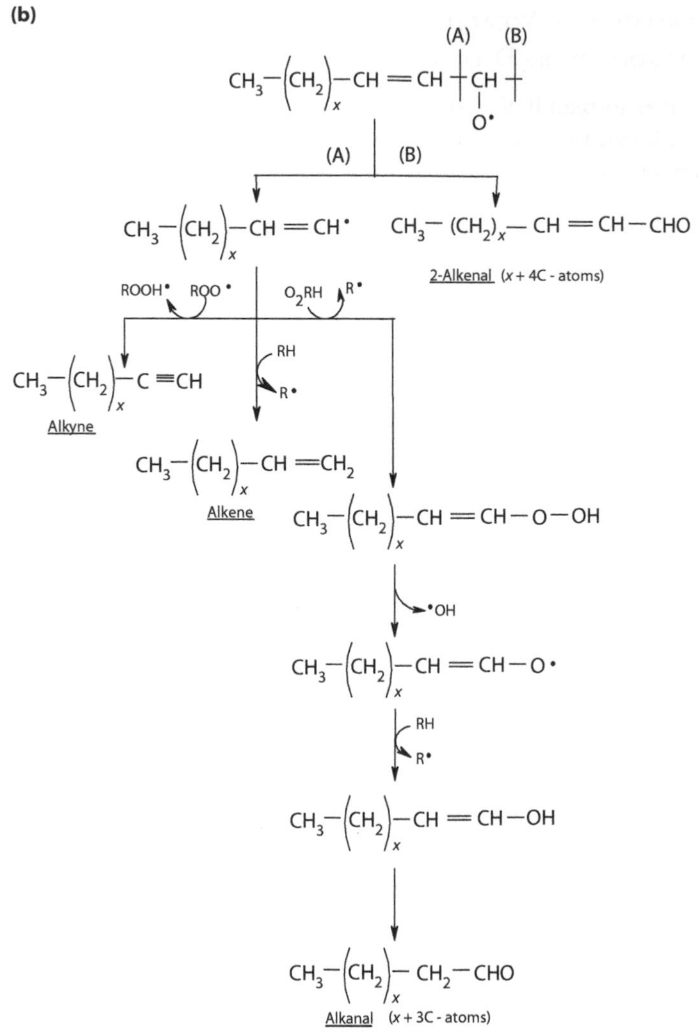

FIGURE 4.4 (Continued)

that takes place [(**A**) or (**B**)] and the presence of oxygen during the breakdown of alkoxy radicals. For saturated chains depending on a scission, alkanal and alkyl radical can yield alkane, or after oxidation an alkanal or alcohol. When an -ene system is present in the chain, (**B**) scission provides 2-alkenal, whereas (**A**) scission followed by subsequent reactions yields alkyne, alkene, or alkanal. For -diene systems, 2,4-alkadienals or 2-alkenals are formed. There is also an opportunity to form alkylfurans and furans. In the last example discussed by Grosch, when an allylic system undergoes decomposition, 3-alkenals are formed, with routes providing formation of sensory important compounds—1-alken-3-ones and 1-alken-3-ols.

Among the compounds identified as lipid oxidation products formed in the autoxidation process, aldehydes play the most important role. As an example, products

(c)

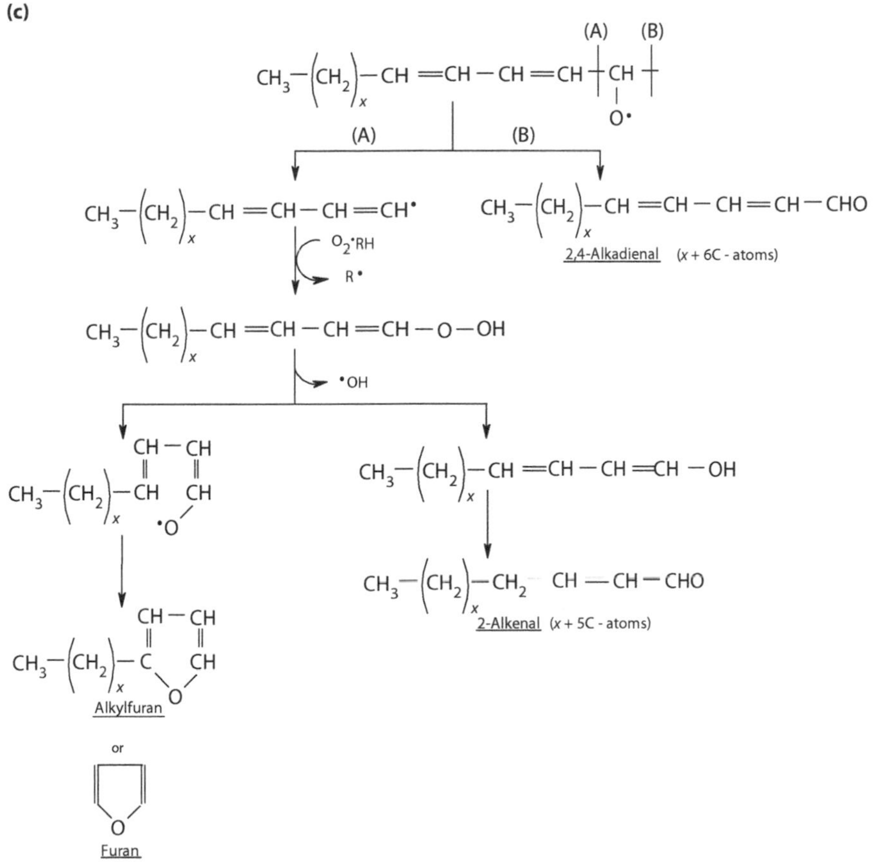

FIGURE 4.4 (Continued)

of the autooxidation of rapeseed oil subjected to accelerated storage test at 60°C for 12 days are shown in Table 4.2. The flavor of aldehydes formed as a result of lipid autooxidation is usually described as rancid, green, tallow, oily, and beany, and contributes to the oxidative rancidity of fat-containing food products. Hexanal is an autooxidation product of special importance as it has been used as an indicator of oxidative deterioration of foods by many researchers. Ketones formed by autooxidation are also important food odorants. Apart from autooxidation, ketones can be formed in β-oxidation and decarboxylation of fatty acids. 1-Octene-3-one can be obtained in an oxidation of arachidonate. 2-Pentylfuran has been identified in many fats, oils, and lipid-containing foods. It is a product of autooxidation of linoleic acid and is known to be one of the compounds responsible for the reversion of soybean oil. It is formed from the conjugated diene radical generated from the cleavage of the 9-hydroxyradical of linoleic acid, which reacts with oxygen to produce vinyl hydroperoxide, which undergoes cyclization via the alkoxy radical to form 2-pentylfuran (Ho and Chen 1994a).

(d)

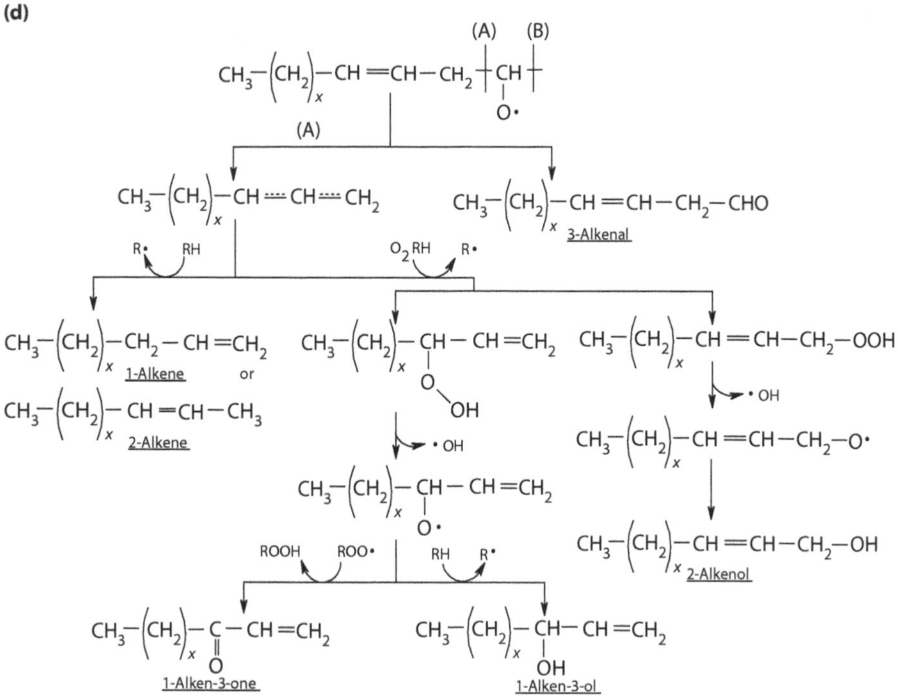
(A)
(B)
(A)
3-Alkenal
R•
RH
O2 RH
R•
1-Alkene
or
2-Alkene
• OH
• OH
ROOH
ROO•
RH
R•
2-Alkenol
1-Alken-3-one
1-Alken-3-ol

FIGURE 4.4 (Continued)

For linolenic acid (E,Z)-2,4-heptadienal dominates, followed by (Z)-3-hexenal, 2,4,7-decatrienal and (E,E)-2,4-heptadienal (Belitz et al. 2009). The importance of volatile lipid autooxidation products for food quality is related to their flavor significance. In many studies, they have been identified as impact odorants. A detailed discussion on particular compounds' importance would exceed the range of this chapter as fatty acids oxidation is a subject of numerous investigations. The importance of lipid oxidation products for flavor of food is not only related with the concentration of a compound, but also with the odor threshold of a particular compound. Table 4.3 shows the threshold values of the main groups of compounds formed during the lipid autooxidation process.

4.3 FLAVOR COMPOUNDS FORMED FROM LIPIDS IN ENZYMATIC REACTIONS

4.3.1 Lipoxygenase Pathway (LOX)

Oxidative enzymes are of interest to food scientists because they influence two important (from sensory point of view) features—color and flavor. Color is influenced by an action of polyphenol oxidases as a result of enzymic browning, also

TABLE 4.2
Main Volatile Compounds Identified by SPME-GC/MS from Rapeseed Oil Subjected to Accelerated Storage Test at 60°C for a Period of 12 Days

Compound	RI	Storage Time at 60°C (Days)						
		Peak Area (TIC units ×10^6)						
		0	2	4	6	8	10	12
Acetic acid*	641	–	–	–	3.30[a]	8.53[b]	12.33[a]	11.74[b]
2-Butenal	665	–	–	1.11[a]	2.66[a]	6.29[b]	7.67[a]	9.83[a]
1-Pentene-3-ol	685	–	–	6.09[b]	13.24[a]	23.72[b]	28.54[b]	30.08[a]
Pentanal*,†	705	–	0.93[c]	6.86[b]	5.52[b]	10.82[a]	12.82[b]	15.34[b]
Propanoic acid	735	–	–	–	1.39[b]	4.29[b]	5.56[b]	5.34[a]
E-2-pentenal*	765	–	–	1.79[b]	5.21[b]	14.51[b]	18.93[b]	23.63[b]
2-Penten-1-ol	781	–	–	–	–	1.1[b]	1.88[b]	2.42[b]
Unidentified	786	–	–	0.24[d]	0.49[d]	0.94[c]	1.51[c]	3.20[a]
Hexanal*	804	0.61[b]	0.80[c]	1.66[c]	4.53[c]	19.68[c]	28.06[c]	45.64[a]
Unidentified	836	–	–	–	0.23[a]	0.98[a]	1.69[c]	3.20[a]
Unidentified	855	–	–	–	0.23[b]	0.37[b]	0.50[b]	0.62[a]
E-2-Hexenal*	861	–	–	–	0.33[a]	1.07[d]	1.94[b]	2.67[a]
1-Hexanol*	878	–	–	–	–	0.51[b]	0.67[a]	0.89[a]
Unidentified	887	–	–	–	–	0.42[a]	0.57[a]	0.88[a]
2-Heptanone	897	–	–	–	0.12[a]	0.37[a]	0.57[a]	0.84[a]
p-Xylene	899	–	–	1.81[a]	0.65[b]	1.76[b]	0.72[b]	2.35[b]
Heptanal*	902	–	–	0.14[b]	0.64[d]	1.43[b]	2.54[a]	3.83[d]
Unidentified	922	–	–	0.29[b]	0.69[b]	1.63[b]	2.24[b]	2.88[a]
Unidentified	947	–	–	2.94[b]	9.62[b]	21.48[b]	29.08[b]	35.59[b]
Unidentified	949	–	–	1.95[b]	6.57[b]	16.66[b]	23.99[b]	31.75[b]
E-2-Heptenal*	964	–	0.52[a]	1.68[b]	4.84[b]	12.60[b]	22.94[b]	32.61[b]
Unidentified	970	–	–	–	–	0.44[d]	1.04[b]	1.43[a]
1-Heptanol*	974	–	–	–	–	0.25[b]	0.58[a]	0.95[d]
1-Octene-3-ol*	983	–	–	–	1.81[b]	4.93[b]	9.09[a]	12.72[b]
6-Methyl 5-hepten-2-one	991	0.36[a]	0.34[b]	0.43[b]	0.70[b]	1.45[a]	2.87[a]	5.64[a]
Hexanoic acid	1001	–	–	–	–	–	–	1.20[b]
2-Octanone	1003	–	–	–	–	0.56[c]	1.19[a]	1.35[a]
2.4-Heptadienal	1007	–	–	1.29[b]	4.62[b]	10.59[b]	18.33[b]	25.53[b]
Oktanal*	1010	–	–	–	0.16[a]	0.78[c]	1.44[c]	2.08[d]
2.4-Heptadienal isomer	1023	–	–	0.72[d]	2.86[b]	8.27[b]	16.27[b]	25.31[b]
Unidentified	1065	–	–	–	–	0.25[b]	0.64[b]	1.02[a]
E-2-Octenal*	1071	–	–	–	0.46[b]	1.61[a]	3.69[a]	7.55[a]
3.5-Octadien-2-one	1098	–	–	0.20[e]	0.58[b]	1.12[b]	1.95[b]	2.58[a]
Nonanal*	1122	–	–	0.11[a]	0.23[c]	0.90[c]	2.25[d]	3.26[c]
E-2-Nonenal*	1174	–	–	–	–	0.19[a]	0.60[d]	0.75[c]

(continued)

TABLE 4.2 (Continued)
Main Volatile Compounds Identified by SPME-GC/MS from Rapeseed Oil Subjected to Accelerated Storage Test at 60°C for a Period of 12 Days

		Storage Time at 60°C (Days)						
Compound	**RI**	**Peak Area (TIC units ×10⁶)**						
E-2-Decenal*	1279	–	–	–	–	–	–	0.67[c]
E,E-2,4-Decadienal*	1315	–	–	–	–	–	0.58[c]	0.83[b]
Total volatiles		0.97	2.59	29.31	71.68	180.50	265.27	358.20

Source: Jeleń, H.H. et al., *J. Am. Oil Chem. Soc.*, 84, 509–517, 2007. With permission.
Note: RI, retention indices on DB-5-type column; TIC, total ion current.[a-e] Relative standard deviations: [a](RSD) < 5%; [b]5 < (RSD) < 15%; [c]15 < (RSD) < 25%; [d]25 < (RSD) < 50%; [e](RSD) > 50%.
* Identification of compounds based on a comparison of their mass spectra and retention times with that of authentic standards; other compounds identified tentatively based on mass spectra library search.
† Other compound coeluting with that identified.

by ascorbic acid oxidase. Flavor changes are associated with action of LOXs. They catalyze the oxidation of PUFAs, and can also catalyze the cooxidation of carotenoids resulting in the loss of essential nutrients and formation of off-odors, forming free radicals that can react with other food constituents such as proteins, vitamins, and colorants.

TABLE 4.3
Odor Threshold Values for Main Groups of Volatile Compounds Generated during Lipid Autooxidation

Compounds	**Concentration (ppm)**
Hydrocarbons	90–2150
Substituted furans	2–27
Vinyl alcohols	0.5–3
1-Alkenes	0.02–9
2-Alkenals	0.04–2.5
Alkanals	0.04–1.0
Trans, *trans*-2,4-alkadienals	0.04–0.3
Isolated alkadienals	0.002–0.3
Isolated *cis*-alkenals	0.0003–0.1
Trans, *cis*-2,4-alkadienals	0.002–0.006
Vinyl ketones	0.00002–0.007

Source: Frankel, E.N., in *Flavor Chemistry of Fats and Oils*, ed. D.B. Min and T.H. Smouse, 1–39, American Oil Chemist's Society, Peoria, III, 1985.

LOXs have been found in animal tissues, mushrooms, fungi, and plants. Damage of plants, from wounding, cutting, and destruction of cell walls as a result of food processing, initiates a series of enzymatic reactions. Enzymes and substrates separated within intact cells, mix and as a consequence, due to enzymic reactions, volatile compounds are formed.

LOX activity is associated with several important pathways in plants. The C-18 hydroperoxides are converted by at least seven different enzyme families and all the reactions are called the LOX pathway (Figure 4.5). From the flavor formation standpoint, the LOX–HPL (lipoxygenase-hydroperoxide lyase) pathway is the predominant one in enzymatic formation of lipid-derived aroma compounds

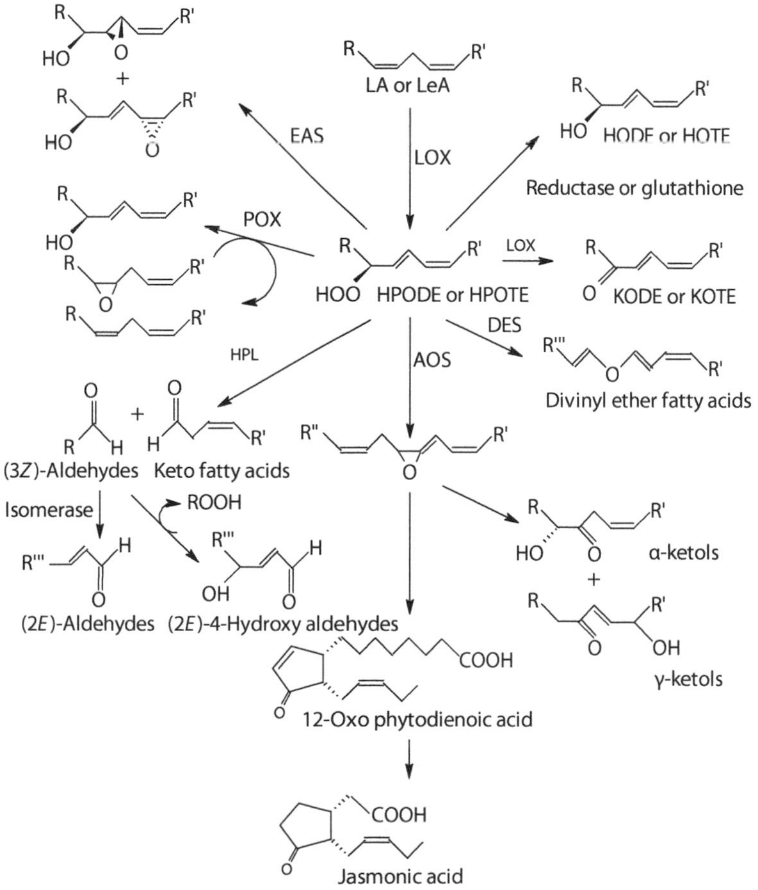

FIGURE 4.5 Metabolism of PUFA in plants—LOX pathway. AOS, allene oxide synthase; DES, divinyl ether syntase; EAS, epoxy alcohol synthase; HPL, hydroperoxide lyase; H(P)O(D/T), hydro(pero)xylinole(n)ic acid; KO(D/T), ketolinole(n)ic acid; POX, peroxygenase (From Weichert, H. et al., *Planta*, 215, 612–619, 2002. With permission.)

taking place in fruits and vegetables. However, aroma compounds produced as a result of LOX activity are not the only metabolites: as a result of LOX activity in oxygenated fatty acids, oxylipins are formed. They can be formed in reactions mediated also by cytochrome P-450 monooxygenases, or cyclooxygenase-like oxygenases. Oxylipins are involved in plant resistance strategies. They are formed in plants from linoleic and linolenic acids, and in mammals from arachidonic acid, where they play an important role in stress responses to infection, allergy, and inflammatory processes. The allene oxide synthase (AOS) pathway is restricted to 13-hydroperoxy isomers of linoleic and linolenic acids that are dehydrated into unstable allene oxide. In the biosynthesis of jasmonates, allene oxide formed by 13-allene oxide synthase (13-AOS) is subsequently cyclized enzymatically into 12-oxo-phytodienoic acid, and further to jasmonic acid (Wasternack and Parthier 1997), or alternatively hydrolyzed into α- and γ-ketols. This pathway plays an important role in plant signaling and defense systems. The peroxygenase pathway in which hydroperoxides are reduced to corresponding alcohols also plays a role in plant resistance since cutin and phytoalexins are formed in this pathway (Blee 1998).

Only a portion of LOX-derived compounds takes part in flavor formation. Aldehydes of C-6 and C-9 type, which are formed by hydroperoxide lyase (HPL) from 13 or 9-hydroperoxides of linoleic or linolenic acids, apart from their considerable role in flavor formation, are reported to be involved in the hypersensitive resistance of plants infected with pathogens and are involved in a defense response again insect herbivores (De Moraes et al. 1998).

4.3.2 Enzymes Involved in Flavor Compounds Formation

LOX (EC 1.13.11.12, linoleate: oxygen oxidoreductase) is an iron-containing dioxygenase that catalyzes the oxidation of PUFAs containing *cis*, *cis*-1,4-pentadiene units to produce conjugated unsaturated fatty acids hydroperoxides (Robinson et al. 1995). LOXs are monomeric proteins that contain a single atom of nonheme iron per molecule. The presence of iron is essential for the catalytic properties of LOX. Lipoxygenase in its "native" form is in the ferrous state and to become catalytically active, the enzyme must be first oxidized into ferric state, by the hydroperoxide product of the enzyme.

In lipid-derived flavor compounds, oxidation of unsaturated C-18 fatty acids (linoleic and linolenic) yields either 9- or 13-hydroperoxyoctadecadi(tri)enoic acids, or a mixture of both depending which of the enzyme source prevails. Arachidonic acid from animal sources also undergoes reaction with LOX. Although mainly associated with oxidation of unsaturated fatty acids, LOX can also oxygenate keto-fatty acids, aldehydes, and has the capability to oxygenate fatty acids in their esterified form, such as in phospholipids (Kondo et al. 1995). LOX can also catalyze free radical reactions involved in decomposition of hydroperoxides to epoxy-hydroxy derivatives or function as HPL-like enzymes (Guerdam et al. 1993).

Various plant LOXs differ by their stereo- and regio-specificity depending on their origin. Soybean LOX-1 exclusively catalyzes the formation of 13(*S*)-hydroperoxide of linoleic or linolenic acids at pH >9, whereas LOX from potato produces 9(*S*)-isomers. Some LOXs are less specific and form both 9- and 13- derivatives in *R* and *S* configurations (Grosch et al. 1976).

HPL (hydroperoxide lyase), and also allene oxide synthase (AOS), and divinyl ether synthase, are the key enzymes in the plant LOX pathway, regarding final products of their activity (Grechkin 2002). HPL was first isolated and characterized from watermelon seedlings (Vick and Zimmerman 1976). It was identified as the enzyme responsible for cleavage of fatty acids hydroperoxides at the carbon–carbon bond adjacent to the hydroperoxide group and their neighboring double bond, generating aldehydes and ω-oxo-acids (Kim and Grosch 1981; Hatanaka et al. 1973). HPLs are widespread in higher plant species. For some plants, activity of the enzyme correlates with the chlorophyll contents, as observed for bell peppers, where HPL is most active in green immature fruits, whereas during maturation, the activity strongly decreases (Matsui et al. 1997). HPL from bell pepper fruit consists of 55-kDa subunits (Shibata et al. 1995; Psylinakis et al. 2001). An HPL subunit of 62 kDa of soybean seedlings was determined and suggested that the lyase was a tetramer of 240–260 kDa (Olias et al. 1990). HPL isolated from green bell pepper (*Capsicum annuum* L.) was identified as a heme protein having heme b (protoheme IX) as a prosthetic group. It has a 40% identity with allene oxide synthase (CYP74A); therefore, it is postulated that the HPL should be a member of a new P-450 subfamily, CYP74B, specialized for the metabolism of lipid peroxides (Matsui et al. 1996). HPLs need no cofactor, such as molecular oxygen or reducing equivalent generally required for most P-450 enzymes (Matsui et al. 2000). It has been proposed that HPO lyase reacts with cytochrome P-450 enzymes. First, the hemolytic cleavage of the oxygen–oxygen bond in the hydroperoxide molecule could result in an alkoxyl radical and ferryl-hydroxo complex. The next step is proton donation to the hydroxyl in the ferryl-hydroxo complex and abstraction of the electron of the alkoxy radical, which forms an intermediate allylic ether cation. The addition of water to the carbocation forms a C-6 aldehyde and C-12-enol, which is transformed into a ω-oxo-acid by keto-enol tautomerization. According to recent findings, the end product of HPL catalysis is a hemiacetal, which serves as an enzyme-bound intermediate, and the aldehydes are produced only as a consequence of the high chemical instability of the hemiacetal (Grechin and Hamberg 2004).

In higher plants, two types of HPL activity exist and are classified according to their substrate specificity. The first is 9-hydroperoxide-specific (9-HPL) as found in pears (Kim and Grosch 1981). Some plants, especially those belonging to the *Curcubitaceae* family, have the ability to form C-9 aldehydes such as (Z)-3-nonenal, or (Z,Z)-3,6-nonadienal from the 9-hydroperoxides of linoleic or linolenic acids, respectively, after HPL cleaving action. The second group of HPL activity, most common and widespread in the plant kingdom, is 13-hydroperoxide-specific (13-HPL) as found in tomato leaves and bell pepper fruits (Shibata et al. 1995) producing C-6 aldehydes. HPL isolated from tea leaves (Matsui et al. 1991), and tomato fruits also have high substrate specificity for the 13-hydroperoxide of α-linolenic acid and nearly zero activity can be found toward 9-hydroperoxide. There is also evidence for nonspecific HPL type cleaving both 9- and 13-hydroperoxides as

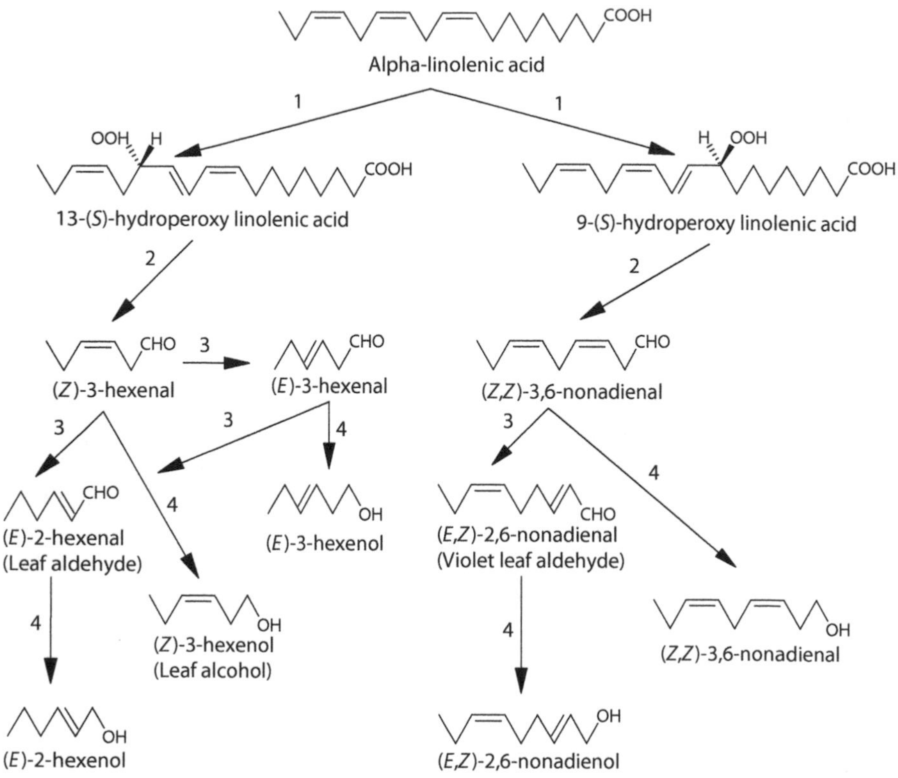

FIGURE 4.6 Pathways indicating formation of flavor compounds originating from linolenic acid oxidation leading to formation of C-6 and C-9 compounds. (1) 13- and 9-Lipoxygenases (LOXs); (2) hydroperoxide lyase (HPL); (3) isomerase; (4) alcohol dehydrogenase.

found in cucumber fruits (Galliard and Philips 1976). Main pathways for the formation of C-6 and C-9 aldehydes as a result of LOX activity in vegetables are shown in Figure 4.6.

4.3.3 Examples of LOX-Derived Flavor Compounds

4.3.3.1 Lipoxygenase Pathway in Tomatoes

The most prominent examples of vegetables where the LOX pathway is responsible for the characteristic aroma are tomatoes, cucumbers, and olives, but LOX is also important in the formation of the characteristic flavor of lettuce, green leaves, and edible kelps (Hatanaka 1996).

Among the crucial aroma compounds in tomatoes are fatty acids-derived short chain aldehydes and alcohols, such as hexanal, (*Z*)-3-hexenal, (*E*)-2-hexenal, and (*Z*)-3-hexenol, which are produced from precursors during maceration, which affects the distribution and quantity of volatile compounds measured (Riley and Thomson 1998). Moreover, competitive pathways sharing substrates or products could affect the accumulation or disappearance of selected volatile compounds. In tomato fruit, linoleic (C18:2) and linolenic (C18:3) acids are the major constituents in the unsaturated fatty acid pool, and production of volatiles from these acids can proceed via the sequential actions of LOX and HPL. (*Z*)-3-Hexenal and hexanal result from LOX/HPL action on C18:3 and C18:2 acids, respectively, and are the two of the most significant fresh tomato volatiles in terms of concentration/odor threshold ratios (Buttery et al. 1989). (*Z*)-3-Hexenal in tomatoes can undergo autooxidation via peracid to 5-ethyl-2(5*H*)furanone (Buttery and Takeoka 2004). The role of five LOX isoforms (TomLoxA, TomLoxB, TomLoxC, TomLoxD, and TomLoxE) in flavor formation was determined using transgenic tomato plants, and suppression of TomLoxA and TomLoxB did not influence the production of flavor compounds (Griffiths et al. 1996). However, for chloroplast-located TomLoxC, Chen et al. (2004) noticed a reduction in the formation of hexanal, hexenal, and hexenol to a level of 1.5% of the wild-type controls.

The highest LOX activities in tomato fruit were observed at the pink stage of ripening; the lowest LOX activity was observed at the mature green stage. Likewise, the highest HPL activities were noted for pink, lowest at mature green, and intermediate at red stage of ripeness, whereas ADH activities generally showed an increase in specific activity as tomato ripened (Yilmaz et al. 2001). However, there are reports of a linear increase of LOX activity throughout ripening (Biacs and Deood 1987) and a lack of change in HPL during ripening (Riley et al. 1996). LOX activity in tomato is localized both in soluble and membranous (microcosmal) compartments. Membrane-associated LOX is the same protein as in soluble form with a charge modification to allow membrane binding (Droillard et al. 1993). The activity of microcosmal HPL does not change significantly during ripening. By contrast, microcosmal LOX activity increases by 50% between the mature green and breaker stages of development, and soluble LOX activity declines by ~50% during the same period (Riley et al. 1996).

LOX catalyzes the oxygenation of PUFA, resulting in hydroperoxide located at carbon 9 or 13, depending on the isozyme. However, flavor compounds that contribute to the tomato flavor are obtained only from the 13-hydroxy products. HPL causes cleavage at the carbon containing a hydroperoxide, resulting in the formation of an aldehyde and an oxoacid. In tomato fruit, the majority of hydroperoxides formed by the LOX activity are the 9-isomers (Smith et al. 1997)—90–95% of LOX products in tomato fruit are thought to be 9-hydroperoxide, the balance being the 13-hydroperoxide isomer, but there is little or no 9-HPL to utilize these substrates, which therefore accumulate. The HPL affinity in tomato is higher for the 13-hydroperoxide product of LOX. It was demonstrated that tomato HPL utilized 85% of 13-hydroperoxide substrate over a 10-min period compared to only 7.1% of the 9-hydroperoxide (Hatanaka et al. 1992). Fatty acid HPLs that cleave 13-hydroperoxy-octadecadienoic acid (13-HPOD) and an α-13-hydroperoxyoctadecatrienoic acid (13-HPOT) were isolated from tomatoes. When incubated with 13-HPOT, it yielded (*Z*)-3-hexenal and (*E*)-2-hexenal, whereas when incubated with 13-HPOD, a small amount of hexanal was formed (Suurmeijer et al. 2000). With the utilization of LOX products by HPL (as described above), the availability of substrate for HPL may be a rate-limiting step in the formation of flavor compounds.

4.3.3.2 Lipoxygenase Pathway in Cucumbers

Lipids of cucumber consist mainly of palmitic acid (16:0; 26%), linoleic acid (18:2; 26.3%), and linolenic acid (18:3, 39.5%). The main volatiles produced as a result of LOX/HPL activity are hexanal originating from linoleic acid-derived 13-LOX activity, followed by (*Z*)-3-hexenal originating from linolenic acid, and (2*E*,6*Z*)-nonadienal originating from linolenic-derived 9-LOX activity, and little (2*E*)-nonenal is produced from linoleic acid (Weichert et al. 2002). Although several LOX forms were distinguished in cucumber (Feussner and Kindl 1994), so far only one HPL of cucumber seedlings has been described (Matsui et al. 2000). It lacks substrate specificity against (9*S*)- or (13*S*)-HPOD. Cucumber fruit LOX oxidizes linoleic acid 1.33 times more readily than linolenic acid forming predominantly 9-hydroperoxides (Galliard and Philips 1976). An optimum pH of 6 for HPL activity toward both the 9-hydroperoxy and 13-hydroperoxy linoleic acids was estimated (Hornostaj and Robinson 1999). 13-Hydroperoxylinoleic acid lysing activity was more sensitive to increased pH losing 50% of its activity at pH 6.5, whereas 9-hydroperoxylinoleic acid lysing activity was 50% at pH 7.5. The enzyme is relatively stable at low temperatures (4°C), and loses activity within a few hours at 50°C.

4.3.3.3 Lipoxygenase Pathway in Olives

The aroma of virgin olive oil is appreciated by consumers, to the same extent as its taste properties. Virgin oil is unique in that it is obtained by pressing without any refining process from olive fruit. The organoleptic properties of olive oil are associated with the by-products of the LOX pathway. The important flavor in olive oil is green notes typical for extra virgin oil and synonymous with its good quality. Analysis of volatile compounds in virgin olive oil reveals more than a hundred compounds, among which aldehydes (both saturated and unsaturated), esters, and alcohols are the dominant groups of compounds (Morales et al. 1994, 1995; Salas et al. 2000; Angerosa et al. 2004; Kalua et al. 2007). The formation of LOX volatiles in olive oil is a unique feature among oil fruits, where usually heat sterilization is performed (as in palm oil fruits) to inactivate engodenous enzymes such as lipases, and LOX enzymes.

LOX in olives is located in the chloroplasts and thylakoids (50% of the activity) and microcosms (35% of the activity) (Salas et al. 1999). LOX activity in olive is in acidic range (pH = 5 to 5.5) similar to tomato and bell pepper; however, the maximum LOX activity was found in alkaline range for linoleic and linolenic acids (pH 8–9) (Williams et al. 2000). 13-Hydroperoxide acids are the main products in olive tissue cultures as well as in olive fruits. Two isoforms of HPL isolated from olive fruit have optimal pH at 6.0 and are active with 13-fatty acids peroxides only, with the highest activity toward 13-hydroperoxy-9(*Z*), 11(*E*), 15(*Z*)-octatrienoic acid (13-*ZEZ*-HOTA), which has a reflection in the formation of volatile compounds as neither nonenal nor derivatives are detectable (Morales et al. 1995). HPL in olives cleaves 13-hydroperoxy fatty acids into C-6 aldehydes and 12-oxo-acids. The C-6 aldehydes include hexanal and (*Z*)-3-hexenal. The latter is unstable and undergoes isomerization to the more stable (*E*)-2-hexenal. (*E*)-2-Hexenal is of special importance, accounting for up to more than 50% of volatiles. Both hexanal and its related aldehydes undergo reduction forming alcohols, which are responsible for the fruity notes of an oil (Morales et al. 1994).

In the case of olive oil, LOX activity takes place mainly during the crushing of olive fruits and the malaxation of the resulting paste that takes place during the extraction process, making the aroma dependent on enzyme levels and activities. At acidic pH, HPL activity is reduced as compared with the optimum level (Salas and Sanchez 1999). However, it is an order-of-magnitude higher than the LOX activity measured under the same conditions, which suggests that HPL activity should not limit the formation of volatile aldehydes during the process of olive oil extraction. The six carbon aldehydes are produced from nonesterified fatty acids, so TAGs lipolytic acylhydrolases need to act. These enzymes include lipases, phospholipases, and galactolipases, as well as enzymes of general specificity. In olives, TAGs can also be hydrolyzed by exogenous lipases produced by microorganisms, such as fungus *Gloesporium olivarum*, which often contaminates the fruit. Hydrolysis of lipids by lipases is caused mainly by fruit damage and poor storage resulting in tissue disruption.

4.3.3.4 Lipoxygenase Pathway in Mushrooms

Characteristic aroma compounds associated with typical mushroom flavor are 1-octene-3-ol, 1-octene-3-one, and a group of related compounds. The literature indicates that the most important flavor compounds of the *Agaricus bisporus* are 1-octene-3-ol, 1-octene-3-one, 3-octanone, benzyl alcohol, and benzaldehyde (Loch-Bonazzi and Wolff 1994). The most important flavor compound (1-octene-3-ol) is formed by physical disruption of tissues of *A. bisporus* and *A. campestris*, and apart from *Basidiomycetes*, it has also been detected in many molds (Tressl et al. 1982; Kamiński et al. 1974). Linoleic acid in aerobic conditions is converted into 1-octene-3-ol and 10-oxo-*trans*-8-decenoic acid as the major products. As intermediates, 13-hydroperoxides, as opposed to 10- and 9-hydroperoxides, were suggested (Tressl et al. 1982). In contrast, the formation of 1-octene-3-ol via 10-hydroperoxide (10-hydroperoxy-*trans*-8-*cis*-12-octadecadienoic acid), not 9- or 13-hydroperoxides, as intermediates, was also proposed. Accumulation of 13-hydroperoxide can inhibit the formation of 1-octene-3-ol and 10-oxo-acid, suggesting that there can be two pathways catalyzed by LOX of different specificities (Wurzenberg and Grosh 1984). Use of LOX-mediated biosynthesis of 1-octene-3-ol from linoleic acid in laboratory conditions can yield 380 μg/g to 2.7 mg/g of mushroom (Husson et al. 2001; Morawicki et al. 2005).

4.3.3.5 Other Examples of LOX-Derived Aromas

Apart from the usual pathways described for C-6 and C-9 aldehydes formation in plants—where these compounds originate from linoleic or linolenic acids—it has been suggested that for some algae such as green kombu (*Laminaria angustata*), C-20 fatty acids might be precursors for C-6 and C-9 aldehydes. The majority of formed C-9 compounds [(*E*)-2- and (*Z*)-3-nonenal in *L. angustata* homogenate] was produced from C20:4(*n*–6) arachidonic acid (Boonprab et al. 2003). It was also demonstrated that in the case of animal tissues, LOX pathway contributes to the formation of specific flavor compounds (Mottram 1998). As demonstrated by Josephson et al. (1984), biogenesis of volatile aroma compounds in fresh fish is associated with formation of prostaglandins and other biologically active LOX-formed hydroxyl compounds. They postulated the formation of 1-octene-3-ol and 1-octene-3-one via the cleavage of prostaglandin H_2 (PGH_2) and the formation of 1,5-octadien-3-ol and 1,5-octadien-3-one as a result of cleavage of prostaglandin H_3 (PGH_3). Starting substrates for the formation of these flavor compounds, which are responsible for the fresh plant-like aroma of fresh fish, are arachidonic (C20:4 ω6) and eicosapentaenoic (C20:5 ω3) acids. The LOX pathways for the formation of fish flavor compounds are shown in Figure 4.7. A similar observation was made by Hsieh and Kinsella (1989), who—using gill homogenate—obtained short-chain aldehydes as a result of lipid enzymatic oxidation. They identified as products of

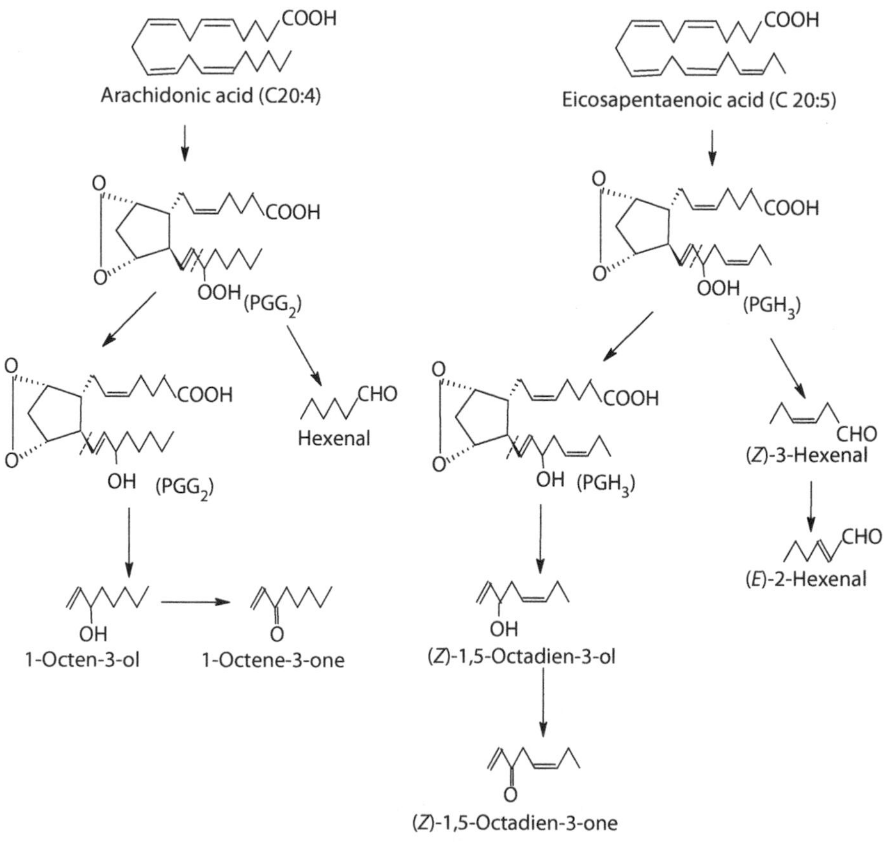

FIGURE 4.7 Pathway for formation of flavor compound in fish. (From Josephson, D.B. et al., *J. Agric. Food Chem.*, 32, 1347–1352, 1984. With permission.)

arachidonic and eicosapentaenoic acids 1-octen-3-ol, 2-octenal, 2-nonenal, 2-nonadienal, 1,5-octadien-3-ol, and 2,5-octadien-1-ol.

The lipid oxidation pathway plays an important role in producing aliphatic esters, alcohols, acids, and carbonyl compounds derived from linoleic and linolenic acids in fruits (Song and Bangerth 2003). In apples, volatile compounds emitted during ripening comprise mainly of esters, which can sum up to 98% of all volatiles (Lopez et al. 1998). Esters, mainly ethyl butanoate, ethyl 2-methylbutanoate, 2-methylbutylbutanoate, 2-methylbutyl acetate, hexyl acetate, hexyl hexanoate, and hexyl 2-methylbutanoate, are main contributors to the apple aroma for the Pink Lady variety (Lopez et al. 2007). Of the two pathways of processing fatty acids precursors—β-oxidation and LOX—the first one prevails in fruits; however, LOX pathway can become active (Villatoro et al. 2008) with high LOX and HPL activities in skin and flesh during ripening of apples.

4.4 LIPOLYSIS, METHYLKETONES, AND LACTONES

Formation of aroma compounds from lipids is not restricted to fatty acids oxidation processes alone. Tryglicerides are a source of free fatty acids, which are released in the process of lipolysis, where subsequently fatty acids can be transformed to methyl ketones and hydroxy fatty acids can be converted to lactones. Short-chain fatty acids have a distinct smell, sharp for butyric and hexanoic acid, which becomes less pronounced for decanoic or dodecanoic (lauric) acid. The formation of free fatty acids in food systems is the most pronounced in metabolism of fatty acids in dairy products. The importance of free fatty acids (and also other groups of compounds) as key odorants in cheese characterized by gas chromatography–olfactometry (GC-O) has been summarized by Curioni and Bosset (2002). There are several important review papers discussing in detail flavor formation in dairy products, describing not only lipid-derived flavors but also those originating from proteolysis and catabolism of amino acids, lactose, and citrate-derived flavors (Marilley and Casey 2004; Collins et al. 2003; McSweeney and Sousa 2000; McSweeney 2004). Also Chapter 12 of this book is devoted to the development of cheese aroma. Fat is essential for the development of characteristic cheese flavor. In dairy products, mainly cheese, lipid oxidation is of lesser importance because of the presence of antioxidants and low redox potential; however, enzymatic hydrolysis (lipolysis) of triglycerides via diacylglycerols, monoacylglycerols yielding free fatty acids is one of the key flavor formation pathways. Lipolytic enzymes are classified as esterases and lipases and are distinguished based on different characteristics and specificity. Esterases hydrolyze ester chains between 2 and 8 carbon atoms, whereas lipases hydrolyze acyl chains of more than 10 carbons. Esterases work in aqueous solutions, whether lipases in emulsified substrates. Moreover, the mode of action differs in these enzymes—esterases exhibit classical Michael–Menten kinetics, whereas lipases are activated only in the presence of hydrophobic/hydrophilic interface. In milk fat, short-chain fatty acids, namely, C4:0 and C6:0, are usually located in the sn-3 position and the sn-1 and sn-3 positions, respectively (Collins et al. 2003). Lipolytic enzymes exhibit region specificity and hydrolyze outer ester bonds in triacylglycerols (sn-1 and sn-3 positions).

Lipolysis in cheeses can be attributed to indigenous lipoprotein lipase, some rennet pastes containing pregastric esterase (Fox and Stepaniak 1993), and a large group of lipases of microbial origin. Microbial lipases and esterases are released by lactic acid bacteria *Lactobacillus delbrueckii* subsp. Lactis, *L. acidophilus*, *Lactococcus lactis* subsp. Cremoris, *L. plantarum*, *Streptococcus thermophilus*, and *Brevibacterium linens*. For mold-ripened cheeses, strains of *Penicillium* have high lipolytic activity. *P. roqueforti* and *P. camemberti* are the dominant *Penicillia* used for this type of cheese. The detailed characteristic of lipolytic enzymes can be found in an excellent review by Collins et al. (2003). The lipolytic activity is exhibited by many fungi, yeasts, and bacteria and used widely in the biotechnological formation of flavors or compounds with enhanced flavor such as enzyme-modified cheese (Regado et al. 2007). Lipolysis is a phenomenon occurring mostly in cheese; however, it is also pronounced in dry fermented sausages and hams. The lipolytic activity in meat products comes mostly from the starter cultures used in manufacturing of fermented meat products, but also lipases from muscular and adipose tissues—endogenous meat lipases, which explain lipolysis process undergoing in inner parts of hams, where microorganisms have a difficult access (Toldrá 1998; Hierro et al. 1997).

Free fatty acids metabolism is also very significant in blue mold cheese, in which they are converted into 2-methyl ketones as a result of β-oxidation followed by decarboxylation. The main fungus associated with this activity is *P. roqueforti* (Larroche et al. 1988; Creuly et al. 1990; Chalier and Crouzet 1998), and also several other fungi, such as *P. camemberti* (Molimard and Spinnler 1996). The most common methylketones that contribute to the aroma of Roquefort and other blue-veined cheeses are 2-pentanone, 2-heptanone, and 2-nonanone (King and Clegg 1979). Methylketones can be subsequently reduced to corresponding secondary alcohols (Figure 4.8). For microorganisms, such as *P. roqueforti*, production of methylketones is probably a defense mechanism in which a fungus utilizes more toxic acids released in lipolysis transforming them into volatile ketones. When *P. roqueforti* is not supplemented during its growth with free fatty acids, its secondary metabolism shifts and it starts to produce sesquiterpene hydrocarbons, of which aristolochene prevails being the volatile intermediate in subsequent production of PR-toxin (Jeleń et al. 2002) (Figure 4.9). This toxin is characteristic for *P. roqueforti*. It is suggested that it undergoes decomposition in the process of cheese making (Scott 1981), but probably due to reasons discussed above, it has not been reported in cheese.

Lactones can be formed from hydroxyl acids. The key process is the intramolecular esterification, and usually γ and σ lactones are formed with five- and six-sided rings, respectively. Lactone aromas are usually associated with peach, coconut, fatty, and fruity notes. They are produced in heated butter (milk fat), but also abundant in fruits and some fermented products. Because they contribute to peach/coconut flavors, they are a valuable flavor material. Lactones are well-explored compounds (bio)synthesized using microorgranisms, usually yeasts, mainly *Yarrovia lipolytica*, for the bioconversion of hydroxy fatty acids into lactones. The main product used in the biotechnological process is γ-decalactone (peach aroma) produced by biotransformation of ricinoleic acid, being the main (90%) acid of hydrolyzed castor oil (Waché et al. 2002, 2003).

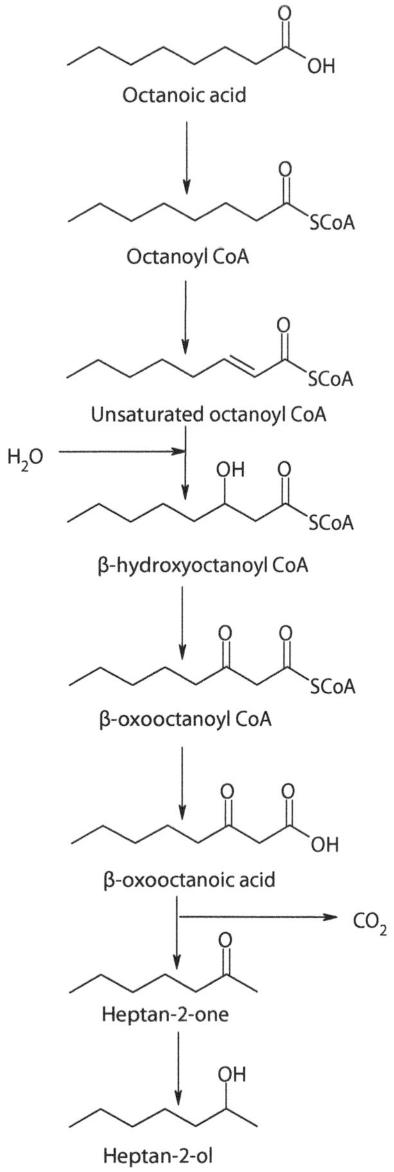

OH
Octanoic acid
SCoA
Octanoyl CoA
SCoA
Unsaturated octanoyl CoA
H2O
OH
SCoA
β-hydroxyoctanoyl CoA
SCoA
β-oxooctanoyl CoA
OH
β-oxooctanoic acid
CO2
Heptan-2-one
OH
Heptan-2-ol

FIGURE 4.8 Scheme of methyl ketones formation from free fatty acids by microbial biotransformation by *P. roqueforti.*

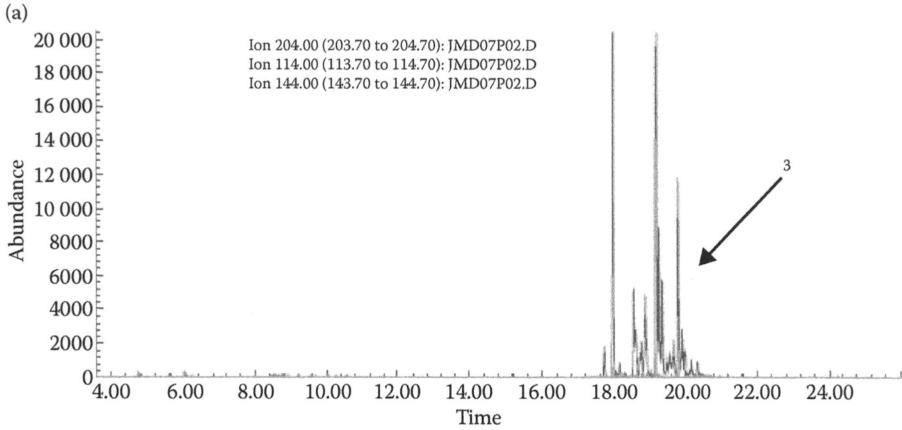

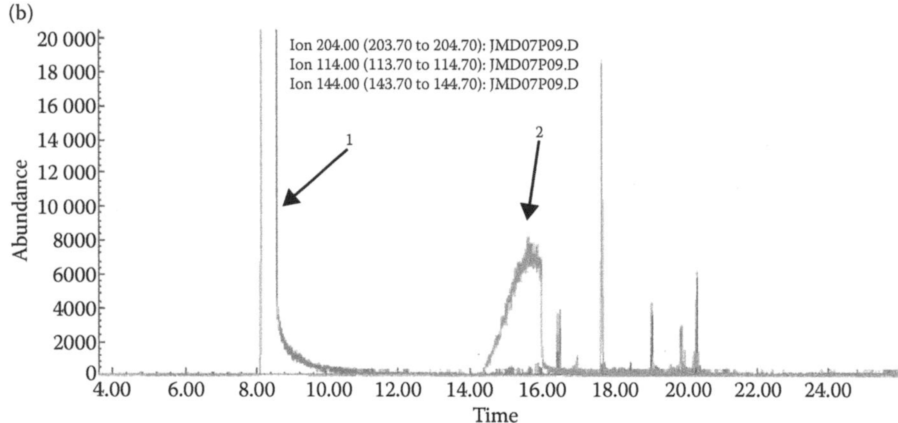

FIGURE 4.9 Extracted ion chromatograms of volatile compounds of *P. roqueforti* strain IBT 16404 grown on autoclaved wheat kernels (a) nonsupplemented and (b) supplemented with octanoic acid. Peaks on chromatograms: (1) 2-heptanone (m/z = 114); (2) octanoic acid (m/z = 144); (3) sesquiterpene hydrocarbons (m/z = 204). (From Jeleń, H. et al., *Lett. Appl. Microbiol.*, 35, 37–41, 2002. With permission.)

4.5 CONCLUSION

Lipid-derived flavor compounds form one of the most important volatiles in food. The autooxidation of lipids is associated with decrease of food quality and formation of off-flavors. However, enzyme-mediated oxidation of fatty acids in plants provides a distinctive aroma of many fruits and vegetables. Flavors formed as a result of triglycerides lipolysis are important for aroma of dairy products, whereas lipid-derived aroma compounds, such as methyl ketones and lactones, are important flavor compounds produced via biotechnological methods.

REFERENCES

Akoh, C.C., and D.B. Min. 2008. Food lipids. *Chemistry, Nutrition and Biotechnology*. Boca Raton, FL: CRC Press, Taylor & Francis Group.

Angerosa, F., M. Servili, R. Selvaggini, A. Taticchi, S. Esposto, and G. Montedoro. 2004. Volatile compounds in virgin olive oil: Occurrence and their relationship with the quality. *J. Chromatogr. A* 1054: 17–31.

Belitz, H.-D., W. Grosch, and P. Schieberle. 2009. *Food Chemistry*. Berlin: Springer-Verlag.

Biacs, P.A., and H. Deood. 1987. Characterization of tomato lipoxygenase. In *The Metabolism, Structure and Function of Plant Lipids*, ed. P.K. Stumpf, J.D. Mudd, and N.D. Nes, 425–429. New York: Plenum Press.

Blee, E. 1998. Phyooxylipins and plant defense reactions. *Prog. Lipid Res.* 37: 33–72.

Boonprab, K., K. Matsui, Y. Akakabe, N. Yotsukura, and T. Kajiwara. 2003. Hydroperoxy-arachidonic acid mediated *n*-hexanal and (*Z*)-3- and (*E*)-2-nonenal formation in *Laminaria angustata*. *Phytochemistry* 63: 669–678.

Buttery, R., and G. Takeoka. 2004. Some unusual minor volatile components of tomato. *J. Agric. Food Chem.* 52: 6264–6266.

Buttery, R.G., R. Teranishi, R.A. Flath, and L.C. Ling. 1989. Fresh tomato volatiles. In *Flavor Chemistry: Trends and Developments*, ed. R. Teranishi, R.G. Buttery, and F. Shahidi, 312–322. Washington, DC: American Chemical Society.

Chalier, P., and J. Crouzet. 1998. Methyl ketone production from kopra oil by *Penicillium roqueforti* spores. *Food Chem.* 63: 447–451.

Chen, G., R. Hackett, D. Walker, A. Taylor, Z. Lin, and D. Grierson. 2004. Identification of a specific isoform of tomato lipoxygenase (TomloxC) involved in the generation of fatty acid-derived flavor compounds. *Plant Physiol.* 136: 2641–2651.

Choe, E., and D.B. Min. 2006. Chemistry and reactions of reactive oxygen species in foods. *Crit. Rev. Food Sci. Nutr.* 46: 1–22.

Collins, Y.F., P.L.H. McSweeney, and M.G. Wilkinson. 2003. Lipolysis and free fatty acid catabolism in cheese: A review of current knowledge. *Int. Dairy J.* 13: 841–866.

Creuly, C., C. Laroche, and J.B. Gros. 1990. A fed batch technique for 2-heptanone production by spores of *Penicillium roqueforti*. *Appl. Microbiol. Biotechnol.* 34: 20–25.

Curioni, P.M.G., and J.O. Bosset. 2002. Key odorants in various cheese types as determined by gas chromatography–olfactometry. *Int. Dairy J.* 12: 959–984.

Damodaran, S., K.L. Parkin, and O.R. Fennema. 2008. *Fennema's Food Chemistry*. Boca Raton, FL: CRC Press, Taylor & Francis Group.

De Moraes, C.M., W.J. Lewis, P.W. Paré, H.T. Alborn, and J.H. Tumlinson. 1998. Herbivore-infested plants selectively attract parasitoids. *Nature* 393: 570–573.

Droillard, M.J., M.A. Rouet-Mayer, J.M. Bureau, and C. Lauriere. 1993. Membrane associated and soluble lipoxygenase isoforms in tomato pericarp. *Plant Physiol.* 103: 1211–1219.

Feussner, I., and H. Kindl. 1994. Particulate and soluble lipoxygenase isoenzymes: Comparison of molecular and enzymatic properties. *Planta* 194: 22–28.

Fox, P.F., and L. Stepaniak. 1993. Enzymes in cheese technology. *Int. Dairy J.* 3: 509–530.

Frankel, E.N. 1985. Chemistry of autooxidation: Mechanism, products and flavor significance. In *Flavor Chemistry of Fats and Oils*, ed. D.B. Min and T.H. Smouse, 1–39. Peoria, Ill: American Oil Chemists' Society.

Galliard, T., and D.R. Philips. 1976. The enzymic cleavage of linoleic acid to C9 carbonyl fragments in extracts of cucumber (*Cucumis sativus*) fruit and the possible role of lipoxygenase. *Biochim. Biophys. Acta* 431: 278–287.

Grechin, A.N., and M. Hamberg. 2004. The heterolytic hydroperoxide lyase is an isomerase producing a short lived fatty acid hemiacetal. *Biochim. Biophys. Acta* 1636: 47–58.

Grechkin, A.N. 2002. Hydroperoxide lyase and divinyl ether synthase. *Prostaglandins Other Lipid Mediat.* 68–69: 457–470.

Griffiths, A., S. Prestage, R. Linforth, J. Zhang, and A. Taylor. 1996. Fruit specific lipoxygenase suppression in antisense-transgenic tomatoes. *Postharvest Biol. Technol.* 17: 163–173.

Grosch, W. 1982. Lipid degradation products and flavor. In *Food Flavours. Part A. Introduction*, ed. I.D. Morton and A.J. Macleod, 325–398. Amsterdam: Elsevier Scientific Publishing Company.

Grosch, W., G. Laskawy, and F. Weber. 1976. Formation of volatile carbonyl compounds and cooxidation of beta-carotene by lipoxygenase from wheat, potato, flax, and beans. *J. Agric. Food Chem.* 24: 456–459.

Guerdam, E., R.H. Andrianarison, H. Rabinowitch-Chable, M. Tixier, and J.L. Beneytout. 1993. Presence of fatty acid-degrading enzyme in a certain variety of peas (*Pisum sativum hortense* cv. Solara). *J. Agric. Food Chem.* 41: 1593–1597.

Hatanaka, A. 1996. The fresh green odor emitted by plants. *Food Rev. Int.* 12: 303–350.

Hatanaka, A., and T. Harada. 1973. Formation of *cis*-3-hexenal, *trans*-2-hexenal and *cis*-3-hexenol in macerated *Thea sinensis* leaves. *Phytochemistry* 12: 2341–2346.

Hatanaka, A., T. Kajiwara, K. Matsui, and A. Kitamura. 1992. Expression of lipoxygenase and hydroperoxide lyase activities in tomato fruits. *Z. Naturforsch.* 47: 369–374.

Hierro E., L. de la Hoz, and J.A. Ordóñez. 1997. Contribution of microbial and meat endogenous enzymes to the lypolysis of dry fermented sausages. *J. Agric. Food Chem.* 45: 2989–2995.

Hsieh, R.J., and J.E. Kinsella. 1989. Lipoxygenase generation of specific volatile flavor carbonyl compounds in fish tissue. *J. Agric. Food Chem.* 37: 279–286.

Ho, C.-T., and Q. Chen. 1994a. Lipids in food flavors. In *Lipids in Food Flavors*, ACS Symposium Series, ed. C. Ho and T.G. Hartman. Washington, DC: ACS.

Ho, C.-T., and T.G. Hartman. 1994b. *Lipids in Food Flavors*, ACS Symposium Series. Washington, DC: ACS.

Hornostaj, A.R., and D.S. Robinson. 1999. Production of hydroperoxide lyase from cucumbers. *Food Chem.* 66: 173–180.

Husson, F., D. Bompas, S. Kermasha, and J.M. Belin. 2001. Biogeneration of 1-octene-3-ol by lipoxygenase and hydroperoxide lyase activities of *Agaricus bisporus*. *Process. Biochem.* 31: 177–182.

Jeleń, H., S. Mildner, and K. Czaczyk. 2002. Influence of octanoic acid addition to medium on some volatile compounds and PR-toxin biosynthesis by *Penicillium roqueforti*. *Lett. Appl. Microbiol.* 35: 37–41.

Jeleń, H.H., S. Mildner-Szkudlarz, I. Jasińska, and E. Wąsowicz. 2007. A headspace–SPME–MS method for monitoring rapeseed oil autooxidation. *J. Am. Oil Chem. Soc.* 84: 509–517.

Josephson, D.B., R.C. Linsday, and D.A. Stuiber. 1984. Biogenesis of lipid derived volatile aroma compounds in the emerald shiner (*Notropis atherinoides*). *J. Agric. Food Chem.* 32: 1347–1352.

Kalua, C.M., M.S. Allen, D.R. Bedgood Jr., A.G. Bishop, P.D. Prenzler, and K. Robards. 2007. Olive oil volatile compounds, flavor development and quality: A critical review. *Food Chem.* 100: 273–286.

Kamiński, E., S. Stawicki, and E. Wąsowicz. 1974. Volatile flavor compounds produced by molds of *Aspergillus*, *Penicillium* and *Fungi imperfecti*. *Appl. Microbiol.* 24: 1001–1004.

Kim, I.S., and W. Grosch. 1981. Partial purification of a hydroperoxide lyase from fruit of pear. *J. Agric. Food Chem.* 29: 1220–1225.

King, R.D., and G.H. Clegg. 1979. The metabolism of fatty acids, methyl ketones and secondary alcohols by *Penicillium roqueforti* in blue cheese slurries. *J. Sci. Food Agric.* 30: 197–202.

Kondo, Y., Y. Hashidoko, and J. Mizutani. 1995. An enzymatic formation of 13-oxo-trideca-9,11-dienoic acid from 13-hydroperoxylinolenic acid by a homolytic hydroperoxide lyase in elicitor-treated soybean cotyledons. *Biochim. Biophys. Acta* 1255: 9–15.

Larroche, C., B. Tallou, and J.B. Gros. 1988. Aroma production by spores of *Penicillium roqueforti* on a synthetic medium. *J. Ind. Microbiol.* 3: 1–8.

Loch-Bonazzi, C.L., and E. Wolff. 1994. Characterization of flavor properties of the cultivated mushroom (*Agaricus bisporus*) and the influence of drying process. *Lebensm.-Wiss. Technol.* 24: 386–289.

Long, C.A., and D.R. Kearns. 1975. Radiationless decay of singlet molecular oxygen in solution: II. Temperature dependence and solvent effects. *J. Am. Oil Chem. Soc.* 97: 2018–2020.

Lopez, M.L., T. Lavilla, I. Recasens, M. Riba, and M. Vendrell. 1998. Influence of different oxygen and carbon dioxide concentrations during storage on production of volatile compounds by "Starking Delicious" apples. *J. Agric. Food Chem.* 46: 634–643.

Lopez, M.L., C. Villatoro, T. Fuentes, J. Graell, I. Lara, and G. Echeverria. 2007. Volatile compounds, quality parameters and consumer acceptance of "Pink Lady®" apples stored in different conditions. *Postharvest Biol. Technol.* 43: 55–66.

Marilley, L., and M.G. Casey. 2004. Flavours of cheese products: Metabolic pathways, analytical tools and identification of producing strains. *Int. J. Food Microbiol.* 90: 139–159.

Matsui, K., Y. Shibata, H. Tateba, A. Hatanaka, and T. Kajiwara. 1997. Changes of lipoxygenase and fatty acid hydroperoxide lyase activities in bell pepper fruits during maturation. *Biosci. Biotechnol. Biochem.* 61: 199–201.

Matsui, K., M. Shibutani, T. Hase, and T. Kajiwara. 1996. Bell pepper fruit fatty acid hydroperoxide lyase is a cytochrome P450 (CYP74B). *FEBS Lett.* 394: 21–24.

Matsui, K., H. Toyota, T. Kajiwara, T. Kakuno, and A. Hatanaka. 1991. Fatty acid hydroperoxide clearing enzyme, hydroperoxide lyase from tea leaves. *Phytochemistry* 30: 2109–2113.

Matsui, K., C. Ujita, S. Fujimoto, J. Wilkinson, B. Hiatt, V. Knauf, T. Kajiwara, and I. Feussner. 2000. Fatty acid 9- and 13-hydroperoxide lyases from cucumber. *FEBS Lett.* 481: 183–188.

McSweeney, P.L.H. 2004. Biochemistry of cheese ripening. *International Journal of Dairy Technology* 57: 127–144.

McSweeney, P.L.H., and M.J. Sousa. 2004. Biochemical pathways for the production of flavor compounds in cheese during ripening: A review. *Lait* 80: 293–324.

Min, D., and T.H. Smouse. 1985. *Flavor Chemistry of Fats and Oils*. Peoria, Ill: American Oil Chemists' Society.

Molimard, P., and H.E. Spinnler. 1996. Compounds involved in the flavor of surface mould-ripened cheeses: Origins and properties. *J. Dairy Sci.* 79: 169–184.

Morales, M.T., M.V. Alonso, J.J. Rios, and R. Aparicio. 1995. Virgin olive oil aroma: Rselationship between volatile compounds and sensory attributes by chemometrics. *J. Agric. Food Chem.* 43: 2925–2931.

Morales, M.T., R. Aparicio, and J.J. Rios. 1994. Dynamic headspace gas chromatographic method for determining volatiles in virgin olive oil. *J. Chromatogr. A.* 668: 455–462.

Morawicki, R.O., R.B. Beelman, D. Petersen, and G. Ziegler. 2005. Biosynthesis of 1-octen-3-ol and 10-oxo-*trans*-8-decenoic acid using a crude homogenate of *Agaricus bisporus*. Optimization of the reaction: Kinetic factors. *Process. Biochem.* 40: 131–137.

Mottram, D.S. 1998. Flavour formation in meat and meat products: A review. *Food Chem.* 62: 415–424.

Olias, J.M., J.L. Rios, M. Valle, R. Zamora, L.C. Sartz, and B. Axelrod. 1990. Fatty acid hydroperoxide lyase in germinating soybean seedlings. *J. Agric. Food Chem.* 38: 624–630.

Privett, O.S., and M.L.J. Blank. 1962. The initial stages of autooxidation. *J. Am. Oil Chem. Soc.* 39: 465–469.

Psylinakis, E., E.M. Davoras, N. Ioannidis, M. Trikeriotis, V. Petrouleas, and D.F. Ghanotakis. 2001. Isolation and spectroscopic characterization of a recombinant bell pepper hydroperoxide lyase. *Biochim. Biophys. Acta* 1533: 119–127.

Rawls, H.R., and P.J. Van Santen. 1970. A possible role of singlet oxygen in the initiation of fatty acids autooxidation. *J. Am. Oil Chem. Soc.* 47: 121–125.

Regado, M.A., B.M. Cristóvão, C.G. Moutinho, V.M. Balcão, R. Aires-Barros, J.P.M. Ferreira, and F.X. Malcata. 2007. Flavour development via lypolysis of milk fats: Changes in free fatty acid pool. *Int. J. Food Sci. Technol.* 42: 961–968.

Riley, J.C.M., and J.E. Thompson. 1998. Ripening induced acceleration of volatile aldehyde generation following tissue disruption in tomato fruit. *Physiol. Plant* 104: 571–576.

Riley, J.C.M., C. Willemot, and J.E. Thompson. 1996. Lipoxygenase and hydroperoxide lyase activities in ripening tomato fruit. *Postharvest Biol. Technol.* 7: 97–107.

Robinson, D.S., Z. Wu, C. Domoney, and R. Casey. 1995. Lipoxygenases and the quality of foods. *Food Chem.* 54: 33–43.

Salas, J., and J. Sanchez. 1999. Hydroperoxide lyase from olive (*Olea europea*) fruits. *Plant Sci.* 143: 19–23.

Salas, J.J., J. Sanchez, U.S. Ramli, A.M. Manaf, M. Williams, and J.L. Harwood. 2000. Biochemistry of lipid metabolism in olive and other oil fruits. *Prog. Lipid Res.* 39: 151–180.

Salas, J., M. Williams, J. Sanchez, and J.L. Harwood. 1999. Lipoxygenase activity in olive oil (*Olea europea*) fruit. *J. Am. Oil Chem. Soc.* 76: 1163–1168.

Scott, P.M. 1981. Toxins of *Penicillium* species used in cheese manufacture. *J. Food Prot.* 44: 702–710.

Shibata, Y., K. Matsui, T. Kajiwara, and A. Hatanaka. 1995. Purification and properties of fatty acid hydroperoxide lyase from Green Bell pepper fruits. *Plant Cell Physiol.* 36: 147–156.

Smith, J.J., R. Linforth, and G.A. Tucker. 1997. Soluble lipoxygenase isoforms from tomato fruit. *Phytochemistry* 45: 453–458.

Song, J., and F. Bangerth. 2003. Fatty acids as precursors for aroma volatile biosynthesis in pre-climacteric and climacteric apple fruit. *Postharvest Biol. Technol.* 30: 113–121.

St. Angelo, A.J. 1996. Lipid oxidation in foods. *Crit. Rev. Food Sci. Nutr.* 36: 175–224.

Suurmeijer, C.N.S.P., M. Perez-Gilabert, D.J. van Unen, T.W.M. van der Hijden, G.A. Veldink, and J.F.G. Vliegenthart. 2000. Purification, stabilization and characterization of tomato fatty acid hydroperoxide lyase. *Phytochemistry* 53: 177–185.

Terao, J., and S. Matsushita. 1977. Products formed by photo-sensitized oxidation of unsaturated fatty acids esters. *J. Am. Oil Chem. Soc.* 54: 234–238.

Toldrá, F. 1998. Proteolysis and lipolysis in flavor development of dry-cured meat products. *Meat Sci.* 49: S101–S110.

Tressl, R.B., D. Bahri, and K.H. Engel. 1982. Formation of eight carbon and ten carbon components in mushroom *Agaricus campestris*. *J. Agric. Food Chem.* 30: 89–93.

Vick, B.A., and D.C. Zimmerman. 1976. Lipoxygenase and hydroperoxide lyase in germinating watermelon seedlings. *Plant Physiol.* 57: 780–788.

Villatoro, C., R. Altisent, G. Echeverria, J. Graell, M.L. López, and I. Lara. 2008. Changes in biosynthesis of aroma volatile compounds during on-tree maturation of "Pink Lady®" apples. *Postharvest Biotechnol. Technol.* 47: 286–295.

Yang W.T., and D.B. Min. 1994. Chemistry of singlet oxygen oxidation of foods. In *Lipids in Food Flavors*, ACS Symposium Series, ed. C.-T. Ho and T.G. Hartman. Washington, DC: ACS.

Yilmaz, E., K.S. Tandon, J.W. Scott, E.A. Baldwin, and R.L. Shewfelt. 2001. Absence of a clear relationship between lipid pathway enzymes and volatile compounds in fresh tomatoes. *J. Plant Physiol.* 158: 1111–1116.

Waché, Y., M. Aguedo, M.-T. LeDall, J.-M. Nicaud, and J.M. Belin. 2002. Optimization of *Yarrowia lipolytica*'s β-oxidation pathway for γ-decalactone production. *J. Mol. Catal. B Enzymol.* 19–20: 347–351.

Waché, Y., M. Aguedo, J.-M. Nicaud, and J.M. Belin. 2003. Catabolism of hydroxyacids and biotechnological production of lactones by *Yarrowia lipolytica. Appl. Microbiol. Biotechnol.* 63: 393–404.

Wasternack, C., and B. Parthier. 1997. Jasmonate signaled plant gene expression. *Trends Plant Sci.* 2: 302–307.

Weichert, H., A. Kolbe, A. Kraus, C. Wasternack, and I. Feussner. 2002. Metabolic profiling of oxylipins in germinating cucumber seedlings: Lipoxygenase-dependent degradation of triacylglycerols and biosynthesis of volatile aldehydes. *Planta* 215: 612–619.

Williams, M., J.J. Salas, J. Sanchez, and J.L. Harwood. 2000. Lipoxygenase pathway in olive callus cultures (*Olea europeae*). *Phytochemistry* 53: 13–19.

Wurzenberg, M., and W. Grosh. 1984. The formation of 1-octene-3-ol from 10-hydroperoxide isomer of linoleic acid by hydroperoxide lyase in mushroom (*Psalliota bispora*). *Biochim. Biophys. Acta* 794: 25–30.

Wacke[illegible], M. [illegible]

Wa[illegible] [illegible]

W[illegible] [illegible]

Welch[illegible] [illegible]

[illegible]

[illegible]

5 Saccharides-Derived Flavor Compounds

Małgorzata Majcher

CONTENTS

5.1 INTRODUCTION

Saccharides are common components of foods as natural components and added ingredients. Together with lipids and proteins, they represent one of the basic nutrients serving as the most important source of energy in human diet. Because of their wide range of chemical and physical properties, saccharides have additionally many valuable functions in foods such as sweetening, gelatinization, thickening, or stabilization properties, and they can also act as precursors for aroma and coloring substances. Saccharides are also called carbohydrates, which suggests that they are all hydrates of carbon, namely, $C_x(H_2O)_y$. Among the most important types of carbohydrates in foods are the sugars, dextrins, starches, celluloses, hemicelluloses, pectins, and certain gums. By now, many other components such as deoxysugars, amino sugars, and sugar carboxylic acids have been included in this class of compounds. Basically, saccharides are divided into monosaccharides, oligosaccharides, and polysaccharides. For the development of food flavor, mostly monosaccharides are involved in the course of caramelization and Maillard reaction (MR). In this chapter, universal mechanisms and pathways in aroma compounds formation of process flavors from saccharides precursors will be discussed.

5.2 CARAMELIZATION

Monosaccharides are carbohydrate molecules that cannot be broken down to simpler carbohydrate molecules by hydrolysis; however, they are the monomeric units joined

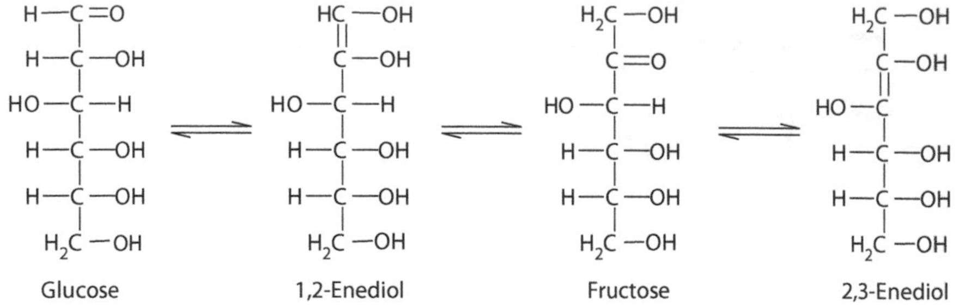

Glucose 1,2-Enediol Fructose 2,3-Enediol

FIGURE 5.1 Enolization reaction of glucose and fructose.

together to form larger structures such as oligo- and polysaccharides. All saccharides have hydroxyl groups (–OH) available for reactions. Sugars possessing free aldehyde or ketone groups are known as reducing sugars. Simple monosaccharides and other low-molecular-weight carbohydrates also have carbonyl groups available for reaction. The reaction involving heating carbohydrates, in particular sucrose and reducing sugars, affects a complex group of reactions called caramelization, giving rise to brown-colored products with typical caramel aroma. Depending on reaction parameters (temperature, catalysts), the process can be directed more toward formation of brown pigments (e.g., heating glucose syrup with sulfuric acid in the presence of ammonia) or toward aroma formation (e.g., heating of sucrose syrup in a buffered solution). During saccharides heating, dehydration of the sugar molecule occurs with introduction of double bonds or formation of anhydro rings. Heating of monosaccharides under acidic conditions gives rise to a large number of furans and pyrones such as furfural, 5-hydroxymethylfurfural, 5-methylfurfural, 2-acetylfuran, isomaltol, or 2-hydroxyacetylfuran. The reaction pathways of these compounds can be explained by the enolization step followed by dehydration (BeMiller and Huber 2007). Enolization reaction, known as the "Lobry de Bruyn–Alberda van Ekenstein transformation," produces enediol anion species, which in cases of hexoses give rise to 1,2-enediol and for pentoses 2,3-enediol as well (Figure 5.1) (Angyal 2001).

Those enediols are important intermediates of caramel aroma development such as formation of 5-hydroxymethyl furfural from 1,2-enediol (Figure 5.2) (Belitz et al. 2004). In the same way, furfural can be formed from 2,3-enediol. Heating of fructose gives rise to wider spectrum of degradation products than with glucose, as in addition to predominant 1,2-enediol also 2,3-enediol can be formed. Examples are introduced in Figure 5.3.

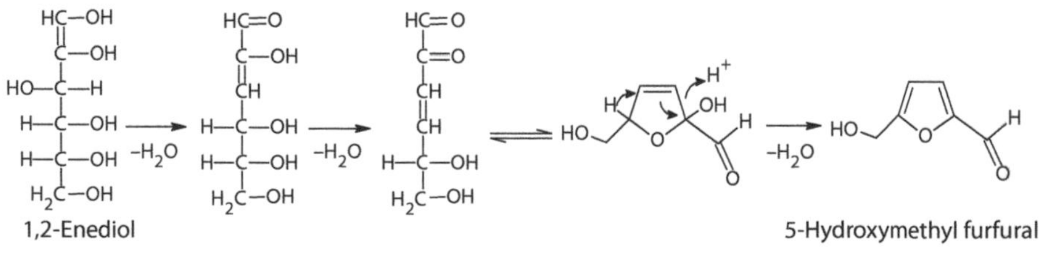

1,2-Enediol 5-Hydroxymethyl furfural

FIGURE 5.2 Formation of 5-hydroxymethyl furfural in a caramelization reaction.

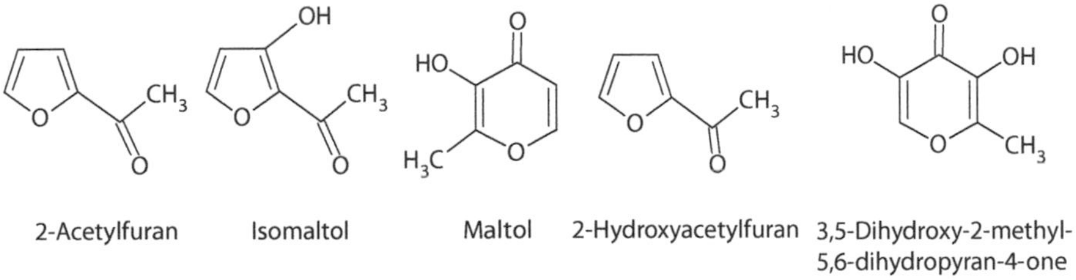

2-Acetylfuran Isomaltol Maltol 2-Hydroxyacetylfuran 3,5-Dihydroxy-2-methyl-5,6-dihydropyran-4-one

FIGURE 5.3 Flavor compounds formed during heating of carbohydrates.

Heating of glucose in pH 8–10 allows for the formation of other group of aroma compounds such as dihydrofuranones, cyclopentenolones, cyclohexenolones, and pyrones. Some of them, such as cyclopentenelones, are typical caramel-like aroma substances. Their formation is understood to follow enolization, dehydration, and aldol condensation. In Figure 5.4, formation of 2-hydroxy-3-methyl-2-cyclopentenone is shown. Substitution of 1-hydroxy-2-butanone or 3-hydroxy-2-butanone for one molecule of hydroxyacetone can give rise analogously to 2-hydroxy-3-ethyl- or 2-hydroxy-3,4-dimethyl-2-cyclopentenone.

5.3 MAILLARD REACTION—MAIN PATHWAYS FOR AROMA COMPOUNDS FORMATION

For aroma compounds development, the most well-known reaction involving saccharides is when they react with compound possessing a free amino group (Maillard 1912). This reaction has been named after the French chemist Louis Maillard, who first described it in 1910, and is called Maillard reaction (MR) (Billaud and Adrian 2003). For as long as food has been cooked, MR has been applied to produce food, for example, fried meat, coffee, and bakery products, which possess the color and flavor

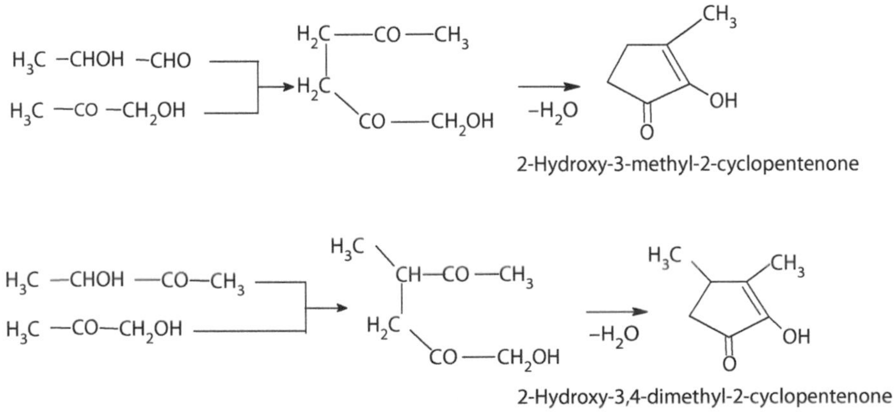

H_3C –CHOH –CHO
H_3C –CO –CH_2OH
H_2C–CO–CH_3 / H_2C / CO–CH_2OH
$-H_2O$
2-Hydroxy-3-methyl-2-cyclopentenone

H_3C –CHOH –CO–CH_3
H_3C –CO–CH_2OH
H_3C / CH–CO–CH_3 / H_2C / CO–CH_2OH
$-H_2O$
2-Hydroxy-3,4-dimethyl-2-cyclopentenone

FIGURE 5.4 Formation of 2-hydroxy-3-methyl-2-cyclopentenone and 2-hydroxy-3,4-dimethyl-2-cyclopentenone. (From Belitz, H.D. et al., eds., *Food Chem.*, Springer, Germany, 2004. With permission.)

demanded by the consumer. At present, the food industry, by applying traditional processes such as roasting, frying, baking, toasting, or cooking, is trying to employ MR to improve the appearance and flavor of foods. For this reason, the understanding of chemical reactions occurring during these processes has been of special interest to many scientists and also to the industry itself. Although this reaction has a very simply structured precursors, namely, reducing sugars and amino acids, it has a very complex pathway reactions depending on many factors: type and combination of precursors, and food processing variables, for instance, temperature, pH, or water activity. The complexity of MR can be also illustrated by the vast number of flavor compounds (greater than 2500) identified in thermally processed foods (Ericsson 1981; Cerny 2008; Poisson et al. 2009). In addition, this reaction is also responsible for development of high-molecular-weight products called melanoidins that are responsible for the browning of foodstuff (Blank and Fay 1996). After Louis Maillard's discovery, the first attempt to understand complexity of those reactions has been made by John Hodge only in 1953 by publishing a consistent scheme (Figure 5.5).

Since then, the knowledge concerning MR has been studied by many authors, resulting in the discovery of many new important pathways, which were not accounted for by the Hodge scheme; however, it still remains widely used to explain the essential steps of MR. Based on this concept, MR can principally be divided

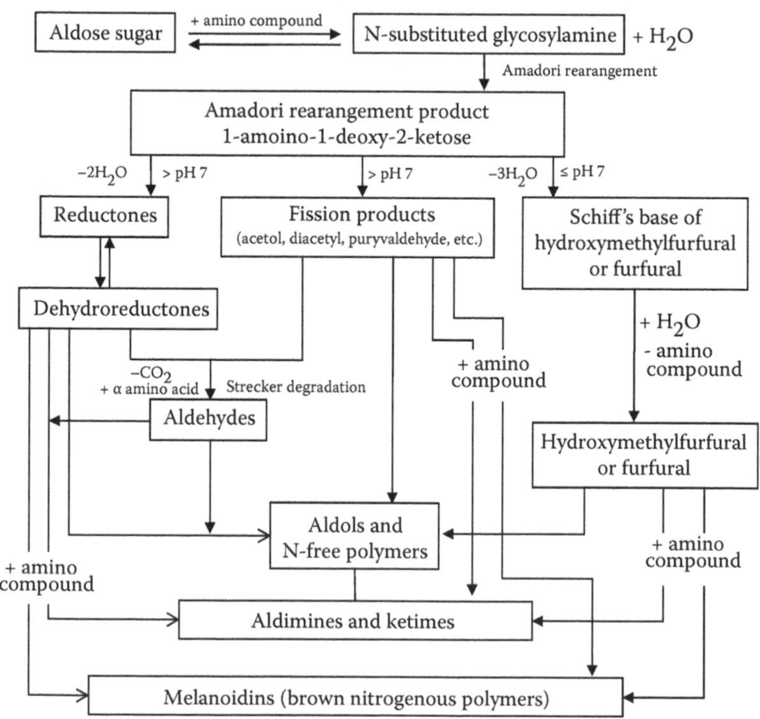

FIGURE 5.5 Maillard reaction scheme. (Adapted from Hodge, J.E., *J. Agric. Food Chem.*, 1, 928–943, 1953.)

into three stages: the early, advanced, and final phases. At the early stage, reducing sugar condenses with a compound possessing a free amino group to give a condensation product N-substituted glycosylamine. The glucosyloamines from aldose sugar undergo *Amadori* rearrangement to yield the so-called *Amadori* compounds (1-amino-1-deoxyketose) where the reaction between ketoses (fructose) and amines usually involves the formation of ketosylamines (Amadori 1931), followed by the *Heyns* rearrangement to form 2-amino-2-deoxyaldose (Figure 5.6).

The following degradation of the *Amadori* and *Heyns* product initiates the advanced stage of MR, in which a series of reactions take place, including cyclizations, dehydrations, retroaldolizations, rearrangements, isomerizations, and further condensations, producing impact food odor compounds (Yaylayan and Huyghues-Despointes 1994; Wedzicha et al. 1994). In the final stage of MR, polymerized protein and brown pigments called melanoidins are formed (Adams et al. 2005).

Amadori products are the only intermediates formed in the course of MR, and they undergo further degradation via several ways to form deoxyosones, which decompose afterward to flavor compounds. Aminoketose (*Amadori* compound) can be converted to 2,3-eneaminol as well as 1,2-eneaminol by enolization; as for aminoaldose (*Heyns* compound), it converts only to 1,2-eneaminol. In the next step, 1,2-eneaminol, by water elimination and hydrolysis, gives 3-deoxyosone, whereas analogous 2,3-eneaminol has two different β-elimination options to give either 1-deoxyosone by retro-Michael reaction or 4-deoxyosone by water elimination. All three deoxyosones can be formed in different cyclic hemiacetal forms (Figure 5.7).

The course of further reactions depends on the pH of the system. At pH 7 or below, it undergoes mainly 1,2-enolization, followed by the formation of 3-deoxyosone to finally form furfural when pentoses are involved, or 5-hydroxymethylfurfural when hexoses are involved (Figures 5.8 and 5.9).

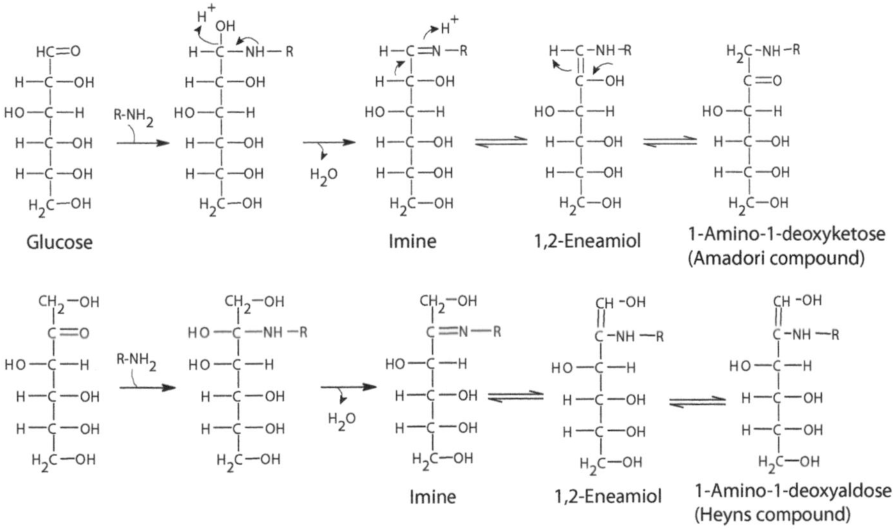

FIGURE 5.6 *Amadori* and *Heyns* rearrangement.

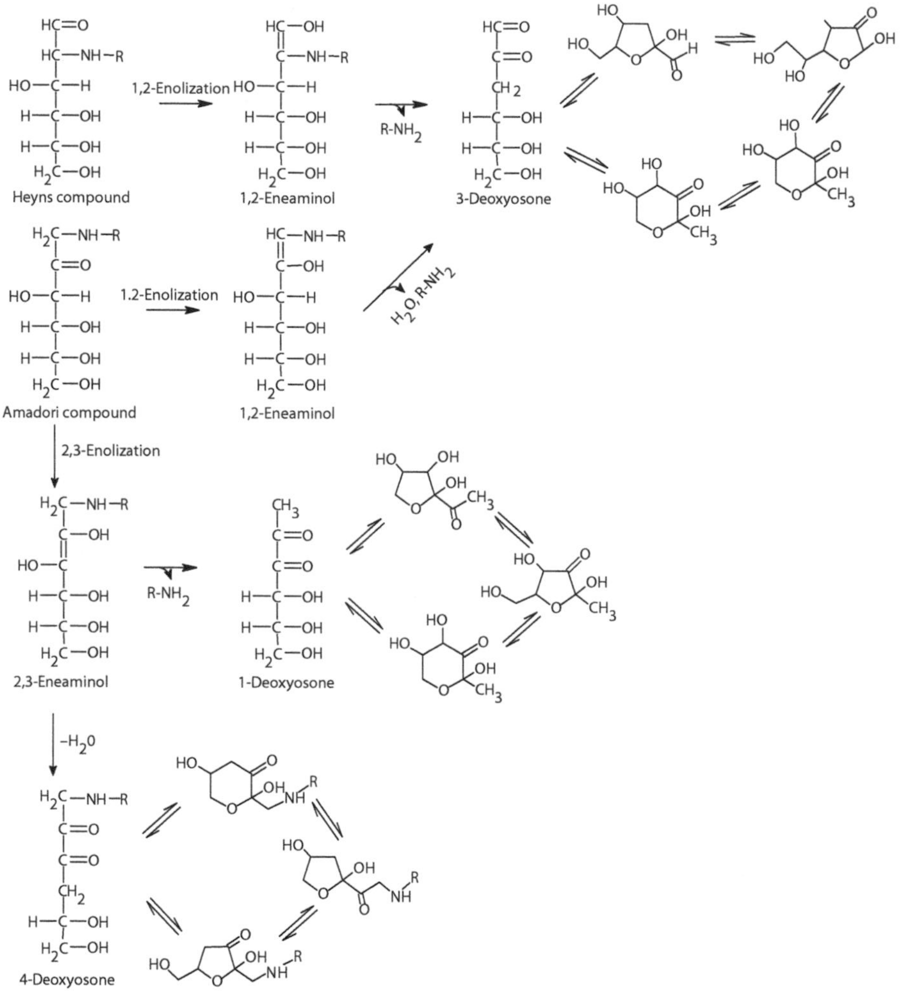

FIGURE 5.7 Rearrangement of *Amadori* and *Heyns* intermediates to form deoxyosones. (From Belitz, H.D., et al., eds., *Food Chem.*, Springer, Germany, 2004. With permission.)

At pH > 7, the degradation of the *Amadori* compound is understood to comprise by the 2,3-enolization, 1-deoxyosone, 1-deoxyreductone, or reductone formation that can turn into development of many flavor compounds such as 4-hydroxy-5-methyl-3(2*H*)-furanone, maltol, isomaltol, and a variety of fission products, including hydroxyacetone, pyruvaldehyde, and diacetyl (Figure 5.8). All these compounds are also highly reactive and take part in further reactions.

Dicarbonyl compounds will react with amino acids with the formation of aldehydes and α-aminoketones. During this reaction, known as Strecker degradation, dicarbonyls degrade amino acids to generate aldehyde with one carbon atom less than the original amino acid (Rizzi 2008). An example of Strecker degradation for L-phenylalanine is shown in Figure 5.10.

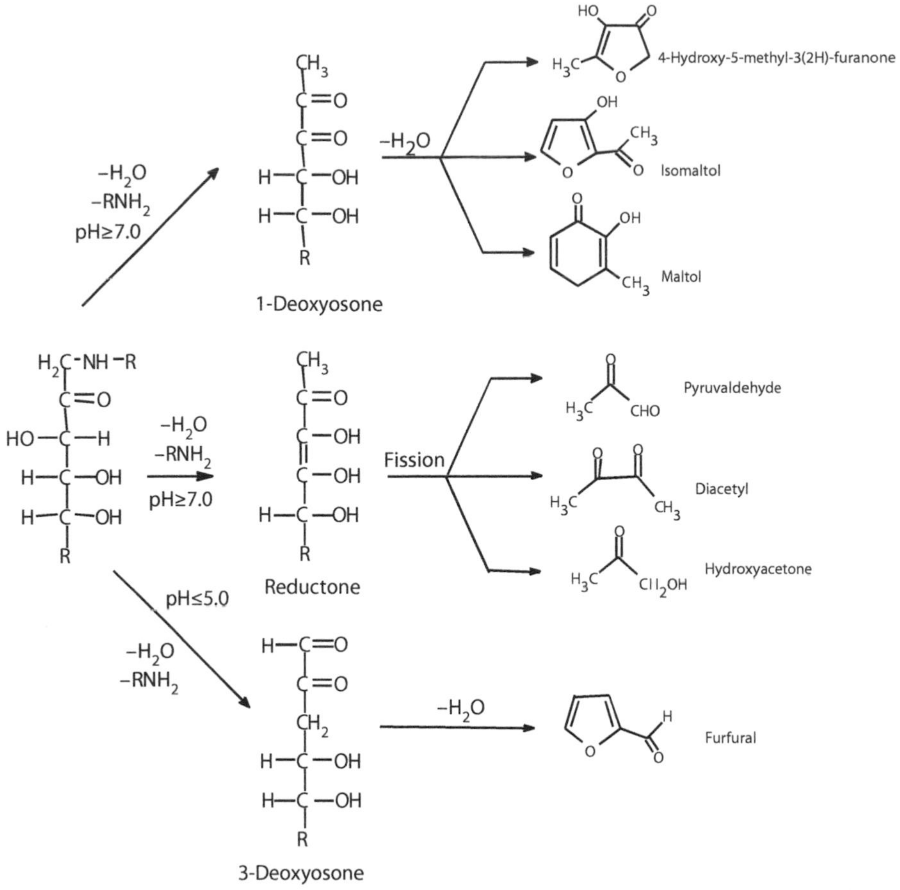

-H2O
-RNH2
pH≥7.0
1-Deoxyosone
4-Hydroxy-5-methyl-3(2H)-furanone
Isomaltol
Maltol
Reductone
Fission
Pyruvaldehyde
Diacetyl
Hydroxyacetone
pH≤5.0
3-Deoxyosone
Furfural

FIGURE 5.8 Deamination and dehydration of *Amadori* compounds to form important food flavor intermediates. (From Bailey, M.E., in *Flavor of Meat and Meat Products*, ed. F. Shahidi, 153–173, Blackie Academic and Professional, Glasgow, UK, 1994. With permission.)

These aldehydes, also called Strecker aldehydes, are often major contributors to the food aroma of thermally produced products. Among the important aroma compounds obtained in this manner are methional obtained from L-methionine, 2- and 3-methylbutanal from L-leucine, and phenylacetaldehyde from L-phenylalanine. As an example, a detailed pathway of methional formation as well as its further conversion into methanethiol and dimethylsulfide is presented in Figure 5.11.

Because of their low odor thresholds, Strecker aldehydes such as 3-methyl butanal and methional (0.2 μg/L) or phenylacetaldehyde (4 μg/L) are known to have significant odor strength, and by application of aroma extract dilution analysis (AEDA), they have been confirmed as key contributors to the aromas of numerous thermally processed foods. For example, 3-methylbutanal has been identified in boiled beef, boiled chicken, and yeast extract; phenylacetaldehyde has been identified in roasted coffee, and both compounds were found as important odorants in wheat and rye

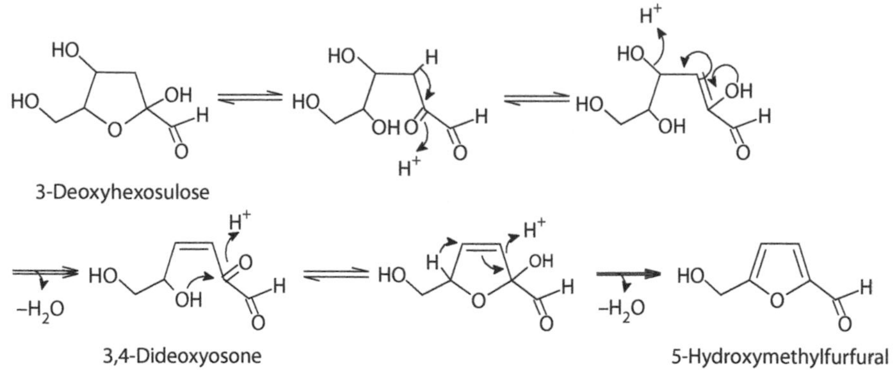

FIGURE 5.9 Conversion of 3-deoxyosone into 5-hydroxymethylfurfural.

bread crust, chocolate, and roasted sesame. Methional has also been characterized as a key contributor to the flavor of boiled potatoes, wheat bread crust, roasted wild mango seeds, or popcorn.

In the same reaction, besides Strecker aldehydes, the corresponding acids are also always formed. They are likewise of great importance to the flavor industry, as many of them (e.g., 3-methylbutanoic acid and phenylacetic acid) are often identified as the key odorants of thermally processed foods, for instance, wheat bread crust, boiled chicken, boiled beef, or caramel. The ratio of the formation of aldehyde and acid is influenced by the reaction parameters, especially by the presence of oxygen but also by the structure of α-dicarbonyl reaction with the amino acid and the pH of the reaction mixture. Using model experiments, Hofmann et al. (2000) have shown that carbohydrate degradation products, such as glyoxal and 2-oxopropanal, preferentially generate acids as opposed to α-dicarbonyl with an intact skeleton, such as 3-deoxy-2-hexosulose or 2-hexosulose, which favors the formation of aldehydes. Figure 5.12 shows a proposed reaction pathway of phenylacetic acid formation in the course of Strecker degradation from 2-oxopropanal and L-phenylalanine. After Schiff base formation, followed by decarboxylation and hydrolysis, a hemiaminal intermediate

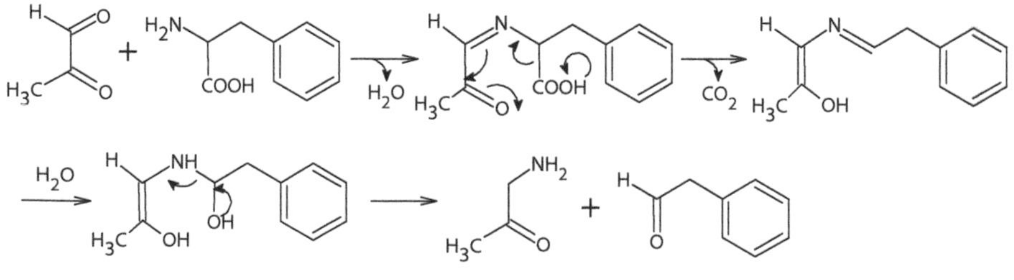

FIGURE 5.10 Strecker degradation of L-phenylalanine initiated by pyruvaldehyde (2-oxopropanal). (Adapted from Hofmann, T. et al., *J. Agric. Food Chem.*, 48, 434–440, 2000.)

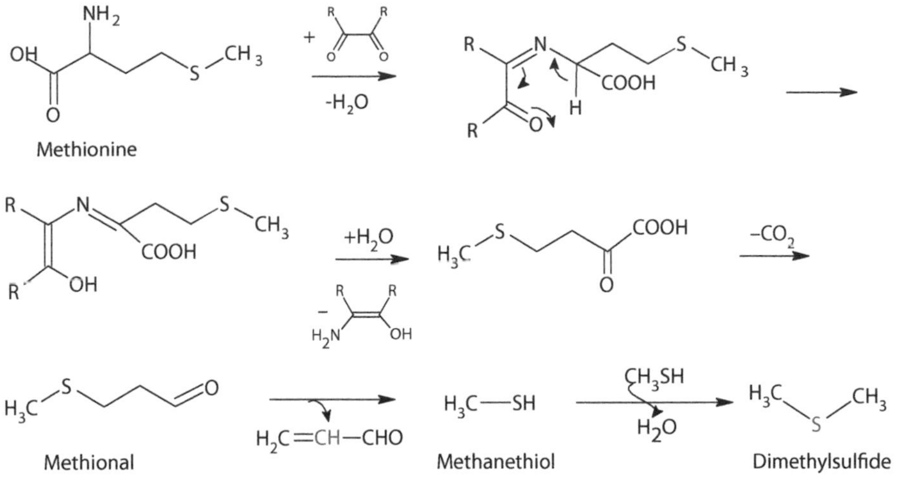

FIGURE 5.11 Strecker degradation of methionine to methional, methanethiol and dimethylsulfide. (From Belitz, H.D. et al., eds., *Food Chem.*, Springer, Germany, 2004. With permission.)

is formed, which may either liberate phenylacetaldehyde or be oxidized to form an iminoketone. After enolization and hydrolysis, phenylacetic acid is formed.

Other compounds that are also formed via Strecker degradation and influence food aroma are hydrogen sulfide, ammonia, 1-pyrroline, and cysteamine. As an example, formation of hydrogen sulfide and 2-mercaptoethanal from corresponding

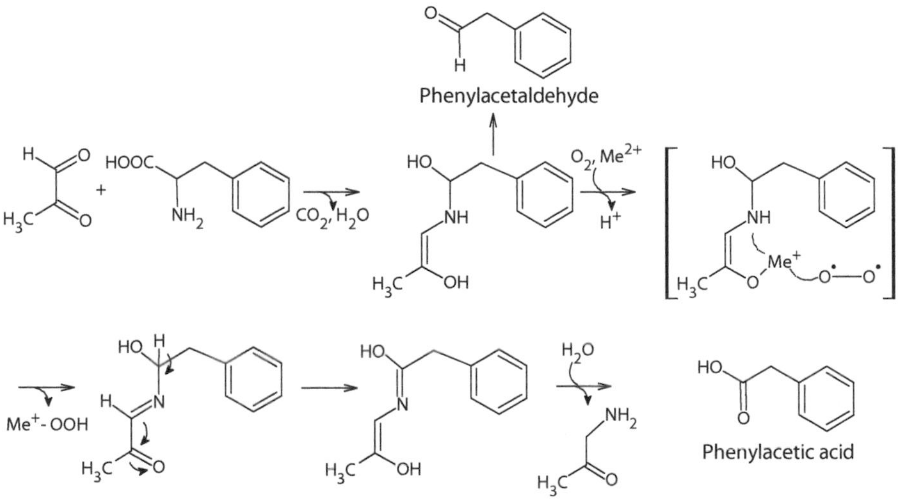

FIGURE 5.12 Reaction pathway of formation of phenylacetic acid from 2-oxopropanal and L-phenylalanine. (Proposed by Hofmann, T. et al., *J. Agric. Food Chem.*, 48, 434–440, 2000.)

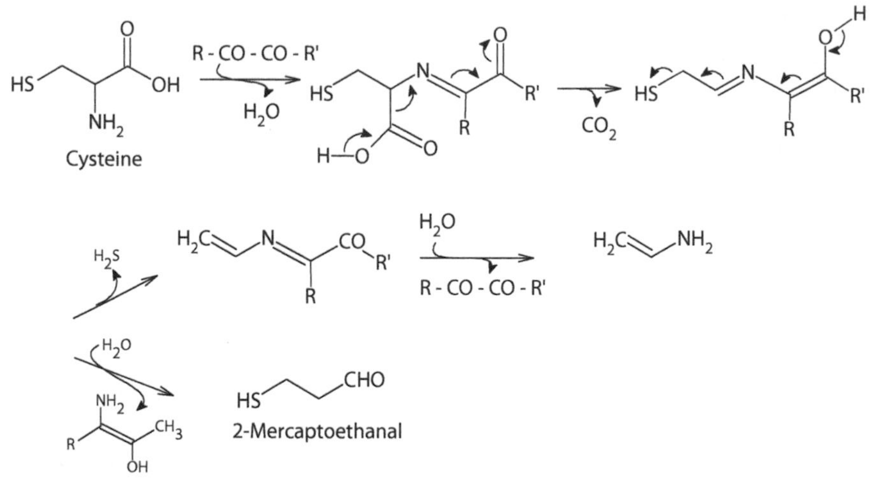

FIGURE 5.13 Formation of hydrogen sulfide and 2-mercaptoethanal in course of Strecker degradation of cysteine.

amino acids cysteine is presented in Figure 5.13. Figure 5.14 shows the decomposition of another amino acid (proline) to form 1-pyrroline, an important food flavor intermediate.

Then again, these compounds can serve as a precursor of another aroma active compounds, for instance, hydrogen sulfide for 2-furfurylthiol, 2-methyl-3-furanthiol, or its disulfide (bis(2-methyl-3-furyl)disulfide) and 1-pyrroline for 2-acetyl-pyrroline or 2-acetyltetrahydropyridine. Those aroma compounds are characterized by very low odor thresholds and odor notes typical for thermally processed foods (Table 5.1).

2-Furfurylthiol with a roasted odor note is the key aroma component of roasted coffee, white-bread crust, or popcorn. This compound can be formed in numerous ways. One reaction pathway, next to hydrogen sulfide, includes furfural as a precursor and leads through water elimination and reductive sulfhydrylation (Figure 5.15).

An isomer of 2-furfurylthiol, 2-methyl-3-furanthiol, has a very different odor quality that gives boiled meat aroma to food products. This compound can be formed from the interaction of norfuraneol and H_2S, through the development of 4-mercapto-5-methyl-3(2*H*)-furanone, or by the hydrolysis of thiamine with the very reactive intermediate 5-hydroxy-3-mercaptopentan-2-one (Figures 5.16 and 5.17).

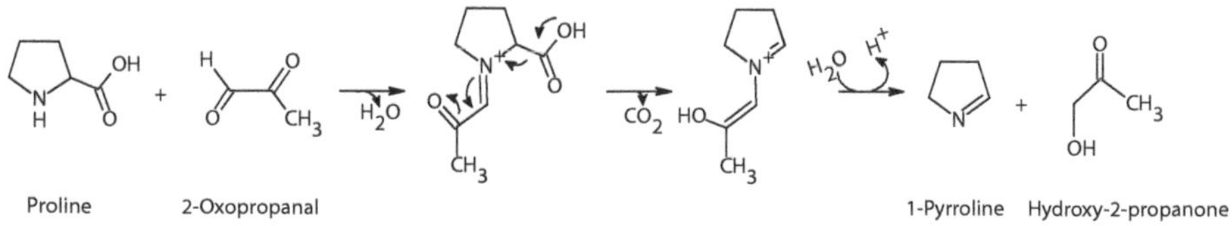

FIGURE 5.14 Formation of 1-pyrroline in course of Strecker degradation of proline. (From Hodge, J.E. et al., *Cereal Sci. Today*, 17, 34–40, 1972. With permission.)

TABLE 5.1
Odor Notes, Odor Thresholds and Occurrence of Some Aroma Active Compounds of Thermally Processed Foods

Compound	Odor Note	Odor Threshold (μg/L, Water)	Occurrence
2-Furfurylthiol	Roasted, like coffee	0.012	Roasted coffee, cooked meat, popcorn
2-Methyl-3-furanthiol	Boiled meat	0.007	Cooked meat
Bis(2-methyl-3-furyl)disulfide	Meat-like	0.00002	Boiled meat
2-Acetyl-1-pyrroline	Popcorn	0.1	White-bread crust, roasted meat, popcorn
2-Acetyltetrahydropyridine	Popcorn	1.6	Popcorn, heated meat

At this point, it is noteworthy that most sulfur-containing compounds, such as thiols, thioethers, di- and trisulfides, thiophenes, or thiazoles, are very powerful aroma components with extremely low odor thresholds. They are formed from cysteine, cystine, thiamine, and methionine by heating foods (Hwang et al. 1997; Hofmann and Schieberle 1997, 1998b; Mottram and Norbega 2002). Although their odor notes are highly appreciated in certain foods such as roasted coffee, fried meat, or toasted bread, they can also lead to very irritating off-flavors, for example, in milk.

Another compound, 2-acetyl-1-pyrroline, has a pleasant popcorn aroma typical for Basmati type of rice, but it has been also reported as a key odorant of several thermally produced foods such as crust of white bread, freshly popped popcorn, or roasted skin of fried chicken (Buttery and Ling 1982). Hoffman and Schieberle (1998a) by application of roasting conditions on the model mixture of proline and $[^{13}C]_6$-glucose, demonstrated that 2-acetyl-1-pyrroline can be formed in foods from 1-pyrroline, the Strecker degradation product of proline (Figure 5.18).

Alternatively, the aminoketone formed in the course of Strecker degradation can generate through the condensation pathway, dihydropyrazines, which are then oxidized to the corresponding pyrazines (Figure 5.19). It is often considered the most direct and important route for pyrazine formation (Adams et al. 2008).

Many types of pyrazines have been identified in processed foods. However, only a small fraction of them play a role in aromas. It is described as the aroma activity of the compound, and it depends on odor threshold values. Compounds can be considered potent odorants of food aroma only when they are produced in foods at a concentration higher than their odor threshold concentration. From the group of alkyl

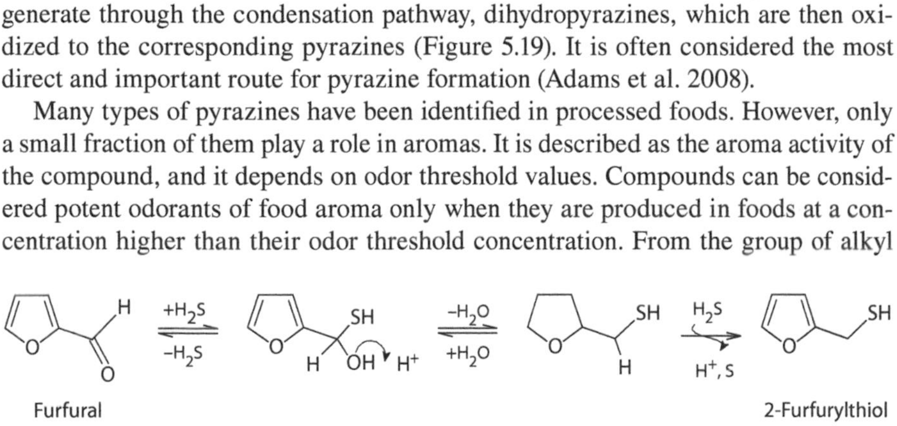

Furfural $\underset{-H_2S}{\overset{+H_2S}{\rightleftharpoons}}$ $\underset{+H_2O}{\overset{-H_2O}{\rightleftharpoons}}$ $\underset{H^+, S}{\overset{H_2S}{\rightarrow}}$ 2-Furfurylthiol

FIGURE 5.15 Formation of 2-furfurylthiol from hydrogen sulfide and furfural. (From Belitz, H.D. et al., eds., *Food Chem.*, Springer, Germany, 2004. With permission.)

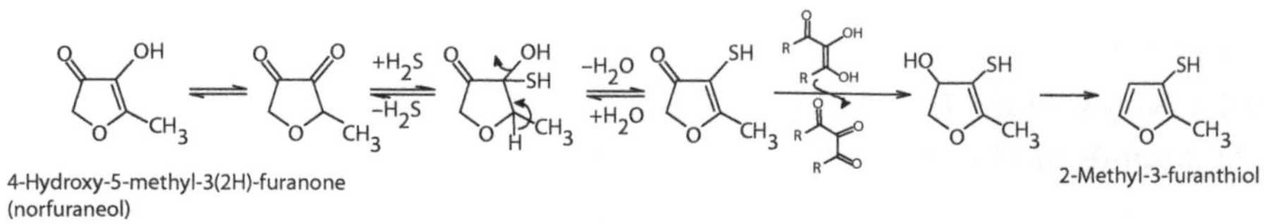

4-Hydroxy-5-methyl-3(2H)-furanone (norfuraneol)

2-Methyl-3-furanthiol

FIGURE 5.16 Formation of 2-methyl-3-furanthiol in a Maillard reaction. (From Hofmann, T., and Schieberle, P., *J. Agric. Food Chem.*, 46, 2721–2726, 1998a. With permission.)

pyrazines, the lowest odor thresholds have been measured for 2-ethyl-3,5-dimethylpyrazine (0.04 μg/L; water) and 2,3-diethyl-5-methylpyrazine (0.09 μg/L; water). They are known to have an impact on flavor of roasted sesame seeds, roasted beef, or coffee. The formation of 2-ethyl-3,5-dimethylpyrazine starts from Strecker reaction of alanine and 2-oxopropanal, followed by condensation of produced aminoacetone and 2-aminopropanal with 3,5-dimethyldihyropyrazine intermediate (Figure 5.20).

Another powerful aroma compounds obtained from carbohydrate degradation are furanones: 3(2*H*)- and 2(5*H*)-furanones. A list of furanones identified in foods with their odor thresholds is presented in Table 5.2.

2,5-Dimethyl-4-hydroxy-3(2*H*)-furanone, known as furaneol, with an intense caramel-like aroma, has been found in various natural and processed foods such as pineapple, tomato, and grape, as well as roasted coffee, roasted almond, and soy sauce. 5-Ethyl-4-hydroxy-2(methyl)-3(2*H*)-furanone (ethylfuraneol), which exists in the tautomeric forms in the ratio of 1:2, has been reported a key odorant of soya and also identified in roasted coffee, melon, and Emmentaler cheese.

Most furans are formed from deoxyosone in the presence or absence of amino acids during intermediate or final stage of MR. Furaneol and ethylfuraneol can be formed by the decomposition of 1-deoxyosone from 2-hydroxypropanal and its oxidation product, that is, 2-oxopropanal in the case of furaneol, and 2-oxobutanal in the case of ethylfuraneol (Figure 5.21). In a similar manner,

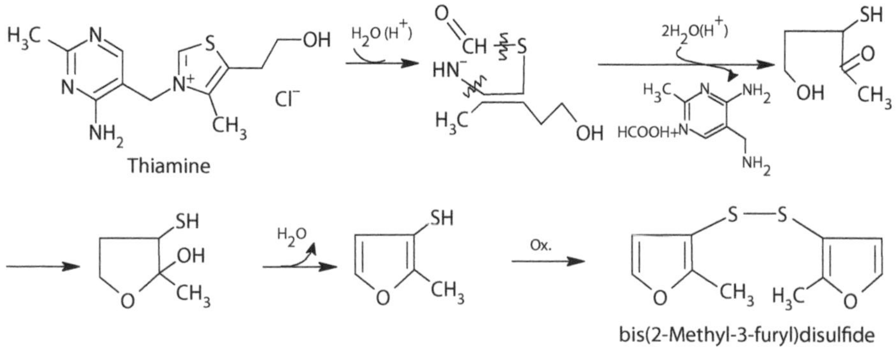

Thiamine

bis(2-Methyl-3-furyl)disulfide

FIGURE 5.17 Formation of 2-methyl-3-furanthiol and bis(2-methyl-3-furyl)disulfide from thiamine.

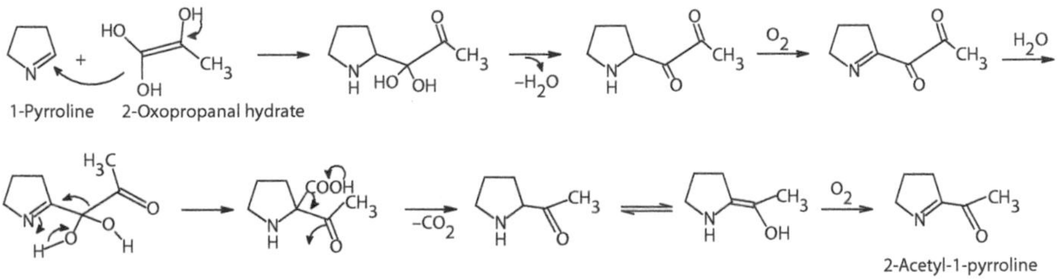

FIGURE 5.18 Formation of 2-acetyl-1-pyrroline from 1-pyrroline and 2-oxopropanal hydrate. (From Hofmann, T., and Schieberle, P., *J. Agric. Food Chem.*, 46, 2721–2726, 1998a. With permission.)

3-hydroxy-4,5-dimethyl-2(5*H*)-furanone (sotolon) can be formed from 2,3-butanedione and glycoaldehyde (Figure 5.22). It is a well-known, very powerful odorant, characterized by burnt sweet note of cane sugar, aged sake, and by a spicy-curry note of fenugreek, lovage, and condiments, as well as the typical nutty-sweet flavor of botrytized and sherry wines. Another aroma active furanone, 5-ethyl-3-hydroxy-4-methyl-2(5*H*)-furanone (abhexon), has an aroma similar to that of sotolon. One way in which it could be formed is by aldol condensation of two molecules of α-oxobutyric acid, the degradation product of threonine.

On the other hand, 4-hydroxy-5-methyl-3(2*H*)-furanone (norfuraneol) does not significantly influence food aroma because of its relatively high odor threshold; however, it serves as a precursor for the formation of key odorants of boiled meat flavor: 2-methyl-3-furanthiol, 3-mercapto-2-pentanone, 2-mercapto-3-pentanone, and 2-methyl-3-thiophenethiol. For the formation of mercaptoalkanones (3-mercapto-2-pentanone and 2-mercapto-3-pentanone), Whitfield and Mottram (1999) proposed

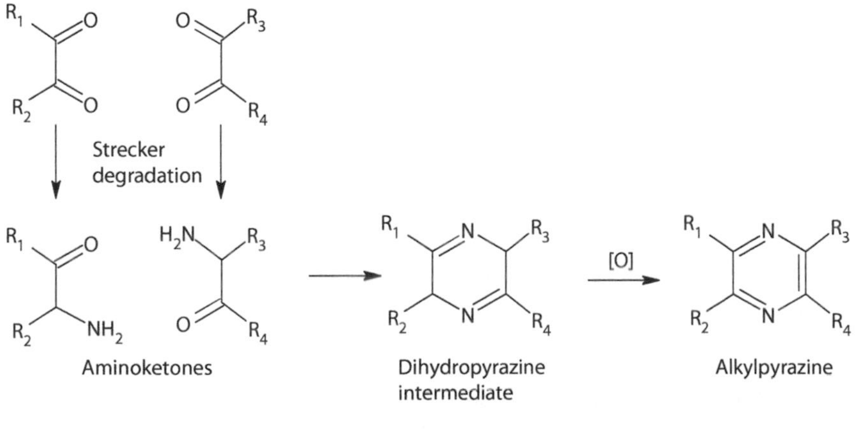

FIGURE 5.19 Pathway for alkylpyrazine formation, involving condensation of two aminoketones to form a dihydropyrazine, with direct oxidation to corresponding alkylpyrazine. (From Low, M.Y. et al., *J. Agric. Food Chem.*, 54, 5976–5983, 2006. With permission.)

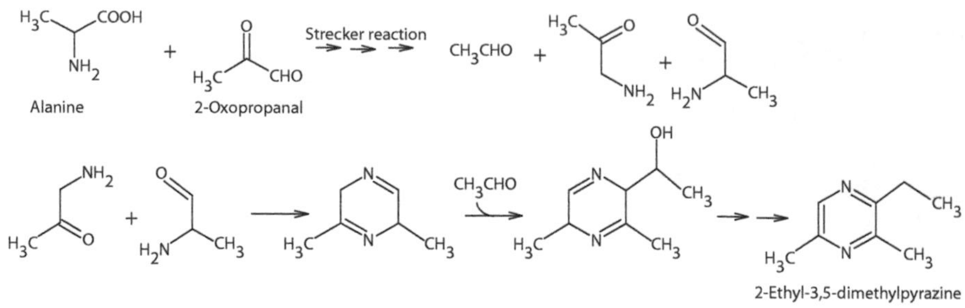

FIGURE 5.20 Formation of 2-ethyl-3,5-dimethylpyrazine from the reaction of alanine and 2-oxopropanal. (From Belitz, H.D. et al., eds., *Food Chem.*, Springer, Germany, 2004. With permission.)

two pathways involving, other than norfuraneol, either hydrogen sulfide or cysteine. During the interaction with hydrogen sulfide, the acid hydrolysis of norfuraneol takes place, yielding 1-deoxypentosone followed by reduction and acid-catalyzed dehydration (Figure 5.23). Formation of the 2-methyl-3-thiophenethiol, illustrated in Figure 5.24, occurs via a route similar to that proposed for the 2-methyl-3-furanthiol (Figure 5.16).

For those looking for specific flavor compounds belonging to a certain chemical group, another scheme of MR, described by Jousse et al. (2002), might be useful. This scheme follows closely the classical mechanism of Hodge but concentrates only on volatiles and is based on different classes of compounds—those that are chemically related and formed generally in a similar manner (Figure 5.25). It categorizes the volatile compounds into four classes: pyrroles and other nitrogen-containing heterocyclic compounds; furans and other oxygen-containing heterocyclic compounds; carbonyls; and pyrazines.

According to Jousse et al. (2002), MR starts with condensation of a sugar with an amino acid to form an *Amadori* or *Heyns* rearrangement product. As an alternative, the sugar can also directly degrade, for example, by a caramelization reaction at high temperature. Afterward, the intermediate can cyclize to form nitrogen-containing

TABLE 5.2
3(2*H*)- and 2(5*H*)-Furanones Identified in Foods

Compound	Odor Threshold (µg/L, Water)
4-Hydroxy-2,5-dimethyl-3(2*H*)-furanone (furaneol)	60
4-Hydroxy-5-methyl-3(2*H*)-furanone (norfuraneol)	23,000
2-(5)-Ethyl-4-hydroxy-5-(2)-methyl-3(2*H*)-furanone (ethylfuraneol)	7.5
4-Methoxy-2,5-dimethyl-3(2*H*)-furanone (mesifuran)	3400
3-Hydroxy-4,5-dimethyl-2(5*H*)-furanone (sotolon)	7
5-Ethyl-3-hydroxy-4-methyl-2(5*H*)-furanone (abhexon)	30

Source: Belitz, H.D. et al., eds., *Food Chem.*, Springer, Germany, 2004. With permission.

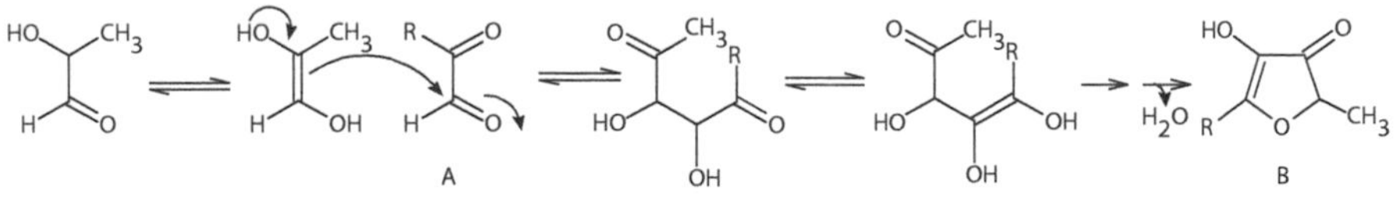

R = methyl and A = 2-oxopropanal; B = furaneol
R = ethyl and A = 2-oxobutanal; B = ethylfuraneol

FIGURE 5.21 Formation of furaneol and ethylfuraneol in Maillard reaction. (From Belitz, H.D. et al., eds., *Food Chem.*, Springer, Germany, 2004. With permission.)

heterocyclic compounds, such as pyrroles or pyridines. In its place, it may also cleave to give rearranged sugars, which contain the intact chain of the starting sugar. These rearranged sugars include the 1-desoxy-2,3-diketones and the 3-desoxy-1,2-diketones, as well as further rearrangement from these via ketoenol tautomerization. This cleavage gives back the original amino acid. Subsequently, the rearranged sugars may cyclize into oxygen-containing heterocyclic compounds, such as furans or furfurals. It can also break up into α-dicarbonyl fragments, which may recombine to give furans or furfurals. Another pathway explains the formation of Strecker aldehydes and pyrazines in the Strecker degradation reaction of dicarbonyl fragments with the amine group of the amino acid. As a substitute, Strecker aldehydes can also represent the nitrogen-containing heterocyclic compounds such as pyrroline and pyrrolidine coming from the reaction of dicarbonyls with proline and hydroxyproline.

5.4 FACTORS INFLUENCING MAILLARD REACTION

The reaction of carbohydrates and amino acids through MR leads to the formation of more than 2500 flavor compounds. Many of these compounds can be formed in several different pathways, depending on the reaction parameters. Control of the MR is of particular industrial and scientific interest, as the aroma profile of a product influences the quality and consumer acceptance of the food products (Ames 1990; Ericsson 1981; Tai and Ho 1998; Ames 1998). A number of factors influence the generation of flavors in the MR. For the characteristics of volatiles and flavor of thermally treated foods obtained in the course of MR, the biggest influence comes from the type of sugars and amino acids involved in the reaction as well as the pH at which reaction is conducted. As for the kinetics of the reaction, it is principally influenced by temperature and time, and water activity, while leaving the nature of volatiles generally unchanged. The food industry, which expects and depends on the MR to produce foods possessing the flavor and color demanded by the consumers, therefore is looking for well-controlled and accurate reaction pathways. Toward this end, the effect of type of reactants, pH, temperature, time, and water activity on MR has been discussed.

5.4.1 Nature of Saccharides Involved

Hofmann and Schieberle (1995, 1997) in their extensive research on the formation of meat flavor, have performed a number of studies on model mixtures containing cysteine

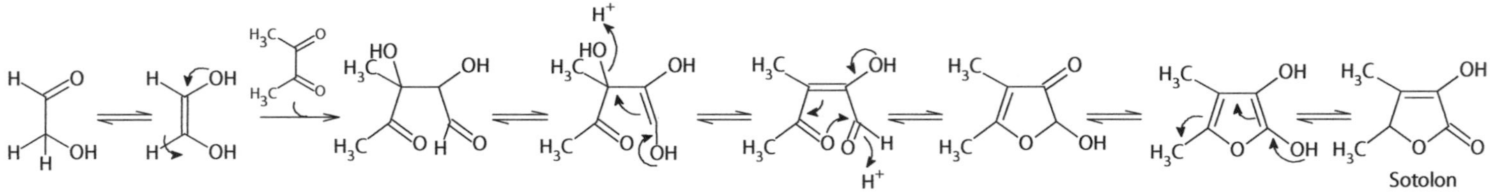

FIGURE 5.22 Formation of sotolon in Maillard reaction.

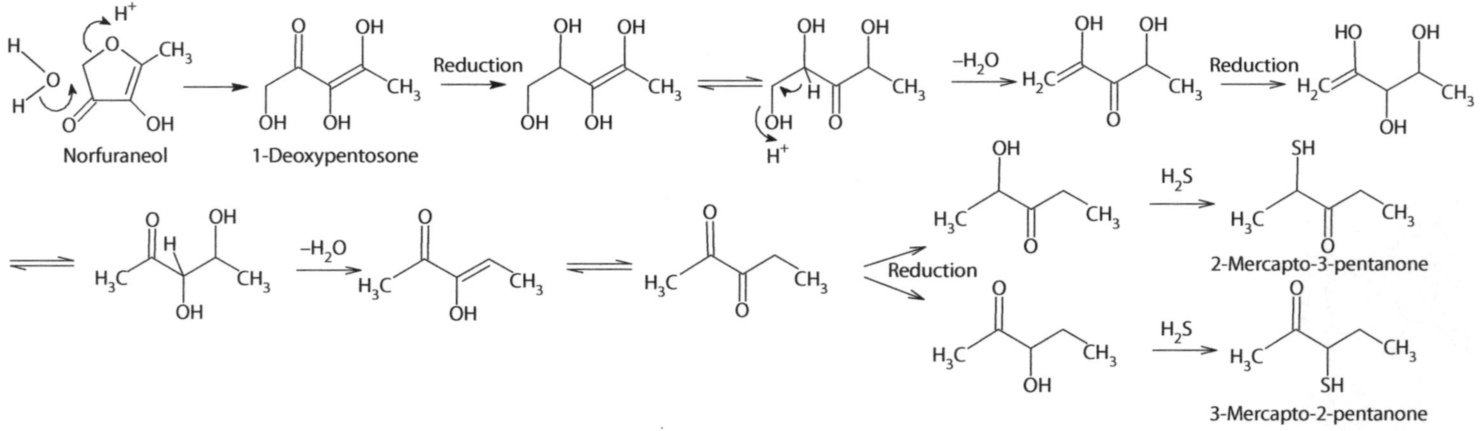

FIGURE 5.23 Formation of 2-mercapto-3-pentanone and 3-mercapto-2-pentanone from norfuraneol and hydrogen sulfide. (From Whitfield, F.B., Mottram, D.M., *J. Agric. Food Chem.*, 47, 1626–1634, 1999. With permission.)

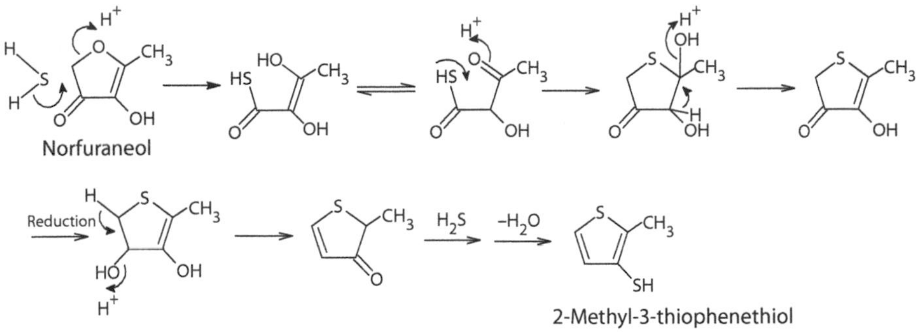

FIGURE 5.24 Formation of 2-methyl-3-thiophenethiol in the reaction of norfuraneol with hydrogen sulfide. (From Whitfield, F.B., and Mottram, D.M., *J. Agric. Food Chem.*, 47, 1626–1634, 1999. With permission.)

and different monohydrates. By application of AEDA, a method combining sensory evaluation with analytical chemistry, they have revealed an impressive diversity of odor active compounds generated from heating cysteine either with ribose, glucose, or rhamnose (Table 5.3).

For example, 2-furfurylthiol, 5-acetyl-2,3-dihydro-1,4-thiazine, and 2-acetyl-2-thiazoline were identified as the key odorants in the presence of each of the three carbohydrates (glucose, rhamnose, or ribose). On the other hand, several odorants

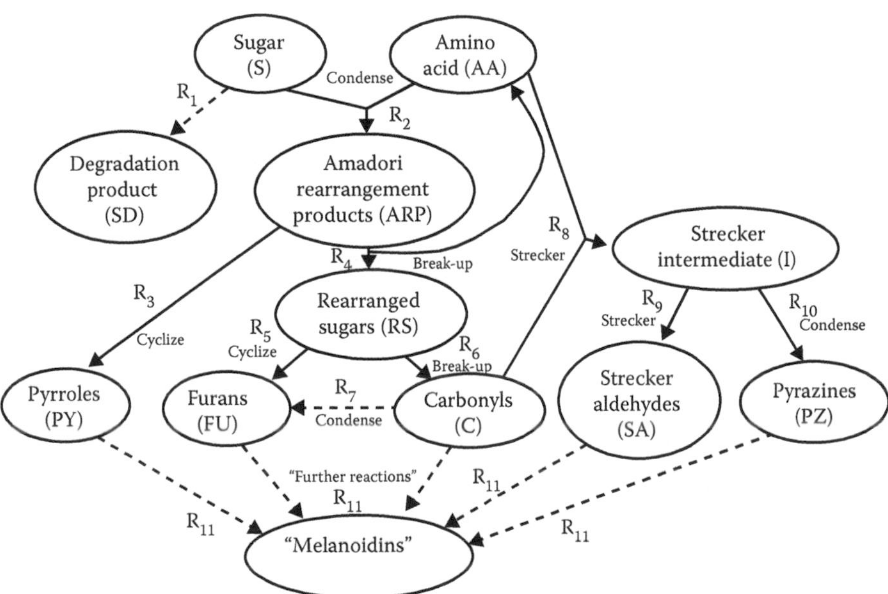

FIGURE 5.25 Simplified kinetic scheme of flavor generation by the Maillard reaction. (From Jousse, F. et al., *Food Chem. Toxicol.*, 67, 2534–2542, 2002. With permission.)

TABLE 5.3
Comparison of the Key Odorants (FD Factors >128 to 65,536) Formed by Reacting Cysteine in the Presence of Carbohydrates

	FD Factor		
Odorant	**Glucose**	**Rhamnose**	**Ribose**
2-Furfurylthiol	1,024	512	4,096
5-Acetyl-2,3-dihydro-1,4-thiazine	1,024	512	1,024
5-Methyl-2-furfurylthiol	<1	2,048	<1
3-Mercapto-2-pentanone	512	128	2,048
3-Mercapto-2-butanone	512	32	512
4-Hydroxy-2,5-dimethyl-3(2*H*)-furanone	512	65,536	128
2-(1-Mercaptoethyl)furan	256	<1	<1
2-Acetyl-2-thiazoline	128	256	256
3-Hydroxy-6-methyl-2(2*H*)-pyranone	<1	16,384	<1
3-Hydroxy-4,5-dimethyl-2(5*H*)-furanone	16	128	64
2-Methyl-3-furanthiol	<1	<1	1,024
4-Hydroxy-5-methyl-3(2*H*)-furanone	<1	<1	256
4-Hydroxy-2,5-dimethyl-3(2*H*)-thiophenone	128	<1	<1

Source: Hofmann, T., and Schieberle, P., *J. Agric. Food Chem.*, 43, 2187–2194, 1995. With permission. Hofmann, T., and Schieberle, P., *J. Agric. Food Chem.*, 45, 898–906, 1997. With permission.

with an intense smell were selectively generated; for example, only rhamnose produced 5-methyl-2-furfurylthiol and 3-hydroxy-6-methyl-2(2*H*)-pyranone, whereas 2-(1-mercaptoethyl) furan was exclusively formed from glucose. Additionally, the comparison of sensory analysis revealed variety in odor attributes of obtained cysteine/carbohydrates mixtures. The panelists described an overall aroma of cysteine/ribose mixture similar to roasted chicken flavor with strongest notes obtained for sulfur-like, meat-like, and roasted. In cysteine/rhamnose mixture, the same notes were noted; however, in addition, caramel-like and seasoning-like odor notes were detected. On the other hand, for cysteine/glucose mixture, a pungent note dominated together with meat-like aroma (Hofmann and Schieberle 1995, 1997).

5.4.2 Influence of pH

Influence of pH on flavor formation in MR has been already indicated on Hodge scheme, where the degradation of the Amadori product has been divided into three ways dependent on the pH of the system (Figure 5.5). Since then, many research groups have also reported the major influence of pH on reaction pathways and consequently on the profile of reaction products.

Quantitative measurements performed by Engel and Schieberle (2002) by means of stable isotope dilution assay on the formation of odor active compounds formed during heating of model aqueous fructose/cysteamine mixture revealed a significant effect of pH. As presented in Figure 5.26, the maximum amount of flavor compound with popcorn-like odor—5-acetyl-3,4-dihydro-2*H*-1,4-thiazine—was observed at pH 7.0, whereas at pH values of 6 or 8, significantly less amount of the compound was generated. Below pH 5.0, it was not formed at all. On the contrary, 2-acetyl-2-thiazoline was preferably generated at pH > 7.0, whereas its formation has been inhibited at pH < 6.0.

Studies performed by Cerny and Briffod (2007) reveal another example of pH influence on the formation of MR flavor compounds. By heating a model mixture of labeled $[^{13}C]_5$-xylose with thiamin and cysteine, at different pH values (between 4.0 and 7.0), they showed great diversity in the generated sulfur volatiles. Compounds such as 2-mercapto-3-pentanone, 3-methyl-1,2-dithian-4-one, 3-acetyl-1,2-dithiolane, and 4,5-dihydro-2-methyl-3(2*H*)-thiophene were detected only when the reaction was carried out at pH 6.0 and 7.0, and not under more acidic conditions. The formation of 4,5-dihydro-2-methyl-3(2*H*)-furanone was also favored at higher pH values. On the other hand, 2-furaldehyde, 2-furfurylthiol, and 2-methyl-3-(methylthio)-furan were formed only at acidic pH levels, and no traceable amounts were detected at pH 7.0. This could be explained by the fact that at alkaline conditions, relatively high amounts of hydrogen sulfide is obtained, which results in the formation of thiophenes instead of furans. Additionally, application of labeled xylose in the mixture with thiamin and cysteine allowed authors to confirm that for mentioned compounds, different formation pathways exist with and without involving xylose as a precursor. 4,5-Dihydro-2-methyl-3(2*H*)-thiophenone as well as 3-methyl-1,2-dithian-4-one and 3-acetyl-1,2-dithiolane were formed only at the higher pH values of 6.0 and 7.0. This finding is in agreement with the stronger involvement of hydrogen sulfide at higher pH, and a formation pathway involving its reaction with 5-hydroxy-3-mercapto-2-pentanone, which is generally accepted as a key intermediate of thiamin degradation (Figure 5.17).

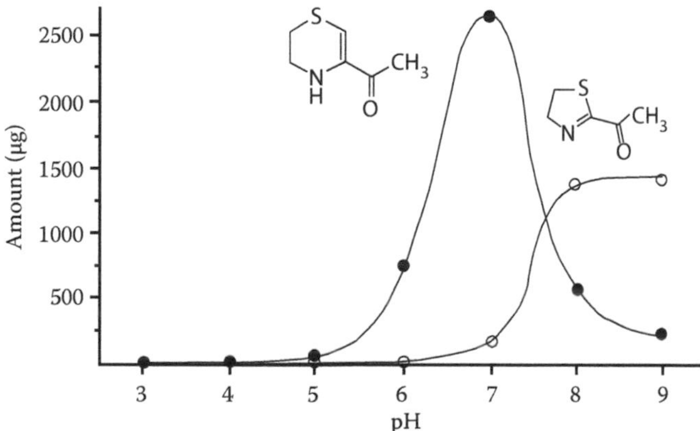

FIGURE 5.26 Influence of pH on formation of 5-acetyl-3,4-dihydro-2*H*-1,4-thiazine (–●–) and 2-acetyl-2-thiazoline (–o–) in model aqueous mixture of fructose/cysteamine). (Adapted from Engel, W., and Schieberle, P., *J. Agric. Food Chem.*, 50, 5394–5399, 2002).

5.4.3 Effect of Temperature and Time

It is evident that the aroma profile of a food product varies with the temperature and time of heating. At any given temperature–time combination, a unique aroma is obtained, which is almost impossible to reproduce by any other heating conditions. The range of possible temperatures used is also very wide. Whereas caramelization occurs only at elevated temperatures (typically >150°C), MR can be initiated also at lower temperatures (<100°C). The effect of temperature and time has been studied by many authors, usually in model studies. In 1989, Shu and Ho showed that heating of cysteine and 2,5-dimethyl-4-hydroxy-2,3-dihydrofuranone in a water/glycerol (75:25) mixture at 160°C gave the post-roasted, meaty aroma with the major reaction products trithiolanes, thiophenones, and 2,4-hexadienone. Temperatures of 100°C or 200°C resulted in different aroma description with more burnt and biting flavors. In more recent studies on formation of thiazoles and thiazolines in foods, Hofmann and Schieberle (1996) showed in model systems different yields of 2-acetyl-2-thiazoline depending on time and temperature of heating. In Figure 5.27, it can be seen that this key odorant of quick fried beef had the highest concentration after 10 min of heating in 100°C. Longer time of heating at 100°C decreased its concentration significantly.

In 2009, Balagiannis and coworkers published a report on the influence of heating time on 3-methylbutanal. At the beginning of heat treatment, the first 15–20 min, usually intermediates are formed and there is little of the final product present. As the intermediate accumulates, the rate of product formation increases, and there is a period of steady growth until finally the precursor material is used up and the levels reach a maximum. Afterward, for 3-methylbutanal, the levels started to decrease, in agreement with the fact that Strecker aldehydes are not end products and they can react further to form other compounds.

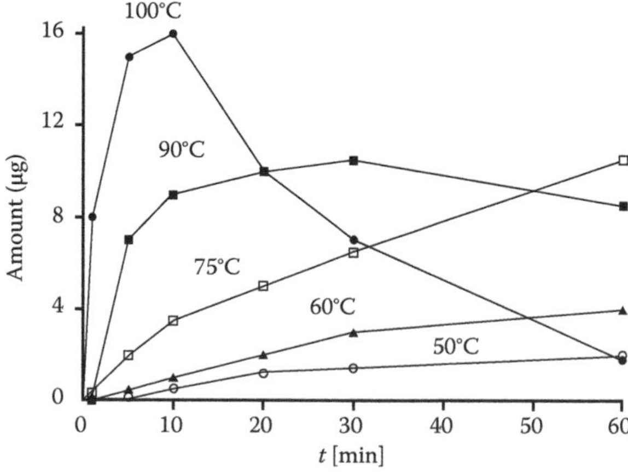

FIGURE 5.27 Dependence on time and temperature of the formation of 2-acetyl-2-thiazoline from 2-(1-hydroxyethyl)-3,5-dihydrothiazole. (From Hofmann, T., and Schieberle, P., in *Flavour Science—Recent Developments*, ed. A.J. Taylor and D.S. Mottram, Royal Society of Chemistry, Cambridge, 1996. With permission.)

TABLE 5.4
Quantitation of Volatile Compounds Generated From 3-Hydroxy-2-Butanone/Ammonium Sulfide Model Systems

	Quantity (mg/g Acetoin)					
Compound	**25°C**	**50°C**	**75°C**	**100°C**	**125°C**	**150°C**
2,4,5-Trimethyl-3-oxazoline	0.284	0.101	0.175	0.346	0.300	0.019
2,4,5-Trimethyloxazole	0.106	0.113	0.111	0.142	0.516	1.846
3-Mercaptobutane	0.698	0.327	0.249	0.172	1.002	1.329
4,5-Dimethylthiazole	0.000	0.000	0.000	0.000	0.070	0.077
2,4,5-Trimethylthiazole	0.093	0.086	0.186	0.596	2.447	3.765
2,4,5-Trimethyl-3-thiazoline	0.000	0.018	0.127	0.325	1.503	1.686
Tetramethylpyrazine	0.000	0.080	1.180	9.776	58.540	85.909
2-(1-Hydroxyethyl)-2,4,5-trimethyl-3-oxazoline	22.441	21.394	10.474	3.258	0.264	0.700
2-(1-Mercaptoethyl)-2,4,5-trimethyl-3-oxazoline	0.491	0.710	0.114	0.154	0.034	0.048
2-(1-Hydroxyethyl)-2,4,5-trimethyl-3-thiazoline	0.174	1.230	2.356	1.270	1.532	0.769
2-(1-Mercaptoethyl)-2,4,5-trimethyl-3-thiazoline	0.300	1.849	3.302	3.052	2.161	1.809

Source: Xi, J. et al., *J. Agric. Food Chem.*, 47, 245–248, 1999. With permission.

An additional research paper describes a variety of flavor compounds obtained from the reaction between 3-hydroxy-2-butanone and ammoniun sulfide at different temperatures: 25°C, 50°C, 75°C, 100°C, 125°C, and 150°C (Xi et al. 1999). The authors noted that at 25°C, four flavor compounds were identified—2,4,5-trimethyloxazole, 2,4,5-trimethyl-3-oxazoline, 2,4,5-trimethylthiazole, and 2,4,5-trimethyl-3-thiazoline—as well as four intermediates—2-(1-hydroxyethyl)-2,4,5-trimethyl-3-oxazoline, 2-(1-mercaptoethyl)-2,4,5-trimethyl-3-oxazoline, 2-(1-hydroxyethyl)-2,4,5-trimethyl-3-thiazoline, and 2-(1-mercaptoethyl)-2,4,5-trimethyl-3-thiazoline. In contrast, when the reaction temperature was higher than 100°C, tetramethylpyrazine was the major product (Table 5.4).

5.4.4 Influence of Water Activity

As Figure 5.28 illustrates, MR (nonenzymatic browning) occurs between water activity 0.3 and 0.9, reaching a maximum at intermediate values 0.6–0.8, which is a typical value for dried and intermediate-moisture foods. However, a characteristic feature of the MR is that it liberates water in the early stages (Figure 5.5); further increases in water activity may inhibit MRs. So, for some samples, measuring and controlling water activity is a good way to control Maillard browning problems.

Development of aroma active flavor compounds depends on water activity of the reaction system as well. Hofmann and Schieberle (1998a) have performed an

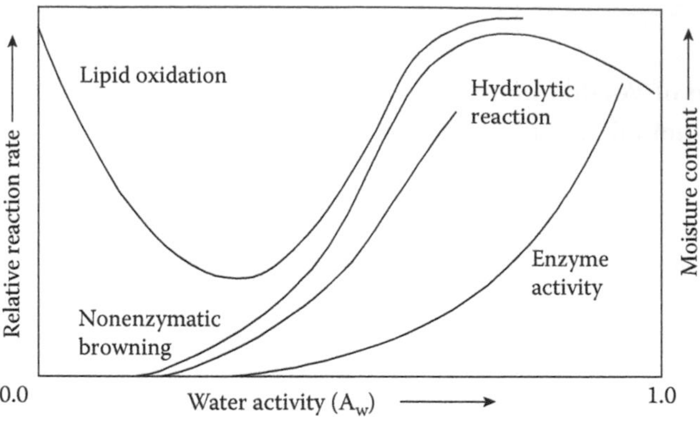

FIGURE 5.28 Reaction rates in foods as a function of water activity.

experiment of heating proline and glucose mixture in two different reaction moisture conditions: aqueous and dry-heating conditions. By application of AEDA, they have identified key odorants in both mixtures. Results showed that although the characteristics of key odorants were the same in both mixtures, their contribution to the overall flavor expressed as FD (flavor dilution factor) differed significantly (Table 5.5). Compared to liquid conditions, the FD factor of 2-acetyl-1-pyrroline increased in dry-heating conditions from 64 to 16,384, becoming the most important odorant in the mixture. Similarly, its homologue, 2-propionyl-1-pyrroline, was notably enhanced. A different effect was observed for four tetrahydropyridine derivatives

TABLE 5.5
Key Odorants Generated in Thermally Treated Proline/Glucose Reaction Mixtures: Aqueous and Dry-Heating

	FD Factor	
Compound	**Aqueous**	**Dry Heating**
2,3-Butanedione	16	16
2-Acetyl-1-pyrroline	64	16,384
2-Propionyl-1-pyrroline	16	512
2-Acetyl-3,4,5,6-tetrahydropyridine	4,096	2,048
4-Hydroxy-2,5-dimethyl-3(2*H*)-furanone	<1	8,192
2-Acetyl-1,4,5,6-tetrahydropyridine	4,096	2,048
2-Propionyl-3,4,5,6-tetrahydropyridine	128	128
2-Propionyl-1,4,5,6-tetrahydropyridine	128	128
Unknown	<1	16

Source: Hofmann, T., and Schieberle, P., *J. Agric. Food Chem.*, 46, 2721–2726, 1998b. With permission.

(2-acetyl-3,4,5,6-tetrahydropyridine, 2-acetyl-1,4,5,6-tetrahydropyridine, 2-propionyl-3,4,5,6-tetrahydropyridine, and 2-propionyl-1,4,5,6-tetrahydropyridine), which remained constant in their odor activities in both aqueous and dry conditions.

5.5 CONCLUSION

Saccharides belong to one of the most important group of precursors involved in flavor formation. They can either take part in the reactions involving simply sugars such as caramelization, or they can react with amino acids in the so-called Maillard reaction and produce an extensive number of flavor compounds recognized as potent aroma components in many food products. Changes caused by those reactions can be very favorable and are desired by consumers in foods such as coffee, fried meat, or baked bread, or they can bring off-flavors, for example, during prolonged storage and exclude food stuff from consumption. Therefore, the reactions described above have been studied by a great number of scientists, resulting in hundreds of research papers with new findings, continually bringing further knowledge and capability of reaction control.

REFERENCES

Adams, A., R.C. Borrelli, V. Fogliano, and N. De Kimpe. 2005. Thermal degradation studies of food melanoidins. *J. Agric. Food Chem.* 53: 4136–4142.

Adams, A., V. Polizzi, M. van Boekel, and N. De Kimpe. 2008. Formation of pyrazines and a novel pyrrole in Maillard model systems of 1,3-dihydroxyacetone and 2-oxopropanal. *J. Agric. Food Chem.* 56: 2147–2153.

Amadori, M. 1931. Condensation products of glucose with *p*-toluidine. *Atti. R. Accad. Naz. Lincei. Mem. Cl. Sci. Fis. Mat. Nat.* 13: 72–78.

Ames, J.M. 1990. Control of the Maillard reaction in food systems. *Trends Food Sci. Technol.* 1: 150–154.

Ames, J. 1998. Application of the Maillard reaction in the food industry. *Food Chem.* 62: 431–439.

Angyal, S.J. 2001. The Lobry de Bruyn Alberda van Ekenstein transformation and related reactions. In *Glycoscience: Epimerisation, Isomerisation and Rearrangement Reactions of Carbohydrates*, Volume 215, 1–14. Berlin: Springer Publishing House.

Bailey, M.E. 1994. Maillard reaction and meat flavor development. In *Flavor of Meat and Meat Products*, ed. F. Shahidi, 153–173. Glasgow, UK: Blackie Academic and Professional.

Balagiannis, D.P., J.K. Parker, D.L. Pyle, N. Desforges, B.L. Wedzicha, and D.S. Mottram. 2009. Kinetic modeling of the generation of 2- and 3-methylbutanal in a heated extract of beef liver. *J. Agric. Food Chem.* 57: 9916–9922.

Belitz, H.D., W. Grosch, and P. Schieberle, eds. 2004. *Food Chemistry*. Germany: Springer.

BeMiller, J.N., and K.C. Huber. 2007. In *Fennema's Food Chemistry*, ed. S. Damodaran, K.L. Parkin, and O.R. Fennema. Boca Raton, FL: CRC Press, Taylor & Francis Group.

Billaud, C., and J. Adrian. 2003. Louis-Camille Maillard, 1878–1936. *Food Rev. Int.* 19: 345–374.

Blank, I., and L.B. Fay. 1996. Formation of 4-hydroxy-2,5-dimethyl-3(2*H*)-furanone and 4-hydroxy-2(or 5)-ethyl-5(or 2)-methyl-3(2*H*)-furanone through Maillard reaction based on pentose sugars. *J. Agric. Food Chem.* 44: 531–536.

Buttery, R.G., and L.C. Ling. 1982. 2-Acetyl-1-pyrroline: An important component of cooked rice. *Chem. Ind.* 23: 958–959.

Cerny, C. 2008. The aroma side of the Maillard reaction. Maillard reaction: Recent advances in food and biomedical sciences. *Ann. N. Y. Acad. Sci.* 1126: 66–71.

Cerny, C., and M. Briffod. 2007. Effect of pH on the Maillard reaction of [$^{13}C_5$]xylose, cysteine, and thiamin. *J. Agric. Food Chem.* 55: 1552–1556.

Engel, W., and P. Schieberle. 2002. Identification and quantitation of key aroma compounds formed in Maillard-type reactions of fructose with cysteamine or isothiaproline (1,3-thiazolidine-2-carboxylic acid). *J. Agric. Food Chem.* 50: 5394–5399.

Ericsson, C., ed. 1981. Maillard reaction in food. *Prog. Food Nutr. Sci.* 5: 1–6.

Frank, B., D. Whitfield, and D.S. Mottram. 1999. Investigation of the reaction between 4-hydroxy-5-methyl-3(2*H*)-furanone and cysteine or hydrogen sulfide at pH 4.5. *J. Agric. Food Chem.* 47: 1626–1634.

Hodge, J.E. 1953. Dehydrated foods: Chemistry of browning reactions in model systems. *J. Agric. Food Chem.* 1: 928–943.

Hodge, J.E., F.D. Mills, and B.E. Fisher. 1972. Compounds derived from browned flavors. *Cereal Sci. Today* 17: 34–40.

Hofmann, T., P. Münch, and P. Schieberle. 2000. Quantitative model studies on the formation of aroma-active aldehydes and acids by Strecker-type reactions. *J. Agric. Food Chem.* 48: 434–440.

Hofmann, T., and P. Schieberle. 1995. Evaluation of the key odorants in a thermally treated solution of ribose and cysteine by aroma extract dilution techniques. *J. Agric. Food Chem.* 43: 2187–2194.

Hofmann, T., and P. Schieberle. 1996. Studies on intermediates generating the flavour compounds 2-methyl-3-furanthiol, 2-acetyl-2-thiazoline, and sotolon by Maillard-type reactions. In *Flavour Science—Recent Developments*, ed. A.J. Taylor, D.S. Mottram, Hrsg., Proc. zum 8, Weurman Flavour Research Symposium, Reading, UK, 23–26 July 1996, pp. S175–S181. Cambrige: Royal Society of Chemistry.

Hofmann, T., and P. Schieberle. 1997. Identification of potent aroma compounds in thermally treated mixtures of glucose/cysteine and rhamnose/cysteine using aroma extract dilution techniques. *J. Agric. Food Chem.* 45: 898–906.

Hofmann, T., and P. Schieberle. 1998a. 2-Oxopropanal, hydroxy-2-propanone, and 1-pyrrolines: Important intermediates in the generation of the roast-smelling food flavor compounds 2-acetyl-1-pyrroline and 2-acetyltetrahydropyridine. *J. Agric. Food Chem.* 46: 2721–2726.

Hofmann, T., and P. Schieberle. 1998b. Identification of key aroma compounds generated from cysteine and carbohydrates under roasting conditions. *Z. Lebensm.-Unters.-Forsch. A* 207: 229–236.

Hwang, C.F., W.E.I. Riha, B. Jin, M.V. Karwe, T.G. Hartman, H. Daun, and C.T. Ho. 1997. Effect of cysteine addition on the volatile compounds released at the die during twin-screw extrusion of wheat flour. *Lebensm.-Wiss. Technol.* 30: 411–416.

Jousse, F., T. Jongen, W. Agterof, S. Russel, and P. Braat. 2002. Simplified kinetic scheme of flavor formation by the Maillard reaction. *Food Chem. Toxicol.* 67: 2534–2542.

Low, M.Y., G. Koutsidis, J.K. Parker, J.S. Elmore, A.T. Dodson, and D.S. Mottram. 2006. Effect of citric acid and glycine addition on acrylamide and flavor in a potato model system. *J. Agric. Food Chem.* 54: 5976–5983.

Maillard, L.C. 1912. Action des acides aminés sur les sucres: Formation des mélanoïdines par voie méthodique (Amino acids action upon sugars: Melanoidins formation according to a methodical route). *C. R. Acad. Sci., Paris* 154: 66–67.

Mottram, D.S., and I.C.C. Norbega. 2002. Formation of sulfur aroma compounds in reaction mixtures containing cysteine and three different forms of ribose. *J. Agric. Food Chem.* 50: 4080–4086.

Poisson, L., F. Schmalzried, T. Davidek, I. Blank., and J. Kerler. 2009. Study on the role of precursors in coffee flavor formation using in-bean experiments. *J. Agric. Food Chem.* 57: 9923–9931.

Rizzi, G.P. 2008. The Strecker degradation of amino acids: Newer avenues for flavor formation. *Food Rev. Int.* 24: 416–435.

Shu, C.-K., and C.-T. Ho. 1989. Parameter effects on the thermal reaction of cystine and 2,5-dimethyl-4-hydroxy-3(2*H*)-furanone. In *Thermal Generation of Aromas*, ed. T.H. Parliment, 229–241. Washington, DC: American Chemical Society.

Tai, C.-Y., and C.-T. Ho. 1998. Influence of glutathione oxidation and pH on thermal formation of Maillard-type volatile compounds. *J. Agric. Food Chem.* 46: 2260–2265.

Wedzicha, B.L., I.R. Bellion., and G. German. 1994. New insight into the mechanism of the Maillard reaction from studies of the kinetics of its inhibition by sulfite. In *Maillard Reaction in Chemistry, Food and Health*, ed. T.P. Labuza, G.A. Reineccius, V.M. Monnier, J. O'Brien, and J.W. Baynes, 82–87. Cambridge: Royal Society of Chemistry.

Whitfield, F.B., and D.M. Mottram. 1999. Investigation of the reaction between 4-hydroxy-5-methyl-3(2*H*)-furanone and cysteine or hydrogen sulfide at pH 4.5. *J. Agric. Food Chem.* 47: 1626–1634.

Xi, J., T.-C. Huang, and C.-T. Ho. 1999. Characterization of volatile compounds from the reaction of 3-hydroxy-2-butanone and ammonium sulfide model system. *J. Agric. Food Chem.* 47: 245–248.

Yaylayan, V., and A. Huyghues-Despointes. 1994. Chemistry of Amadori rearrangement products: Analysis, synthesis, kinetics, reactions, and spectroscopic properties. *Crit. Rev. Food Sci. Nutr.* 34(4): 321–369.

6 Flavors from Amino Acids

Henry-Eric Spinnler

CONTENTS

6.1 INTRODUCTION

Protein breakdown is the source of many compounds involved in food sensory properties. The first step of protein catabolism is done by proteases. They produce a variety of peptides that are then hydrolyzed into amino acids by peptidases (aminopeptidases and carboxypeptidases). Peptides and amino acids are not volatile, and consequently cannot be perceived by the olfactive receptors; however, they are perceived by the gustative nerve and constitute a very important factor in food palatability (Jinap and Hajeb 2010).

Peptides can have sour, sweet, salty, umami, or bitter flavors. Bitterness of the peptides can be an issue in certain products. In fermented foods, such as cheese, it is possible to formulate the association of flora to equilibrate the protease activity of certain microflora (e.g., *Penicillium*) having strong proteolytic activities with flora having peptidase activities such as *Geotrichum* (Molimard et al. 1994) to avoid the occurrence of the bitter defect.

The breakdown of amino acids leads directly to a large diversity of volatile compounds. Many plants and microorganisms (filamentous fungi, yeasts, and bacteria)

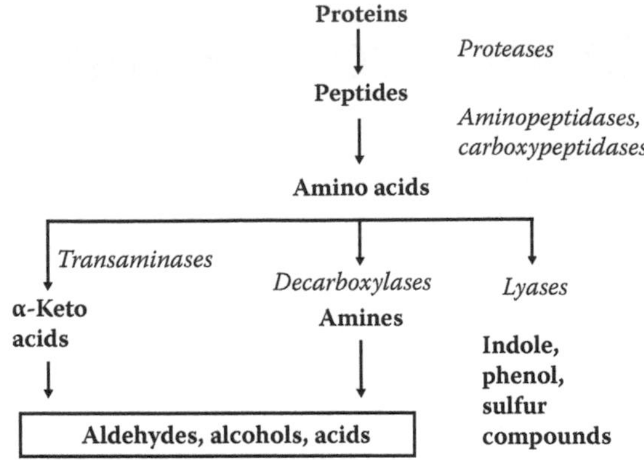

FIGURE 6.1 General scheme of protein breakdown.

are very efficient for the amino acid breakdown. Because they are easy to handle and because of their importance in fermented foods and beverages, the knowledge of the pathways leading from the amino acids to the flavor compounds has often progressed quicker in microorganisms (bacteria and fungi) than in plants. However, they give good base to emit hypothesis and are helpful in understanding what could happen in plants.

The amino acid catabolism can mainly lead to branched chains or benzenic aldehydes, alcohols, acids, or esters (Figure 6.1). They can also lead to nitrogen-containing compounds such as amines, pyrazines, and pyridines. The sulfur amino acids methionine and cysteine are the origins of a large family of sulfur compounds including thiols, polysulfides, and thioacetals, but also very diverse heterocyclic compounds such as thiophenes, thiazoles, or lanthionines. They can also be the source of metabolites that will disturb the metabolism of the cell, for example, alanine, which induces the synthesis of terpene compounds in bacteria, fungi, and chloroplasts (Lichtenthaler et al. 1997).

In some cases, amino acids can fix volatile compounds and be the source of new volatile compounds such as cysteinyl precursors. In such cases, they can lead to the synthesis of volatile precursors like in passion fruit, grapes, and wines.

Finally, they are also able to produce a large diversity of volatile compounds through reactions with sugars and carbonyls (see the chapter devoted to Maillard reactions). This chapter, however, will focus on the impact of amino acids involved in the synthesis of flavor compounds by biological reactions or by reactions involving enzymes in at least at one step.

6.2 DIRECT CONVERSION OF AMINO ACIDS

6.2.1 Ehrlich–Neubauer Pathway

The most common pathway used by microorganisms for amino acid breakdown is Erhlich–Neubauer's pathway, which leads (1) through a transamination process

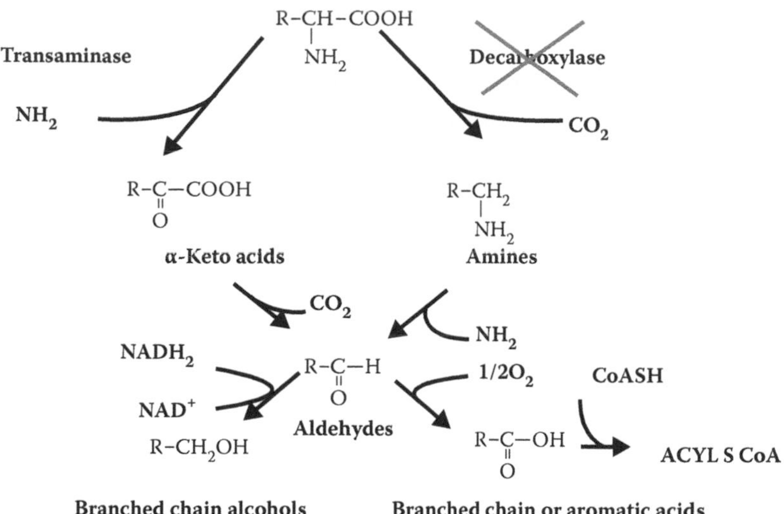

FIGURE 6.2 General scheme of Ehrlich–Neubauer pathway.

to an α-keto acid, (2) followed by a decarboxylation step leading to the production of aldehydes and the corresponding alcohols and acids having one carbon less than the starting amino acid (Figure 6.2). It has been shown in many models that the deamination/transamination step is very often a limiting step in amino acid catabolism and especially for lactic acid bacteria in dairy products (Yvon et al. 1998, 2000).

The acids obtained are easily esterified and permit the synthesis of a series of esters with the alcohols produced by the same metabolism or from the one issued from other pathways (Table 6.1).

6.2.1.1 Aldehydes from Ehrlich–Neubauer Pathway

Aldehydes originate from amino acid catabolism either by transamination, leading to an α-keto acid followed by decarboxylation, or by decarboxylation, and then transamination. These aldehydes can also be produced through an anabolic pathway during amino acid synthesis (Maloney et al. 2010). Aldehydes are often transitory compounds because they are transformed to alcohols or to the corresponding acids.

Aldehydes have flowery (benzene acetaldehyde), fruity, or malty flavors (isobutanal, 2-methylbutanal, and 3-methylbutanal). Some may have vegetable flavors such as methional, which reminds one of cooked potatoes. In oxidative conditions, aldehydes are converted into acids, whereas in reductive conditions they are reduced in alcohols with flowery or fruity flavors. Aldehydes are present in many fruits and vegetables such as olives (Collin et al. 2008) or tomato (Tieman et al. 2007). In tomato, a recent work (Maloney et al. 2010) investigated the transaminase activities of the plants that could be involved in the synthesis of some of these aldehydes. In flowers, 2-phenylacetaldehyde has been identified as the major volatile associated with hyacinth and lilac (Knudsen et al. 1993).

TABLE 6.1
Major Compounds Issued from Ehrlich–Neubauer Pathway, Issued from Oxidation of α-Keto Acids, and Issued from Amino Acids by Mn^{2+} or Mg^{2+} (Bold Characters)

Amino Acid	Aldehyde	Alcohols	Acids
Leu	3-Methylbutanal	3-Methylbutanol	3-Methylbutanoic acid or isovaleric acid
Ileu	2-Methylbutanal	2-Methylbutanol	2-Methylbutanoic acid
Val	Isobutanal	Isobutanol	Isobutanoic acid
Ala	Acetaldehyde	Ethanol	Acetic acid
Met	Methional or 3-methylthio-propionaldehyde	Methionol or 3-methylthio-propyl alcohol	3-Methylthio-propionic acid
Met	**Methylthioacetaldehyde**[a]	**2-Methylthioethanol**	**Methylthioacetic acid**
Cys	Mercaptoethanal	2-Mercaptoethanol	Mercaptoacetic acid
Phe	Phenylacetaldehyde	2-Phenylethanol	Phenylacetic acid
Phe	**Benzaldehyde**	**Benzyl alcohol**	**Benzoic acid**
Glu	4-Oxo-butyric acid	4-Hydroxybutyric acid or butyrolactone	butyric diacid or succinic acid
Tyr	(4-Hydroxyphenyl) acetaldehyde	2-(4-Hydroxyphenyl) ethanol or tyrosol	(4-Hydroxyphenyl) acetic acid
Tyr	Indyl acetaldehyde	2-Indyl ethanol	Indyl acetic acid

[a] Bonnarme et al. 2004.

Aldehydes are considered to be very important compounds in fermented foods such as cheese or alcoholic beverages (Molimard and Spinnler 1996; Gijs et al. 2000; Spitzke and Fauhl-Hassek 2010). Detection thresholds in malt culture media of 0.1, 0.13, and 0.06 mg/kg, respectively, for 2-methylpropanal, 2-methylbutanal, and 3-methylbutanal, have been reported (Margalith 1981).

The role of methional should be specifically mentioned. This compound is particularly important in fermented foods. In beer it is provided by malt toasting, whereas in wine and cheese, it is provided by the amino acid breakdown. It has been shown (Gijs et al. 2000) that this compound can provide dimethyltrisulfide (DMTS) by reduction. This compound has a particular role in wines. Methional, cited for the first time as off-odor in wine in 2000 (Escudero et al. 2000, 2002), is responsible for a cooked potato aromatic note, which is very characteristic of oxidized wine. The speed of the production of this compound in oxidized red wine is quicker than the one observed in white wine (Bueno et al. 2010). Moreover, methional is important in many fruits such as lychee (Mahattanatawee et al. 2007) or raspberry (Klesk et al. 2004).

6.2.1.2 Alcohols from Ehrlich–Neubauer Pathway

Among primary alcohols, 3-methylbutan-1-ol has an alcoholic, floral note. 2-Phenylethanol is well known to have an odor of faded roses. It has recently been shown, in two populations of genetically characterized roses, that 2-phenylethanol displayed quantitative variations in the progeny, and six quantitative trait loci from the rose marker map were shown to influence the amounts of this volatile compound in the studied populations (Spiller et al. 2010). It is present in many fruits such as tomato or kiwi (Tatsuka et al. 1990).

In tomato, it has been shown that 2-phenylethanol is present as a glycosylated precursor. During the ripening from green to red, in the Raf variety, the amount of precursors increased from 311 to 894 µg/kg, whereas the free 2-phenylethanol only increased from 116 to 280 µg/kg (Ortiz-Serrano and Gil 2010). This alcohol is one of the major compounds in Camembert after 7 days of ripening, at a concentration of 1.15 mg/kg. Its concentration stabilizes at approximately 1 mg/kg at the end of ripening. It is lower than the detection threshold in cheese (9 mg/kg), but is close to the detection threshold of the most sensitive panelist of the panel used in the study of Roger et al. (1988).

In cheese, 2-phenylethanol is largely present during the first week of ripening, because it is mainly a metabolic product of yeasts, which are the only microorganisms that can grow at the pH corresponding to the one in soft curd pH (pH = 4.6) (Lee and Richard 1984). This alcohol can be methylated, as is probably the case for the 1-methoxy-3-methylbutanc that has been found produced by *Proteus vulgaris* (Deetae et al. 2009).

6.2.1.3 Acids from Ehrlich–Neubauer Pathway

Aldehydes can also be oxidized into isobutyric, 2-methylbutyric, and isovaleric acids. These acids are described as having a mild odor, sometimes unpleasant and reminiscent of sweat (Molimard and Spinnler 1996). 3-(Methylthio)propionic acid is characterized by chocolate and roasted odors, and contributes significantly to the lactic aromas of red wines (Pripis-Nicolau et al. 2004).

6.2.1.4 Esters and Thioesters from Ehrlich–Neubauer Pathway

There is a great diversity of esters coming from these acids and alcohols in food and especially in fermented foods. They have generally fruity flavors. The acids are activated into acyl-*S*-coenzyme A (Helinck et al. 2000). The coenzyme is then substituted with a short-chain alcohol to produce alkyl esters or by thiols to produce thioesters.

2-Phenylethylacetate and 2-phenylethylpropanoate are qualitatively important in the flavor of Camembert cheese. On the seventh day of ripening, 2-phenylethylacetate is the principal compound in the aromatic profile, at a concentration of 4.6 mg/kg (Roger et al. 1988). Most of the esters found in cheeses are described as having fruity, floral notes. The most cited aromatic notes of these compounds are pineapple, banana, apricot, pear, floral, rose, honey, and wine. Some of these esters have a very low perception threshold, for example, isoamylacetate, which is detectable in water at a concentration of 2 µg/kg (Piendl and Geiger 1980). Esters with a low carbon number have a perception threshold approximately 10 times lower than the corresponding alcohols (for a recent review, see Liu et al. 2004).

Esterification reactions are well-known detoxification reactions in media, enabling the elimination of toxic alcohols and carboxylic acids. A wide variety of enzymes are involved in esterification reactions including carboxylesterases, which have a very wide range of substrates, and arylesterases, which are present in most plants and microorganisms. Ester formation has been widely studied in fermented beverages, in which they play an important aromatic role. The contributions of ethyl 2-methylbutyrate, ethyl isobutyrate, ethyl octanoate, ethyl hexanoate, and 3-methylbutyl acetate have been shown to be important in Scheurebe and Gewürztraminer wines (Guth 1997a, 1997b). Their production is related to yeast activity (Pretorius 2000).

In all cheeses, microorganisms involved in ester production seem to be mainly yeasts. Production of esters occurs early during ripening. *Geotrichum candidum* is capable of producing numerous esters, some of which have a very pronounced melon odor (Latrasse et al. 1987), but it depends on the strain of *G. candidum. Kluyveromyces lactis*, another yeast found on cheese at the early stages of cheese ripening, is also well known to produce esters and to develop a typical fruity flavor (Martin et al. 2001).

6.3 OXIDATION OF α-KETO ACIDS BY DIVALENT CATIONS

The α-keto acids issued from the transamination of amino acids can lose two carbons. The biosynthesis pathway for benzaldehyde was determined by Nierop-Groot and de Bont (1999). It was shown that a chemical breakdown of phenylpyruvic acid was catalyzed by divalent cations such as Mn^{2+}. Benzaldehyde is described as having an aromatic note reminiscent of bitter almond. Its detection threshold in water is 350 μg/kg (Buttery et al. 1988).

This pathway also seems to be used for other amino acids such as methionine, producing 2-(methylthio)ethanal, which has a green apple flavor (Yvon et al. 2001). In fact, it is probable that all the α-keto acids when oxidating metals are present in the environment as precursors of aldehydes having a carbon chain whose length is shorter (two carbons) than the corresponding amino acid. The carbons eliminated are the one bearing the carboxylic function and the other bearing the amino group.

6.4 AMINES

For some microorganisms, the breakdown of amino acids starts by decarboxylation with the production of amines. Decarboxylation of amino acids leads to the production of CO_2 and amines. This reaction requires the presence of pyridoxalphosphate and coenzyme. Decarboxylation of leucine gives isobutylamine, phenylalanine gives phenylethylamine, and tyrosine gives tyramine. A low oxygen pressure favors these reactions. Amines are not the final products but are subjected to oxidative deamination to form aldehydes. They can also be the starting point of compounds such as *N*-isobutylacetamide encountered in Camembert, presumably by reaction with acetic acid. Volatile amines such as isopropylamine, iso-amylamine, anteiso-amylamine, phenethylamine, spermine, spermidine, putrescine, and cadaverine can have amino acids as precursors. Many volatile amines are described as having a fruity, alcoholic, or varnish-like aroma notes (Laivg et al. 1978).

6.5 CATABOLISM OF AMINO ACIDS SIDE CHAIN

6.5.1 Indole Ring

Degradation of the side chain of tyrosine and of tryptophan by tyrosine–phenol–lyase and by tryptophan–indole–lyase, respectively, leads to the formation of phenol and indole. Parliment et al. (1982) considered that phenol found in Limburger results from degradation of tyrosine by *Brevibacterium linens*. The catabolism of tryptophan by *B. linens* was recently studied by Ummadi and Weimer (2001). In model media, tryptophan was broken down into anthranilic acid at a high rate. However, the physicochemical environment of ripening cheese is quite far from optimal conditions, and these authors concluded that it is unlikely that *B. linens* could be responsible for fecal, putrid, or meaty–brothy defects in Cheddar cheese.

A high concentration of indole has been linked to "plastic-like" off-flavor in wines, predominantly in wines produced under sluggish fermentation conditions (Arevalo-Villena et al. 2010). Tryptophan was required for the accumulation of indole in chemically defined medium, and all yeast and bacteria fermentations were able to accumulate indole. *Candida stellata* showed the greatest potential for indole formation (1033 mg/L), and among the bacteria, the highest concentration was generated by *Lactobacillus lindneri* (370 mg/L).

6.5.2 Sulfur Compounds

During their work on the identification of minor components present in aromatic extracts of Pont l'Evêque and Camembert, Dumont et al. (1976a, 1976b) isolated four sulfur compounds from a fraction with a garlic flavor note: 2,4-dithiapentane, diethyldisulfide, 2,4,5-trithiahexane, and 3-methylthio-2,4-dithiapentane. They also identified traces of a sulfur-containing alcohol, 3-methylthiopropanol (or methionol), and ethyldisulfide. Other sulfur compounds are also found in Camembert cheese (Leclercq-Perlat et al. 2004). Disulfides are generally absent from young cheeses. In these cheeses, a low level of proteolysis yields only a low level of sulfur amino acids, which are precursors of disulfides. In late ripening, sulfur compounds are quantitatively reduced and even disappear in some products. This can be explained by their high volatility. Sulfur compounds found in cheeses are described as having a strong garlic or "very ripe cheese" odor. Furthermore, these compounds have a very low detection threshold in water, from 0.02 μg/kg for methanethiol to 0.3 μg/kg for dimethylsulfide (Shankaranarayma et al. 1974).

In cheese, sulfur compounds originate principally from methionine degradation resulting from a carbon–sulfur bond cleavage by a methionine-γ-demethiolase. This amino acid is a precursor of methanethiol, which is itself the starting point for some other compounds, including dimethyldisulfide and dimethyltrisulfide obtained by methane thiol oxidation. Many microorganisms are able to produce methanethiol from methionine especially coryneform bacteria and yeasts (Spinnler et al. 2001). Among the ripening fungi, many have this potential, such as *Penicillium camemberti*, *G. candidum*, and *Yarrowia lipolytica* (Bonnarme et al. 2000). Molimard et al. (1997) have shown that some strains of *G. candidum*, although its growth was quite early in the ripening process, were able to change the typicality of a Camembert cheese. One strain of *G. candidum* caused the development of cabbage and cowshed notes. It was then shown that *G. candidum* growing in a curd medium enriched with methionine was able to accumulate a large variety of sulfur compounds, including various thioesters such as methylthioacetate, methylthiopropionate, methylthiobutyrate, methylthioisobutyrate, methylthioisovalerate (MTIV), and methylthiohexanoate (MTH) (Berger et al. 1999a). These thioesters have various flavor notes: from cheesy (MTIV) to fruity (MTH) (Berger et al. 1999b). The metabolism of *G. candidum* was explored and it was shown that this species, unlike *B. linens*, was able to accumulate α-keto-4-methylthiobutyric acid as an intermediate in catabolism (Bonnarme et al. 2001).The origins of different sulfur flavor compounds are summarized in Figure 6.3.

Smear bacteria have also been studied, but mainly *B. linens* or *Arthrobacter* spp. (Bonnarme et al. 2000). Among these microorganisms, coryneform bacteria, especially *B. linens*, are considered key agents in the production of sulfur compounds in cheeses in which they grow. It has been shown that dimethylsulfide is produced by *G. candidum* using a separate pathway than methanethiol production from methionine (Demarigny et al. 2000).

These sulfur compounds are also very important in wines and fruits where their origin is not as well known. Some of them are present as precursors and liberated during the ripening of the fruits (see below).

6.5.3 Volatiles Issued from the Activity of Phenylalanine Ammonia Lyase

The cleavage of the amino chain of the phenylalanine by phenylalanine ammonia lyase provides cinnamic acid. This compound has never been detected in cheese but

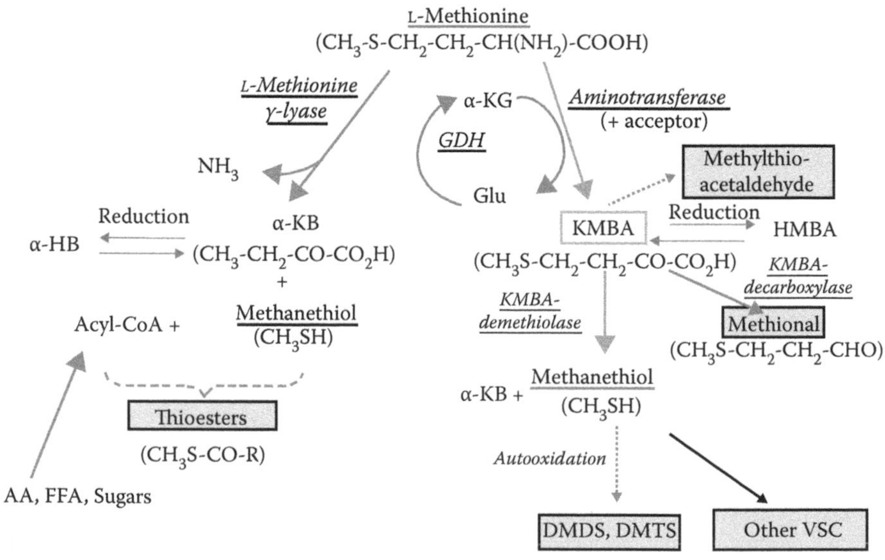

FIGURE 6.3 Volatile compounds issued from methionine.

its methyl ester, methylcinnamate, has been identified in Camembert by Moinas et al. (1975). It seems to be particularly important in the aroma of this cheese. When varying the concentration of this compound in a neutral cheese base, to which heptan-2-one, heptan-2-ol, oct-1-en-3-ol, nonan-2-ol, phenol, and butyric acid had been added, these authors developed a characteristic Camembert note.

P. camemberti is able to decarboxylate the cinnamic acid and to produce styrene (Pagot et al. 2007). Styrene has a very strong plastic-like odor. Its perception threshold in cream is 5 μg/kg. This compound has been described as a trace element in several cheeses including Camembert (Dumont et al. 1974). Adda et al. (1989) found an abnormally high quantity of styrene (5 mg/kg) in Camembert with a pronounced celluloid taste. These authors demonstrated the role played by *P. camemberti* in the production of these compounds. Spinnler et al. (1992) observed a correlation between the production of styrene and oct-1-en-3-ol in a minimal medium. These two compounds were produced after 15 days of culture, when there is no more glucose in the

TABLE 6.2
Thiols Found in Wines, Concentration Detected, Description, and Olfactive Threshold of Detection

Compounds	Concentration (min–max)/Threshold (ng/L) in Model Alcoholic Solution	Olfactory Description	Wines
2-*SH*-Ethyl acetate	40–134 μg/L / 65 μg/L	Roasted meat	Semillon, Sauvignon[a]
3-*SH*-Propyl acetate	3–32 μg/L / 35 μg/L	Roasted meat	Semillon, Sauvignon[a]
4-*SH*-4-Methylpentan-2-one	4–44/0.8	Box tree, broom	Sauvignon[b], Scheurebe wines[c]
3-*SH*-Hexyl acetate	0–777/4	Box tree Passion fruit	Sauvignon[b], Cabernet Sauvignon and Merlot[d]
3-*SH*-Hexanol	600–12,800/60	Grapefruit	Sauvignon[b], Cabernet Sauvignon[d]
4-*SH*-4-Methylpentan-2-ol	1–111/55	Citrus zest	Sauvignon[b]
3-*SH*-3-Methylbutan-2-ol	34–128/1500	Cooked leeks	Sauvignon[b]
2-Methyl-3-furane thiol	No data/2–8	Empyreumatic	Cabernet Sauvignon, Merlot[e]
2-Furanmethane thiol	2–39/0.4	Roast coffee, toasty	Petit Manseng, Red Bordeaux[f]

[a] Lavigne et al. 1998.
[b] Tominaga et al. 1998.
[c] Guth 1997b.
[d] Bouchilloux et al. 1998a.
[e] Bouchilloux et al. 1998b.
[f] Tominaga et al. 2000.

growth medium. Oct-1-en-3-ol is produced 2 to 3 days before styrene. ^{13}C styrene is produced from ^{13}C-phenylalanine (Pagot et al. 2007), suggesting that this amino acid is the precursor of styrene.

6.6 CYSTEINYLATED PRECURSORS OF FLAVORS

Different thiols are present under the form of precursors fixed on cysteine. It is the case for some ketones or unsaturated compounds that can react with cysteine. One of the first examples published was the conversion of pulegone into P-8 menthenthiol (Kerkenaar et al. 1988). The molecule on which the volatile compound is fixed can be the free cysteine or the glutathione (Roland et al. 2010).

It is probably in wines that these cysteinylated compounds have been studied the most. At least some of the free thiols are liberated during fermentation by the yeasts probably through their lyase activities. These thiols have very low olfactive thresholds as presented in Table 6.2. In Sauvignon Blanc and in several other types of grapes, they determine the typicality of the wine. However, the amount of precursors and their yield of conversion do not permit to explain the whole amount of thiols in wine.

It has recently been demonstrated that with a mixture of yeast, it was possible to obtain different flavor perceptions in a Sauvignon Blanc and that the yeast mixture used to realize the fermentation affects consumer acceptance (King et al. 2010).

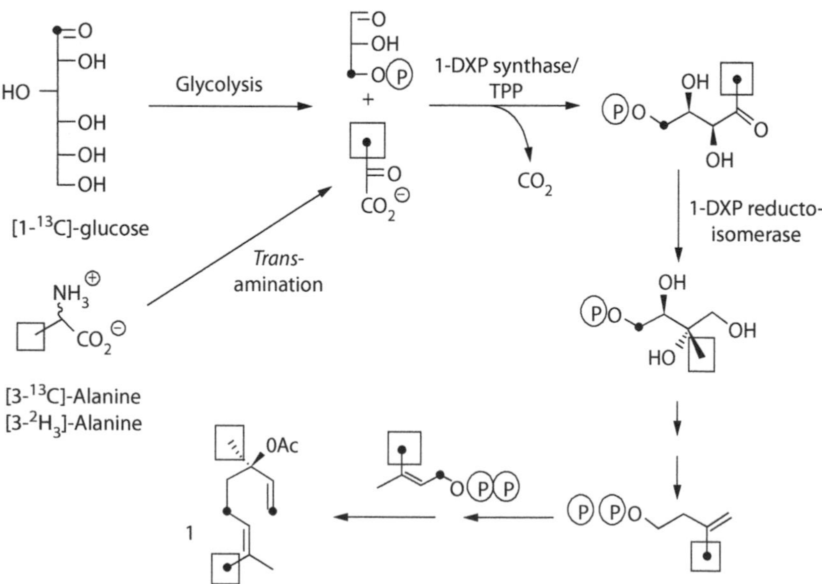

FIGURE 6.4 Pathway leading to synthesis of terpene from alanine. (From Lichtenthaler, H.K. et al., *Physiol. Plant*, 101, 643–652, 1997. With permission.)

6.7 AMINO ACID PRECURSORS OF TERPENES

Because alanine by deamination leads to pyruvate (see Ehrlich–Neubauer reaction), it considerably disturbs the metabolism. In 1997, Lichtenthaler et al. discovered that this overconcentration of pyruvate first led to 1-deoxy-D-xylulose and then to methyl-D-erythritol-4-phosphate. This was a new pathway and leads to the synthesis of linalool. The different trials with labeled precursors were consistent with this hypothesis (Figure 6.4).

Since it has been shown that this pathway was present in chloroplast, bacteria, and fungi, Hampel et al. (2005) have shown in *Daucus carota* that monoterpene was

TABLE 6.3
Pyrazines Identified in the Headspace of *Bacillus cereus* Surface Cultures with Their Flavor Properties and Linear Retention Indices (LRI)

Compound	Structure	Odor Description	Odor Threshold (ng/l, Air)[a]	LRI (EC-5)
2 Acetyl 1 pyroline	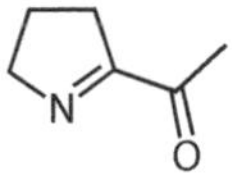 1	Popcorn, scented rice	0.02[b]	920
Pyrazine	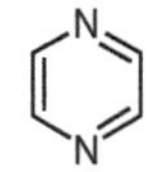 2	Nutty	>2000[c]	724
Methylpyrazine	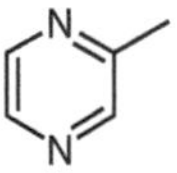 3	Nutty, green	>2000	818
2,5-Dimethylpyrazine	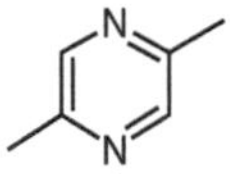 4	Nutty	1720	910
Trimethylpyrazine	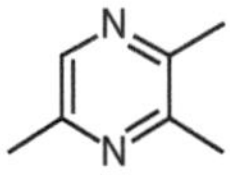 5	Roasted	50	1000
3-Ethyl-2,5-dimethylpyrazine	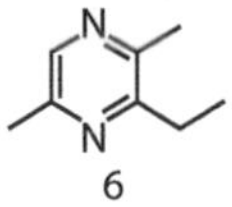 6	Earthy, roasted	0.011	1089

Source: Adams, A. and de Kimpe, N., *Food Chem.*, 101, 1230–1238, 2007.
[a] Wagner et al. (1999).
[b] Schieberle (1991).
[c] Odor threshold (pyrazine) > odor threshold (methylpyrazine) (Buttery et al. 1999).

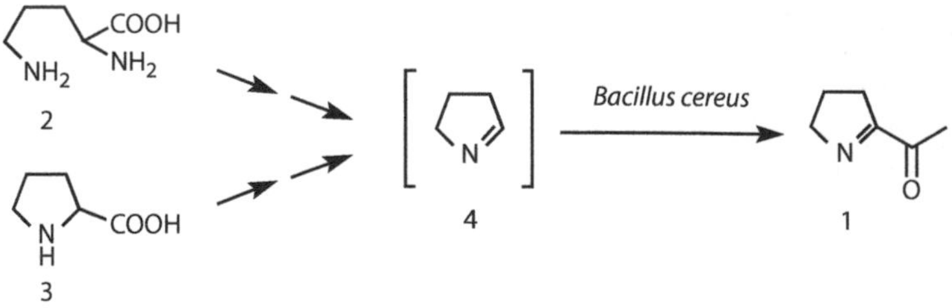

FIGURE 6.5 Production of 2-acetyl pyroline (1) from ornithine (2) and proline (3) through pyrolidine (4) by *Bacillus cereus*. (From Adams and de Kimpe 2007.)

built through this route in chloroplasts, whereas sesquiterpenes were synthetized in the cytoplasm and this organization needs a flow of the intermediate isopentyl allyl pyrophosphate from the plastids into the cytoplasm.

6.8 AMINO ACID PRECURSORS OF PYRAZINES AND PYROLINES

Pyrazines are very powerful flavoring compounds leading to nutty flavors or vegetable flavors. They are produced by the Maillard reactions, and also by microorganisms (Table 6.3) and fruits.

Many microorganisms have been reported to produce pyrazines. They are well known to give their typicality to some traditional fermented foods like natto in Japan (Demyttenaere et al. 2002). Among them, *Bacillus subtilis*, *Pseudomonas* sp., and *Corynebacterium glutamicum* have been the more studied. It is mainly polyalkylpyrazines and methoxypyrazines that are biosynthesized. The role of diketones like diacetyl, glyoxal, and methylglyoxal in the reaction is greatly suspected. It is possible to produce pure pyrazines from bacteria to be used as natural flavorings. Recently, Zhu et al. (2010) have shown that it was possible to produce an average of 4 g/L of tetramethylpyrazine through *B. subtilis*. Other heteroatomic cycles can be produced from amino acids like pyroline (Figure 6.5).

6.9 CONCLUSIONS

Amino acids are the source of a large diversity of flavor compounds. The Ehrlich pathway is mainly the source of branched chains or aryl aldehydes, alcohols, acids, and esters very important in fruit flavors and fermented foods and beverages. In this reaction, the transamination step is often the limiting step of the chain of reactions.

Lyase activity will liberate precursors of aryl compounds from phenylalanine or tyrosine (cinnamic acid or 4-hydroxycinnamic acid) and sulfur amino acids (methane thiols and H2S). These precursors are the base of a large diversity of very odorous compounds responsible of food typicality, but also of taints when they are present in high concentrations as compared with their olfactive thresholds.

Finally, the nitrogen of the amino acids is a key element for the occurrence of amines, pyrazines, pyroline, and diverse odorous nitrogen-containing compounds.

REFERENCES

Adams A., and N. de Kimpe. 2007. Formation of pyrazines and 2-acetyl-1-pyrroline by *Bacillus cereus*. *Food Chem.* 101: 1230–1238.

Adda, J., J. Dekimpe, L. Vassal, and H.E. Spinnler. 1989. Production de styrène par *Penicillium camemberti* Thom. *Lait* 69: 115–120.

Arevalo-Villena, M., E.J. Bartowsky, D. Capone, and M.A. Sefton. 2010. Production of indole by wine-associated microorganisms under oenological conditions. *Food Microbiol.* 27: 685–690.

Berger, C., J.A. Khan, P. Molimard, N. Martin, and H.E. Spinnler. 1999a. Production of sulfur flavors by 10 strains of *G. candidum*. *Appl. Environ. Microbiol.* 65: 5510–5551.

Berger, C., N. Martin, S. Collin, L. Gijs, J.A. Khan, G. Piraprez, H.E. Spinnler, and E.N. Vulfson. 1999b. Combinatorial approach to flavor analysis: II. Olfactory investigation of a library of *S*-methyl thioesters and sensory evaluation of selected components. *J. Agric. Food Chem.* 47: 3274–3279.

Bonnarme, P., K. Arfi, C. Dury, S. Helinck, M. Yvon, and H.E. Spinnler. 2001. Sulfur compound production by *G. candidum* from methionine: Importance of the transamination step. *FEMS Microbiol. Lett.* 205: 247–252.

Bonnarme, P., L. Psoni, and H.E. Spinnler. 2000. Diversity of L-methionine catabolism pathways in cheese-ripening bacteria. *Appl. Environ. Microbiol.* 66: 5514–5517.

Bonnarme, P., F. Amarita, E. Chambellon, E. Semon, H.E. Spinnler, and M. Yvon. 2004. Methylthioacetaldehyde, a possible intermediate metabolite for the production of volatile sulfur compounds from l-methionine by *Lactococccus lactis*. *FEMS Microbiol. Lett.* 236: 85–90.

Bouchilloux, P., P. Darriet, and R. Henry. 1998a. Identification of volatile and powerful odorous thiols in Bordeaux red wine varieties. *J. Agric. Food Chem.* 46: 3095–3099.

Bouchilloux, P., P. Darriet, and D. Dubourdieu. 1998b. Identification of a very odoriferous thiol, 2 methyl-3-furanthiol, in wines. *Vitis* 37: 177–180.

Bueno, M., L. Culleré, J. Cacho, and V. Ferreira. 2010. Chemical and sensory characterization of oxidative behavior in different wines. *Food Res. Int.* 43: 1423–1428.

Buttery, R.G., J.G. Turnbaugh, and L.C. Ling. 1988. Contribution of volatiles to rice aroma. *J. Agric. Food Chem.* 36: 1006–1009.

Collin, S., S. Nizet, S. Muls, R. Iraqi, and A. Bouseta. 2008. Characterization of odor-active compounds in extracts obtained by simultaneous extraction/distillation from Moroccan black olives. *J. Agric. Food Chem.* 56: 3273–3278.

Deetae, P., P. Bonnarme, H.E. Spinnler, and S. Helinck. 2009. The growth and aroma contribution of *Microbacterium foliorum*, *Proteus vulgaris* and *Psychrobacter* sp. during ripening in a cheese model medium. *Appl. Microbiol. Biotechnol.* 82: 169–177.

Demyttenaere, J., K.A. Tehrani, and N. de Kimpe. 2002. The chemistry of the most important Maillard flavor compounds of bread and cooked rice. In *Heteroatomic Aroma Compounds*, ed. G.A. Reineccius and T.A. Reineccius. *ACS Symposium Series* 826, 150–165.

Demarigny, Y., C. Berger, N. Desmazures, M. Gueguen, and H.E. Spinnler. 2000. Flavour sulphides produced from methionine by two different pathways by *G. candidum*. *J. Dairy Res.* 67: 371–380.

Dumont, J.P., C. Degas, and J. Adda. 1976a. L'arôme du Pont l'Evêque. Mise en évidence de constituants volatils quantitativement mineurs. *Lait* 56: 177–180.

Dumont, J.P., S. Roger, and J. Adda. 1976b. L'arôme du camembert: Autres composés mineurs mis en évidence. *Lait* 56: 595–599.

Dumont, J.P., S. Roger, P. Cerf, and J. Adda. 1974. Etude des composés volatils neutres présents dans le Vacherin. *Lait* 54: 243–251.

Escudero, A., E. Asensio, J. Cacho, and V. Ferreira. 2002. Sensory and chemical changes of young white wines stored under oxygen. An assessment of the role played by aldehydes and some other important odorants. *Food Chem.* 77: 325–331.

Escudero, A., P. Hernandez-Orte, J. Cacho, and V. Ferreira. 2000. Clues about the role of methional as character impact odorant of some oxidized wines. *J. Agric. Food Chem.* 48: 4268–4272.

Forquin, M.P. 2010. Étude de *Brevibacterium aurantiacum*, une bactérie d'affinage de fromage: de son métabolisme du soufre à son interaction avec *Kluyveromyces lactis*. PhD thesis, AgroParisTech. Paris, France.

Gijs, L., P. Perpete, A. Timmermans, and S. Collin. 2000. 3-Methylthiopropionaldehyde as precursor of dimethyl trisulfide in aged beers. *J. Agric. Food Chem.* 48: 6196–6199.

Guth, H. 1997a. Identification of character impact odorants of different white wine varieties. *J. Agric. Food Chem.* 45: 3022–3026.

Guth, H. 1997b. Quantitation and sensory studies of character impact odorants of different white wine varieties. *J. Agric. Food Chem.* 45: 3027–3032.

Hampel, D., A. Mosandl, and M. Wüst. 2005. Biosynthesis of mono- and sesquiterpenes in carrot roots and leaves (*Daucus carota* L.): Metabolic cross talk of cytosolic mevalonate and plastidial methylerythritol phosphate pathways. *Phytochemistry* 66: 305–311.

Helinck, S., H.E. Spinnler, S. Parayre, M. Dame-Cahagne, and P. Bonnarme. 2000. Enzymatic versus spontaneous *S*-methyl thioester synthesis in *Geotrichum candidum*. *FEMS Microbiol. Lett.* 193: 237–241.

Jinap, S., and P. Hajeb. 2010. Glutamate. Its applications in food and contribution to health. *Appetite* 55: 1–10.

Kerkenaar, A., D.J.M. Schmedding, and J. Berg. 1988. European patent application, 0 277 688 A2.

King, E.S., R.L. Kievit, C. Curtin, J.H. Swiegers, H. Jan, I.S. Pretorius, S.E.P. Bastian, and I.L. Francis. 2010. The effect of multiple yeasts co-inoculations on Sauvignon Blanc wine aroma composition, sensory properties and consumer preference. *Food Chem.* 122: 618–626.

Klesk, K., M. Qian, and R.R. Martin. 2004. Aroma extract dilution analysis of cv. Meeker (*Rubus idaeus* L.) red raspberries from Oregon and Washington. *J. Agric. Food Chem.* 52: 5155–5161.

Knudsen, J.T., L. Tollsten, and L.G. Bergstron. 1993. Floral scents—A checklist of volatile compounds isolated by head-space techniques. *Phytochemistry* 33: 253–280.

Laivg, D.G., H. Panhuber, and R.I. Baxter. 1978. Olfactory properties of amines and *n*-butanol. *Chem. Sens. Flavour* 3: 149–166.

Latrasse, A., P. Dameron, M. Hassani, and T. Staron. 1987. Production of a fruity aroma by *Geotrichum candidum* (Staron). *Sci. Aliments* 7: 637.

Lavigne, V., R. Henry, and D. Dubourdieu. 1998. Identification and determination of sulfur compounds responsible for "grilled" aroma in wines. *Sciences des Aliments* 18: 175–191.

Leclercq-Perlat, M.N., E. Latrille, G. Corrieu, and H.E. Spinnler. 2004. Controlled production of Camembert-type cheeses: Part II. Changes in the concentration of the more volatile compounds. *J. Dairy Res.* 71: 355–366.

Lee, C.W., and J. Richard. 1984. Catabolism of L-phenylalanine by some microorganisms of cheese origin. *J. Dairy Res.* 51: 461–469.

Lichtenthaler, H.K, M. Rohmer, and J. Schwender. 1997. Two independent biochemical pathways for isopentenyl diphosphate and isoprenoid biosynthesis in higher plants. *Physiol. Plant* 101: 643–652.

Liu, S.Q., R. Holland, and V. Crow. 2004. Esters and their biosynthesis in fermented dairy products: A review. *Int. Dairy J.* 14: 923–945.

Mahattanatawee, K., P. Ruiz Perez-Cacho, T. Davenport, and R. Rouseff. 2007. Comparison of three lychee cultivar odor profiles using gas chromatography–olfactometry and gas chromatography–sulfur detection. *J. Agric. Food Chem.* 55: 1939–1944.

Maloney, G.S., A. Kochevenko, D.M. Tieman, T. Tohge, U. Krieger, D. Zamir, M.G. Taylor, A.R. Fernie, and H.J. Klee. 2010. Characterization of the branched-chain amino acid aminotransferase enzyme family in tomato. *Plant Physiol.* 153: 925–936.

Margalith, P.Z. 1981. Dairy products. In *Flavor Microbiology*, ed. R.A. Magalitz, 32–118. Springfield, IL: C.C. Thomas Publisher.

Martin, N., C. Berger, C. Le Du, and H.E. Spinnler. 2001. Aroma compound production in cheese curd by coculturing with selected yeast and bacteria. *J. Dairy Sci.* 84: 2125–2135.

Moinas, M., M. Groux, and I. Horman. 1975. La flaveur des fromages: III. Mise en évidence de quelques constituants mineurs de l'arôme du Camembert. *Lait* 55: 414–417.

Molimard, P., I. Lesschaeve, I. Bouvier, L. Vassal, P. Schlich, S. Issanchou, and H.E. Spinnler. 1994. Amertume et fractions azotées de fromages à pâte molle de type Camembert: Rôle de l'association de *Penicillium camemberti* avec *Geotrichum candidum*. *Lait* 74: 361–374.

Molimard, P., I. Lesschaeve, S. Issanchou, M. Brousse, and H.E. Spinnler. 1997. Effect of the association of surface flora on the sensory properties of mould ripened cheese. *Lait* 77: 181–187.

Molimard, P., and H.E. Spinnler. 1996. Review: Compounds involved in the flavor surface mold ripened cheeses: Origins and properties. *J. Dairy Sci.* 79: 169–184.

Nierop-Groot, M.N., and J.A.M. de Bont. 1999. Involvement of manganese in conversion of phenylalanine to benzaldehyde by lactic acid bacteria. *Appl. Environ. Microbiol.* 65: 5590–5593.

Ortiz-Serrano, P., and J.V. Gil. 2010. Quantitative comparison of free and bound volatiles of two commercial tomato cultivars (*Solanum lycopersicum* L.) during ripening. *J. Agric. Food Chem.* 58: 1106–1114.

Pagot, Y., J.M. Belin, F. Husson, and H.E. Spinnler. 2007. Metabolism of phenylalanine and biosynthesis of styrene in *Penicillium camemberti*. *J. Dairy Res.* 74:180–185.

Parliment, T.H., M.J. Kolo, and D.J. Rizzo. 1982. Volatile components of Limburger cheese. *J. Food Chem.* 30: 1006–1008.

Piendl, A., and E. Geiger. 1980. Technological factors in the formation of esters during fermentation. *Brew. Dig.* 55: 26–38.

Pretorius, I.S. 2000. Tailoring wine yeast for the new millennium: Novel approaches to the ancient art of wine making. *Yeast* 16: 675–729.

Pripis-Nicolau, L., G. De Revel, A. Bertrand, and A. Lonvaud-Funel. 2004. Methionine catabolism and production of volatile sulfur compounds by *Oenococcus oeni*. *J. Appl. Microbiol.* 96: 1176–1184.

Roger, S., C. Degas, and J.C. Gripon. 1988. Production of phenyl ethyl alcohol and its esters during ripening of traditional Camembert. *J. Food Chem.* 28: 129–140.

Roland, A., R. Schneider, A. Razungles, C. Le Guernevé, and F. Cavelier. 2010. Straightforward synthesis of deuterated precursors to demonstrate the biogenesis of aromatic thiols in wine *J. Agric. Food Chem.* 58: 10684–10689.

Shankaranarayma, M.L., B. Raghavan, K.O. Abraham, and C.P. Natarajan. 1974. Volatile sulfur compounds in food flavors. *CRC Crit. Rev. Sci. Technol.* 4: 395–435.

Spiller, M., R.G. Berger, and T. Debener. 2010. Genetic dissection of scent metabolic profiles in diploid rose populations. *Theor. Appl. Genet.* 120: 1461–1471.

Spinnler, H.E., C. Berger, C. Lapadatescu, and P. Bonnarme. 2001. Production of sulfur compounds by several yeasts of technological interest for cheese ripening. *Int. Dairy J.* 11: 245–252.

Spinnler, H.E., O. Grosjean, and I. Bouvier. 1992. Effect of culture parameters on the production of styrene (vinyl benzene) and 1-octene-3-ol by *Penicillium caseicolum*. *J. Dairy Res.* 59: 533–541.

Spitzke, M.E., and C. Fauhl-Hassek. 2010. Determination of the $^{13}C/^{12}C$ ratios of ethanol and higher alcohols in wine by GC-C-IRMS analysis. *Eur. Food Res. Technol.* 231: 247–257.

Tatsuka, K., S. Suekane, S. Sakai, and H. Sumitani. 1990. Volatile constituents of kiwi fruit flowers: Simultaneous distillation and extraction versus headspace sampling. *J. Agric. Food Chem.* 38: 2176–2180.

Tieman, D.M., H.M. Loucas, J.Y. Kim, D.G. Clark, and H.J. Klee. 2007. Tomato phenylacetaldehyde reductases catalyze the last step in the synthesis of the aroma volatile 2-phenylethanol. *Phytochemistry* 68: 2660–2669.

Tominaga, T., M.L. Murat, and D. Dubourdieu. 1998. Development of a method for analyzing the volatile thiols involved in the characteristic aroma of wines made from *Vitis vinifera* L. cv. Sauvignon Blanc. *J. Agric. Food Chem.* 46: 1044–1048.

Tominaga, T., L. Blanchard, P. Darriet, and D. Dubourdieu. 2000. A powerful aromatic volatile thiol, 2-furanmethanethiol, exhibiting roast coffee aroma in wines made from several *Vitis vinifera* grape varieties. *J. Agric. Food Chem.* 48: 1799–1802.

Ummadi, M., and B.C. Weimer. 2001. Tryptophan catabolism in *Brevibacterium linen*s as a potential cheese flavor adjunct. *J. Dairy Sci.* 84: 1773–1782.

Yvon, M., S. Berthelot, and J.C. Gripon. 1998. Adding aketoglutarate to semi-hard cheese curd highly enhances the conversion of amino acids to aroma compounds. *Int. Dairy J.* 8: 889–898.

Yvon, M., E. Chambellon, A. Bolotin, and F. Roudot-Algaron. 2000. Characterization and role of the branched-chain aminotransferase (BcaT) isolated from *Lactococcus lactis* subsp *cremoris* NCDO 763. *Appl. Env. Microbiol.* 66: 571–577.

Zhu B.F., Y. Xu, and W.L. Fan. 2010. High-yield fermentative preparation of tetramethylpyrazine by *Bacillus* sp. using an endogenous precursor approach. *J. Ind. Microbiol. Biotechnol.* 37: 179–186.

7 Binding and Release of Flavor Compounds

Elisabeth Guichard

CONTENTS

7.1 INTRODUCTION

Flavor perception is mainly related to the release in the gas phase of aroma compounds present in food and is of major importance for acceptability of foods by consumers. However, food products are very complex matrices because of their chemical composition and structure. The nature and the amount of the different components such as proteins, lipids, and carbohydrates greatly influence aroma release and perception (Druaux and Voilley 1997; Guichard 2002). Understanding the behavior of aroma compounds in the food matrix and of the strength and nature of interactions between aroma compounds and ingredients of food matrix is therefore of major significance for improving the overall aroma and thus the quality of food products. The objective of this chapter is to give several physicochemical characteristics of aroma compounds and to discuss some of the main factors that influence aroma binding and release in foods in order to better explain flavor perception.

7.2 THERMODYNAMIC AND KINETIC ASPECTS OF BINDING AND RELEASE OF FLAVOR COMPOUNDS

There are two major factors that control the rate of aroma release from products: volatility of aroma compounds in the product (thermodynamic factor) and resistance to mass transfer from product to air (kinetic factor) (Voilley and Souchon 2006). Examples of flavor compounds are presented in Table 7.1 along with their main physicochemical parameters (hydrophobicity, solubility in water, saturated vapor pressure, etc.).

The volatility of a flavor compound is most conveniently expressed as the proportion of its concentrations in the air and product phase under equilibrium conditions:

$$K = \frac{C_G}{C_M} \tag{7.1}$$

where C_G is the concentration of volatile compound in the headspace and C_M is the concentration of volatile compound in the matrix. The partition coefficient can be expressed using the activity coefficient (γ_i), mass fraction (k_m), molar fraction (K_i), or molar concentration (k_i).

Therefore, to compare the values obtained in different units, some conversions are necessary, using either standard constants (R, T, P, M_a, M_{liq} and d_{liq}) or the saturation pressure of the pure compound at the corresponding temperature (p_i^s).

Conversion of partition coefficient k_i expressed in molar concentration, or in mass concentration to K_i expressed in molar fraction

$$K_i = k_i \times \frac{RTd_{liq}}{PM_{liq}} \times 10^6 \tag{7.2}$$

TABLE 7.1
Main Physicochemical Characteristics of Aroma Compounds

	Diacetyl	Ethyl Acetate	Ethyl Butanoate	Ethyl Hexanoate	2-Nonanone
Molecular mass	86	88	116	144	142
Log P[a]	−1.34	0.73	1.9	2.83	3.14
Solubility in water[b]	1000	29.93	2.7	0.31	0.17
Saturated vapor pressure p_i^s 25°C (mm Hg)	56.8[b]	101[c]	15.6[c]	1.8[b]	0.62[b]
$T_{eb}(K)$[d]	361	350	394	440	468
ΔH^0_{vap} (kJ mol^{-1})[d]	38.7	35.0	42	52.0	56.4

[a] Hydrophobicity (log P) calculated using the method described by Rekker (1977).
[b] Estimation program EPI Suite™.
[c] Calculated with the Antoine's equation parameters.
[d] Available values on National Institute of Standard and Techonology (NIST) Chemistry WebBook (http://webbook.nist.gov/chemistry/).

Conversion of partition coefficient k_m expressed in mass fraction to K_i expressed in molar fraction

$$K_i = k_m \times \frac{M_a}{M_{liq}} \tag{7.3}$$

Conversion of partition coefficient k_m expressed in mass fraction to K_i expressed in molar fraction

$$\gamma_t = K_i \times \frac{P}{p_i^s} \tag{7.4}$$

Conversion of partition coefficient k_i expressed in molar concentration to γ_t activity coefficient

$$\gamma_t = k_i \times \frac{RTd_{liq}}{p_i^s M_{liq}} \times 10^6 \tag{7.5}$$

where:

P = pressure (Pa)
T = temperature (K)
R = gas constant (8.314 J mol^{-1} K^{-1})
d_{liq} = liquid density (kg L^{-1})
M_{liq} = molar mass of liquid (g mol^{-1})
M_a = molar mass of air (28.8 g mol^{-1})
p_i^s = saturated vapor pressure of compound i (Pa).

Partition coefficients between the gas phase and the food matrix at the thermodynamic equilibrium provide quantitative information on the retention of aroma compounds by the food matrix. They are the function of the composition of the product and the temperature, but are not affected by the texture and structure of the product except in case of crystallization. Data in the literature are obtained by different methods, in different conditions, and expressed in different units and they have to be converted into the same unit in order to be compared as shown in Table 7.2 for ethyl hexanoate.

A compilation of gas/water partition coefficients obtained for a same aroma compound at different temperatures allowed the calculation of standard thermodynamic equilibrium values (Kopjar et al. 2010). A structure–property relationships study was thus performed using molecular descriptors on aroma compounds to better understand their vaporization behavior. Quantitative Structure–Activity Relationship (QSAR) methods attempt to find relationships between the properties of molecules and an experimental response; the assumption is that changes in molecular properties elicit different responses (Selassie 2003; Dudek et al. 2006). In addition to the role of polarity for vapor–liquid equilibrium of compounds in aqueous solution, the

TABLE 7.2
Conversion of Gas/Water Partition Coefficient for Ethyl Hexanoate at 25°C

Activity Coefficient	Mass Fraction	Molar Fraction	Molar Concentration
1.62×10^4[a]	*2.11×10^1*	*3.85×10^1*	*2.83×10^{-2}*
1.26×10^4	**1.80×10^1[b]**	*2.88×10^1*	*2.12×10^{-3}*
1.54×10^4	*2.23×10^1*	**3.58×10^1[a]**	*2.63×10^{-2}*
1.94×10^4	*2.87×10^1*	*4.60×10^1*	**3.40×10^{-2}[c]**

Note: **Bold**—literature data; *italic*—converted data.
[a] Le Thanh et al. 1993.
[b] Covarrubias-Cervantes et al. 2004.
[c] Landy et al. 1996.

structure–property study points out the role of chain length and branching (Tromelin et al. 2010a). Application of this approach to apparent thermodynamic liquid–vapor equilibrium study helps improve our understanding of the behavior of aroma compounds in aqueous media, which is of fundamental interest. Indeed, these equilibria are crucial for aroma compounds perception, insofar that aroma compounds are submitted to successive liquid–vapor changes in their transport from the food matrix to the olfactory receptors, through intermediate media such as saliva in the mouth and olfactory mucus in the olfactory epithelium. However, aroma release and perception are time-dependent phenomena, and kinetic parameters give complementary information to help us better understand the behavior of volatiles in food matrices.

Diffusion is caused by random molecular motion that leads to complete mixing. This is a spontaneous process by which matter is transported from one part of a system to another by random molecular movements. These molecules in solution move with rotational and translational movements. Two main factors could impact on the diffusion process: (1) obstructions or entrapment effects due to macromolecule nature and structural organization and (2) the strength and nature of specific interactions (chemical or nonchemical such as hydrogen bonding) between small solutes, including water molecules and ions, and large food molecules (Tavel et al. 2008b). Several methods are available to determine diffusion coefficients. They are characterized by the type of diffusion coefficient measured (diffusion according to a concentration gradient or self-diffusion), by the method used, by the scale at which diffusion is measured (macroscopic or microscopic), and by the means of detection of the diffusing molecule (Cayot et al. 2008). Each one has its specificity and its application fields. The concentration profiles consist of bringing into contact two cylinders of solid or semisolid products, both of which contain a different initial concentration of the studied substance and can be used to estimate apparent diffusivity in direct contact situations (Voilley and Bettenfeld 1985; Rega et al. 2002). The rotating diffusion cell method is designed hydrodynamically so that stationary diffusion layers of known thickeners are created on each side of an oil layer. It enables the calculation of diffusion coefficients in different liquid phases and of the mass transfer coefficients of solutes (Rogacheva et al. 1999).

TABLE 7.3
Diffusion Coefficients (Molar Concentration) of Aroma Compounds in Different Food Matrices at 25°C, Obtained by DOSY-NMR[a,b,c] or by Concentration Profile[d]

	Ethyl Acetate	Ethyl Butanoate	Ethyl Hexanoate
D_2O	10.5×10^{-10}[c]	8.7×10^{-10}[a,c]	8.5×10^{-10}[b,c]
D_2O + sucrose (35%)	3.5×10^{-10}[c]	2.9×10^{-10}[c]	2×10^{-10}[c]
Water			7.9×10^{-10}[d]
Water + sucrose (60%)			1.2×10^{-10}[d]
Water + sucrose (60%) + pectin (0.1%)			0.47×10^{-10}[d]

[a] Gostan et al. 2004.
[b] Juteau-Vigier et al. 2007.
[c] Savary et al. 2006b.
[d] Rega et al. 2002.

The nuclear magnetic resonance (NMR) spectroscopic technique using the pulsed field gradient (PFG) spin–echo method is a nondestructive and noninvasive way to measure the self-diffusion coefficient of small molecules by detecting the mobility of their protons (Stilbs 1987). This method is therefore perfectly suitable for high water content media, such as water or aroma compounds in gels (Rondeau-Mouro et al. 2004). Likewise, the PFG-DOSY technique (for Diffusion Ordered SpectroscopY) also gives self-diffusion coefficients through the identification of protons assigned to a given molecule, as for aroma compounds in a carrageenan matrix (Gostan et al. 2004) or sucrose-based solutions (Savary et al. 2006a). Results obtained in D_2O for ethyl hexanoate are comparable with those obtained in water by the concentration profile method (Table 7.3). A decrease in diffusion coefficient is observed by increasing the size of the aroma compound, from ethyl acetate to ethyl hexanoate. Moreover, transfer of small molecules can occur between two heterogeneous phases of the product. The mass transfer coefficient corresponds to the resistance to the transfer between the product and the vapor phase. The calculation of the mass transfer h_D needs the determination of diffusion coefficient D: h_D is a function of D. Depending on experimental conditions, mass transfer coefficients vary between 10^{-4} and 10^{-9} m s^{-1} (Voilley and Souchon 2006).

The determination of relevant thermodynamic and kinetic parameters such as air/product partition coefficient and diffusion or mass-transfer coefficients is a way to characterize the impact of product composition and structure on aroma mobility in the food matrix and release in the vapor phase.

7.3 COMPOSITION OF FOOD MATRIX

The composition and complexity of the food matrix influence the gas/matrix partition coefficients of aroma compounds in function of their hydrophobicity and, to a lower extent, enthalpy of vaporization (Kopjar et al. 2010). A small but noticeable decrease in enthalpy of affinity is observed for ethyl butyrate and ethyl hexanoate between water and the food matrices, suggesting that the energy needed for the volatilization is lower in matrices than in water. In function of the composition of the food matrix, different types of interactions occur between aroma compounds and macromolecules, which induce either retention or a salting out effect, depending on the nature of the aroma compound and the macromolecule (Guichard 2002). However, the effect of fat on the retention of hydrophobic aroma compounds is higher than that of proteins (Seuvre et al. 2000) or that of carbohydrates (Burseg et al. 2009).

7.3.1 Influence of Lipids

The presence of lipids influences the partitioning of aroma compounds between the food matrix and gas phase in function of their respective affinity for each phase (Table 7.4). For a polar compound such as diacetyl (log $P = -1.34$), few differences are observed between air/water and air/oil partition coefficients, whereas for hydrophobic compounds such as esters, a lower partition coefficient is obtained in oil and oil/water emulsions than in water; this decrease in aroma release is more pronounced for the more hydrophobic compounds (Table 7.4). The volatility of aroma compounds in emulsions also depends on the amount of fat, but also on the type of fat (Relkin et al. 2004). In emulsions, the influence of the droplet size is not significant or rather small (Landy et al. 1996; Carey et al. 2002) and seems to be different in function of

TABLE 7.4
Partition Coefficients (Molar Concentration) of Aroma Compounds in Different Food Matrices

	Diacetyl (25°C)	Diacetyl (37°C)	Ethyl Butanoate (37°C)	Ethyl Hexanoate (25°C)	2-Nonanone (25°C)
Olive oil/water	0.412[a]			380[a]	644[a]
Sun flower oil/water	0.418[a]			406[a]	687[a]
Air/water	4.5×10^{-4}[b]	1.8×10^{-3}[c]	18.1×10^{-3}[c]	3.4×10^{-2}[d]	1.5×10^{-2}[e]
Air/oil	6.3×10^{-4}[b]	2.7×10^{-3}[c]	0.58×10^{-3}[c]	3×10^{-5}[d]	
Air/emulsion (O50/W50)	6.4×10^{-4}[b]				
Air/emulsion (O40/W60)		1.9×10^{-3}[c]	1.4×10^{-3}[c]		
Air/emulsion (O65/W35)		3.9×10^{-3}[c]	0.9×10^{-3}[c]		

[a] Guichard 2002.
[b] Salvador et al. 1994.
[c] Van Ruth et al. 2002.
[d] Landy et al. 1996.
[e] Buttery et al. 1969.

both the aroma compounds and the type of emulsifier (Miettinen et al. 2002; Van Ruth et al. 2002). Even if a good correlation is often found between volatility of aroma compounds in emulsions and their log *P* value (Piraprez et al. 1998; Roberts et al. 2003), a better correlation was obtained by addition of a quadratic term $(\log P)^2$ in the equation (Carey et al. 2002).

In inhomogeneous systems, flavor retention is at least partly determined by the microenvironment of the flavor. For example, in sugar syrup, a lower rate of volatilization is obtained for aroma compounds dissolved in propylene glycol rather than in vegetable oil. This cannot be explained by the volatility level, which is much higher in propylene glycol than in vegetable oil, but by the presence of a biphasic system with oil, which induces a volatilization from the oil phase rather than from the total product phase, as is the case with propylene glycol due to a better mixing with sugar syrup (de Roos 2003).

7.3.2 Influence of Proteins

The influence of different types of proteins on aroma compounds has been widely studied, pointing out the existence of molecular interactions, ionic bonding, hydrogen bonding, and hydrophobic bonding (Landy et al. 1995; Guichard 2006; Tan and Siebert 2008). In the case of dairy products, milk proteins (β-lactoglobulin or caseins) are known to interact with aroma compounds by hydrophobic interactions (Landy et al. 1995; Hansen 1997; Lubbers et al. 1998; Andriot et al. 2000; Tromelin et al. 2006). The strength of the interactions depends on the physicochemical properties of the aroma compound. For instance, the strength of hydrophobic interactions with β-lactoglobulin increases with the hydrophobicity of the compounds (O'Neill and Kinsella 1987; Sostmann et al. 1997). In the case of esters in model systems with protein, long chain compounds are more retained than short chain esters (Jouenne and Crouzet 2000). Investigations of binding sites using Fourier transform infrared and two-dimensional nuclear magnetic resonance (2D NMR) spectroscopies suggest the existence of at least two binding behaviors between β-lactoglobulin in its monomeric form and aroma compounds in function of the chemical class, the hydrophobicity, or the shape of ligands (Lübke et al. 2002). The binding within the central cavity of aroma compounds that have an elongated structure involves the entrance of the calyx and Trp19. The binding onto the protein surface of aroma compounds that have or adopt a compact structure occurs in a site located between strand β G, α helix, and strand β I of β-lactoglobulin (Tavel et al. 2008a). The existence of at least two binding sites was also suggested using a 3D QSAR study with CATALYST software on binding constants between aroma compounds and β-lactoglobulin: one binding site for ketones, esters, lactones, and alcohols, and the other one for terpenes, some cyclic compounds, and aromatic compounds (Tromelin and Guichard 2003). The use of CATALYST for modeling interaction also underlined the fact that hydrophobicity was not the only important feature, but also that the topology of hydrocarbon chain and hydrogen bonding should be essential in the binding involved between aroma and protein.

However, in real foods, proteins are not always in their native state, for example, the application of a thermal treatment to the system induces denaturation or aggregation of the protein, which changes the nature of the interactions (O'Neill 1996; Famelart et al. 2004). Interactions between two aroma compounds (β-Ionone and guaiacol) and β-lactoglobulin in a molten globule state were studied by 2D NMR spectroscopy. The less tightly packed structure of the molten globule favored ligand binding, in particular within the central cavity. The greater flexibility of the calyx entrance, and the conformational change of loop EF induced an easier access of the central cavity after thermal treatment (Tavel et al. 2010).

7.3.3 Influence of Carbohydrates

The impact of sugar on aroma release has recently been reviewed (Delarue and Giampaoli 2006). The influence on the addition of sugars on aroma release depends on the aroma compound and the tasting agent studied and on their concentration (Nahon et al. 2000). Kieckbusch and King (1979) have, for instance, observed a higher volatility of alkyl acetates (C1 to C5) in the presence of saccharose. This can be due to a salting out effect attributed to a decrease in free water because of disaccharide hydration (Voilley et al. 1977).

Carbohydrates are commonly used as thickening agents. However, they often induce a decrease in aroma perception, because of their effect on aroma retention in the matrix. Among the carbohydrates, starch has been the subject of numerous studies because of its wide use in food technology. Its interactions with volatile compounds have already been reviewed (Escher et al. 2000). Starch consists of linear amylose and branched amylopectin, which in the native state are packed in well-organized starch granules. As shown in binary model systems, amylose binds flavor compounds by formation of inclusion complexes (Rutschmann and Solms 1990a, 1990b, 1990c). The type of the complex between aroma and amylose can influence aroma release and thus aroma perception (Heinemann et al. 2005; Pozo-Bayon et al. 2008). However, aroma retention by starch is not always explained by complexation with amylose and thus is supposed to be also due to interaction with amylopectin (Langourieux and Crouzet 1994). Very few studies deal with the interactions between amylopectin and flavors. In other cases, aroma retention does not correlate with the amylose content of native starch (Cayot et al. 1998), which suggests that aroma starch interactions mainly result from adsorption involving hydrogen bonds and not from inclusion complexes (Boutboul et al. 2002).

Other types of carbohydrates can play an important role in aroma compounds retention (Lubbers 2006). Even if some interactions occur between pectines and aroma compounds, their main effect seems to be an increase in aroma retention due to a lower diffusion in the gel network (Godshall 1997; Hansson et al. 2001). The addition of carrageenan seems to induce only few effects on aroma release. Under equilibrium, no overall effect of lambda-carrageenan was found, except with the most hydrophobic compounds. Analysis of flavor release under nonequilibrium conditions revealed a suppressing effect of lambda-carrageenan on the release rates of aroma compounds, which was dependent on the physicochemical characteristics of the aroma compounds, with the largest effect for the most volatile compounds. This effect was therefore attributable to the thickener and not to the physical properties of the increasingly viscous systems (Bylaite et al. 2004). In model systems with iota-carrageenan, a small increase in the retention of esters was only observed under gelling conditions (Juteau et al. 2004). In fact, the presence of carrageenan polymers only modulates, but does not change, the interaction of aroma compounds with water molecules in a saline solution (NaCl 0.34%). This was shown by using a Quantitative Structure–Property Relationships (QSPR) approach to evaluate the influence of the chemical structure of aqueous matrices over the partition coefficient of 12 aroma compounds between the gas phase and the matrix (saline solution and iota-carrageenan gel) (Chana et al. 2006). A similar study using the QSPR approach on partition coefficients of aroma compounds in water and pectin gels suggests that interactions with pectin involve positive charged surface area in the retention phenomenon (Tromelin et al. 2010b). A QSPR study on β-glucan matrices also emphasizes the importance of multivariate approaches to establish the connections between the release phenomenon and molecular descriptors (Christensen et al. 2009). Polysaccharides are often used in mixture, and, for example, xanthan is often associated with guar or caroube. The release of limonene from xanthane/guar and xanthane/carob water-based mixed solutions has been investigated in the dilute and semidilute polymer concentration regimes (Secouard et al. 2003). When considering the limonene release as a function of the concentration regime, a dramatic decrease in limonene release was observed for the xanthane/carob mixtures in the neighborhood and above the critical overlap concentration C^*. This result, combined with kinetic experiments, allowed researchers to state the occurrence of specific interactions between limonene and polymeric junction zones. A different mechanism for synergistic interactions between xanthane and guar was observed. The synergy appears quite lower than xanthane/carob mixtures and can be explained by phase separation occurring in the xanthane/guar mixture, favored by ordered conformation of xanthane.

In model dairy gels differing in their content in texturing agents (starch, pectin, and locust bean gum), aroma retention was strongly dependent on the nature of polyoside (Decourcelle et al. 2004). An increase in starch concentration induced an overall decrease in aroma release, whereas only pectin concentrations higher than 0.04% caused an increase in aroma release. No significant effect was observed with locust bean gum. The differences observed between the texturing agents were higher for esters than for alcohols (Lubbers et al. 2007). However, in most studies, it seems difficult to understand the role of the interactions because of the composition and the physical modification of the structure (Cayot et al. 1998).

7.4 PHYSICAL STATE AND STRUCTURE OF FOOD MATRIX

The physicochemical interactions between proteins or texturing agents and aroma compounds are not the only factor responsible for a modification of aroma perception (Guichard 2002). Indeed, even if the macromolecules do not bind aroma compounds, their addition can affect the mobility and the release of aroma compounds by their effect on the structure of the matrix. The best impression of the effect of texture and microstructure on flavor retention is obtained by measuring the released amounts from (semi-)solid systems under equilibrium and dynamic conditions. Under further identical circumstances, the differences between the equilibrium and dynamic flavor release can be attributed to the effects of texture and microstructure (de Roos 2003). For example, aroma perception decreased rapidly when the guar gum concentration was higher than the critical concentration c^* (Baines and Morris 1987). This concentration, c^*, is the one from which the viscosity of the system abruptly increases and corresponds to the transition of a solution where macromolecules can move easily to a solution where molecules get tangled. The decrease in aroma perception in harder gels would be attributable to a reduced transport of the aroma compounds in the system and thus a lower rate of release in the mouth as shown in different types of gels (Baek et al. 1999; Boland et al. 2006).

Working on gelled systems such as yogurts, Saint-Eve et al. (2006) showed that for the same protein concentration, a decrease in the viscosity induced by the application of a mechanical treatment resulted in an increase in the intensity of aroma perception. A comparison of partition coefficients of esters between yogurt and water showed that the more hydrophobic esters were more retained in yogurt than the less hydrophobic ones (Table 7.5). Gierczynski et al. (2007) also showed an increase in the retention of ethyl hexanoate and nonan-2-one in gels (model of fresh cheeses) in comparison with the milk used for the preparation of these gels and a decrease in retention for ethyl butanoate. The modification of interactions between aroma compounds and proteins, due to the modification of protein structure during acidification, may be responsible for this interaction. However, the formation of a three-dimensional network resulting from proteins aggregation could also be responsible for a trapping of aroma compounds in the network. By comparing retention in yogurts, polyoside gels, and cheeses, it was observed that most of the esters are more retained in yogurts than in cheeses or polyoside gels (Table 7.5). Yogurts containing lipids and proteins induced a higher retention of aroma compounds than the other matrices. The observed effects strongly depend on the hydrophobicity of aroma compounds showing a retention for ethyl hexanoate and a salting out effect for ethyl acetate.

In the case of model dairy desserts with different types of texturing agents, very few effects of changes in texture were observed on aroma release, whereas substantial

TABLE 7.5
Percentages of Retention of Aroma Compounds in Different Food Matrices at 10°C and 30°C in Comparison with Water (Kopjar et al. 2010)

	Ethyl Acetate		Ethyl Butanoate		Ethyl Hexanoate	
Food matrix	10°C	30°C	10°C	30°C	10°C	30°C
Yogurt[a]	−35	−30	18	44	90	95
Water + sucrose (35%) + starch (1.4%) + carragenan (0.05%)[b]	−53	−50	−60	−14	−50	9
Model cheese[c]				−28		48

[a] Saint-Eve et al. 2006.
[b] Savary et al. 2006b.
[c] Gierczynski et al. 2007.

differences in aroma perception were noticed (Lethuaut et al. 2004). However, the effects of the same changes in texture on sweetness perception could be explained by a modification of sugar diffusion (Brossard et al. 2006). More recently, Koliandris et al. (2008) also observed only a small effect of the structure of model hydrocolloids gels on aroma release but a greater effect on the release of tastants.

In more complex systems such as biscuits, fat was shown to be a key component to explain aroma release, but also to modify the structure of the biscuit and thus play an important role in the rate of hydration and rate of aroma release during eating (Burseg et al. 2009).

It is thus difficult to draw general conclusions on the effect of the structure of the product because of the diversity of food models used and of the conditions by which the texture has been modified. Moreover, food composition and food structure are not the only parameters that affect aroma release in the mouth. For solid foods such as cheeses, the chewing process including masticatory behavior and saliva composition has to be taken into account to explain aroma release and then aroma perception (Tarrega et al. 2007). These aspects will not be detailed in the present review (for more information, see Salles et al. 2011).

7.5 MECHANISTIC MODELING OF FLAVOR RELEASE

To achieve a better understanding of the release mechanisms of aroma and taste compounds from well-characterized food products, there is a need to use a mechanistic approach. Modeling of phase partitioning and aroma transfer has been extensively studied and allows a pretty good description of experimental data (de Roos 2006). In this review, we focus mainly on the mechanistic models developed for in *vitro* and in *vivo* aroma release. Simple models were first developed to predict the behavior of volatile compounds in oil–water systems on the basis of oil–water air partitioning, and then validated using a homologous series of alcohols (Mc Nulty and Karel 1973) or used to predict odor thresholds (Buttery et al. 1973). Overbosch et al. (1991) presented a number of useful mathematical relationships that relate release to both partition and diffusion coefficients. These models were further used by some authors as a predictive model for partition coefficients in emulsions (Landy et al. 1996; Carey et al. 2002; Roberts et al. 2003). A rigorous physicochemical model was also developed (de Roos and Wolswinkel 1994), which takes into account the composition of food and resistance to mass transfer. A good agreement was noted between observed and calculated data on flavor retention, and it was shown that the food constituents having the most pronounced effect on the equilibrium headspace concentration of flavor compounds are water, lipids, and alcohol.

However, aroma release is a time-dependent phenomenon and partitioning is not a sufficient key for an overall understanding of the behavior of volatiles in food matrices. Marin et al. (1999) proposed and validated a mechanistic model that mimics the dilution of volatiles in the surrounding air and showed that the hydrodynamic regime in the gas phase could have a significant effect on the release of certain volatile compounds. By modeling theoretical physicochemical data of aroma compounds in miglyol/water model emulsions, Rabe et al. (2004a) found that the rate-limiting factor determining initial flavor release was the dynamic partitioning from the aqueous phase into the gas phase. The obtained results were experimentally confirmed by real-time measurements of dynamic flavor release.

Beyond the simple prediction of release kinetics, the focus of such approaches is the possible evaluation of some parameters that are difficult to modify or to determine experimentally. As an example, a mechanistic mathematical model was developed for the prediction of aroma release from dairy emulsions and for the calculation of apparent diffusion properties (Deleris et al. 2008, 2009). It was demonstrated that the most influent parameters of the apparent diffusivity in dairy fat emulsions are the diffusivity in the aqueous phase, the liquid fat fraction, and the partition between the fat and the aqueous phase. Since the final objective is a better understanding of aroma compound release from food products during consumption, other mechanistic models were developed for *in vivo* aroma release and *in vitro* aroma release simulating *in vivo* conditions. Modeling flavor release from foods in the mouth requires detailed knowledge of the mastication process. The more simple cases were those applied on liquid samples or chewing gums, which did not involve any mastication process. In the case of liquid samples, the release equations only require adjustment to account for the effect of dilution with saliva (de Roos and Wolswinkel 1994); the correction factors mainly depend on the fat content of the liquid. More recently, novel mathematical models for flavor release during drinking have been described, based on the physiology of breathing and swallowing (Normand et al. 2004). In the case of chewing gums, de Roos and Wolswinkel (1994) assumed that flavor compounds are released as a result of a series of subsequent extractions of the gum phase. A computer program was developed to calculate the flavor concentration to be added in a gum for a desirable flavor perception. However, this program did not take into account the individual variations.

The first attempts to develop a simulation of mastication was restricted to foods that fragment during chewing, and the simulation program modeled chewing and swallowing as periodic events with characteristic frequencies (Harrison et al. 1998), which could not make use of real mastication data for real subjects. Wright et al. (2003) and Wright and Hills (2003) proposed a probabilistic model to describe the masticatory cycles with the aim of predicting the generation of in-mouth exchange area in relation with aroma release. They assumed that transfer of flavor from the saliva into the headspace is very fast compared to the transfer from the bolus into the saliva. This model takes into account neither the effect of breathing and swallowing nor the adhesion phenomena. Another model proposed by Rabe et al. (2004b) using *in vitro* simulation only takes into account a few physiological parameters such as salivary flow rate and the presence of mucosa. Until now, models have not correctly incorporated the whole physiology of eating and focused more on the product than on the subject since chewing conditions were often imposed (Lian et al. 2004). In addition, all these approaches need to be confronted with experimental data. More recently, a model of flavor release during the eating process was established, initiated when the product is introduced into the mouth up to the end of signal acquisition, including the swallowing events (Trelea et al. 2008). The deglutition process was observed and segmented in different steps according to Buettner's description of swallowing (Buettner et al. 2001). In each step, the mechanistic model was based on mass balance in each compartment (mouth, pharynx, nasal cavity, product in mouth, or in pharynx) including ingoing and outgoing mass fluxes (Normand et al. 2004). The model also integrates the duration of each step of deglutition, the volume variation in the compartments through time, the saliva flow, the breathing flow, and the concentration of aroma compounds in each compartment. The model has been validated using *in vivo* aroma release data (Saint-Eve et al. 2006) and showed that volatile compound concentration profiles in the nasal cavity are highly dependent on breath rate, mouth volume, and the moment of the velopharyngeal closure between two swallows.

7.6 CONCLUSION

In the literature, much scientific research has been published on binding and release of flavor in foods. The physicochemical properties of aroma compounds must be quantified correctly, and their thermodynamic and kinetic parameters must be known as a function not only of the composition but also of the physical state and structure of the food matrix. Even if a comparison of the different results obtained in the literature is difficult, general tends of binding and release can be observed taking into account the molecular characteristics of aroma compounds by using QSPR approaches. The recent developments of mechanistic approaches including parameters from the food together with physiological parameters will help us to better understand the limited steps of mass transfer and to better control flavor release during consumption.

REFERENCES

Andriot, I., M. Harrison, N. Fournier, and E. Guichard. 2000. Interactions between methyl ketones and β-lactoglobulin: Sensory analysis, headspace analysis, and mathematical modeling. *J. Agric. Food Chem.* 48: 4246–4251.

Baek, I., R.S.T. Linforth, A. Blake, and A.J. Taylor. 1999. Sensory perception is related to the rate of change of volatile concentration in-nose during eating of model gels. *Chem. Sens.* 24: 155–160.

Baines, Z.V., and E.R. Morris. 1987. Flavour/taste perception in thickened systems: The effect of guar gum above and below c. *Food Hydrocolloids* 1: 197–205.

Boland, A.B., C.M. Delahunty, and S.M. van Ruth. 2006. Influence of the texture of gelatin gels and pectin gels on strawberry flavour release and perception. *Food Chem.* 96: 452–460.

Boutboul, A., P. Giampaoli, A. Feigenbaum, and V. Ducruet. 2002. Influence of the nature and treatment of starch on aroma retention. *Carbohydr. Polym.* 47: 73–82.

Brossard, C., L. Lethuaut, A.E.M. Boelrijk, F. Mariette, and C. Genot. 2006. Sweetness and aroma perceptions in model dairy desserts: An overview. *Flavour Fragrance J.* 21: 48–52.

Buettner, A., A. Beer, C. Hannig, and M. Settles. 2001. Observation of the swallowing process by application of videofluoroscopy and real-time magnetic resonance imaging—Consequences for retronasal aroma stimulation. *Chem. Sens.* 26: 1211–1219.

Burseg, K., R.S.T. Linforth, J. Hort, and A.J. Taylor. 2009. Flavor perception in biscuits: Correlating sensory properties with composition, aroma release, and texture. *Chemosens. Percept.* 2: 70–78.

Buttery, R., D.G. Guadagni, and L.C. Ling. 1973. Flavor compounds: Volatilitites in vegetable oil and oil–water mixtures. Estimation of odor thresholds. *J. Agric. Food Chem.* 21: 198–201.

Buttery, R.G., L.C. Ling, and D.G. Guadagni. 1969. Food volatiles. Volatilities of aldehydes, ketones, and esters in dilute water solution. *J. Agric. Food Chem.* 21: 385–389.

Bylaite, E., Z. Ilgunaite, A.S. Meyer, and J. Adler-Nissen. 2004. Influence of lambda-carrageenan on the release of systematic series of volatile flavor compounds from viscous food model systems. *J. Agric. Food Chem.* 52: 3542–3549.

Carey, M.E., T. Asquith, R.S.T. Linforth, and A.J. Taylor. 2002. Modeling the partition of volatile aroma compounds from a cloud emulsion. *J. Agric. Food Chem.* 50: 1985–1990.

Cayot, N., C. Dury-Brun, T. Karbowiak, G. Savary, and A. Voilley. 2008. Measurement of transport phenomena of volatile compounds: A review. *Food Res. Int.* 41: 349–362.

Cayot, N., C. Taisant, and A. Voilley. 1998. Release and perception of isoamyl acetate from a starch-based food matrix. *J. Agric. Food Chem.* 46: 3201–3206.

Chana, A., A. Tromelin, I. Andriot, and E. Guichard. 2006. Flavor release from iota-carrageenan matrix: A quantitative structure–property relationships approach. *J. Agric. Food Chem.* 54: 3679–3685.

Christensen, N.J., S.M.D. Leitao, M.A. Petersen, B.M. Jespersen, and S.B. Engelsen. 2009. A quantitative structure–property relationship study of the release of some esters and alcohols from barley and oat beta-glucan matrices. *J. Agric. Food Chem.* 57: 4924–4930.

Covarrubias-Cervantes, M., D. Champion, F. Debeaufort, and A. Voilley. 2004. Aroma volatility from aqueous sucrose solutions at low and subzero temperatures. *J. Agric. Food Chem.* 52: 7064–7069.

de Roos, K.B. 2003. Effect of texture and microstructure on flavour retention and release. *Int. Dairy J.* 13: 593–605.

de Roos, K.B. 2006. Modeling aroma interactions in food matrices. In *Flavour in Food*, ed. A. Voilley and P. Etievant, 229–259. Cambridge, England: Woodhead Publishing Limited and CRC Press LLC.

de Roos, K.B., and K. Wolswinkel. 1994. Non-equilibrium partition model for predicting flavour release in the mouth. In *Trends in Flavour Research*, ed. H. Maarse, and D.G. van der Heij, 15–32. Amsterdam, The Netherlands: Elsevier Science.

Decourcelle, N., S. Lubbers, N. Vallet, P. Rondeau, and E. Guichard. 2004. Effect of thickeners and sweeteners on the release of blended aroma compounds in fat-free stirred yoghurt during shear conditions. *Int. Dairy J.* 14: 783–789.

Delarue, J., and P. Giampaoli. 2006. Carbohydrate–flavour interactions. In *Flavour in Food*, ed. A. Voilley and P. Etievant, 208–228. Cambridge, UK: Woodhead Publishing Limited and CRC Press.

Deleris, I., S. Atlan, I. Souchon, M. Marin, and L.C. Trelea. 2008. An experimental device to determine the apparent diffusivities of aroma compounds. *J. Food Eng.* 85: 232–242.

Deleris, I., I. Zouid, I. Souchon, and I.C. Trelea. 2009. Calculation of apparent diffusion coefficients of aroma compounds in dairy emulsions based on fat content and physicochemical properties in each phase. *J. Food Eng.* 94: 205–214.

Druaux, C., and A. Voilley. 1997. Effect of food composition and microstructure on volatile flavour release. *Trends Food Sci. Technol.* 8: 364–368.

Dudek, A.Z., T. Arodz, and J. Galvez. 2006. Computational methods in developing quantitative structure–activity relationships (QSAR): A review. *Comb. Chem. High Throughput Screening* 9: 213–228.

Escher, F., J. Nuessli, and B. Conde-Petit. 2000. Interactions of flavor compounds with starch in food processing. In *Flavour Release*, ed. D.D. Roberts, and A.J. Taylor, 230–245. Washington, DC: American Chemical Society.

Famelart, M.H., J. Tomazewski, M. Piot, and S. Pezennec. 2004. Comprehensive study of acid gelation of heated milk with model protein systems. *Int. Dairy J.* 14: 313–321.

Gierczynski, I., H. Laboure, E. Semon, and E. Guichard. 2007. Impact of hardness of model cheese on aroma release: *In vivo* and *in-vitro* study. *J. Agric. Food Chem.* 55: 3066–3073.

Godshall, M.A. 1997. How carbohydrates influence food flavor. *Food Technol.* 51: 63–67.

Gostan, T., C. Moreau, A. Juteau, E. Guichard, and M.-A. Delsuc. 2004. Measurement of aroma compound self-diffusion in food models by DOSY. *Magn. Reson. Chem.* 42: 496–499.

Guichard, E. 2002. Interactions between flavor compounds and food ingredients and their influence on flavor perception. *Food Rev. Int.* 18: 49–70.

Guichard, E. 2006. Flavour retention and release from protein solutions. *Biotechnol. Adv.* 24: 226–229.

Hansen, A. 1997. A review of the interactions between milk proteins and dairy flavor compounds. In *Food Proteins and Lipids*, ed. Damodaran, 67–75. New York: Plenum Press.

Hansson, A., J. Andersson, and A. Leufven. 2001. The effect of sugars and pectin on flavour release from a soft drink-related model system. *Food Chem.* 72: 363–368.

Harrison, M., S. Campbell, and B. Hills. 1998. Computer simulation of flavor release from solid foods in the mouth. *J. Agric. Food Chem.* 46: 2736–2743.

Heinemann, C., M. Zinsli, A. Renggli, F. Escher, and B. Conde-Petit. 2005. Influence of amylose-flavor complexation on build-up and breakdown of starch structures in aqueous food model systems. *Lebensm.-Wiss. Technol.* 38: 885–894.

Jouenne, E., and J. Crouzet. 2000. Effect of pH on retention of aroma compounds by β-lactoglobulin. *J. Agric. Food Chem.* 48: 1273–1277.

Juteau-Vigier, A., S. Atlan, I. Deleris, E. Guichard, I. Souchon, and I.C. Trelea. 2007. Ethyl hexanoate transfer modeling in carrageenan matrices for determination of diffusion and partition properties. *J. Agric. Food Chem.* 55: 3577–3584.

Juteau, A., J.L. Doublier, and E. Guichard. 2004. Flavor release from iota-carrageenan matrices: A kinetic approach. *J. Agric. Food Chem.* 52: 1621–1629.

Kieckbusch, T.G., and C.J. King. 1979. Partition coefficients for acetates in food systems. *J. Agric. Food Chem.* 27: 504–507.

Koliandris, A., A. Lee, A.L. Ferry, S. Hill, and J. Mitchell. 2008. Relationship between structure of hydrocolloid gels and solutions and flavour release. *Food Hydrocolloids* 22: 623–630.

Kopjar, M., I. Andriot, A. Saint-Eve, I. Souchon, and E. Guichard. 2010. Retention of aroma compounds: An interlaboratory study on the effect of the composition of food matrices on thermodynamic parameters in comparison with water. *J. Sci. Food Agric.* 90: 1285–1292.

Landy, P., J.L. Courthaudon, C. Dubois, and A. Voilley. 1996. Effect of interface in model food emulsions on the volatility of aroma compounds. *J. Agric. Food Chem.* 44: 526–530.

Landy, P., C. Druaux, and A. Voilley. 1995. Retention of aroma compounds by proteins in aqueous solution. *Food Chem.* 54: 387–392.

Langourieux, S., and J. Crouzet. 1994. Study of aroma compounds–polysaccharides interactions by dynamic exponential dilution. *Lebensm.-Wiss. Technol.* 27: 544–549.

Le Thanh, M., T. Lamer, A. Voilley, and J. Jose. 1993. Détermination des coefficients de partage vapeur-liquide et d'activité de composés d'arôme à partir de leurs caractéristiques physico-chimiques. *J. Chem. Phys.* 90: 545–560.

Lethuaut, L., K.G.C. Weel, A.E.M. Boelrijk, and C.D. Brossard. 2004. Flavor perception and aroma release from model dairy desserts. *J. Agric. Food Chem.* 52: 3478–3485.

Lian, G.P., M.E. Malone, J.E. Homan, and I.T. Norton. 2004. A mathematical model of volatile release in mouth from the dispersion of gelled emulsion particles. *J. Controlled Release* 98: 139–155.

Lubbers, S. 2006. Texture–aroma interactions. In *Flavour in Food*, ed. A. Voilley, and P. Etievant, 327–344. Cambridge, CB1 6AH (GBR): Woodhead Publishing Limited and CRC Press LLC.

Lubbers, S., N. Decourcelle, D. Martinez, E. Guichard, and A. Tromelin. 2007. Effect of thickeners on aroma compound behavior in a model dairy gel. *J. Agric. Food Chem.* 55: 4835–4841.

Lubbers, S., P. Landy, and A. Voilley. 1998. Retention and release of aroma compounds. *Food Technol.* 52: 68, 70, 72, 74, 208, 210, 212, 214.

Lübke, M., E. Guichard, A. Tromelin, and J.-L. Le Quéré. 2002. Nuclear magnetic resonance spectroscopic study of beta-lactoglobulin interactions with two flavor compounds, gamma-decalactone and beta-ionone. *J. Agric. Food Chem.* 50: 7094–7099.

Marin, M., I. Baek, and A.J. Taylor. 1999. Volatile release from aqueous solutions under dynamic headspace dilution conditions. *J. Agric. Food Chem.* 47: 4750–4755.

Mc Nulty, P.B., and M. Karel. 1973. Factors affecting flavour release and uptake in o/w emulsion: III. Scale-up model and emulsion studies. *J. Food Technol.* 8: 415–427.

Miettinen, S.M., H. Tuorila, V. Piironen, K. Vehkalahti, and L. Hyvönen. 2002. Effect of emulsion characteristics on the release of aroma as detected by sensory evaluation, static headspace gas chromatography, and electronic nose. *J. Agric. Food Chem.* 50: 4232–4239.

Nahon, D.F., M. Harrison, and J.P. Roozen. 2000. Modeling flavor release from aqueous sucrose solutions, using mass transfer and partition coefficients. *J. Agric. Food Chem.* 48: 1278–1284.

Normand, V., S. Avison, and A. Parker. 2004. Modeling the kinetics of flavour release during drinking. *Chem. Sens.* 29: 235–245.

O'Neill, T.E. 1996. Flavor binding by food proteins: An overview. In *Flavor–Food Interactions*, ed. J.V. Leland, and R.J. McGorrin, 59–74. Washington, DC: American Chemical Society.

O'Neill, T.E., and J.E. Kinsella. 1987. Binding of alkanone flavors to beta-lactoglobulin: Effects of conformational and chemical modification. *J. Agric. Food Chem.* 35: 770–774.

Overbosch, P., W.G.M. Afterof, and P.G.M. Haring. 1991. Flavor release in the mouth. *Food Rev. Int.* 7: 137–184.

Piraprez, G., M.F. Herent, and S. Collin. 1998. Flavour retention by lipids measured in a fresh cheese matrix. *Food Chem.* 61: 119–125.

Pozo-Bayon, M.-A., B. Biais, V. Rampon, N. Cayot, and P. Le Bail. 2008. Influence of complexation between amylose and a flavored model sponge cake on the degree of aroma compound release. *J. Agric. Food Chem.* 56: 6640–6647.

Rabe, S., U. Krings, and R.G. Berger. 2004a. Dynamic flavour release from miglyol/water emulsions: Modelling and validation. *Food Chem.* 84: 117–125.

Rabe, S., R.S.T. Linforth, U. Krings, A.J. Taylor, and R.G. Berger. 2004b. Volatile release from liquids: A comparison of *in vivo* APCI-MS, in-mouth headspace trapping and *in vitro* mouth model data. *Chem. Sens.* 29: 163–173.

Rega, B., E. Guichard, and A. Voilley. 2002. Flavour release from pectin gels: Effects of texture, molecular interactions and aroma compounds diffusion. *Sci. Aliments* 22: 235–248.

Rekker, R.F., ed. 1977. *The Hydrophobic Fragmental Constant. Its Derivation and Application.* Amsterdam, The Netherlands: Elsevier Scientific Publishing Co.

Relkin, P., M. Fabre, and E. Guichard. 2004. Effect of fat nature and aroma compound hydrophobicity on flavor release from complex food emulsions. *J. Agric. Food Chem.* 52: 6257–6263.

Roberts, D.D., P. Pollien, and B. Watzke. 2003. Experimental and modeling studies showing the effect of lipid type and level on flavor release from milk-based liquid emulsions. *J. Agric. Food Chem.* 51: 189–195.

Rogacheva, S., M.A. Espinosa-Diaz, and A. Voilley. 1999. Transfer of aroma compounds in water–lipid systems: Binding tendency of β–lactoglobulin. *J. Agric. Food Chem.* 47: 259–263.

Rondeau-Mouro, C., A. Zykwinska, S. Durand, J.L. Doublier, and A. Buléon. 2004. NMR investigations of the 4-ethyl guaicol self-diffusion in iota (iota)-carrageenan gels. *Carbohydr. Polym.* 57: 459–468.

Rutschmann, M.A., and J. Solms. 1990a. Formation of inclusion complexes of starch with different organic compounds: II. Study of ligand binding in binary model systems with decanal, 1-naphthol, monostearate and monopalmitate. *Lebensm.-Wiss. Technol.* 23: 70–79.

Rutschmann, M.A., and J. Solms. 1990b. Formation of inclusion complexes of starch with different organic compounds: III. Study of ligand binding in binary model systems with (–)limonene. *Lebensm.-Wiss. Technol.* 23: 80–83.

Rutschmann, M.A., and J. Solms. 1990c. Formation of inclusion complexes of starch with different organic compounds: IV. Ligand binding and variability in helical conformations of v amylose complexes. *Lebensm.-Wiss. Technol.* 23: 84–87.

Saint-Eve, A., N. Martin, H. Guillemin, E. Sémon, E. Guichard, and I. Souchon. 2006. Flavored yogurt complex viscosity influences real-time aroma release in the mouth and sensory properties. *J. Agric. Food Chem.* 54: 7794–7803.

Salles, C., M.-C. Chagnon, G. Feron, E. Guichard, H. Laboure, M. Morzel, E. Semon, A. Tarrega, and C. Yven. 2011. In-mouth mechanisms leading to flavor release and perception. *Crit. Rev. Food Sci. Nutr.* 51: 67–90.

Salvador, D., J. Bakker, K.R. Langley, R. Potjewijd, A. Martin, and J.S. Elmore. 1994. Flavour release of diacetyl from water, sunflower oil and emulsions in model systems. *Food Qual. Preference* 5: 103–107.

Savary, G., E. Guichard, J.-L. Doublier, N. Cayot, and C. Moreau. 2006a. Influence of ingredients on the self-diffusion of aroma compounds in a model fruit preparation: A nuclear magnetic resonance-diffusion-ordered spectroscopy investigation. *J. Agric. Food Chem.* 54: 665–671.

Savary, G., E. Guichard, J.L. Doublier, and N. Cayot. 2006b. Mixture of aroma compounds: Determination of partition coefficients in complex semi-solid matrices. *Food Res. Int.* 39: 372–379.

Secouard, S., C. Malhiac, M. Grisel, and B. Decroix. 2003. Release of limonene from polysaccharide matrices: Viscosity and synergy effect. *Food Chem.* 82: 227–234.

Selassie, C.D. 2003. History of quantitative structure–activity relationships. In *Burger's Medicinal Chemistry and Drug Discovery*, Vol 1: Drug Discovery, ed. D.J. Abraham. New York: John Wiley & Sons, Inc.

Seuvre, A.M., M.A. Espinosa Diaz, and A. Voilley. 2000. Influence of the food matrix structure on the retention of aroma compounds. *J. Agric. Food Chem.* 48: 4296–4300.

Sostmann, K., B. Bernal, I. Andriot, and E. Guichard. 1997. Flavour binding by β-lactoglobulin: Different approaches. In *Flavor Perception. Aroma Evaluation*, ed. H.-P. Kruse, and M. Rothe, 425–434. Eisenach: Eigenverlag Universität Postdam.

Stilbs, P. 1987. Fourier transform pulsed-field gradient spin-echo studies of molecular diffusion. *Prog. NMR Spectrosc.* 19: 1–45.

Tan, Y., and K.J. Siebert. 2008. Modeling bovine serum albumin binding of flavor compounds (alcohols, aldehydes, esters, and ketones) as a function of molecular properties. *J. Food Sci.* 73: S56–S63.

Tarrega, A., C. Yven, E. Semon, and C. Salles. 2007. Aroma release and chewing activity during eating different model cheeses. *Int. Dairy J.* 18: 849–857.

Tavel, L., I. Andriot, C. Moreau, and E. Guichard. 2008a. Interactions between beta-lactoglobulin and aroma compounds: Different binding behaviors as a function of ligand structure. *J. Agric. Food Chem.* 56: 10208–10217.

Tavel, L., E. Guichard, and C. Moreau. 2008b. Contribution of NMR spectroscopy to flavour release and perception. *Ann. Rep. NMR Spectros.*, ed. G.A. Webb. 64: 173–188.

Tavel, L., C. Moreau, S. Bouhallab, E.C.Y. Li-Chan, and E. Guichard. 2010. Interactions between aroma compounds and beta-lactoglobulin in the heat-induced molten globule state. *Food Chem.* 119: 1550–1556.

Trelea, I.C., S. Atlan, I. Deleris, A. Saint-Eve, M. Marin, and I. Souchon. 2008. Mechanistic mathematical model for *in vivo* aroma release during eating of semiliquid foods. *Chem. Sens.* 33: 181–192.

Tromelin, A., I. Andriot, and E. Guichard. 2006. Protein–flavour interactions. In *Flavour in Food*, ed. A. Voilley and P. Etievant, 172–207. Cambridge, UK: Woodhead Publishing Limited and CRC Press LLC.

Tromelin, A., I. Andriot, M. Kopjar, and E. Guichard. 2010a. Thermodynamic and structure–property study of liquid–vapor equilibrium for aroma compounds pure and in aqueous solution. *J. Agric. Food Chem.* 58: 4372–4387.

Tromelin, A., and E. Guichard. 2003. Use of catalyst in a 3D-QSAR study of the interactions between flavor compounds and β-lactoglobulin. *J. Agric. Food Chem.* 51: 1977–1983.

Tromelin, A., Y. Merabtine, I. Andriot, S. Lubbers, and E. Guichard. 2010b. Retention-release equilibrium of aroma compounds in polysaccharide gels: Study by quantitative structure–activity/property relationships approach. *Flavour Fragrance J.* 25: 431–442.

Van Ruth, S.M., G. de Vries, M. Geary, and P. Giannouli. 2002. Influence of composition and structure of oil-in-water emulsions on retention of aroma compounds. *J. Sci. Food Agric.* 82: 1028–1035.

Voilley, A., and M. Bettenfeld. 1985. Diffusivities of volatiles in concentrated solutions. *J. Food Eng.* 4: 313–323.

Voilley, A., D. Simatos, and M. Loncin. 1977. Gas phase concentration of volatiles in equilibrium with a liquid aqueous phase. *Lebensm.-Wiss. Technol.* 10: 45–49.

Voilley, A., and I. Souchon. 2006. Flavour retention and release from the food matrix: An overview. In *Flavour in Food*, ed. A. Voilley and P. Etievant, 117–132. Cambridge, UK: Woodhead Publishing Limited and CRC Press LLC.

Wright, K.M., and B.P. Hills. 2003. Modelling flavour release from a chewed bolus in the mouth: Part II. The release kinetics. *Int. J. Food Sci. Technol.* 38: 361–368.

Wright, K.M., J. Sprunt, A.C. Smith, and B.P. Hills. 2003. Modelling flavour release from a chewed bolus in the mouth: Part I. Mastication. *Int. J. Food Sci. Technol.* 38: 351–360.

8 Flavor Suppression and Enhancement

Jakob Ley, Katharina Reichelt, and Gerhard Krammer

CONTENTS

8.1 INTRODUCTION

Because of the global mega trend in modern food applications to improve the healthiness of our diet, certain new taste and flavor challenges have been reported. Two main categories of processed food products show severe taste or flavor deficits: "food minus" and "food plus" products. "Food minus" products can be categorized as foods reduced in sugar, salt, fat, or monosodium glutamate (MSG), whereas "food plus" products include all foods enriched with healthy but unpleasant tasting ingredients such as high amounts of polyunsaturated fatty acids or oils, certain polyphenols from tea or grape skin or seed, and certain vitamins found in the B group, to name a few. In addition to these taste-related deficits, off-flavors can occur in food or processed foods. Frequently, the presence of off-flavors is a sign of food spoilage but they can, however, also be intrinsically present in healthier raw materials, which is unappreciated by consumers. For example, soy proteins often show beany, green, or cardboard flavor profiles, and do not fit with the intended profile for dairy products in western markets. Another example is the replacement of butterfat in ice creams with healthier plant-based oils. These oils often have a green, herbal character, which needs masking to gain consumer acceptance. Other highly problematic consumables are pharmaceutical actives that are often adverse in taste (e.g., bitter, astringent, metallic) and some oral care products that can contain large amounts of unpleasant tasting detergents or antimicrobials.

Some of these challenges are tackled using high potency sweeteners for sugar reduction; potassium chloride for sodium chloride replacement; thickening agents based on proteins or polysaccharides for fat replacement; encapsulation of fish oil or polyphenols; or by using specific and strong masking flavors for soy, vegetable oils, and pharmaceutical actives. It is generally acknowledged that these obvious solutions in creating "light" or "healthy" products do not always meet with consumer acceptance. Consequently, there is a demand for ingredients with flavor-modifying capabilities that have minor or weak intrinsic taste and aroma characteristics at the intended dosage level. Flavor modifiers should be able to enhance the positive while suppressing the negative hedonic experiences of the consumer, thereby resulting in "lighter" or "healthier" foods no longer being distinguishable from the original products.

8.2 DEFINITION OF FLAVOR MODIFIERS

The difference between a pure flavor and a flavor-modifying compound requires in-depth understanding of the performance of flavoring substances. Classically, a flavor compound is defined as being a chemical substance that imparts a certain taste or flavor via direct contribution to a flavor profile of a mixture. Consequently, a flavor modifier is able to influence the intensity or change the quality of the flavor or taste profiles of another flavor or taste compound and acts therefore as an indirect contributor. Flavor modifiers can show an intrinsic flavor or taste profile that is often different from the influenced direction. Since ancient times, glutamates, applied in the form of fish or soy sauce, sodium chloride, and sweet carbohydrates, have been described as "flavor modifiers" because of their inherent strong taste, but considering the definition of modifiers they are in fact primary tastants.

For a better understanding of flavor modification, one needs to understand flavor composition. A flavor can consist of up to three main parts: aroma compounds, taste compounds, and trigeminally active compounds (also called chemosensates or somatosensorial actives). Perception of aroma compounds occurs ortho- or retronasally via the olfactory system, taste compounds elicit effects on the tongue and soft palate, and trigeminally active compounds cause the activation of free nerve endings in the nose and mouth cavity mainly responsible for pain detection. Consider the fact that numerous flavor compounds show not only one type of effect but can impart aroma, taste, and/or trigeminal effects in combination. Correlated in the brain with additional nonchemosensory signals such as visual (Gottfried and Dolan 2003) and acoustical stimuli, these compounds combined with experience and memory result in a flavor recognition pattern growing into the human awareness. At present, there is an incomplete understanding of these processes and effects. Research to elucidate the processing pathways and mechanisms of the brain is being conducted using brain imaging technologies such as functional magnetic resonance imaging (fMRI) or positron emitting tomography (Small et al. 2004).

Whereas whole flavor modification, in general, is very complex because of the involvement of the olfactory, trigeminal, and taste systems, singular effects to taste modification are easier to detect. Until the end of the 20th century, only rare cases of real taste modification were described. There are some nucleotides that can substantially increase the umami taste of glutamic acid (Zhang et al. 2008). These include also the protein miraculin, which changes sour to sweet taste (Gibbs et al. 1996), lactisole (Schiffmann et al. 1999) or gymnemic acid (Suttisri et al. 1995), which suppress sweet taste, and herba santa fluid extract (*Sirupus eriodictyonis*), which has been described as being able to reduce quinine bitter taste (Lewin 1894). There are not only modifications in the single flavor subgroups, but all these different parts of flavor can influence each other. Therefore, in this chapter, the following flavor modification types are discussed:

- Aroma/aroma interactions
- Trigeminal/aroma interactions
- Aroma/taste interactions
- Taste/taste interactions
- Taste/trigeminal interactions
- Flavorless flavor modifiers

It is now accepted on codex level that flavoring substances are also characterized by modulation or modifying effects. In this context, the dose–effect correlation is a parameter of major importance, which can be monitored by detailed concentration-dependent studies obtained from sensory panels. At the same time, molecules showing only sweet, sour, or salty taste do not fall under the definition of flavorings according to the Codex Alimentarius (WHO 2008).

8.3 METHODS OF DETECTING FLAVOR MODIFICATION

Sensory methodology plays a vital role in the detection of flavor modification, and using a sensory panel is the most reliable method to determine flavor modification. However, additional methodologies exist such as "artificial nose," "artificial tongue," and cell biological and animal models to elucidate flavor modification effects or combinations of such methods.

8.3.1 Sensory Methods

Broad varieties of sensory methods using human assessors, as particularly sensitive testing systems or biosensors, have been established. These include: the degree of difference test, the aroma extract dilution analysis (AEDA) (Steinhaus and Schieberle 2007), the comparative taste dilution analysis (cTDA) (Ottinger et al. 2003), the dose over threshold determination (Stark and Hofmann 2005), and the half-tongue test

(McMahon et al. 2001). For the detection of taste-modifying compounds, a duo test, also called paired comparison test, is probably the easiest way to gain insight into a possible effect (Jellinek 1966). Ranking tests and duo–trio tests, also referred to as triangle tests, can be used to identify taste-modifying effects. These tests can also confirm the results of the previous duo tests, if necessary. Comparative taste dilution analysis (cTDA), as described by Ottinger et al., is a novel screening protocol for the identification of taste-modifying compounds from mixtures. Use of cTDA led to the identification of the Maillard derived taste enhancer alapyridaine (Ottinger et al. 2003). Computer-based time–intensity measurement is a tool to monitor the intensity of a certain taste attribute over a period of time (Rossetti et al. 2009) and allows for the observation of different factors such as flavor release or persistence in aftertaste of single compounds or complex products. Moreover, it is possible to obtain data about the influence on taste-modifying compounds, for example, sweet or bitter intensity. Another approach to find novel flavor-modifying compounds uses olfactometers or gustometers, such as Olfactoscan® by NIZO (Burseg and de Jong 2009). This method allows screening for aroma–aroma or aroma–taste interactions by presenting a continuous flow of flavor or taste pulses to a tester in combination with a stream of different aroma compounds. Another example of instrument supported sensory screening is the so-called LC Taste® system. This method combines separation of complex mixtures, such as plant extracts, by high-temperature liquid chromatography using water/ethanol gradients and direct sensory evaluation of the obtained fractions (Roloff et al. 2006, 2009). Blending selected fractions with a tastant such as sucrose or caffeine allows a fast screening for taste-modifying effects (Reichelt et al. 2010).

8.3.2 Cell Biological Methods

Screening protocols using cell cultures expressing the responsible receptors are available for the most important flavor targets. This methodology originated within the pharmaceutical industry which developed high throughput screening assays based on G-protein coupled receptors (GPCR) or ion channels to screen chemical libraries (Milligan 2006). In most cases, immortal cancerous cells such as human embryonic kidney (HEK) cells are used as base and the receptors as well as some of the transduction proteins (e.g., PLC-β2, α-gustducin, or other G-proteins) are stably expressed for screening purposes. Transient transfected cells, which are easier to produce, are also used but not preferred because they show problems in high throughput assays. For ion channels such as the ENaC responsible for salt taste transduction, frequently transfected oocytes of the *Xenopus* frog are used and the activation of the cell is measured by electrophysiological experiments using patch clamp technology (Stähler et al. 2008).

A typical setup for the detection of taste modifiers was described for sweet enhancers (Zoller 2009) using the sweet receptor couple T1R2 and T1R3 stably expressed in HEK cells. Slack et al. (2009a) described the setup for an assay to screen for antagonists of the hTAS2R44, hTAS2R43, hTAS2R46 GPCR responsible for the bitter taste of certain high intensity sweetener. Methods to search agonists or antagonists for trigeminal effects were also reported, for example, TRPM8, the receptor responsible for cooling effects (Stucchi 2008). Olfactory receptors (ORs) are used as targets for the screening of potential agonists (Shirokova et al. 2005) and agonist–antagonist couples such as bourgeonal and undecanal (Spehr et al. 2003). Because of the large number of ORs and the role of additional proteins and mechanisms on a higher level of olfactory processing, it seems difficult to receive a clear correlation between these results and those of sensory experiments (Reisert and Restrepo 2009).

8.3.3 Biological Routes to Artificial Noses and Tongues

Nowadays, electronic noses are widely available for the detection of single volatiles or classes of volatiles and for the detection of more complex aroma mixtures (Röck et al. 2008), but will not reviewed in the context of this chapter. Chapter 20 describes the specificity of electronic noses flavor analysis. Most of the systems are not used to describe aroma or flavor modification effects because of the high complexity of interpretation. Systems dealing with complex aroma mixtures generate a large and complex set data points that have to be compressed using neuronal networks including a pattern recognition process to yield a simpler description resembling those used by human individuals. For known descriptor–aroma composition relationships, there is a high probability for the artificial nose to recognize the pattern, but for unknown variations, it is difficult to derive predictions of the correlation between sensory and data structure. Even in cases when the systems can detect and "describe" a new pattern, conclusions with regard to preference are not possible because of the lack of hedonic evaluation data. A general problem of most of these systems is the use of artificial, nonbiological "receptors" that do not exhibit the original binding pattern of the aroma compounds. Trials to use new hybrid systems using olfactory receptors combined with electrochemical sensors were recently described (Benilova et al. 2008; Casuso et al. 2008; Yoon et al. 2009), but for real electronic nose systems, the whole set of all approximately 340 human olfactory receptors will have to be used.

For the detection of bitter taste and its modification, sensor systems based on the measurement of electrochemical potential in solutions with special polymer membranes were described (Miyanag et al. 2003). The authors studied the modification of some bitter compounds by the use of phospholipids and lipoprotein mixtures. Because of the polymeric character of the modifiers, it was possible to use the general membrane adsorption effects for characterization of masking effects. More specific membranes using enzymes such as tyrosinase as coatings were described for the detection of polyphenols in olive oil and correlations to the bitter and pungent taste could be found (Busch et al. 2006). Additionally, optical methods are used to measure precipitation effects of certain proteins (Edelmann and Lendl 2002), which is discussed as one possible cause for the astringency perception of tannins and other mouth-drying compounds. Recently, the first trials to combine whole taste receptor cells with electrodes to construct a more reliable artificial tongue were described (Chen et al. 2009). This method did not determine the modification effects caused by the interaction of different tastants or modifiers, but it might be the most interesting approach in the future. Unfortunately, artificial sensor systems to describe trigeminal sensations are not yet discussed in the literature.

In general, artificial nose and tongue systems show promising development for direct detection of flavor compounds and also for single point modification (i.e., enhancement or suppression of one quality), but the currently available systems are not able to reflect the much more complex interaction pattern between taste and aroma or even different taste and aroma qualities.

8.3.4 Animal Models

Animals are not able to express their flavor impressions. Consequently, indirect parameters need to be defined. Hedonic responses of rats or mice can be determined by two bottle preference tests (Riera et al. 2009) or the so-called licking frequency experiments (Kawai et al. 2008). Animals cannot override the hedonic response and stereotypically reject negative tasting food and prefer positive tasting food. As a result, a certain correlation between human and animal taste perceptions could deduced from these experiments. As a strong support of these hypotheses, genetically engineered mice expressing bitter receptors in sweet taste cells showed a positive hedonic response, that is, an increased licking frequency for bitter (and sometimes finally toxic) compounds (Chandrashekar et al. 2006).

Another, more invasive method is based on recording nerve signals especially the chorda tympani nerve. Studies using hamsters showed a clear suppression of the sweet response by acids with increasing titratable acidity (Formaker et al. 2009), which corresponds to the results from sensory experiments with humans.

8.4 AROMA/AROMA INTERACTIONS

In general, the interaction between different aroma compounds in a flavor is very complex and is not necessarily linear. Because of the transduction of olfactory signals from receptor neurons to the brain via various levels of complexity, an aroma impression is very difficult to predict from the single compounds. Similarly, the mixing of two odorants can generate new activation patterns in the olfactory cortex of mammals that cannot be detected for the single compounds, as found by fMRI experiments (Zou and Buck 2006). However, assumption of linear and independent responses is very helpful for identification of the most active flavor compounds as demonstrated by the widely used AEDA method (for an actual example, see Steinhaus and Schieberle 2007). The concepts of aroma or odor activity value and taste activity value are also commonly used, that is, the assumption that aroma and taste compounds present in an amount lower than their thresholds do not contribute to the overall aroma or taste. The latter concept is often used to recombine flavors starting from AEDA and TDA analyses, but in some cases, the resulting partial recombinants do not match the original flavor. As already noted, aroma prediction is not as simple as assumed and the combination of certain aroma compounds can yield somewhat surprising results: the nonfishy smelling aroma compounds (*Z*)-1,5-octadien-3-on (geranium-like, leaf-like, 0.1 ppb) and methional (cooked potato 10.0 ppb), mixed in their original concentration, occurring in freshly cooked cod indeed smell like fish (Belitz et al. 2001). The threshold itself can also be influenced by other aroma compounds: β-damascenone (baked plum) shows a threshold in water of 0.002 ppb, whereas it is 90 times higher when the threshold is measured in a solution also containing 6.75 ppm 4-hydroxy-2,5-dimethyl-3(2*H*)-furanon (caramellic, fruity), a

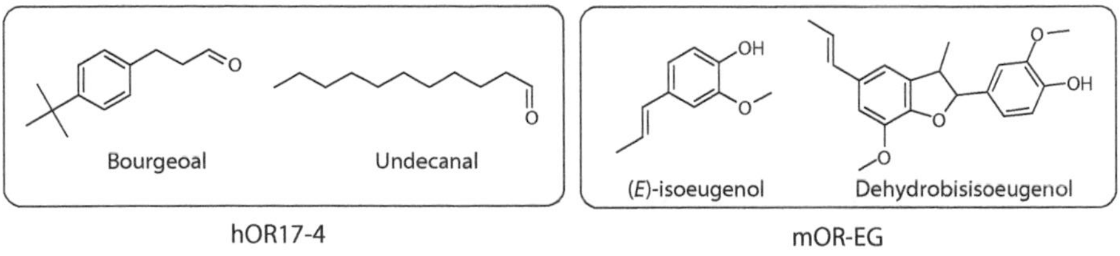

Bourgeoal Undecanal

hOR17-4

(*E*)-isoeugenol Dehydrobisisoeugenol

mOR-EG

FIGURE 8.1 Aroma compounds with modulating effects on receptor level.

concentration present in brewed coffee (Belitz et al. 2001). In sensory experiments with hop flavor compounds, a synergistic enhancement of the intensity of 3-sulfanyl-4-methylpentyl acetate (below thresholds) by 3-sulfanyl-4-methylpentan-1-ol (concentration two times of its threshold) could be detected (Takoi et al. 2009). A similar phenomenon was reported by Schlutt et al. (2007); they investigated the increase in perceived creaminess and creamy flavor by using various δ-lactones in subthreshold concentrations. The authors found that only δ-tetradecalactone was able to enhance the retronasal creamy flavor, whereas the higher homologues only influenced the melting behavior of the fat phase and therefore the flavor release. These flavor-modifying effects caused by subthreshold concentrations of further aroma compounds are under discussion as reviewed by Ryan et al. (2008).

Regarding olfactory receptor research, until now only a few agonist/antagonist findings are described: undecanal is able to block the response of the fragrance chemical bourgeonal (Spehr et al. 2003). This phenomenon can be also detected by sensory experiments (Brodin et al. 2009). Further examples are the inhibition of the isoeugenol activation by its dehydrodimer on the mouse olfactory receptor mOR-EG (Oka et al. 2004) and the reduction of citral response by octanal on mouse receptor mORI7 (Figure 8.1) (Reisert and Restrepo 2009). Therefore, in the future, it might be possible to design dedicated blockers for certain unpleasant smelling flavor compounds; however, it will remain a complex task because of the possible additional binding of one or more of the remaining olfactory receptors in the nose.

8.5 TRIGEMINAL/AROMA INTERACTIONS

There is little information available in literature regarding the influence of trigeminally active compounds on aroma perception and vice versa. The effects are used by flavorists on a more phenomenological level. Some aroma compounds also show strong intrinsic trigeminal characteristics. Typical examples are (–)-menthol, which shows a strong fresh aroma and at the same time a strong cooling effect, and 1,8-cineol, which exhibits a camphoreous smell and a sharp, pungent sensation. The threshold of the aroma versus the trigeminal effect is normally very different with (–)-menthol, for example, the cooling effect can be perceived only at higher concentrations (Ottinger et al. 2001). The activation of the trigeminal or pain sense is important for the overall description of the aroma compound; therefore, there is an intrinsic modification of the aroma value when the trigeminal effect can be reduced or vice versa. An investigation into color–coolant–aroma interactions showed that congruent mixtures of green–coolant–melon caused increases in the fruit flavor perception by the cooling compounds (Petit et al. 2007). In a further investigation regarding cooling, the use of mint aromas increased the cooling perception, whereas the use of a peach flavor showed no effect at all (Labbe et al. 2008).

8.6 AROMA/TASTE INTERACTIONS

It is well known that aroma/taste interactions exist: there are large differences between tasting a neutral and unsweetened aqueous solution containing an aroma compound with a sweetened or salted solution of the same compound. When comparing (2*E*)-nonenal in 5% sucrose and 0.5% salt solution, the sweet solution reminds most people of watermelons, whereas the salty solution is described as typically cucumber-like. The concept of congruent and incongruent aroma compounds is frequently discussed as already demonstrated for some aroma/trigeminal effects. For most people, vanillin is connected or "congruent" with sweet flavors and they will expect sweetness if they perceive the vanillin smell, which is indeed often described as "sweet." Panelists exposed to vanillin in a salty application usually report a reduction of the savory character. In several other cases, the increase in sweet taste response using sub- or near-threshold concentrations of volatiles is described. As an example, ethyl butyrate in a subthreshold concentration was able to enhance the sweet impression of sucrose, but surprisingly maltol, which is often described as having a sweet flavor character, did not (Labbe et al. 2007). In a sensory study investigating fruity aroma compositions—sweet taste interactions, the level of sweetness strongly influenced the fruity aroma, whereas this was not the inverse case (King et al. 2006).

Sweet aroma types were reported to reduce the bitterness of branched-chain amino acids such as L-leucine, L-isoleucine, and L-valine (Mukai et al. 2007). The taste inactive butyl phthalide found in celery increased the umami and sweet notes of chicken broth (Kurobayashi et al. 2008). The aroma of a dried bonito stock (Manabe et al. 2009) or the application of the sotolon occurring, for example, in soy sauce (Busch et al. 2009) was able to increase saltiness of sodium chloride in sensory tests. Most of these interactions are phenomenologically described but not fully understood.

Examples of aroma molecules that can directly influence a taste perception by binding to the appropriate receptors are anisole, which shows an intrinsic sweet taste and sweet-enhancing effect (Prakash and Dubois 2007), and 4-ethoxybenzaldehyde, which is able to substantially increase the sweetness of sucrose (Slack et al. 2009b). The latter has shown to be a true agonist or modifier of the sweet receptor.

8.7 TASTE/TASTE INTERACTIONS

In general, the basic tastes are known to influence each other. Examples are the reduction of sweet taste at lower pH, the suppression of bitterness by sweeteners, and the reduction of sourness using sweeteners. Besides the general overwhelming effect of high sweetness, some sweeteners are also active in much lower concentrations showing only limited sweetness. One example is neohesperidin dihydrochalcone, which is used as a bitter-reducing flavor compound (Cano et al. 2000). The use of sodium chloride to improve bitter-tasting food is well known in the past centuries and was thoroughly studied through sensory experiments (Keast et al. 2001). Acidic peptides are able to reduce the bitterness of certain molecules. It was shown on receptor level that addition of acidic peptides or organic acids, but not of neutral amino acids, can reduce the response of the bitter receptor hTAS2R16 to the bitter β-glucoside salicin and the hTAS2R38 to the bitter *N*-phenylthiourea (Sakurai et al. 2009).

Discrimination of the different taste qualities is possible but it is less difficult for younger than for elder people (Mojet et al. 2004). A thorough sensory study on eight different bitter tastants (denatonium benzoate, quinine, sucrose octaacetate, urea, L-tryptophan, L-phenylalanine, ranitidine, and tetralone) by Keast et al. (2003) showed only linear interactions between these compounds. Neither synergies nor antagonistic effects could be detected. The same was found during experiments with binary bitter-compound mixtures and additional amounts of sucrose or sodium salts. Some bitter compounds could be reduced in taste intensity by using sodium salts, whereas others such as tetralone were not reduced in taste and the effects of binary bitter mixtures with sodium salts showed a predictable linear reduction in bitterness (Keast et al. 2004). Salty and umami taste interaction may also be linear (Simons and Albin 2009): at constant NaCl level and increasing monosodium glutamate (MSG) concentration, the saltiness and umami intensity increased, but in contrast at constant MSG level and increasing NaCl content, only the saltiness increased. The saltiness enhancement, however, was probably only caused by the sodium ion, because by comparison of NaCl and NaCl plus MSG solutions with the same Na^+ concentration, the increase of saltiness by MSG was not detectable.

Without sound sensory evidence for the above-mentioned linear behavior, a general conclusion might be misleading. Although research regarding bitter compounds is limited, it remains possible that an antagonist of a certain bitter receptor can also activate another bitter receptor. McLaughlin et al. (2008) showed that hop polyphenols were able to significantly increase the bitterness and some negative attributes of iso-α-acids found in hops.

A more or less trivial phenomenon is the additive activity of at least two sweet molecules as shown for neohesperidin dihydrochalcone and sucrose (Kroeze 2000) or the sweet protein thaumatin in subthreshold concentrations (van der Loo and Wiener 1983). Because real synergistic effects (i.e., much more than additive) are difficult to define and to measure by sensory methods because of the nonlinear dose responses, most hints in the literature regarding "synergistic" sweet mixtures, need to be judged carefully. A good example showing clear synergistic effects is the mixture of lactose with D-tryptophane, whereas other interactions of lactose with amino acids or carbohydrates were only linear (Williams and Bernhard 1981).

A surprising effect is caused by the protein miraculin from the fruits *Richadella dulcifica* (Gibbs et al. 1996): after rinsing the tongue with a protein solution, aqueous sour solutions were described as sweet only. A similar effect was reported for the proteins neoculin (Shirasuka et al. 2004), curculin (Suzuki et al. 2004), and mabinlin (Kant et al. 2005). Because of the high price of the protein and the total suppression of sour taste for several minutes, the application remains of limited use.

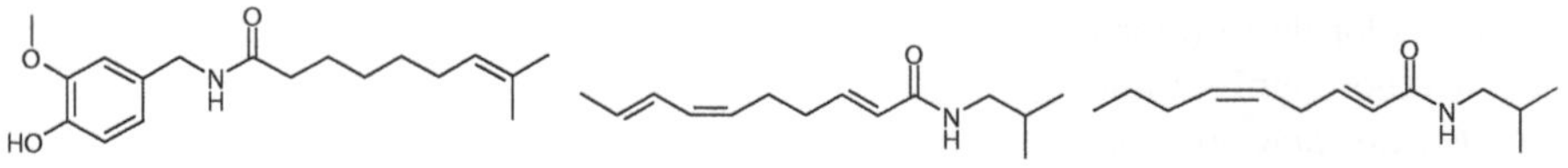

FIGURE 8.2 Trigeminal active taste modulators.

8.8 TASTE/TRIGEMINAL INTERACTIONS

The influence of temperature on taste is well known and an example can be seen in how higher temperatures strongly increase the sweet taste. The same TRP ion channels responsible for temperature sensation in the body trigger trigeminal effects. Indeed, it could be shown that not only temperature but also chemical cold and warm elicitors can influence perception of taste sensations in vivo and in vitro (Talavera et al. 2007). Capsaicin can reduce bitter response (Simons et al. 2003). This effect is not limited to bitterness; it affects sweet and umami taste as well (Simons et al. 2003). Some trigeminally active alkamides such as spilanthol (Miyazawa et al. 2006) and *N*-isobutyl (2*E*,6*Z*)-nonadienamides (Dewis et al. 2005) (see Figure 8.2) are also claimed to enhance saltiness.

8.9 FLAVORLESS FLAVOR MODIFIERS

Additionally, weak- or neutral-tasting molecules can influence the basic taste qualities, and complex taste patterns occur. Some lipids such as linoleic acid can influence sweet and bitter perception (Mattes 2007). In the past decade, the identification and description of new, weak-, or neutral-tasting taste modifiers has increased tremendously. Only the most relevant developments will be summarized in this chapter.

8.9.1 Bitter Taste

A thorough review regarding the modification of bitter taste using flavor modifiers was given by Ley in 2008. Therefore, only some highlights and updates will be presented here. Before 2000, only rare cases of bitter-masking effects of small, neutral-tasting molecules were described: gymnemic acid and herba santa extract were reported to reduce quinine bitterness (Lewin 1894) and neodiosmine to lower quinine and caffeine taste (Figure 8.3) (Guadagni et al. 1979). Zinc ions are able to reduce bitterness, but they show also some astringency in applications and reduce sweetness as well (Keast 2003). Lactisole (see next subchapter) was described as able to reduce the bitterness of potassium chloride (Johnson et al. 1994). γ-Amino butyric acid (<100 ppm) reduces caffeine and quinine bitterness (Ley et al. 2005a), and the bitter taste of peptide hydrolysates as well as for brucine and caffeine was eliminated by the dipeptide L-Glu-L-Glu (Belikov and Gololobov 1986). Some pyridinium betain derivatives based on amino acids, isolated from Maillard reaction mixtures, demonstrate bitter-masking effects (Soldo and Hofmann 2005). The flavanones homoeriodictyol, eriodictyol, and to some extent sterubin can reduce the bitterness of several bitter compounds such as caffeine, quinine, and amarogentine

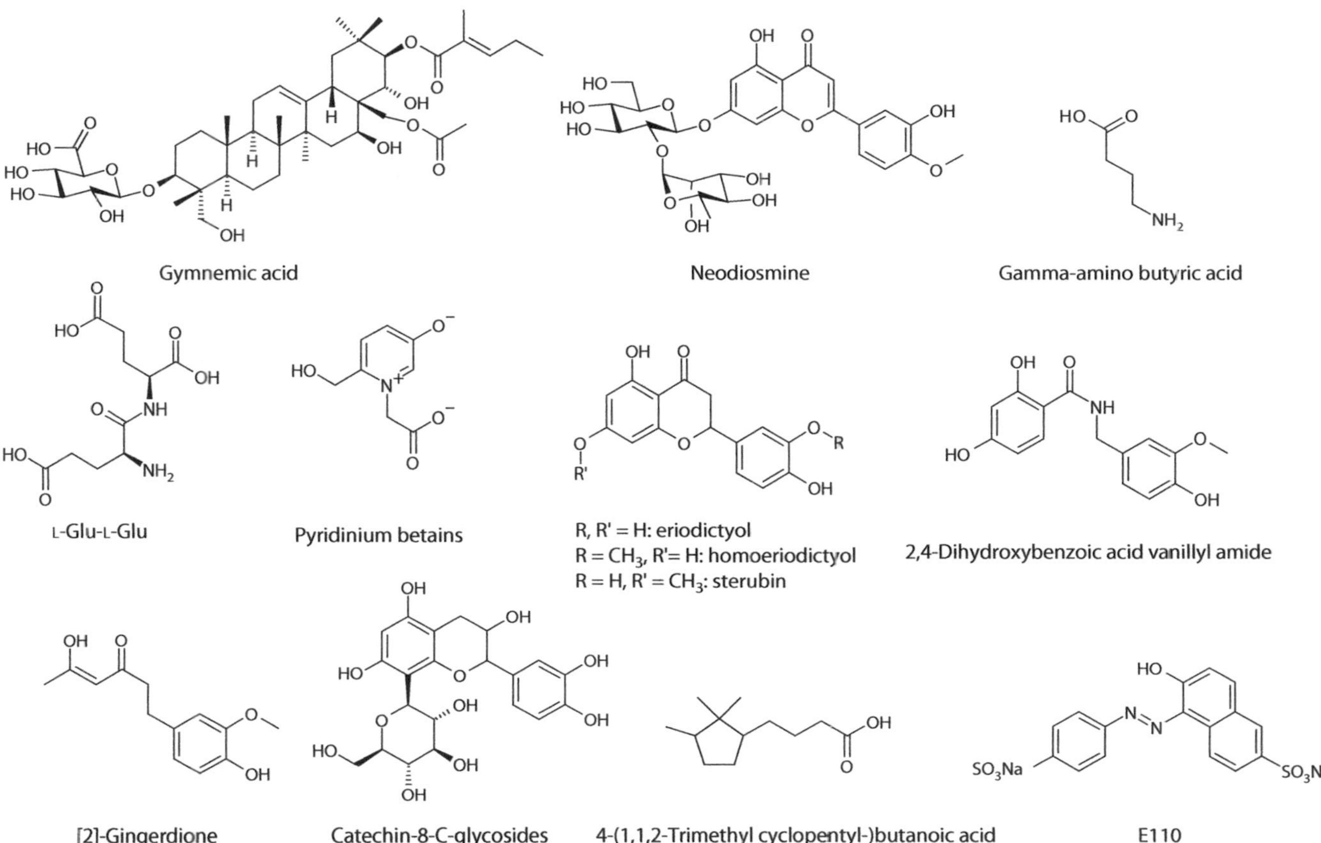

FIGURE 8.3 Neutral tasting bitter suppressors.

(Ley et al. 2005b). Several structurally related polyphenols such as hydroxybenzoic acid vanillyl amides (Ley et al. 2006), and gingerdiones showed a similar modification pattern (Ley et al. 2008). Some catechin-C-glucosides were described as able to reduce bitterness and to combat astringency (Stark and Hofmann 2006). Recent additions include specific bitter inhibitors, for example, 4-(1,1,2-trimethyl cyclopentyl-)butanoic acid (Slack et al. 2009a) or some azoic dyes in combination with carotene preparations allowed as food colorings in the European Union such as E110 (Matuschek et al. 2008) against the bitter (after)taste of sweeteners such as saccharine or acesulfame K. Both were found by extensive screening of chemical libraries using heterologous expression systems containing the responsible hTAS2 variants.

8.9.2 Sweet Taste

Sweet taste reduction is of limited value. In most cases, sugar will not be added more than the level necessary for sweetening effects. In rare cases, a demand for reduction of sweetness may arise when the mouth feel of sucrose is important in non-sweet applications. As mentioned earlier, sweet taste can be largely reduced by lactisole (Figure 8.4) (Schiffmann et al. 1999). Lactisole acts also on the sweet receptor, and the binding site was identified in between (Jiang et al. 2005). Other sweet taste reducers include the gymnemic acids (Gent et al. 1999) (see Figure 8.3); the taste of sweet proteins such as thaumatin, monellin, and lysozyme could be selectively suppressed by riboflavin-binding protein (Maehashi et al. 2007).

A so-called sweet water taste can be induced by several molecules such as cynarin, a compound isolated from artichokes: after swallowing the solution and subsequent rinsing with distilled water, a sweet taste can be detected (Bartoshuk et al. 1972). A

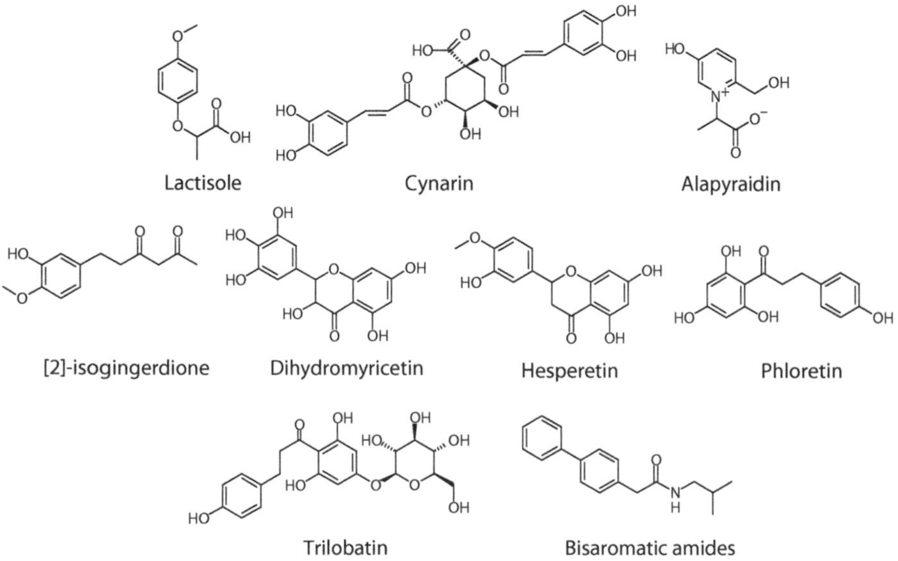

Lactisole Cynarin Alapyraidin

[2]-isogingerdione Dihydromyricetin Hesperetin Phloretin

Trilobatin Bisaromatic amides

FIGURE 8.4 Neutral sweet taste modulators.

similar effect was described for saccharine after rinsing with water (Winnig et al. 2008). In the latter case, it was shown that the sweet water taste was caused by an autoinhibition of sweet taste of saccharine by saccharine at higher concentrations by itself, but after dilution, the saccharine was again able to elicit only the sweet taste (Galindo-Cuspinera and Breslin 2007).

Much progress can be seen in the development of new sweetness enhancers, a very important category. Whereas the use of synergistic mixtures of different sweeteners is a common practice, the use of non- or only weak sweet flavor molecules to gain synergistic sweet-enhancing effects is relatively new. From Maillard reaction products, Ottinger et al. (2003) were able to isolate alapyridine, which shows a sweet as well as an umami or even general taste-enhancing effect (Soldo et al. 2004). Some isogingerdiones such as [2]-isogingerdione exhibit sweet-enhancing effects (Ley et al. 2008). These compounds show only a very weak sweet taste, which is very limited in intensity even at higher concentrations. It could be shown that the enhancing effect is synergistic. Other weak- or neutral-tasting sweet taste enhancers described in the literature include dihydromyricetin (Sugita et al. 2003), hesperetin (Ley et al. 2007), phloretin (Krammer et al. 2007), trilobatin (Jia et al. 2008), combinations of 3-hydroxy- and 2,4-dihydroxybenzoic acid (Bingley et al. 2007), certain bisaromatic amides (Tachdjian et al. 2009), and naringin dihydrochalcone (Hansen et al. 2008).

8.9.3 Umami/Salty/Sour Taste

For meat or culinary preparations, in particular, the interaction of taste compounds and the influence of non- or weak-tasting compounds seem to be important for the overall flavor. Typical examples are the complex salty and umami taste patterns in morels containing morelid (Rotzoll et al. 2006), or in seafood such as scallop caused by strombine (Figure 8.5) (Starkenmann 2009). Some nucleotides such as inositol monophosphate (IMP), guanosyl monophosphate (GMP), and adenosyl monophosphate (AMP) are long known for their synergistic umami-enhancing effect (Morita et al. 2006), and the receptor mechanism was elucidated recently (Zhang et al. 2008). Several lactic acid derivatives were isolated from complex cheese matrices and described as umami/salt enhancing, for example, lactoylguanosine 5′-monophosphate or lactoyltyramine (Winkel et al. 2008). Glycoconjugates from Maillard mixtures also showed umami-potentiating activities (Beksan et al. 2003). Using a virtual *in silico* screening, Grigorov et al. (2003) were able to identify *N*-acetyl glycine as an umami- and salt-enhancing compound. Some neutral peptides also show umami-enhancing effects (Lioe et al. 2005). Other molecules described as umami or salt taste enhancers are theogalline (Kaneko et al. 2006), *N*-(1-methyl-4-hydroxy-3-imidazolin-2,2-ylidene)alanine (Shima et al. 1998), alkyldienamides (Dewis et al. 2004; Pei et al. 2007), certain amino acid derivatives of dicarboxylic acids, and some glutathione derivatives (Dunkel et al. 2007).

It was shown through sensory experiments as well as cellular models that the amino acids L-arginine and L-lysine, and choline chloride are able to significantly increase saltiness (Stähler et al. 2008). Cetylpyridinium chloride and capsaicin can increase saltiness at low dosage (DeSimone et al. 2001), as do some Maillard

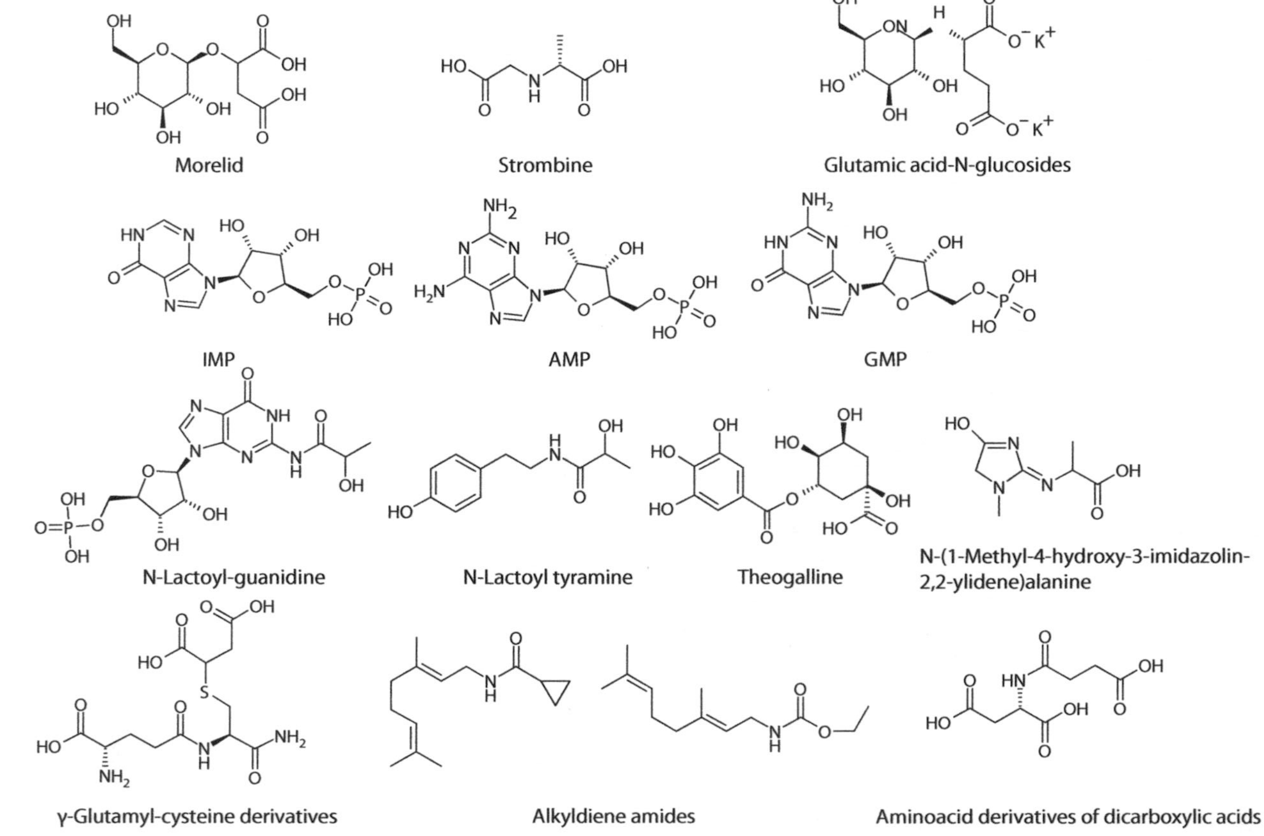

FIGURE 8.5 Weak or neutral tasting umami or salt taste modulators.

reaction compounds (Katsumata et al. 2008; Rhyu et al. 2006). Chlorhexidine can selectively suppress the salty taste perception in humans for a longer period (Wang et al. 2009). Chlorhexidine itself is bitter, but the effect is also seen for nonbitter dosages of the antiseptic. Neutral taste modifiers for sour taste are not widely known; some peptides isolated from beef were also described as able to suppress sourness (Nishimura et al. 2008).

8.10 CONCLUSION

The elucidation of the very complex interactions between the three main components of a flavor—aroma chemicals, tastants, and trigeminally active chemosensates—has evolved tremendously. Nevertheless, a general understanding and prediction of such interactions remains challenging. Recently, much progress occurred in the area of taste-modifying compounds and some were legally applied and approved as flavor compounds. In the future, cell biological models will strongly contribute to the identification and development of further, specifically active compounds in the areas of bitter, sour, and salt modification. Their main advantage is the reduction of required tasting sessions to identify hits from the chemical space, but the experienced nose and tongue will still play a major role in the evaluation and judgment of the complex interactions of aroma, taste, and chemosensory perceptions.

REFERENCES

Bartoshuk, L.M., C.H. Leyy, and R. Scarpellino. 1972. Sweet taste of water induced by artichoke (*Cynara scolymus*). *Science* 178: 988–990.

Beksan, E., P. Schieberle, F. Robert, et al. 2003. Synthesis and sensory characterization of novel umami-tasting glutamate glycoconjugates. *J. Agric. Food Chem.* 51: 5428–5436.

Belikov, V.M., and M.Y. Gololobov. 1986. Plasteins. Their preparation, properties, and use in nutrition. *Nahrung* 30: 281–287.

Belitz, H.D., W. Grosch, and P. Schieberle. 2001. *Lehrbuch der Lebensmittelchemie*. Berlin: Springer.

Benilova, I.V., J. Minic Vidic, E. Pajot-Augy, A.P. Soldatkin, C. Martelet, and N. Jaffrezic-Renault. 2008. Electrochemical study of human olfactory receptor OR 17-40 stimulation by odorants in solution. *Mater. Sci. Eng. C Biomim. Supramol. Syst.* 28: 633–639.

Bingley, C.A., G. Olcese, and K.C. Darnell. 2007. Taste potentiator compositions and beverages containing sam. US 2007 54,023. USA: Cadbury Adams USA LLC.

Brodin, M., M. Laska, and M.J. Olsson. 2009. Odor Interaction between bourgeonal and its antagonist undecanal. *Chem. Sens.* 34: 625–630.

Burseg, K., and C. de Jong. 2009. Application of the olfactoscan method to study the ability of saturated aldehydes in masking the odor of methional. *J. Agric. Food Chem.* 57: 9086–9090.

Busch, J., M. Batenburg, R. van der Velden, and G. Smit. 2009. Reviewing progress towards finding an acceptable natural flavor alternative to salt. *Agro Food Ind. Hi-Tech* 20: 66–68.

Busch, J.L.H.C., K. Hrncirik, E. Bulukin, C. Boucon, and M. Mascini. 2006. Biosensor measurements of polar phenolics for the assessment of the bitterness and pungency of virgin olive oil. *J. Agric. Food Chem.* 54: 4371–4377.

Cano, J., H. Mintijano, F. Lopez-Cremades, and F. Borrego. 2000. Masking the bitter taste of pharmaceuticals. *Manuf. Chem.* 71: 16–17.

Casuso, I., M. Pla-Roca, G. Gomila, et al. 2008. Immobilization of olfactory receptors onto gold electrodes for electrical biosensor. *Mater. Sci. Eng. C Biomim. Supramol. Syst.* 28: 686–691.

Chandrashekar, J., M.A. Hoon, N.J.P. Ryba, and C.S. Zuker. 2006. The receptors and cells for mammalian taste. *Nature* 444: 288–294.

Chen, P., B. Wang, G. Cheng, and P. Wang. 2009. Taste receptor cell-based biosensor for taste specific recognition based on temporal firing. *Biosens. Bioelectron.* 25: 228–233.

DeSimone, J.A., V. Lyall, G.L. Heck, et al. 2001. A novel pharmacological probe links the amiloride-insensitive NaCl, KCl, and NH_4Cl chorda tympani taste responses. *J. Neurophysiol.* 86: 2638–2641.

Dewis, M.L., M.E. Huber, M.V. Cossette, and D.O. Agyemang. 2004. *Alkyldienamides Exhibiting Taste and Sensory Effect in Flavor Compositions*. US 2004 202,760. USA: International Flavours & Fragrances Inc.

Dewis, M.L., T.V. John, M.A. Eckert, J.H. Colstee, N.C. Da Costa, and T. Pei. 2005. *Conjugated Dienamides from Piper Species for Imparting Aroma, Taste, and Chemesthetic Effects*. US 2005 075,368. USA: Int. Flavours & Fragrances.

Dunkel, A., J. Koester, and T. Hofmann. 2007. Molecular and sensory characterization of gamma-glutamyl peptides as key contributors to the kokumi taste of edible beans (*Phaseolus vulgaris* L.). *J. Agric. Food Chem.* 55: 6712–6719.

Edelmann, A., and B. Lendl. 2002. Toward the optical tongue: Flow through sensing of tannin–protein interactions based on FTIR spectroscopy. *J. Am. Chem. Soc.* 125: 14741–14747.

Formaker, B.K., H. Lin, T.P. Hettinger, and M.E. Frank. 2009. Responses of the hamster chorda tympani nerve to sucrose+acid and sucrose+citrate taste mixtures. *Chem. Sens.* 34: 607–616.

Galindo-Cuspinera, V., and P.A.S. Breslin. 2007. Taste after-images: The science of "water-tastes". *Cell. Mol. Life Sci.* 64: 2049–2052.

Gent, F.J., T.P. Hettinger, M.E. Frank, and L.E. Marks. 1999. Taste confusion following gymnemic acid rinse. *Chem. Sens.* 24: 393–403.

Gibbs, B.F., I. Alli, and C. Mulligan. 1996. Sweet and taste-modifying proteins: A review. *Nutr. Res. (N.Y.)* 16: 1619–1630.

Gottfried, J.A., and R.J. Dolan. 2003. The nose smells what the eye sees: Crossmodal visual facilitation of human olfactory preception. *Neuron* 39: 375–386.

Grigorov, M.G., H. Schlichtherle-Cerny, M. Affolter, and S. Kochhar. 2003. Design of virtual libraries of umami-tasting molecules. *J. Chem. Inf. Comp. Sci.* 43: 1248–1258.

Guadagni, D.G., R.M. Horowitz, B. Gentili, and V.P. Maier. 1979. *Method of Reducing Bitter and Off-After-Taste*. US 4,154,862. The United States of America as represented by the Secretary of Agriculture, Washington, DC, USA.

Hansen, C.A., J.P. Slack, and C.T. Simons. 2008. *Preparation of Sweetened Consumables Comprising Sweeteners and Sweetness Enhancer*. WO 2008 049,256. Switzerland: Givaudan SA.

Jellinek, G. 1966. Moderne Verfahren der sensorischen Analyse. *Planta Med.* 14: 21–33.

Jia, Z., X. Yang, C.A. Hansen, et al. 2008. *Sweeteners Containing Dihydrochalcones*. WO 2008 148,239. Switzerland: Givaudan SA.

Jiang, P., M. Cui, B. Zhao, et al. 2005. Lactisole interacts with the transmembrane domains of human T1R3 to inhibit sweet taste. *J. Biol. Chem.* 280: 15238–15246.

Johnson, C., G.G. Birch, and D.B. MacDougall. 1994. The effect of the sweetness inhibitor 2-(4-methoxyphenoxy)propanoic acid (sodium salt) (NA-PMP) on the taste of bitter stimuli. *Chem. Sens.* 19: 349–358.

Kaneko, S., K. Kumazawa, H. Masuda, A. Henze, and T. Hofmann. 2006. Molecular and sensory studies on the umami taste of Japanese green tea. *J. Agric. Food Chem.* 54: 2688–2694.

Kant, R., M.B. Rajasekaran, and R. Suryanarayanarao. 2005. Sweet and taste modifying proteins—comparative modeling and docking studies of curculin, mabinlin, miraculin with the T1R2 T1R3 receptor. *Internet Electron. J. Mol. Des.* 4: 106–123.

Katsumata, T., H. Nakakuki, C. Tokunaga, et al. 2008. Effect of Maillard reacted peptides on human salt taste and the amiloride-insensitive salt taste receptor (TRPV1t). *Chem. Sens.* 33: 665–680.

Kawai, T., Y. Kusakabe, and T. Ookura. 2008. Behavioral estimation of enhancing effects in taste enhance. *Chem. Sens.* 33: J6.

Keast, R.S.J. 2003. The effect of zinc on human taste perception. *J. Food Sci.* 68: 1871–1877.

Keast, R.S.J., M.M.E. Bournazel, and P.A.S. Breslin. 2003. A psychophysical investigation of binary bitter-compound interactions. *Chem. Sens.* 28: 301–313.

Keast, R.S.J., P.A.S. Breslin, and G.K. Beauchamp. 2001. Surpression of bitterness using sodium salts. *Chimia* 55: 441–447.

Keast, R.S.J., T.M. Canty, and P.A.S. Breslin. 2004. The Influence of sodium salts on binary mixtures of bitter-tasting compounds. *Chem. Sens.* 29: 431–439.

King, B.M., P. Arents, N. Bouter, et al. 2006. Sweetener/sweetness-induced changes in flavor perception and flavor release of fruity and green character in beverages. *J. Agric. Food Chem.* 54: 2671–2677.

Krammer, G., J. Ley, T. Riess, M. Haug, and S. Paetz. 2007. Use of 4-hydroxydihydrochalcones and their salts for enhancing an impression of sweetness. WO 2007 107,596. Germany: Symrise GmbH & Co. K.G.

Kroeze, J.H.A. 2000. Neohesperidin dihydrochalcone is not a taste enhancer in aqueous sucrose solutions. *Chem. Sens.* 25: 555–559.

Kurobayashi, Y., Y. Katsumi, A. Fujita, Y. Morimitsu, and K. Kubota. 2008. Flavor enhancement of chicken broth from boiled celery constituents. *J. Agric. Food Chem.* 56: 512–516.

Labbe, D., F. Gilbert, and N. Martin. 2008. Impact of olfaction on taste, trigeminal, and texture perceptions. *Chemosens. Percept.* 1: 217–226.

Labbe, D., A. Rytz, C. Morgenegg, S. Ali, and N. Martin. 2007. Subthreshold olfactory stimulation can enhance sweetness. *Chem. Sens.* 32: 205–214.

Lewin, L. 1894. Ueber die Geschmacksverbesserung von Medicamenten und über Saturationen. *Berl. Klin. Wochenschr.* 28: 644–645.

Ley, J.P. 2008. Masking bitter taste by molecules. *Chemosens. Percept.* 1: 58–77.

Ley, J.P., M. Blings, S. Paetz, G.E. Krammer, and H.-J. Bertram. 2006. New bitter-masking compounds: Hydroxylated benzoic acid amides of aromatic amines as structural analogues of homoeriodictyol. *J. Agric. Food Chem.* 54: 8574–8579.

Ley, J., G. Kindel, G. Krammer, T. Hofmann, and N. Rotzoll. 2005a. *Use of Gamma-Aminobutanoic Acid for Masking or Reducing an Unpleasant Flavor Impression, and Preparations Containing Gamma-Aminobutanoic Acid.* WO 2005 096,841. Germany: Symrise GmbH & Co. K.G.

Ley, J., G. Kindel, S. Paetz, et al. 2007. *Use of Hesperetin for Enhancing the Sweet Taste.* WO 2007 014,879. Germany: Symrise GmbH & Co. K.G.

Ley, J.P., G. Krammer, G. Reinders, I.L. Gatfield, and H.-J. Bertram. 2005b. Evaluation of bitter masking flavanones from herba santa (*Eriodictyon californicum* (H. & A.) Torr., Hydrophyllaceae). *J. Agric. Food Chem.* 53: 6061–6066.

Ley, J.P., S. Paetz, M. Blings, P. Hoffmann-Lücke, H.-J. Bertram, and G.E. Krammer. 2008. Structural analogues of homoeriodictyol as flavor modifiers: Part III. Short chain gingerdione derivatives. *J. Agric. Food Chem.* 56: 6656–6664.

Lioe, H.N., A. Apriyantono, K. Takara, K. Wada, and M. Yasuda. 2005. Umami taste enhancement of MSG/NaCl mixtures by subthreshold L-alpha-aromatic amino acids. *J. Food Sci.* 70: S401–S405.

Maehashi, K., M. Matano, A. Kondo, Y. Yamamoto, and S. Udaka. 2007. Riboflavin-binding protein exhibits selective sweet suppression toward protein sweeteners. *Chem. Sens.* 32: 183–190.

Manabe, M., S. Ishizaki, T. Yoshioka, and N. Oginome. 2009. Improving the palatability of salt-reduced food using dried bonito stock. *J. Food Sci.* 74: S315–S321.

Mattes, R.D. 2007. Effects of linoleic acid on sweet, sour, salty, and bitter taste thresholds and intensity ratings of adults. *Am. J. Physiol.* 292: G1243–G1248.

Matuschek, M., M.B. Jager, A. Kleber, M. Krohn, and H. Zinke. 2008. *Method for Modulating the Taste of Material Compositions Containing at Least One High Intensity Sweetener (HIS).* WO 2008 102,018. Germany: BASF SE.

McLaughlin, I.R., C. Lederer, and T.H. Shellhammer. 2008. Bitterness-modifying properties of hop polyphenols extracted from spent hop material. *J. Am. Soc. Brew. Chem.* 66: 174–183.

McMahon, D.B.T., H. Shikata, and P.A. Breslin. 2001. Are the human taste thresholds similiar on the right and left sides of the tongue? *Chem. Sens.* 26: 875–883.

Milligan, G. 2006. G-protein-coupled receptor heterodimers: Pharmacology, function and relevance to drug discovery. *Drug Discovery Today* 11: 541–549.

Miyanag, Y., N. Inoue, A. Ohnishi, E. Fujisawa, M. Yamaguchi, and T. Uchidam. 2003. Quantitative pediction of the bitterness suppression of elemental diets by various flavors using a taste sensor. *Pharm. Res.* 20: 1932–1938.

Miyazawa, T., T. Matsuda, S. Muranishi, and K. Miyake. 2006. *Salty Taste Enhancers Containing Spilanthol, Flavoring Materials Containing Them, Foods and Beverages Containing the Flavoring Materials, and Taste-Enhancing Method.* JP 2006 296,357. Japan: Ogawa and Co. Ltd.

Mojet, J., J. Heidema, and E. Christ-Hazelhof. 2004. Effect of concentration on taste–taste interactions in foods for elderly and young subjects. *Chem. Sens.* 29: 671–681.

Morita, K., M. Narukawa, and Y. Hayashi. 2006. Evaluation of taste intensity and quality of umami substances and the synergistic effect with nucleotide. *Nippon Aji to Nioi Gakkaishi* 13: 319–322.

Mukai, J., E. Tokuyama, T. Ishizaka, S. Okada, and T. Uchida. 2007. Inhibitory effect of aroma on the bitterness of branched-chain amino acid solutions. *Chem. Pharm. Bull.* 55: 1581–1584.

Nishimura, T., Y. Fujita, and Y. Furukawa. 2008. *Sourness-Suppressing Peptides in Beef.* AChemS 2008 Abstract Book #172.

Oka, Y., A. Nakamura, H. Watanabe, and K. Touhara. 2004. An odorant derivative as an antagonist for an olfactory receptor. *Chem. Sens.* 29: 815–822.

Ottinger, H., A. Bareth, and T. Hofmann. 2001. Characterization of natural "cooling" compounds formed from glucose and L-proline in dark malt by application of taste dilution analysis. *J. Agric. Food Chem.* 49: 1336–1344.

Ottinger, H., T. Soldo, and T. Hoffmann. 2003. Discovery and structure determination of a novel Maillard-derived sweetness enhancer by application of the comparative taste dilution analysis (cTDA). *J. Agric. Food Chem.* 51: 1035–1041.

Pei, T., M.L. Dewis, and A.J. Janczuk. 2007. Unsaturated cyclic and acyclic carbamates exhibiting taste and flavor enhancement effect in flavor compositions. US 2007 134,389. USA: International Flavors & Fragrances Inc.

Petit, C.E.F., T.A. Hollowood, F. Wulfert, and J. Hort. 2007. Colour-coolant–aroma interactions and the impact of congruency and exposure on flavor perception. *Food Qual. Prefer.* 18: 880–889.

Prakash, I., and G.E. Dubois. 2007. High-Potency Sweetener Composition With Antioxidant and Compositions Sweetened Therewith. US 2007 116,838. USA: Coca Cola Co.

Reichelt, K.V., R. Peter, S. Paetz, et al. 2010. Characterization of flavor modulating effects in complex mixtures via high temperature liquid chromatography. *J. Agric. Food Chem.*, 54: 458–464.

Reisert, J., and D. Restrepo. 2009. Molecular tuning of odorant receptors and its implication for odor signal processing. *Chem. Sens.* 34: 535–545.

Rhyu, M., M. Ogasawara, M. Egi, et al. 2006. Effect of Maillard peptides (MPS) on TRPV1 variant salt taste receptor (TRPV1T). *Chem. Sens.* 31: A105.

Riera, C.E., H. Vogel, S.A. Simon, S. Damak, and J. Le Coutre. 2009. Sensory attributes of complex tasting divalent salts are mediated by TRPM5 and TRPV1 channels. *J. Neurosci.* 29: 2654–2662.

Röck, F., N. Barsan, and U. Weimar. 2008. Electronic nose: Current status and future trends. *Chem. Rev.* 108: 705–725.

Roloff, M., H. Erfurt, G. Kindel, C.O. Schmidt, and G. Krammer. 2006. *Process for the Separation and Sensory Evaluation of Flavors*. WO 2006 111,476. Germany: Symrise GmbH & Co. KG.

Roloff, M., R. Peter, and M. Luetkenhaus. 2009. *Vorrichtung, Verwendung der Vorrichtung und Verfahren zur Herstellung eines flüssigen Geschmackstoffkondensates*. EP 2,113,771. Germany: Symrise GmbH & Co. KG.

Rossetti, D., J.H.H. Bongaerts, E. Wantling, J.R. Stokes, and A.-M. Williamson. 2009. Astringency of tea catechins: More than an oral lubrication tactile percept. *Food Hydrocolloids* 23: 1984–1992.

Rotzoll, N., A. Dunkel, and T. Hofmann. 2006. Quantitative studies, taste reconstitution, and omission experiments on the key taste compounds in Morel mushrooms (*Morchella deliciosa* Fr.). *J. Agric. Food Chem.* 54: 2705–2711.

Ryan, D., P.D. Prenzler, A.J. Saliba, and G.R. Scollary. 2008. The significance of low impact odorants in global odour perception. *Trends Food Sci. Technol.* 19: 383–389.

Sakurai, T., T. Misaka, T. Nagai, et al. 2009. Acidic substances added in the oral cavity reduce our bitter taste sensation by pH-dependent inhibition of hTAS2R response. *Chem. Sens.* 34: A81.

Schiffmann, S.S., B.J. Booth, E.A. Sattely-Miller, B.G. Graham, and K.M. Gibes. 1999. Selective inhibition of sweetness by the sodium salt of ±2-(4-methoxyphenoxy)propanoic acid. *Chem. Sens.* 24: 439–447.

Schlutt, B., N. Moran, P. Schieberle, and T. Hofmann. 2007. Sensory-directed identification of creaminess-enhancing volatiles and semivolatiles in full-fat cream. *J. Agric. Food Chem.* 55: 9634–9645.

Shima, K., N. Yamada, E.-I. Suzuki, and T. Harada. 1998. Novel brothy taste modifier isolated from beef broth. *J. Agric. Food Chem.* 46: 1465–1468.

Shirasuka, Y., K.-I. Nakajima, T. Asakura, et al. 2004. Neoculin as a new taste-modifying protein occurring in the fruit of *Curculigo latifolia. Biosci. Biotech. Biochem.* 68: 1403–1407.

Shirokova, E., K. Schmiedeberg, P. Bedner, et al. 2005. Identification of specific ligands for orphan olfactory receptors: G Protein-dependent agonism and antagonism of odorants. *J. Biol. Chem.* 280: 11807–11815.

Simons, C.T., and K. Albin. 2009. Understanding the relationship between saltiness and umami. *Chem. Sens.* 34: A116.

Simons, C.T., Y. Boucher, and E. Carstens. 2003. Suppression of central taste transmission by oral capsaicin. *J. Neurosci.* 23: 978–985.

Slack, J., A. Brockhoff, B. Claudia, et al. 2009a. Inhibition of bitter taste receptors. *Chem. Sens.* 34: A82.

Slack, J.P., C.T. Simons, and C.A. Hansen. 2009b. *Sweetness Enhancement and Screening for Enhancers by Using Taste Receptor Subunit Transmembrane Domain*. WO 2007 121,604. Switzerland: Givaudan SA.

Small, D.M., J. Voss, Y.E. Mak, K.B. Simmons, T. Parrish, and D. Gitelman. 2004. Experience dependent neural integration of taste and smell in the human brain. *J. Neurophysiol.* 92: 1892–1903.

Soldo, T., O. Frank, H. Ottinger, and T. Hofmann. 2004. Systematic studies of the structure and physiological activity of alapyridaine. A novel food-borne taste enhacer. *Mol. Nutr. Food Res.* 48: 270–281.

Soldo, T., and T. Hofmann. 2005. Application of hydrophilic interaction liquid chromatography/comparative taste dilution analysis for identification of a bitter inhibitor by a combinatorial approach based on Maillard reaction chemistry. *J. Agric. Food Chem.* 53: 9165–9171.

Spehr, M., G. Gisselmann, A. Poplawski, et al. 2003. Identification of a testicular odorant receptor mediating human sperm chemotaxis. *Science* 299: 2054–2058.

Stähler, F., K. Riedel, S. Demgensky, et al. 2008. A role of the epithelial sodium channel in human salt taste transduction? *Chemosens. Percept.* 1: 78–90.

Stark, T., and T. Hofmann. 2005. Structures, sensory activity, and dose/response functions of 2,5-diketopiperazines in roasted cocoa nibs (*Theobroma cacao*). *J. Agric. Food Chem.* 53: 7222–7231.

Stark, T., and T. Hofmann. 2006. Application of a molecular sensory science approach to alkalized cocoa (*Theobroma cacao*): Structure determination and sensory activity of nonenzymatically C-glycosylated flavan-3-ols. *J. Agric. Food Chem.* 54: 9510–9521.

Starkenmann, C. 2009. Contribution of (*R*)-strombine to dry scallop mouthfeel. *J. Agric. Food Chem.* 57: 7938–7943.

Steinhaus, P., and P. Schieberle. 2007. Characterization of the key aroma compounds in soy sauce using approaches of molecular sensory science. *J. Agric. Food Chem.* 55: 6262–6269.

Stucchi, M. 2008. Application of HTS for the identification of modulators of an ancient pain target, the menthol receptor TRPM8. *Abstracts of Papers, 236th ACS National Meeting, Philadelphia, PA, United States, August 17–21, 2008*. AGFD-265.

Sugita, T., K. Yoshida, and S. Shimura. 2003. *Dihydromyricetin for Modification of Taste and Food Containing the Taste Modifier*. JP 2003 38,121. Japan: Lotte Co. Ltd.

Suttisri, R., I.S. Lee, and A.D. Kinghorn. 1995. Plant-derived triterpenoid sweetness inhibitors. *J. Ethnopharmacol.* 47: 9–26.

Suzuki, M., E. Kurimoto, S. Nirasawa, et al. 2004. Recombinant curculin heterodimer exhibits taste-modifying and sweet-tasting activities. *FEBS Lett.* 573: 135–138.

Tachdjian, C., A.P. Patron, M. Qi, et al. 2009. Aromatic Amides and Ureas and their Uses as Sweet and/or Umami Flavor Modifiers, Tastants and Taste Enhancers. US 2006 045,953. USA: Senomyx.

Takoi, K., M. Degueil, S. Shinkaruk, et al. 2009. Identification and characteristics of new volatile thiols derived from the hop (*Humulus luplus* L.) cultivar Nelson Sauvin. *J. Agric. Food Chem.* 57: 2493–2502.

Talavera, K., Y. Ninomiya, C. Winkel, T. Voets, and B. Nilius. 2007. Influence of temperature on taste perception. *Cell. Mol. Life Sci.* 64: 377–381.

van der Loo, H.E., and C. Wiener. 1983. *Flavor Potentiated Oral Compositions Containing Thaumatin or Monellin*. US 4,412,984. USA: Talres Dev (Nl).

Wang, M.-F., L.E. Marks, and M.E. Frank. 2009. Taste coding after selective inhibition by chlorhexidine. *Chem. Sens.* 34: 653–666.

WHO. 2008. Guidelines for the use of flavorings. *Codex Alimentarius Guideline CAC/GL* 66: 1–3.

Williams, J.G., and R.A. Bernhard. 1981. Amino acid–lactose interactions and their sensory consequences. *J. Food Sci.* 46: 1245–1251.

Winkel, C., A. de Klerk, J. Visser, et al. 2008. New developments in umami (enhancing) molecules. *Chem. Biodivers.* 5: 1195–1203.

Winnig, M., C. Kuhn, O. Frank, et al. 2008. Saccharin: Artificial sweetener, bitter tastant, and sweet taste inhibitor. In *Sweetness and Sweeteners—Biology, Chemistry and Psychophysics*, ed. D.K. Weerasinghe and G. DuBois, 230–240. New York: Oxford University Press.

Yoon, H., S.H. Lee, O.S. Kwon, et al. 2009. Polypyrrole nanotubes conjugated with human olfactory receptors: High-performance transducers for FET-type bioelectronic noses. *Angew. Chem.* 121: 2793–2796.

Zhang, F., B. Klebansky, R.M. Fine, et al. 2008. Molecular mechanism for the umami taste synergism. *Proc. Natl. Acad. Sci. USA* 105: 20930–20934.

Zoller, M. 2009. The discovery and function of sweet taste enhancers. *Abstracts of Papers, 237th ACS National Meeting, Salt Lake City, UT, United States, March 22–26, 2009*. MEDI-138.

Zou, Z., and L.B. Buck. 2006. Combinatorial effects of odorant mixes in olfactory cortex. *Science* 311: 1477–1481.

9 Legislation, Safety Assessment, and Labeling of Food Flavors and Flavorings

Manfred Lützow

CONTENTS

9.1 FOOD SAFETY AND FLAVORINGS—AN INTRODUCTION

Food is a "substance taken into the body to maintain life and growth"—a definition proposed by the 2nd edition of the *Oxford Illustrated Dictionary* in 1975 that does not only refer to nutrition but also to other aspects such as the pleasure associated with the process of taking food into the body. Any food item, whether a basic ingredient or a processed food that consists of several ingredients, has distinctive organoleptical properties that appeal to all five human senses and especially to sight, smell, and taste. The *flavor* of a food is an experience of the individual consumer eating that food, and the perceived flavors will depend not only on the substances present in the food and its structure, but also on the individual's genetic and sociocultural background. According to the Codex Alimentarius, "flavour is the sum of those characteristics of any material taken in the mouth, perceived principally by the senses of taste and smell, and also the general pain and tactile receptors in the mouth, as received and interpreted by the brain." Whereas flavor is therefore rather the result of an interaction between the consumer and a food, "flavourings are products that are added to food to impart, modify, or enhance the flavour of food." Both definitions were agreed upon recently by the Codex Alimentarius Commission (CAC), an international body representing more than 190 members who lay down global food standards, as part of the *Guidelines for the Use of Flavourings* (CAC 2008).

Modern food regulatory frameworks address the addition of *flavorings* to food and the identification of *flavors* in descriptions of foods (e.g., on labels) as separate areas. The former requires an assessment of the risk of such a practice; the latter strives to assure that consumers receive truthful information about the food they intend to consume. For regulators, the question of whether strawberry-flavored yogurt is safe to eat is different from labeling it as "strawberry yogurt." And both questions are today of equal importance, whereas until the last decade of the twentieth century, most countries were mainly concerned with the labeling aspects of flavored foods. Regulators provided specific guidance under what conditions, for which recipe, "strawberry flavor" was considered a "true" or a "false" statement, but were rather vague about what flavoring substances were safe to be added to food and what were the criteria to establish such safety. Of importance was whether a product called *strawberry yogurt* would possibly mislead the consumer rather than the safety of the flavorings imparting such a flavor to the yogurt. This brief characterization of the regulatory approach until recently should not be understood as if foods containing flavorings had been unsafe. Food industry is marketing processed foods under the general obligation that they should be safe to the consumer, and the safety of flavorings used by this industry was and continues to be established in most markets mainly by its suppliers, the flavoring or flavor industry. The producers of flavorings assure the safety of their products, often by an organized self-regulating activity, the FEMA GRAS assessment program being the most prominent and influential example.

This chapter is not the place to discuss the pros and cons of self-regulation vs. premarket authorization schemes; as a matter of fact, industry and governments have supported the development of a more coherent and modern approach to assure the safety of the flavorings used by the food industry and have therefore increased the regulatory control by official authorities at the cost of self-regulation. This is notably true for two important international bodies, the CAC and the European Union, which are discussed in this chapter as the leading international regulatory bodies that apply modern risk analysis thinking to flavorings.

9.2 RISK ANALYSIS—A COMPREHENSIVE APPROACH TO FOOD SAFETY

The CAC, managed jointly by the Food and Agriculture Organization of the United Nations (FAO) and the World Health Organization (WHO), provides for almost 50 years, guidance in the area of food safety and quality to regulators, industry, trade, and consumers in developed and developing countries via a growing set of food standards and guidance documents, the *Codex Alimentarius*. After 30 years of work that had addressed all major food commodities, most chemicals added to or occurring in foods such as additives, contaminants, residues of pesticides, and other important aspects of food trade such as import/export conditions and labeling of food sold to the consumer, the Commission recognized—in view of the growing international trade with foods and the challenge posed by the World Trade Organization in 1995—the need to establish for its work a formal risk analysis framework that is based on science. The adopted concept consists of three separate but connected activities: *risk assessment*, which establishes the probability of a food-related hazard to affect human health; *risk management*, which prevents or minimizes the exposure of consumers to such hazards; and *risk communication*, the information provided to all stakeholders including food industry, trade, and consumers concerning the safe production, distribution, and consumption of food. When applying this risk analysis framework, any risk management measure should be based on a risk assessment that itself should meet generally recognized criteria for scientific work such as using all available data of sufficient quality, involving experts representing the required scientific disciplines, publishing transparently the data, and applying modern assessment methodologies.

In the area of food flavorings, little systematic work had been done by the CAC until 1995; the guiding bodies were rather the Council of Europe (CoE), an international governmental organization based in Strasbourg, and the Flavor and Extracts Manufacturers Association (FEMA), a Washington-based nongovernmental organization representing the U.S. industry involved in the production and trade of flavorings. However, this picture has changed: whereas FEMA continues to be operative for the U.S. market and is recognized as an authoritative source abroad, the CoE was replaced by the work of the European Commission and the European Food Safety Authority (EFSA), and in parallel, the CAC, together with FAO and WHO, has been working for 15 years on a comprehensive flavoring evaluation program. Much of this is work in progress and therefore it may be more important to describe the common and distinguishing elements of all three approaches rather than dealing with them separately.

9.3 RISK ANALYSIS/ASSESSMENT OF FLAVORINGS—CHALLENGE BY THEIR COMPLEXITY

The flavor of a food may result from the use of an ingredient (e.g., strawberries), an extract from a raw material (e.g., oregano essential oil), a pure chemical compound (e.g., cinnamaldehyde), a mixture of compounds produced for flavoring purposes (e.g., reaction flavor, smoking flavorings), or from reactions that take place during food processing (e.g., Maillard reaction, smoking). Usually, several such elements will be used at the same time and lead to the complex flavor of a processed food; furthermore, the borders between the categories are not always clear-cut. At least from a risk assessment point of view, there is no big difference between an oregano essential oil with a content of ~80% of carvacrol and a chemically synthesized carvacrol containing not less than 96%: both will be used in food at levels that result in similar intakes of carvacrol by the consumer. In a similar manner, processes applied to food with the intention to create flavor such as smoking will result in the same array of chemicals being present at the same levels if added using a prefabricated preparation, in this case a smoking flavoring.

However, regardless of the difficulties to clearly distinguish—from a scientific viewpoint—between the different ways that impart flavor to a food, the categories sketched out above are the main ones applied by regulators to lay down requirements for risk assessment and to establish subsequent risk management measures. These categories shall be discussed in detail in the following sections.

9.4 CHEMICALLY DEFINED FLAVORINGS

The Codex Alimentarius defines *flavoring substances* as being chemically defined; they may be produced by chemical synthesis or are obtained from materials of plant or animal origin. This category is subdivided accordingly into *natural flavoring substances* and *synthetic flavoring substances*. Most of these substances are pure compounds with purity well above 95% or consist of few defined constituents that are closely related to each other (usually isomers). Probably more than 4000 such substances are being used and listed in inventories of flavoring substances.

Meanwhile, there is general agreement that such substances shall only be used if an evaluation has concluded that their use in food does not pose a risk to the consumer at the estimated levels of intake. Such evaluations are undertaken currently by three programs/bodies: the FEMA GRAS program, the Joint FAO/WHO Expert Committee on Food Additives (JECFA) evaluation of flavoring groups, and the EFSA flavoring group evaluations. The work of all three programs results in comprehensive lists of chemically defined flavorings that are considered to be safe for use in food; however, the legal status of these lists differ and therefore the programs behind them need to be characterized.

9.4.1 FEMA GRAS

In the United States, certain categories of food ingredients may be used safely in food if they are "generally recognized as safe" (GRAS). One way to achieve such a status is to ask a panel of experts to assess the ingredient and the proposed conditions of use and to conclude, based on all available data, whether its use would pose a risk to the consumer. In 1960, FEMA established such a panel, the FEMA Expert Panel, which is *de facto* recognized by the Food and Drug Administration (FDA) as an appropriate expert body to assess the safety of flavoring substances. The history and development of this program were reviewed by Smith et al. (2005a), an excellent starting point for understanding the details of this program, the work of the panel, and how the results integrate into the legal framework of the U.S. Federal Food, Drug, and Cosmetic Act, the Code of Federal Regulation (CFR) and the work of the FDA.

Working for almost five decades, it is of no surprise that the assessments performed by the FEMA GRAS Panel have influenced significantly the scientific and regulatory debates about the safety evaluation of flavoring substances. Some of the principles such as the threshold of concern concept, the use of intake data, the association of structure and exposure to toxic potential were developed by FEMA in parallel with JECFA and the European Scientific Committee for Food (SCF).

The FEMA panel has published since 1965 (in 24 publications) lists of more than 2000 GRAS flavorings considered to be safe "under conditions of intended use in food." References to the single publications, the list of GRAS flavorings, and additional information about the FEMA GRAS program are available from the association's webpage (see Table 9.2). FEMA assigns also numbers to flavoring substances, the FEMA No., which is one of the reference numbering systems used by industry and trade.

9.4.2 JECFA/Codex Alimentarius

JECFA was founded by FAO and WHO in 1956 as an international expert body tasked to provide advice on several classes of chemicals used in food such as food additives, processing additives, flavorings, or occurring in food such as natural contaminants or residues of veterinary drugs. During the first four decades, the committee occasionally evaluated flavorings, for example, anethol or limonene, applying the same approach as for food additives that requires sufficiently robust toxicological data obtained with the substance to be evaluated resulting in an acceptable daily intake (ADI).

At the 44th meeting, a paper presented by Ian Munro (WHO 1996), who had also been a long-serving member of the FEMA Gras Panel, was the trigger for a new approach that build on several concepts that had been under discussion for some time: the threshold of concern concept assumed that there was a level of intake at which, taking into consideration the chemical structure of a substance and its probable toxicological properties, no hazard was expected to occur. The assignment of a substance to one of three structural chemical classes for which default *No Observed Effect Levels* had been determined based on a comprehensive database of more than 600 substances for which toxicological tests had been done (Cramer et al. 1978). The grouping of substances according to their chemical properties was based on the understanding that substances sharing functional groups and properties would be metabolized and possibly excreted similarly, and toxicological data from tested substances could be extrapolated within a chemical group to those not tested. Levels of intake would be derived using the annual poundage data collected by the industry for the U.S. or the European market. All data and assumptions were combined in a decision tree that resulted in several evaluation scenarios.

JECFA evaluated until 2010 more than 2060 individual flavoring substances, published for the flavoring groups comprehensive toxicological monographs and proposed specifications for the commercial materials for which the evaluation is applicable. This information can be accessed from the committee's FAO and WHO secretariat webpages (see Table 9.2). As noted earlier, this work was undertaken in response to requests from the CAC, the resulting list of JECFA-evaluated chemically defined flavorings (identified by a JECFA No.) is not adopted by the CAC and does not constitute an official positive list; however, the specifications (purity criteria) for flavorings proposed by JECFA are adopted by the CAC as so-called Codex Specifications and have thereby become international standards for trade.

9.4.3 EFSA/European Commission

As a consequence of the enhanced efforts by the end of the 1980s to harmonize the European market for goods including foods, legislation on the use of flavorings in foodstuffs was adopted in June 1988, and soon afterward the SCF was asked to develop guidelines for the safety evaluation of flavorings. These guidelines, published in 1992, took into account work by the CoE, the FEMA GRAS assessments, and JECFA. Based on these guidelines, a first batch of chemically defined flavorings was evaluated in 1995; however, work was interrupted until 2000 when eventually the framework for an evaluation program of these substances was adopted. The core elements were an evaluation by 34 chemical groups (see Table 9.1), specifications for the commercially used flavorings, use levels for 16 food categories, and an electronic submission format for the data. Evaluations would be processed by a small expert team working initially at the Danish Veterinary and Food Administration, a team that runs the EU Flavour Information System (FLAVIS). The draft evaluations should be reviewed by the SCF and eventually adopted as opinions that provide guidance whether flavorings were safe under the proposed conditions of use. The program took into consideration work done by the CoE, the SCF, and JECFA: substances that had been evaluated already before 1999 were in principle excluded from the program unless there was new information that would possibly change the previous evaluation. Substances evaluated by the JECFA after 1999 should only be reviewed briefly whether there was any disagreement. The approximately 2700 flavoring substances to be covered by the program were published as a community register and received a unique identifier, FLAVIS or FL-No, that consists of two numbers (the first reflecting a grouping previously assigned by the CoE).

The European Flavour and Fragrance Association (EFFA), the European sister organization to FEMA (both are organized in the International Organization of the Flavour Industry), started to submit documentations in 2000, and the SCF adopted its first opinions in December 2002. The following year, the new EFSA became operative and one of its scientific panels, the *Panel on food additives, flavourings, processing aids and materials in contact with food* (AFC Panel) continued the work until 2008, when it was replaced by the *Panel on food contact materials, enzymes, flavourings, and processing aids* (CEF). Whereas the regulation foresaw 34 chemical groups (like the ones applied by JECFA and the FEMA GRAS panel), the FLAVIS working group deviated from this grouping and started a different system that numbered the opinions from the SCF and the EFSA panels as Flavouring Group Evaluations (FGE.No). With time, the number of groups grew and as the program is not yet finished, it is unclear how many FGE reports will eventually be published.

TABLE 9.1
Chemical Groups in Which Flavorings Are Grouped for Safety Evaluations

1. Straight-chain primary aliphatic alcohols/aldehydes/acids, acetals, and esters with esters containing saturated alcohols and acetals containing saturated aldehydes. No aromatic or heteroaromatic moiety as a component of an ester or acetal.
2. Branched-chain primary aliphatic alcohols/aldehydes/acids, acetal, and esters with esters containing branched-chain alcohols and acetals containing branched-chain aldehydes. No aromatic or heteroaromatic moiety as a component of an ester or acetal.
3. α, β-unsaturated (alkene or alkyne) straight-chain and branched-chain aliphatic primary alcohols/aldehydes/acids, acetals, and esters with esters containing α, β-unsaturated alcohol and acetal containing α, β-unsaturated alcohols or aldehydes. No aromatic or heteroaromatic moiety as a component of an ester or acetal.
4. Nonconjugated and accumulated unsaturated straight-chain and branched-chain aliphatic primary alcohols/aldehydes/acids, acetals, and esters with esters containing unsaturated alcohols and acetals containing unsaturated alcohols or aldehydes. No aromatic or heteroaromatic moiety as a component of an ester or acetal.
5. Saturated and unsaturated aliphatic secondary alcohols/ketones/ketals/esters with esters containing secondary alcohols. No aromatic or heteroaromatic moiety as a component of an ester or ketal.
6. Aliphatic, alicyclic, and aromatic saturated and unsaturated tertiary alcohols and esters with esters containing tertiary alcohols. Esters may contain any acid component.
7. Primary alicyclic saturated and unsaturated alcohols/aldehydes/acids/acetals/esters with esters containing alicyclic alcohols. Esters/acetals may contain aliphatic acyclic or alicylic acids or alcohol component.
8. Secondary alicyclic saturated and unsaturated alcohols/ketones/ketals/esters with ketals containing alicyclic alcohols or ketones and esters containing secondary alicyclic alcohols. Esters may contain aliphatic acyclic or alicyclic acid component.
9. Primary aliphatic saturated or unsaturated alcohols/aldehydes/acids/acetals/esters with a second primary, secondary, or tertiary oxygenated functional group including aliphatic lactones.
10. Secondary aliphatic saturated or unsaturated alcohols/ketones/ketals/esters with a second secondary or tertiary oxygenated functional group.
11. Alicyclic and aromatic lactones.
12. Maltol derivatives and ketodioxane derivatives.
13. Furanones and tetrahydrofurfuryl derivatives.
14. Furfuryl and furan derivatives with and without additional side-chain substituents and heteroatoms.
15. Phenyl ethyl alcohols, phenylacetic acids, related esters, phenoxyacetic acids, and related esters.
16. Aliphatic and alicyclic ethers.
17. Propenylhydroxybenzenes.
18. Allylhydroxybenzenes.
19. Capsaicin-related substances and related amides.
20. Aliphatic and aromatic mono- and di-thiols and mono-, di-, tri-, and polysulfides with or without additional oxygenated functional groups.
21. Aromatic ketones, secondary alcohols, and related esters.
22. Aryl-substituted primary alcohol/aldehyde/acid/ester/acetal derivatives, including unsaturated ones.
23. Benzyl alcohols/aldehydes/acids/esters/acetals. Benzyl and benzoate esters included. May also contain aliphatic acyclic or alicyclic ester or acetal component.

(continued)

TABLE 9.1 (Continued)
Chemical Groups in Which Flavorings Are Grouped for Safety Evaluations

24. Pyrazine derivatives.
25. Phenol derivatives containing ring-alkyl, ring-alkoxy, and side-chains with an oxygenated functional group.
26. Aromatic ethers including anisole derivatives.
27. Anthranilate derivatives.
28. Pyridine, pyrrole, and quinoline derivatives.
29. Thiazoles, thiophene, thiazoline, and thienyl derivatives.
30. Miscellaneous substances.
31. Aliphatic and aromatic hydrocarbons.
32. Epoxides.
33. Aliphatic and aromatic amines.
34. Amino acids.

TABLE 9.2
Internet Resources on Flavoring Regulations and Related Information

EFFA	European Flavour Association provides information and access to the European regulatory approach for flavorings used in foods as viewed by the flavoring industry: http://www.effa.eu
European Commission	Legislation, regulations, and guidance on flavorings in the European Union: http://ec.europa.eu/food/food/chemicalsafety/flavouring Online version of the list of flavorings substances to be evaluated for their safety with the intention to authorize them: http://ec.europa.eu/food/food/chemicalsafety/flavouring/database/dsp_search.cfm
European Food Safety Authority	Information of the European safety evaluation program: http://www.efsa.europa.eu/en/panels/cef.htm
FEMA	Flavor and Extracts Manufacturers Association provides information and access to the U.S. regulatory approach for flavorings used in foods: http://www.femaflavor.org/
IOFI	The International Organization of the Flavor Industry maintains a global database on flavorings used by member organizations; access is currently restricted: http://www.iofi.org/Iofi/English/Home/IOFI-database/page.aspx/59
JECFA	Access to the safety evaluation of flavorings performed by JECFA: http://apps.who.int/ipsc/database/evaluations/search.aspx All documents published by JECFA: http://www.fao.org/ag/agn/agns/jecfa_archive_en.asp
RIFM	RIFM/FEMA Fragrance and Flavor Database provides the comprehensive database relevant for the occurrence, use and safety of flavorings; access by subscription only: http://rifm.org/nd/Login.cfm

The safety evaluation of chemically defined flavorings by European experts of both EFSA panels differs in some aspects from the work done by JECFA; however, a brief review is not the place to do full justice to these differences. The most important ones relate to the exposure assessment, the alerts for genotoxicity, and the possible presence of isomers.

EFSA experts stressed the need to base intake calculations also on normal use levels in food (modified Theoretical Added Maximum Daily Intake), whereas JECFA based the evaluations mainly on the per capita consumption derived from market disappearance data (Maximal Survey Derived Intake).

The EFSA evaluation includes predictions for possible genotoxic activity based on in vitro models for structure–activity relationships; if these models predict such an activity, a genotoxic potential is not excluded and they are not evaluated by the group evaluation procedure. For a significant number of flavorings—among them, important ones such as maltol or, for some time, cinnamaldehyde—the evaluations by EFSA resulted in the opinion that safety could not be established; for most of these flavorings, EFFA intends to submit additional data, which means that the evaluations are not yet concluded. From an international point of view, problems may arise from the fact that most of these flavorings have been determined to be safe by JECFA.

For nature-identical substances such as limonene, by default the naturally occurring isomer is the one used by industry, a fact sometimes not clearly spelled out by the specifications defining the commercial products. Whereas JECFA accepted frequently the proposed descriptions by industry as being sufficient, EFSA panels tended to be more critical and to request more data and a clear and unambiguous chemical characterization of the flavoring.

EFSA does not maintain a publicly available database or index of the substances that have been evaluated by the expert panels. Since the work by EFSA is still ongoing, the future Community List of authorized chemically defined flavorings cannot be fully envisaged. The current electronic version of the register of flavoring substances (available at the Commission webpage, see Table 9.2) contains some information about the status of a substance including its Flavouring Group Evaluation (FGE), from which its fate may be deduced, but it is not regularly updated and is sometimes ambiguous in its descriptions.

9.5 NATURAL FLAVORING COMPLEXES

The work of flavorists is aimed at creating an organoleptic impression of a food that meets the taste expectation of the consumer for that food. Using an extract from that food or another source material serving the same purpose will, naturally, be a good starting point to meet this objective. In addition, consumers in many markets value highly the natural origin of food and prefer food products with "clean" labels. Substances of natural origin, specifically more complex mixtures, are difficult to standardize, but they gain market share and are also of interest because of potential nonflavor-related functional properties.

One may be surprised that regulators put much less focus on regulating such extracts that are now defined by the Codex Alimentarius as *natural flavoring complexes* than on chemically defined flavorings. In this area with few exceptions, the traditional self-control of industry prevails as the following analysis of the situation in the United States, at Codex Alimentarius, and in the European Union reveals.

9.5.1 United States/FEMA

Most of the natural flavoring complexes are essential oils or other extracted preparations, many of which have been in use since the nineteenth century and will therefore, since they had been in use before 1958, fall under the so-called grandfather clause and are by default accepted for use in food. The U.S. FDA published in addition lists of the most common herbs and spices and other *natural seasonings and flavorings* that had been used traditionally in food.

The FEMA GRAS Expert Panel proposed recently a safety evaluation procedure for natural flavoring complexes that builds on the approach developed for chemically defined flavorings. Since many of the more popular extracts are essential oils or similar preparations composed of volatile substances that, to a large extent, are also used as chemically defined flavorings, it should be possible to assess the risk of a natural flavoring complex used in food by referring to the existing evaluations of its constituents (Smith et al. 2005b). This reductionist approach has not been used yet by FEMA in an evaluation program of natural flavoring complexes similar to the one addressing chemically defined flavorings.

9.5.2 JECFA/Codex Alimentarius

The flavoring guidelines adopted by the CAC in 2008 provided the first comprehensive definition of natural flavoring complexes, which are "preparations that contain flavoring substances obtained by physical processes that may result in unavoidable but unintentional changes in the chemical structure of the flavoring (e.g., distillation and solvent extraction), or by enzymatic or microbiological processes, from material of plant or animal origin. Such material may be unprocessed, or processed for human consumption by traditional food-preparation processes (e.g., drying, torrefaction (roasting) and fermentation). Natural flavoring complexes include the essential oil, essence, or extractive, protein hydrolysate, distillate, or any product of roasting, heating, or enzymolysis." With the exception of the general requirement that the use of such flavorings in food should not lead to unsafe levels of their intake, no further requirements about the risk assessment of natural flavoring complexes were made.

JECFA discussed at its 63rd meeting in 2004 the "reductionist" approach to the evaluation of natural flavoring complexes using three essential oils as examples and agreed that such an approach was feasible (WHO 2004). The committee defined more closely the analytical data that would be necessary to embark on such an exercise and emphasized the need for appropriate product specifications that define the material(s) for which the safety evaluation was applicable. Evaluations of naturally flavoring complexes have not been requested to be performed by JECFA since this initial discussion, and it remains unclear whether JECFA/Codex will strive to develop a list of acceptable natural flavoring complexes similar to the one for chemically defined flavorings.

9.5.3 European Union

For more complex mixtures used as flavorings in food, the recently adopted European Union legislation Regulation (EC) No. 1334/2008 (which entered into force in January 2011) applies a regulatory approach that is distinctly different to chemically defined flavorings: premarket authorization and risk assessment by the competent authority EFSA are by default not required. *Flavoring preparations* (the term under which natural flavoring complexes are regulated in the European Union) may be obtained from food or nonfood sources by appropriate physical, enzymatic, or microbiological processes or by a process listed as a traditional food preparation process. All flavorings derived from foods may be used without an evaluation and approval as long as "they do not, on the basis of the scientific evidence available, pose a safety risk to the health of the consumer, and their use does not mislead the consumer." This waiver for a risk assessment applies according to an explanatory note also to many nonfood materials "if it can be sufficiently demonstrated that they have hitherto been used for the production of flavorings, [they] are considered to be food materials for this purpose, even though some of these source materials, such as rose wood and strawberry leaves, may not have been used for food as such. They do not need to be evaluated."

All other flavoring preparations that are of nonfood origin or produced by nonfood processing procedures need to be evaluated by EFSA before marketing. EFSA recently published guidelines for submission of an application for flavorings that shall define the data requirements also for such products.

9.6 PROCESS AND SMOKING FLAVORINGS

Processing of foodstuffs may lead to significant and intended changes of their organoleptic properties. Some processes such as frying and roasting will involve complex chemical reactions, such as the Maillard reactions, that will change not only the structure/texture of food but also its chemical composition, creating many components that are originally not present in food. Being characteristic for the taste of certain food categories, it is no surprise that the flavor industry produces flavorings using the same or similar processes that consist eventually of a complex mixture of substances. Since it is quite difficult to qualitatively and quantitatively define the composition of the thereby created flavoring preparation, it is no surprise that little emphasis has been applied on regulating these products with the exception of smoking flavorings. They contain polycyclic aromatic hydrocarbons (PAH), notably benzo[*a*]pyrene, which are known mutagens and suspected carcinogens for which human exposure may be controlled by setting maximum limits on their presence in flavoring preparations. Such limits were agreed upon by Codex Alimentarius and the European Union, and adopted also by many other countries.

The European Union adopted even more specific legislation on smoke flavorings in 2003: any product requires a premarket approval that is granted only after a safety evaluation by EFSA has determined that it is safe under the proposed conditions of use. The program covered at least eight products that so far have been evaluated, and corresponding legislation for their eventual authorization is pending.

For other flavorings produced by chemical reactions that result in rather complex mixtures used for flavoring purposes, the European Union is the first major entity providing a regulatory definition that is based on the idea that any flavoring preparation produced utilizing "traditional" food processing methods is acceptable, and that only processing methods that go beyond such long-term use require attention from a regulatory point of view. The criteria for assigning a flavoring preparation as "acceptable" or "to be evaluated prior to marketing" are laid down in an Annex to the new European flavoring regulation 1334/2008. The example given as "Heating, cooking, baking, frying (up to 240°C at atmospheric pressure) and pressure cooking (up to 120°C)" is a traditional food preparation process, whereas the example "flavorings which are obtained by heating oil or fat to an extremely high temperature for a very short period of time, resulting in a grill-like flavor…may be used in and on foods [only] after they have undergone an evaluation and approval procedure" is not.

For thermal process flavorings, specific conditions for their production were laid down by European legislation that include maximum level for two polycyclic heteroaromatic amines. If a flavoring preparation is produced from food starting materials, according to these conditions, no premarket authorization is required; however, using nonfood raw materials and productions conditions outside the range of these criteria will require submission of a dossier for evaluation by EFSA.

9.7 SUBSTANCES OF SPECIFIC CONCERN

For risk assessors, flavoring substances present a number of challenges, which have been discussed in other sections of this chapter. One of these challenges is that a number of characteristic compounds are classified by toxicologists as substances that have serious toxic effects for which no human intake (even very low ones) without such effects can be estimated. This would normally not be a problem since the intentional addition of such substances to food is anyhow prohibited by most countries. However, among them there are a number of substances that are characteristic constituents imparting the specific taste to popular ingredients and foods such as estragol in basil (used for *Pesto Genovese*) or menthofuran in mint leaves (used in mint sweets). Estragol is identified as mutagen that may cause cancer—and for such substances, exposure at any dose, even a very low one, may lead to a certain (low) probability of developing this disease; and menthofuran is hepatotoxic. Banning pesto or peppermint sweets is not that popular with consumers; such substances are prohibited in many countries as pure flavoring substances but tolerated as part of natural flavoring complexes. Until very recently, the Codex Alimentarius contained a list of substances that should be present only in a limited number of foods at defined maximum levels. In 2008, this list was abolished until a new risk-based list is put in place. In the same year, the European Union adopted its own list of 15 substances that should not be added to food, of which 12 are tolerated in some foods at maximum levels (e.g., estragol, menthofuran, coumarin, pulegone) and one at least is tolerated without any maximum level being set (capsaicin).

9.8 LABELING OF FLAVORINGS AND FLAVORED FOODS

Adding flavorings to food serves one ultimate goal: to modify smell and taste of food in such a way that the consumer enjoys the food. The flavorist will use all his skills to create for a food the one flavor the consumer expects; for this purpose, he will use food ingredients, natural flavoring complexes, chemically defined flavorings of natural or synthetic origin, and the processes that are known to create in situ flavor. And by doing so, he/she is possibly straying into an area which, during the entire history of human food supply, has been the main reason for mistrust by consumers toward people producing and selling food: fraud. That food shall be safe is currently the parole of the day, and it is quickly forgotten that fraudulent practices that deceive consumers about the nature of food items they buy in the market are, from their viewpoint, just as important.

Labeling of flavored food is governed by a number of principles, which will be discussed in the following paragraphs. It is beyond the scope of this chapter to provide specific guidance for specific foods; there are too many special cases and exemptions, and no attempt is made to address those.

9.8.1 Labeling of Ingredients and Flavorings

A common denominator of food labels in most regulatory frameworks is a list of ingredients in which, as the Codex Alimentarius stipulates, "all ingredients shall be listed in descending order of ingoing weight at the time of the manufacture of the food." "All ingredients" means every single ingredient of the food is listed whether it is an ingredient (e.g., strawberries), a food additive (e.g., ascorbic acid), or a flavoring. Whereas ingredients need to be listed individually, the presence of flavorings (usually several starting materials used in one flavoring preparation) may be indicated by listing them collectively as *flavorings* (or *flavor*). Ingredients that are themselves food items or derived from food items and which impart flavor to food need to be declared as such—for example, a *strawberry concentrate* will be identified as one ingredient, whereas a mixture of flavorings will just be *flavorings*.

9.8.2 Referring to a Flavor

Using a specific flavor designation in the name of a food shall lead to associations and expectations by consumers who may link the expected taste of a food to its composition: a *lemon tart* shall not only taste of the fruit but also contain it in the amount necessary to impart this characteristic flavor. Correspondingly, there are rules limiting or even prohibiting the addition of flavorings to foods that are known for a flavor resulting from the use of a characteristic ingredient. Such rules may also require a specific statement about the origin of the flavor if not caused by a characteristic ingredient. As examples, in the European Union certain chocolate products may contain added flavorings but not those that "mimic the taste of chocolate," and in the United States the label of a smoked food may not imply that a food flavored with artificial smoke flavor has been smoked or has a true smoked flavor resulting from smoking.

9.8.3 Using the Claim "Natural"

Consumers in today's societies appreciate if their food is *natural*. However, most of them probably never contemplate critically about what properties of a food or what processes applied in its production justify the statement that it is natural. Although a biologist—the author of this chapter is from that profession—might consider any kind of human skillful intervention as not being natural but cultural (or share the opposite view that humans are part of nature and their activities are therefore natural), the term natural is loaded in most markets with consumers' expectations that the concerned product is uncontaminated, healthy, and safe. In their view, the opposite of natural is artificial or chemical, and the latter term carries the burden of all efforts by industry to modify food in order to deceive them and putting their health at risk.

Not surprisingly, the term natural when used in relation to flavoring substances added to food is defined by regulatory authorities rather to meet consumer expectations. The most recent example is the legislation adopted by the European Union 2008, which provides a definition for natural that includes traditional and even modern food processing techniques: all these human inventions many of which lead to significant chemical changes may be used when producing flavorings and flavorings preparations and still the resulting flavoring is natural and the whole flavored food may be called natural (if other ingredients meet corresponding criteria).

The borderline between enjoying the status natural or missing it is a fine one and differ between markets; flavorists are advised to consult experts familiar with the legal requirements in the main markets before envisaging a formula that is intended to be accepted globally as natural.

In many regulations including the Codex Alimentarius, rules for food labeling the hybrid descriptor "nature identical" is still permitted, a term that tries to evoke specific properties natural substances may have in contrast to "artificial" or "synthetic" ones. However, risk assessors, and specifically the experts from JECFA stressed from the beginning of food chemical risk assessment that neither origin of raw materials, applied production processes, nor structure of a molecule are eventually good predictors of safety. More recently adopted legislation among them, the Codex Alimentarius rules for flavorings and the European Union regulation therefore no longer permit the use of the term "nature identical."

9.9 CONCLUSION

The safety and regulatory aspects of flavorings used in foods were discussed in this chapter by painting the big picture, analyzing the fundamentals, giving possibly the impression that European Union, United States, and the Codex Alimentarius are the only regulatory frameworks addressing flavorings. This is certainly not true, and a recent overview by Salzer (2007) provides information about other markets and their regulatory systems. However, in recent years, the work of the Codex Alimentarius is being increasingly accepted by countries that implement the list of flavorings evaluated by JECFA (and adopted by the CAC) into their national legislation.

During recent years, printed books are increasingly replaced by Internet resources, and Table 9.2 provides a short list of suitable web references; the selection focuses on pages that provide points of entry for regulatory information also from other sources. Valuable information on the regulatory status of many flavorings is also available from *Fenaroli's Handbook of Flavor Ingredients* (Burdock 2009).

The safety evaluation of food flavorings and the subsequent readjustment of regulatory frameworks are ongoing processes that will continue in the coming years; any professional interested in this field is advised to regularly consult the available sources of information in order to understand the status of a flavoring preparation and its conditions of use.

REFERENCES

Burdock, G.A. 2009. *Fenaroli's Handbook of Flavor Ingredients*, 6th edn. Boca Raton, FL: CRC Press.

Cramer, G.M., R.A. Ford, and R.L. Hall. 1978. Estimation of toxic hazard—A decision tree approach. *Food Cosmet. Toxicol.* 16(3): 255–276.

Codex Alimentarius Commission. 2008. Guidelines for the Use of Flavourings (CAC/GL 66-2008). Available at http://www.codexalimentarius.net/.

Salzer, J.U. 2007. Legislation/Toxicology. In *Flavourings*, ed. H. Ziegler. Weinheim, Germany: Wiley-VCH.

Smith, R.L., S.M. Cohen, J. Doull, V.J. Feron, J.I. Goodman, L.J. Marnett, I.C. Munro, P.S. Portoghese, W.J. Waddell, B.M. Wagner, T.B. Adams, and The Expert Panel of the Flavor and Extract Manufacturers Association. 2005a. Criteria for the safety evaluation of flavoring substances. *Food Chem. Toxicol.* 43(8): 1141–1177.

Smith, R.L., S.M. Cohen, J. Doull, V.J. Feron, J.I. Goodman, L.J. Marnett, P.S. Portoghese, W.J. Waddell, B.M. Wagner, R.L. Hall, N.A. Higley, C. Lucas-Gavin, and T.B. Adams. 2005b. A procedure for the safety evaluation of natural flavor complexes used as ingredients in food: Essential oils. *Food Chem. Toxicol.* 43(3): 345–363.

World Health Organization. 1996. Food Additives and Contaminants. WHO Food Additives Series 35. Annex 5. A procedure for the safety evaluation of flavouring substances (Paper by Dr. I.C. Munro, CanTox Inc., Mississauga, Ontario, Canada).

World Health Organization. 2004. Evaluation of certain food additives: Sixty-third report of the Joint FAO/WHO Expert Committee on Food Additives. WHO Technical Report Series, 928. Geneva, Switzerland.

pages that provide points of entry [illegible]
Valuable information on [illegible]
from [illegible]
The safety evaluation [illegible]
lator, framework [illegible]
any professional [illegible]
aspects of [illegible]
and a [illegible]

[illegible]

10 Essential Oils and Spices

Danuta Kalemba and Anna Wajs

CONTENTS

10.1 INTRODUCTION

Plants have been accompanying mankind since ancient times. Plants are people's first food. Subsequently, people began to discover that plants had other uses aside from their nutritional values. People found out that some plants improve the taste and fragrance of food, prolong food stability, or exert a favorable influence on digestion. Others appeared to repel some insects, whereas other plants seemed to cause beneficial medicinal activity. All ancient cultures treated aromatic plants not only as food flavorings but also as medicines, preservatives, and perfumes. Many cultures believed that spices had magical properties and used them in religious and ceremonial functions. In ancient and medieval times, spices had an enormous commercial value and were status symbols. Nowadays, in parallel with the demand for natural and organic food manufactured without the addition of salt, sugar, or chemical preservatives, the demand for spices, especially organic ones, is growing rapidly. Today's consumers are looking for new and exotic ingredients that can be found in ethnic cuisines. The ready-for-use spice blends as well as seasonings that cut the time and facilitate meal preparation are becoming more and more popular, giving the possibility of creating flavor and variety in meals available for everyone.

Spices are different parts of plants, called aromatic plants, used in food preparation to impart food odor, taste, and color. Besides these primary functions, spices also provide secondary effects, such as salt and sugar reduction, improvement of texture, as well as preservative (antimicrobial and antioxidant) and health functions. A majority of spices have been traditionally used for their medicinal properties, from which the capability to stimulate digestion is the most crucial in food. It is estimated that approximately 400 spices are being used around the world, although only 112 plant species are listed as spices by the International Organization for Standardization (ISO).

All parts of plants are used for seasoning purposes: leaves, or leafy stems called sometimes herbs (e.g., mint, basil, marjoram, oregano, thyme, savory, dill, lovage, rosemary, sage, curry, bay); stems (lemongrass); seeds (cardamom, nutmeg, sesame, mustard, nuts); fruits (anise, caraway, cumin, coriander, fennel, coriander, cumin, allspice, citrus, chili, fenugreek, peppers, star anise, vanilla, juniper, cranberry); bark (cinnamon, cassia); roots (angelica, horseradish); rhizomes (ginger, turmeric); flowers or blossoms (rose, orange, jasmine, paprika); flower buds (clove, elder); bulbs (onion, garlic); and wood (camphor). Spices provide a broad range of flavors to food and beverages: herbal (basil, marjoram, sage, dill, rosemary, parsley, etc.); spicy (allspice, cinnamon, clove, coriander, pepper); citrus (lemon, orange); minty (peppermint, spearmint); thymol-like (savory, thyme); alliaceous (garlic, onion, mustard); floral (bergamot, lemongrass, sweet basil, rose, elder, lavender); fruity (fennel, savory, star anise); woody (cassia, cardamom, juniper); and earthy (saffron, turmeric, black cumin). Some of them can give basic taste perception: sweet (anise, fennel, allspice); sour (sumac, caper); and bitter (fenugreek, clove, thyme, oregano, bay). The most important spices providing pungency to food are black and white pepper, capsicum, and ginger, and those providing color are saffron (yellow to orange), paprika (red), turmeric (orange yellow to reddish brown), and ginger (yellow).

Spices are used in different forms: fresh or dried; in whole, crushed, or powdered. Some spices have to be subjected to special treatment, for example, fermentation (black pepper, vanilla) in order to release their flavor. When buying small amounts of spices, on the domestic scale, it is preferable to buy whole spices and grind them just before use to obtain a more intense aroma. Dried ground spices lose their strength in flavor and color faster as a rule.

The nomenclature in spices and products derived from them is not identical in meaning. Sometimes the word "herbs" is used alternatively although it is preferably reserved for whole air part of medicinal plants. Spice blends, flavorings, and seasonings are sometimes also called spices. However, the soy sauce, nuts (almond, sesame), cocoa, chocolate, coffee and tea, as well as some special fruits (orange peel, olives) or vegetables (pickled mushrooms, dehydrated onion and garlic) are rather recognized as flavorings. On the other hand, seasonings are spice blends or formulations containing other ingredients, such as salt, sugar, vinegar, wine, soy sauce, and fish sauce. It is worth noting that ethnic spice blends and seasonings that are becoming more and more popular nowadays can differ significantly in both ingredients and flavoring properties depending on the manufacturer: not all curry, salsa, and pestos taste the same.

10.2 ESSENTIAL OILS

Spices are the source of different forms of extractives that are sometimes preferred to spices because there is a better possibility of standardization and incorporation into food, and are especially suitable for flavoring of beverages because of their more convenient form. Essential oils obtained by water or steam distillation are the most important and popular. However, fragrance- and flavor-producing substances can be isolated by many other methods. It should be stressed that products obtained by other methods shall not be considered essential oils.

Essential oils are complex mixtures of volatile compounds obtained by water or steam distillation of plant material, with the exception of those from citrus peels obtained by cold pressing. Expressed citrus oils have superior odor compared to distilled oils and contain nonvolatile compounds, for example, natural antioxidants that make them more stable to oxidation. Sometimes products obtained by dry distillation of plant exudates such as resins are also classified as essential oils. Essential oils have several physical properties in common. They are usually colorless or pale yellow liquids with strong odor resembling the source plant material. Essential oils are immiscible with water but soluble in alcohol, plant oils, and most organic solvents. Their density is usually smaller than that of water, they are characterized by a high refractive index, and most of them are optically active.

Essential oils are biosynthesized and accumulate in various parts of plants in special organs: oil cells, secretory ducts, and cavities located inside different plant tissues, or in glandular hairs situated in the outer cell layer, mainly of leaves or petals.

Essential oils are obtained from different parts of plants: leaves, flowers, flower buds, fruits, stems, seeds, roots, rhizomes, bark, and wood. They are also found in plant exudates such as resins and balsams. Some plants produce essential oils in different botanical parts. These oils can be similar in composition (e.g., angelica root and seed, clove buds and leaves) or are entirely different (e.g., bitter orange peel, flowers and leaves, coriander immature leaves and seeds, cinnamon bark, and leaves).

Water and steam distillation enable operators to isolate the plant constituents that boil at 150°C–300°C in temperatures below 100°C. Sometimes water distillation at an elevated pressure or more favorably at reduced pressure is used. For most plant materials, with the exception of leaves and flowers, comminution is necessary before distillation. The mode of distillation (water or steam) and the duration of the process depend mainly on the properties of plant material and fluctuates from 40 min for flowers through 90–120 min for herbs, up to 300 min for seeds and 500 min for roots. Essential oils are obtained with a broad range of yield, starting from 0.02% for rose oil, through about 1% for herbs and 3%–5% for seeds and fruits, and up to 18% for clove oil. After water or steam distillation, the essential oil is separated from the water phase that contains some amounts of volatiles and can be also a valuable product called hydrosol.

All essential oils obtained from spices are used not only in food preparation but also in perfumery. On the other side, a few of so-called perfumery oils that were derived from plants not normally classified as food, are found to be indispensable in creating special notes in certain food aroma, for example, cedarwood oils and patchouli oil.

Essential oils are usually used as such. However, for some purposes, raw essential oils are processed by different ways from which vacuum rectification is the most frequently used. It enables us to remove a substantial part of monoterpene hydrocarbons and is used in obtaining terpeneless oils. Such citrus oils are more resistant to oxidation and have better solubility in water. Fractionation is usually carried out for obtaining sesquiterpeneless oils or for separation of desirable oil components. Isolates obtained through this process or in other ways are utilized for special purposes (e.g., menthol form peppermint oil is bubble gum and tobacco flavoring) or for creation of natural flavorings (e.g., linalool from rosewood oil is a component of natural apricot flavors and eugenol from clove leaf oil of banana flavors). Essential oils, their fractions, and their isolates are utilized in food, perfumery, flavors, and fragrances, cosmetics and toiletries, fine chemicals, and pharmaceutical industries, as well as in therapy and aromatherapy.

10.2.1 Composition of Essential Oils

Essential oils are multicomponent mixtures. In a majority of essential oils, about 100 components have been identified and in some important oils, more than 300 constituents have been known (rose oil, lavender oil, bergamot oil). The content of individual components fluctuates from traces (α-thioterpineol in grapefruit oil) to above 90% (limonene in lemon oil, eugenol in clove oil). Usually, the flavor of the essential oil derives from its main constituents (menthol in peppermint oil, thymol in thyme oil, linalool and linalyl acetate in lavender oil). However, in many cases, the very characteristic flavor notes of essential oils are created by minor compounds (geranial and neral in lemon oil, β-damascenone and β-ionone in rose oil, nootkatone in grapefruit oil). Chirality is an important aspect of essential oil compounds because enantiomers may possess different smell and taste. The most spectacular example is monoterpene ketone carvone, in which (*S*)-(+)-isomer is a main constituent of caraway oil and has a caraway flavor, whereas (*R*)-(–)-isomer possessing a mint flavor is the major component of spearmint oil.

Since essential oils are natural products, their chemical composition as well as physicochemical properties cannot be precisely quantified. Correct botanical description of the plant material is precious. The genus *Mentha* L. comprises about 25 species and even 900 taxons; there are hundreds of eucalyptus species and varieties. However, such factors as cultivation conditions (growing region, climate), harvesting time, methods of preparing materials, distillation parameters, and many others influence the final product. Plants of the same species grown wild or cultivated in different parts of the world generally have different percentages of the same components. Numerous species produce nevertheless several chemotypes with different dominant flavor constituents. Hence, it is important not only to know the botanical name of the plant, but also its origin and main constituents. The most important example is common thyme, *Thymus vulgaris* L. Thymol and carvacrol types are the most relevant, but others such as geraniol, cineole, or linalool types grow in different regions. What is more, different species can be traded under the same name. This is the case for thyme, where not only common thyme but also species such as *T. capitatus*, *T. serpyllum*, or their mixtures are universally accepted as thyme. Similarly with sage, which is defined as *Salvia officinalis*, there are some 300 species and some of them, for example, *S. trilobula* and *S. tomatosa*, are traded and accepted throughout the world at present as sage.

Although essential oils are not strictly defined, their standardization is crucial, especially when they are used in food and pharmaceutical industries. Specifications for many essential oils and other plant-derived flavors have been published in different sources. The main are: International Standards Organization (ISO), Flavor and Extract Manufacturers Association (FEMA), and the national pharmacopoeias. Criteria established by these sources include the information on the origin (botanical source, geographical source, plant part used, degree of maturity, and method of isolation), the physical analysis (specific gravity, refractive index, optical rotation, color, solubility, etc.), and chemical analysis (e.g., specific numbers, GC profile, the range or upper limit of concentration required of the target constituents). The majority of essential oils from herbs and spices are classified as Generally Recognized as Safe (GRAS).

10.2.1.1 Constituents of Essential Oils

Essential oils mainly comprise volatile compounds that have carbon, hydrogen, and oxygen as their building blocks. These are subdivided into two groups: hydrocarbons that are made up almost exclusively of terpenes (monoterpene, sesquiterpene, and diterpene) and oxygenated compounds that—besides all mentioned groups of terpenes—consist of compounds with phenylpropanoid and aliphatic skeletons. Some compounds may also contain nitrogen (anthranilates, pyrazines) or sulfur (thiols, sulfides). Structures of the main and characteristic flavor compounds are shown in Figure 10.1 (terpenes) and Figure 10.2 (nonterpenes).

A great diversity of terpenes can be found as essential oil constituents. More than 30 monoterpene hydrocarbon skeletons (acyclic, mono-, bi-, and tricyclic) have been described. Because of the wide variety of constitutional isomers and stereoisomers

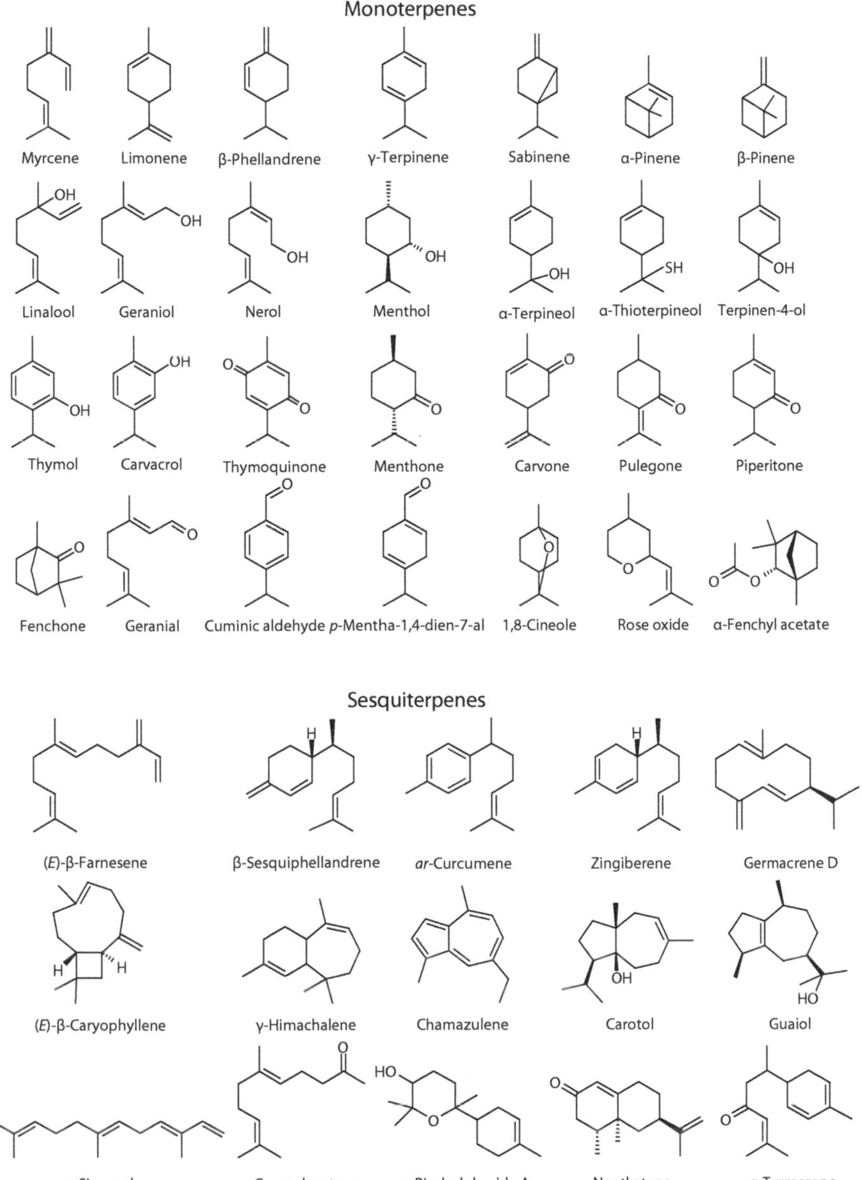

FIGURE 10.1 Terpene flavor compounds.

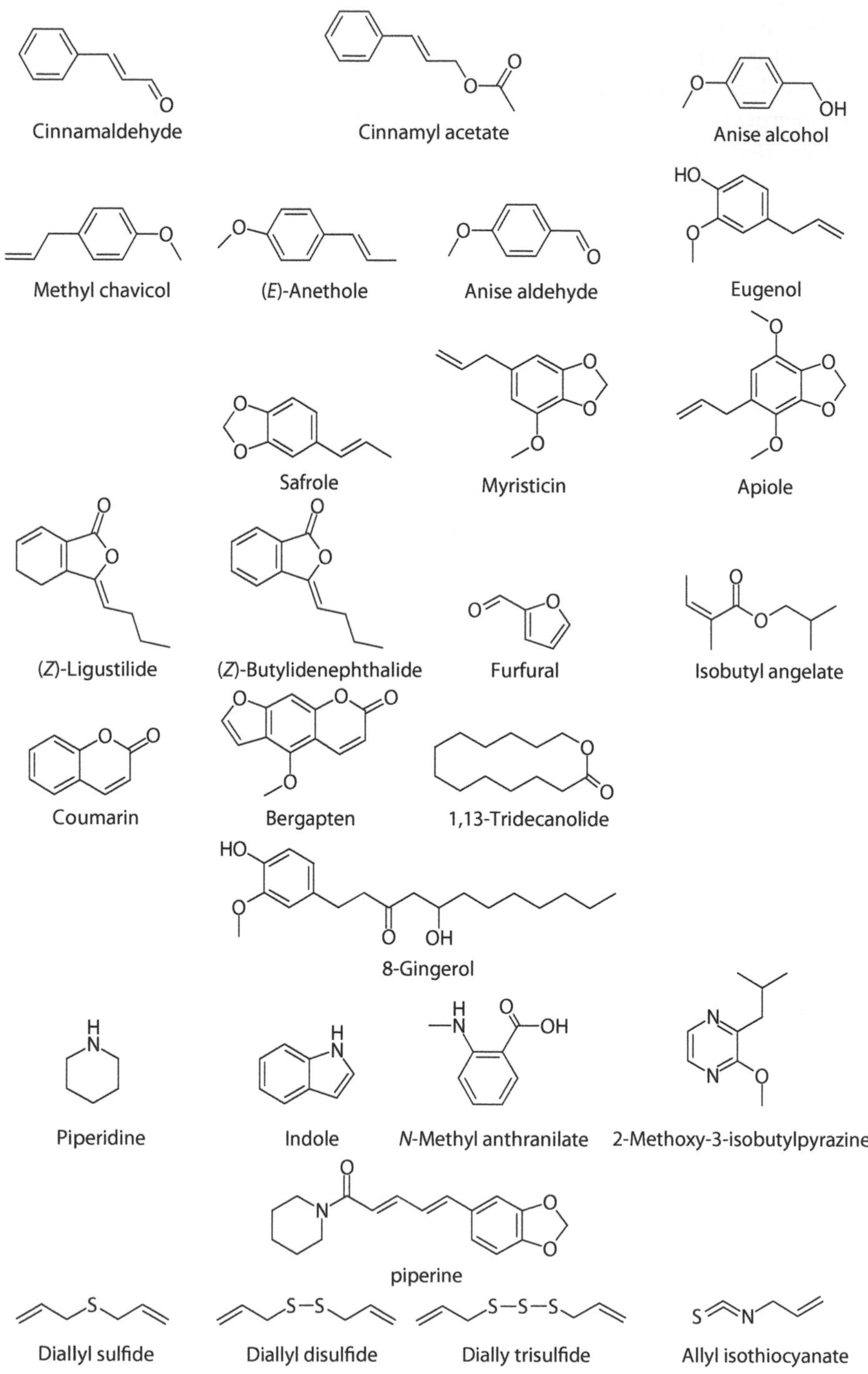

FIGURE 10.2 Selected nonterpene flavor compounds in spices.

combined with different functional groups, each of these skeletons represent dozens of chemical individuals. The number of sesquiterpene hydrocarbon skeletons (from acyclic to fourcyclic) have exceeded 100, and the variety of their isomers is many times higher than that for monoterpenes. For example, there are two enantiomeric forms of pinane, and each has only two unsaturated hydrocarbons: α- and β-pinene. On the other hand, each of the four stereoisomers with cadinane skeleton can result in several unsaturated hydrocarbons, for example, α-, β-, γ-, δ-, ε-, ω-cadinene, and some other cadinadienes are found as natural compounds. From each of these compounds, dozens of oxygenated components can be derived such as alcohols, esters, or lactones, ethers or epoxides, aldehydes, ketones, and acids. This results in a huge number of essential oil constituents estimated to reach 5000. Since many of these components are common to many essential oils, it is evaluated that the ability to identify as few as 500 compounds would enable researchers to identify more than 90% of the essential oils of most species.

The compounds most frequently found as essential oil constituents are monoterpenes. Although oxygenated derivatives are the major contributors to the aromatic sensation of spices, the participation of hydrocarbons cannot be passed over. Monoterpene hydrocarbons that can be the most frequently found as essential oil constituents are: limonene, pinenes (α- and β-), phellandrenes (α- and β-), terpinenes (α- and γ-), sabinene, camphene, and myrcene. β-Caryophyllene is the most common sesquiterpene hydrocarbon. The most important group of compounds are terpene alcohols, which include the following: linalool, menthol, geraniol, nerol, and citronellol, borneol, α-terpineol, terpinen-4-ol, farnesols, as well as esters of these alcohols, mainly acetates. A special group of constituents are phenols with thymol and carvacrol. Other oxygenated terpenes include aldehydes: citronellal, geranial, and neral; ketones: camphor, thujones (α- and β-), carvone, menthone, and pulegone; and oxides: 1,8-cineole, ascaridole, limonene oxides, bisabolol oxides, and lactones (ligustilides). The most common phenylpropanes are: a phenolic compound eugenol, and some methyl ethers—methyleugenol, anethole, estragole (methyl chavicol), and myristicin. To the most important aliphatic constituents belong oct-1-en-3-ol, (*E*)- and (*Z*)-hex-3-enols.

10.2.2 Hydrolates

Hydrolates—known also as hydrosols, aromatic waters, or floral waters—are by-products obtained together with essential oil during water or steam distillation. They contain small amounts of plant volatiles that remained dissolved in water. Their composition is different from that of essential oil. They are free of lipophilic substances such as terpene hydrocarbons. On the other hand, due to their better solubility in water, the content of hydrophylic components is higher in the hydrolate than in the respective oil. Hydrolates are mainly used in food and aromatherapy.

10.2.3 Oleoresins

Oleoresins are—besides essential oils—the second important group of spice-derived products used as food flavorings. Oleoresins are liquid or semisolid extracts obtained from plants by solvent extraction followed by removal of the solvent. Prepared oleoresins contain volatile and nonvolatile flavoring components of spice as well as other nonvolatile ingredients that include fixed oils, antioxidants, pigments, and other extractives soluble in the particular solvent used. Oleoresins have the full aroma and pungency of spice and—what is more and valid—provide the color. The most important commercial oleoresins are produced from red pepper (paprika), black or white pepper, ginger, garlic, and rosemary. The name oleoresins—together with resins—is also ascribed to natural plant exudates known also as balsams (Peru Balsam) or oleo-gum resins (myrrh, olibanum).

The yield and composition of prepared oleoresin depends strongly on the solvent. Recently, liquefied gases or supercritical fluids are being used to increase the extraction rate and yields. Extraction of plants with carbon dioxide is now done on a commercial scale. Depending on extraction conditions, different products can be obtained. Operating in subcritical parameters produces a volatile oil with composition close to the respective essential oil. Using supercritical conditions, it is viable not only to produce oleoresins but also to fractionate them. Carbon dioxide extraction offers numerous advantages. The composition of products can be modified by varying the temperature and pressure and adding modifiers to increase solvent polarity, for example, ethanol. Resulting products are free of solvent residues. The main disadvantage of this process is the high cost.

10.2.4 Other Spice Extractives

Fruit juice concentrates and distillates also known as essences are sometimes used as flavorings. Concentration can be done by vacuum distillation or by freezing.

For the soft drink and alcoholic beverages industry, alcoholic tinctures of spices are produced.

For perfumery and cosmetic purposes, other plant extractives are obtained. Extraction of plant material with apolar solvents, for example, hexane, followed by solvent recovery by distillation provides concrete. Concretes are solid or semisolid mixtures of essential oil and fixed oils. Extraction of resins with a hydrocarbon solvent results in resinoids. Solution of concrete or resinoid in ethyl alcohol and subsequent chilling and filtering eliminates waxes, a part of mono- and sesquiterpene hydrocarbons, and after removing alcohol by distillation produces absolute. These two techniques involve the treatment of plant material that is too sensitive for distillation, for example, jasmine flowers or violet leaves (0.1% of concrete with hexane that yields 40% of absolute or 8% of volatile oil). Extraction via cold fats (tallow, lard), which is named *enfleurage*, is applied to very delicate plant materials especially flowers such as jasmine and tuberose.

10.3 FLAVORING FORMS

Because of their pronounced lipophilic tendencies, essential oils are not soluble in water and well soluble in organic solvents and plant oils. Additionally, essential oil constituents that have low molecular weight are more or less volatile and are prone to oxidization, polymerization, etc., that change their color and viscosity. Therefore, in order to prevent undesirable changes, essential oil should be stored in fully filled steel or dark glass containers with nitrogen in headspace. The cited features limit the use of essential oils, especially in thermally processed foods and food with high content of water (e.g., beverages). To some extent, the same may be applicable to spices, oleoresins, and other plant extractives. In order to protect flavorings from high temperature and to improve their incorporation and solubility in food matrices, different processes are used. Encapsulated and emulsified products belong to the most popular segment. Encapsulated flavorings are prepared form essential oils and oleoresins mainly by spray-drying with modified starch derivatives and/or gum acacia. Such products have extended shelf life, and they are more convenient to handle in dry applications. Emulsions of essential oils and oleoresins are prepared by using permitted emulsifying agents (Reineccius 1994; Raghavan 2007).

10.4 PLANTS USED AS SPICES

The following plants (listed alphabetically) are selected for their importance as spices and as raw materials for isolating essential oil and other flavor products. Monographs of these plants and their flavor products are described according to Ashurst (1999), Burdock (2002), Cseke et al. (2006), Lawrence (1985), Lawrence (1985–2007), Morton et al. (1990), Rowe (2005), Rovira (2008), and Wright (2004).

10.4.1 Angelica (*Angelica archangelica* L.; Apiaceae)

Angelica has been cultivated as a vegetable and medicinal plant, mainly in France, Belgium, and Germany. The essential oil is isolated from the roots (yield 0.35%–1.3%), seeds (yield 0.6%–1.5%), and stems. The root oil is preferred to seed oil, but the composition is quite variable. The most important compound of essential oil is 1,15-pentadecanolide of musk flavor (0.9% root, 0.4% seed), but the main compounds are β-phellandrene (20% root, 60% seed), α-pinene (20%), limonene (10%), borneol (1%), bornyl acetate (1%), and 1,13-tridecanolide (0.5%). The dried roots and the corresponding derivatives are used in the formulation of liqueurs, aquavits, and bitters, as well as in foods: omelettes, trout, and as jam. The long bright green stems are also candied and used as decoration. Angelica oil acts as a key component (because of the musk scent) in many berry flavors: blackberry, raspberry, loganberry. Angelica seeds and roots are sometimes used in making absinthe, whereas angelica roots with juniper berries are used to make gin.

Derivatives (FEMA): [1] angelica root oil (2088); [2] angelica seed oil (2090); [3] angelica stem oil; [4] fluid extract from seeds (2089) and roots; [5] tincture.

10.4.2 Anise (*Pimipinella anisum* L.; Apiaceae; FEMA 2093)

This herbaceous annual plant is native to the eastern Mediterranean region and Minor Asia, and now extensively cultivated in Europe, Russia, and India. It is known for its sweet, soft, and mild flavor with rich effects that resemble licorices, fennel, and tarragon. Anise should not be confused with Chinese (*Illicium verum*) or Japanese star anise (*Illicium anisatum*). The anise seeds are rich in essential oil (yield 2.5%–3%). The most important producer is Turkey, although plants of better quality come from Spain. By 1999, the global annual production of this essential oil was only 8 tons, compared to 400 tons for star anise. The main compound responsible for its characteristic taste and smell as well as for its medicinal properties is *trans*-anthole (85%–95%). The others are: methyl chavicol (2%–4%), γ-himachalene (2%), anise aldehyde, and anise alcohol (<1%). An unusual compound is the ester, 4-methoxy-2-(prop-1-enyl)phenol 2-methyl-butyrate, which is characteristic for anise (<5%). Anise oil is used in beverages, seasoning, and confectionery applications, and in small quantities in natural berry flavors.

Derivatives (FEMA): [1] anise seed oil (2094); [2] fluid extract; [3] tincture; [4] oleoresin.

10.4.3 Balm, Lemon Balm (*Melissa officinalis* L.; Lamiaceae; FEMA 2111)

This plant grows in the submontane areas of Southern and Central Europe and North Africa. Parts used are leaves and flowering tops. Balm leaves, which are harvested before flowering, have a lemon odor and a tonic-like flavor. Fresh leaves add a delicate flavor to many dishes, oils, vinegar, and liqueurs. The yield of leaf oil is 0.1%–0.3%, but the oil is produced in small amounts. The essential oil composition depends widely on the origin of the plant and its harvesting time. The major compounds are geranial, neral, citronallal, β-caryphyllene, linalol, nerol, and geraniol. Methyl citronellate is characteristic for Melissa leaf oil and distinguishes it from the lemongrass oil.

Derivatives (FEMA): [1] balm leaf oil (2113); [2] balm leaf extract (2112); [3] tincture.

10.4.4 Basil (*Ocimum basilicum* L.; Lamiaceae; FEMA 2118)

Basil is a herbaceous plant native to tropical regions of Asia and Pacific Islands. It is cultivated as a culinary herb in Europe. Sweet basil has a warm, spicy odor with fresh, mint-like, and camphor-like flavor. Most culinary and ornamental basils are cultivars of the species *O. basilicum*, but other species are also grown and there are many hybrids between species. Essential oil is isolated from leaves, stems, and flowers with the low yield of about 0.1%. Fourteen tons of the oil is annually produced in Comoros, Madagascar, and France. Major components are methyl chavicol (75%–86% in Comoros type, 25% in sweet European type), linalool (1%–5% in Comoros, 45% in sweet type), 1,8-cineole (1%–5%), and eugenol (1%–3%). The oil is mainly used in seasoning blends but can be useful in small quantities in a wide range of natural flavors.

Derivatives (FEMA): [1] basil oil (2119); [2] oleoresin; [3] infusion; [4] tincture.

10.4.5 Bergamot (*Citrus aurantium* L., syn. *Citrus bergamia* Risso & Poit.; Rutaceae)

This is a small tree similar to other citrus plants. Although bergamot plant is native to Asia, it is cultivated extensively in Calabria (Italy). Essential oil is mainly produced from fresh rind by cold pressing (yield 0.5%) and in small quantities by steam distillation from leaves and twigs. Some of the bergamot oil is distilled to produce oil free from bergaptene (skin sensitizer) and terpenes. Bergamot oil is a complex mixture of more than 300 compounds. Major components are limonene (30%–45%), linalyl acetate (22%–36%), linalool (3%–15%), γ-terpinene (6%–10%), β-pinene (6%–10%); minor compounds are geranial, neral, neryl acetate, geranyl acetate, and bergaptene. Most of the annual total of 200 tons of bergamot oil is produced by Italy and Ivory Coast. Other producers are Guinea, Brazil, Argentina, and Spain. Bergamot has a fragrant, sweet fruit odor so the major use of bergamot oil has imparted a citrus flavor to food and beverages (citrus soft drink), flavored Earl Grey, Lady Grey teas, and confectionery. The oil is also used in many natural fruit flavors, especially apricot. Bergamot oil was also a component of the original Eau de Cologne developed in the seventeenth century, and is now very common in perfumery.

Derivatives (FEMA): [1] bergamot rind oils (2153); [2] bergamot rectified rind oil; [3] bergamot leaf/twig oil.

10.4.6 Camphor (*Cinnamomum camphora* L.; Lauraceae; FEMA 2230)

This tall evergreen tree is native to the Far East (China, Japan). Almost whole camphor oil is obtained by steam distillation of the wood (yield 2.2%) classified as *C. camphora*. L. Sieb and a number of related varieties. A total of 2000 tons of camphor oils is produced in Taiwan, China, and Japan. The original crude oil (including many acids) is semicrystalline. About 70% of camphor is removed from crude oil to give camphor oil. This oil is further fractionated to obtain three separate oils: white camphor oil (13% of the original oil: 1,8-cineole 46%, α-pinene 22%, camphor 21%), which can be rectified to give an oil with some similarity to eucalyptus oil, brown camphor oil (14% of the original oil; isosafrole 32%, safrole 14%), and blue camphor oil (0.7% of the original oil: azulenes).

Derivatives (FEMA): [1] camphor white oil (2231); [2] camphor leaf oil; [3] oils from varieties of *C. camphora*.

10.4.7 Caraway (*Carum carvi* L.; Apiaceae; FEMA 2236)

This herbaceous plant is very common in Europe, Asia, and Africa. Ten tons of caraway oil is produced each year mainly in Egypt and the Netherlands. The parts used are fruits, containing about 17% of fixed oil and 3%–7% of essential oil of warm biting flavor with strong, fatty undernote. The main components are *S*-(+)-carvone (50%–70%) and limonene (40%). Small quantities of caraway oil is used in natural bread and yeast flavors. Caraway seeds are used as a spice in breads, casseroles, curry, and other foods. It is also used as a flavoring agent in some cheeses, aquavits, and several liqueurs.

Derivatives (FEMA): [1] caraway seed oil crude or double rectified (2238); [2] oleoresin; [3] infusion; [4] tincture.

10.4.8 Cardamom (*Elettaria cardamomum* Maton; Zingiberaceae; FEMA 2240)

The plant bears an aromatic green fruit containing many seeds. Cardamom seeds have a warm, spicy, and aromatic odor, and are used not only as spice but also as essential oil (yield 1%–5%). Only 8 tons of this oil is produced in Guatemala, India, and Sri Lanka. The composition of cardamom seed oil from India differs strongly from its counterparts found in Sri Lanka or Guatemala. The main constituents of Indian essential oil are limonene, α-terpinyl acetate, cineole, and α-terpineol. The name cardamom is used not only for *Elettaria* genus (commonly called cardamom, green cardamom, or true cardamom), but also for *Amomum* genus (commonly known as black cardamom). Both forms of cardamom seeds are used as flavorings in foods, drinks, as cooking spices, and as a medicine, whereas *E. cardamomum* oil is used in some tea flavors and in seasoning blends.

Derivatives (FEMA): [1] cardamom seed oil (2241); [2] oleoresin; [3] fluid extract; [4] tincture.

10.4.9 Carrot (*Daucus carota* L.; Apiaceae)

Carrot is a biennial, herbaceous plant common in Europe and many other countries. The parts used are roots and seeds. Carrot roots can be eaten in a variety of ways. Carrot seed oil (yield 0.05%–0.6%) has a pleasant aromatic odor and warm, spicy, sweet piquant flavor. Two tons of this oil is annually produced, mainly in France. The composition of the oil is as follows: carotol (18%–30%), sabinene (9%–11%), geranylacetone (10%–15%), and α- and β-pinene (<13%). The oil is mainly used in vegetable flavors.

Derivatives (FEMA): [1] carrot seed oil (FEMA 2244); [2] carrot root oil; [3] decoction; [4] infusion; [5] tincture.

10.4.10 Celery (*Apium graveolens* L.; Apiaceae; FEMA 2268)

This annual herb is native to Europe, so it is cultivated in many countries there, but also in Asia, United States, and Egypt. The parts mainly used are seeds, and sometimes roots and leaves. Celery seed essential oil (1.5%–2.5% yield) and solid seed extract (22% yield) has long-lasting, spicy–warm and burning taste, and is one of the most diffusive odors and one of the most penetrating flavors. Most of the annual production of 25 tons originates from India. The major component is limonene (60%–72%), but most of the celery scent is derived from minor components: 3-butyl 4,5-dihydrophthalides, 3-*n*-butyl-phthalides, and 3-butyl 3,4,5,6-tetrahydrophthalides. The seed oil is used in seasoning blends and in some natural savory flavors. It adds a unique character to beef, maple, and caramel flavors, but lovage oil is better in maple flavors. About 500 kg of celery herb oil is produced in the United States and Egypt. Only the qualitative composition of herb oil is similar to seed oil. The material is used in seasoning blends.

Derivatives (FEMA): [1] celery seed oil (2271); [2] seed extract (2269); [3] seed solid extract (2270); [4] tincture (from roots—mainly for pharmaceutical use).

10.4.11 Chamomile (German) (*Matricaria chamomilla* L.; Asteraceae)

Only 1 ton of this oil, variously described as "blue," "Hungarian," or "German," is still produced. The yield of flower oil is 0.4%. It contains chamazulene (6%), resulting in the characteristic blue color of the oil, and other compounds: α-bisabolol oxide (40%) and farnesenes (20%). It is much less useful in flavor terms than the Roman chamomile oil. German chamomile oil is FEMA 2273.

10.4.12 Chamomile (Roman) (*Chamaemelum nobile* L.; Asteraceae)

The essential oil is isolated from flowers (yield 0.7%). Three tons of "Roman" or "English" chamomile oil is produced annually in Europe. Major components are isobutyl angelate (30%), isoamyl angelate (12%–22%), and other esters. The oil is used in many natural fruit flavors, particularly apple, pear, peach, apricot, mango, and passion fruit. Roman chamomile oil is FEMA 2272.

10.4.13 Cinnamon (*Cinnamomum zeylanicum* Blume, syn. *C. verum* J.Presl; Lauraceae; FEMA 2289)

Ceylon cinnamon (from Sri Lanka) is a tree growing to 10 m. The part used as a spice is the outer bark, which is peeled from the trees every second year and sun-dried. Cinnamon bark is widely used as a powdered spice. It is generally used in cookery as a condiment and flavoring material. Products isolated from bark are essential oil (yield 0.5%)—5 tons annually and solid extract (yield 10%). Major components of the oil are cinnamaldehyde (65%–80%), cinnamyl acetate (5%), eugenol (4%), linalool (2%), and also coumarin (0.7%), which is limited in food and beverages by EU. Essential oil is also isolated from Ceylon cinnamon leaves (yield 1%)—100 tons annually. The leaf oil has a quite different flavor as compared to bark oil. It has spicy cinnamon, clove-like odor and taste, whereas cinnamon bark oil has a bitter flavor, slightly pungent, and burning. The main components of leaf oil are eugenol (70%), β-caryophyllene (6%), cinnamaldehyde (3%), isoeugenol (2%), linalool (2%), and cinnamyl acetate (2%). It is used as an alternative to clove oil in seasoning blends, and can be blended with cinnamaldehyde to approximate the character of cinnamon bark oil. "Saigon cinnamon" (*C. burmanni*) is also common. "Chinese cinnamon" is much cheaper that cinnamon because this is dried bark of *C. cassia*, which is a very common large tree in Southeast Asia. Both cinnamon bark and leaf oil are used in seasoning blends. Leaf oil can be blended with cinnamaldehyde to approximate the character of cinnamon bark oil.

Derivatives (FEMA): [1] cinnamon bark oil (2291); [2] cinnamon bark extract (2290); [3] cinnamon leaf oil (2292); [4] tincture; [5] fluid extract; [6] oleoresin.

10.4.14 Citronella (*Cymbopogon nardus* L.—Ceylon citronella; *Cymbopogon winterianus* Jowitt—Java citronella; Poaceae)

These are perennial grasses originating in tropical Asia. The fresh cut or dried herb is steam distilled with about 0.5% yield. The oil has a characteristic lemon- and rose-like odor. A total of 150 tons of citronella oil is produced each year. Major components of the oil are citronellal (35%), geraniol (23%), and citronellol (10%). Most of the oil is used in fragrances, but small quantities are used in natural citrus, especially lemon, and fruit flavors. Citronella oil is FEMA 2308.

10.4.15 Clove (*Eugenia caryophyllata* Thunb.; Myrtaceae; FEMA 2327)

This evergreen tree is native to the tropical regions of Asia and South Africa. The parts used are dried unopened flower buds and twig tips, harvested by hand and sun-dried. The highest quality buds are sold as whole cloves. Lower quality clove parts are used in the production of essential oil. The clove bud oil yield is 15%–20% (50 tons is produced in Madagascar), stem oil yield is 5% (100 tons is isolated in Tanzania, Madagascar, and Indonesia), and leaf oil yield is 2% (2000 tons is produced annually worldwide). The quality of the stem oil is intermediate between bud and leaf oils. Approximately 15 tons of bud solid extract (oleoresin) is produced each year. The major component of clove products is eugenol. The oils have a characteristic clove-like aroma and burning, spicy flavor and is used in seasoning blends and in some natural flavors, especially banana, blackberry, cherry, and smoke.

Derivatives (FEMA): [1] clove bud oil (2323); [2] clove bud oleoresin (2324); [3] clove bud extract (2322); [4] clove leaf oil (2325); [5] clove stem oil (2328); [6] oleoresin (from buds only); [7] tincture.

10.4.16 Coriander (*Coriandrum sativum* L.; Apiaceae; FEMA 2333)

This is an annual herbaceous plant that grows in the Middle East. The parts used are ripe dry fruits known as coriander seeds. Fruits reminiscent lemony citrus flavor when crushed, due to linalool and α- and β-pinene. The yield of essential oil from fresh herb is only 0.02%, thus only small quantities are produced mainly in France and Egypt. Seven hundred tons of seed oil is distributed by Russia (yield 0.3%–1.1%). The main seed oil components are linalool (65%–78%), α-pinene (3%–7%), camphor 4%–6%, γ-terpinene 2%–7%, limonene 2%–5%, and geranyl acetate 1%–4%. Coriander seed oil is used as a key component of gin and other alcoholic drink flavors and as a component of natural apricot, peach, and other fruit flavors. Coriander is used in seasoning and curry blends, for pickling vegetables, and making sausages. Leaves have a different taste from seeds, with citrus overtones. Fresh coriander leaves are often used in salads in Russia. Coriander roots have a deeper, more intense flavor than the leaves and are used in a variety of Asian cuisines.

Derivatives (FEMA): [1] coriander seed oil (FEMA 2324); [2] coriander herb oil; [3] tincture; [4] fluid extract; [5] oleoresin (FEMA 2334); [6] infusion.

10.4.17 Cornmint (*Mentha arvensis* L.; Lamiaceae)

It is a herbaceous perennial plant that is native to the polar regions of Europe, Asia (Himalaya, Siberia), and North America. Six subspecies are known. Seven hundred ten tons of essential oil from herb is produced annually in China (yield 0.5%–2%). The major compound is menthol (60%). This oil is often dementholized—amount of menthol is reduced to 35%. The other important components of such oil are: menthone (38%), isomenthone (8%), piperitone (3%), and pulegone (0.5%), which is limited by EU. The oil is used as a cheap alternative to peppermint oil, but is easily recognized organoleptically because of its harsh flavor. Cornmint oil can be used in herbal blends and liquor flavors.

Derivatives (FEMA): [1] cornmint oil; [2] dementholized cornmint oil.

10.4.18 Cumin (*Cuminum cyminum* L.; Apiaceae; FEMA 2340)

This small annual herb is native to Egypt, but is cultivated in Morocco, Cyprus, India, China, Iran, and south Europe because of its aromatic seeds. Today, cumin is one of the most popular spices in the world after black pepper. Cumin seeds are extensively used in cooking to flavor commercial food products. It is a major component of the curry and chilli powder. Cumin has a strong, distinctive odor of cuminaldehyde and spicy, faintly pungent flavor. Seeds are also the source of essential oil (yield 3%–5%). Most of the annual total of 12 tons of the oil is produced from Spain and Egypt. The main components of the oil are cuminic aldehyde (16%–33%), γ-terpinene (16%–22%), β-pinene (12%–18%), *p*-mentha-1,4-dien-7-al (<15%), and *p*-cymene (3%–8%). The odor of this oil is extremely powerful, diffusive, green-spicy, slightly fatty, but at the same time almost soft. This softness is one of the characteristics of cumin oil and is difficult to reproduce artificially with cumin aldehyde. Important aroma compounds of toasted cumin are the substituted pyrazines, 2-ethoxy-3-isopropylpyrazine, 2-methoxy-3-*sec*-butylpyrazine, and 2-methoxy-3-methylpyrazine. The oil is used in natural fruit flavors, especially mandarin, grapefruit, lemon, and orange.

Derivatives (FEMA): [1] cumin seed oil (2343); [2] seed extract.

10.4.19 Cumin Black (*Nigella sativa* L.; Ranunculaceae; FEMA 2237)

This annual plant is widely distributed and cultivated in Mediterranean countries, middle Europe, and western Asia, for its seeds. Material is used in food seasoning, as a bread or cheese flavoring, and as a spice in various kinds of meals. Seed essential oil is isolated in very low amounts, and the major oil compounds are *p*-cymene (30%–80%), γ-terpinene (2%–13%), and α-thujene (2%–15%), whereas in seed volatile oil (obtained by hydrodistillation of seed extract), the main compounds are thymoquinone (25%–60%) and carvacrol (5%–11%).

10.4.20 Dill (*Anethum graveolens* L.; Apiaceae; FEMA 2382)

Dill is a short-lived perennial aromatic plant that grows wild almost everywhere and cultivated in many countries. Fresh and dried dill leaves are used as herbs. The essential oil is isolated from seeds (seed oil) or from fresh herb: stalks, leaves, and seeds (weed oil). Only 2.5 tons of dill seed oil (yield 3.5%) is produced in Russia, Hungary, Bulgaria, and Egypt. The major compounds are (+)-carvone (20%–50%), limonene (6%–40%), and apiole (<50%). Because of high amounts of carvone, this oil has a strong caraway-like odor and flavor in comparison to the weed oil. One hundred tons of weed oil per year is isolated mainly in United States, Hungary, and Bulgaria (yield 0.7%). Major components are similar to those in seed oil, with the exception of β-phellandrene, which is one of the main components (25%). The main use of both oils is in seasoning blends, particularly in pickles.

Derivatives (FEMA): [1] dill weed oil (2383); [2] dill seed oil (2384); [3] seed solid extract.

10.4.21 Eucalyptus (*Eucalyptus globulus* Labill.; Myrtaceae)

This tree, like about 700 other species of *Eucalyptus*, is native to Australia. Species of *Eucalyptus* are cultivated throughout the tropics and subtropics including Asia, the Americas, Europe, and Africa. The part used are leaves of the mature tree. Oil is isolated from fresh or partially dried leaves and reminiscent characteristic camphoraceous odor and pungent, spicy, cooling taste. This oil has aromatic and medicinal uses (herbal blends). Small amount of the oil is added to natural blackcurrant flavors. The main oil component is 1,8-cineole (eucalyptol: 60%–80%). One thousand tons of cineole type oil (yield 1.5%) is produced in Portugal, South Africa, and Spain. "Chinese eucalyptus oil" is generally not true eucalyptus oil and is produced from white camphor oil fraction. It is noteworthy that essential oils are also produced from numerous varieties of *Eucalyptus* but in small quantities.

Derivatives (FEMA): [1] eucalyptus oil (2466); [2] tincture; [3] infusion; [4] fluid extract.

10.4.22 Fennel Sweet (*Foenicilum vulgare* Mill.; Apiaceae; FEMA 2482)

Fennel sweet is a hardy, perennial, herb that usually grows near the seacoast and on riverbanks, especially around the Mediterranean. The essential oil is isolated mainly from seeds (yield 3%–6%), but also from stalks and roots. The oil has a very sweet spicy and anise-like odor. Twenty-five tons of seed oil is produced annually in Spain. Major components are anethole (60%–80%), limonene (<9%), and fenchone (<7%). The oil is used in seasoning blends, liquor, and other natural flavors (often as a terpeneless oil).

Derivatives (FEMA): [1] sweet fennel oil (2483); [2] terpeneless sweet fennel oil; [3] solid extract; [4] fluid extract; [5] tincture.

10.4.23 Galbanum (*Ferula galbaniflua* Boiss. & Bushe; Apiaceae; FEMA 2502)

This is a herbaceous plant growing in northern Persia; other *Ferula* species are widespread in south Persia, Turkey, and Lebanon. The plant yields a resinous exudate occurring commercially in two types: Persian galbanum (hard)—oleoresin and Levant galbanum (soft)—essential oil isolated from galbanum resin (yield 16%). Galbanum has an aromatic odor, a bitter warm, acrid taste. Galbanum oleoresin contains resin acids and many oxygenated terpenes (α-terpinyl acetate, α-fenchyl acetate, guaiol, bulnesol, dihydrofarnesols). The most important essential oil compounds ar: 2-methoxy-3-isobutyl pyrazine (0.05%) and undeca-l,3,5-triene (1%). The materials are used in fragrances, but are also very useful in natural fruit and vegetable flavors, especially bell pepper, blackcurrant, and blueberry formulations.

Derivatives (FEMA): [1] galbanum oil (2501); [2] resin (2502).

10.4.24 Garlic (*Allium sativum* L.; Liliaceae)

This perennial plant has a Mediterranean origin, but is nowadays cultivated in many countries as a spice. The part used are bulbs of pungent, acridic, garlic-like odor. Garlic can be used as fresh, powdered, or as a garlic salt. *A. sativum* is known to contain the highest sulfur content of any member of the *Allium* genus. Garlic oil (yield 0.1%–0.2%) has extremely strong odor with mercaptan-like note and is composed of sulfur-containing compounds (diallyl sulfide, diallyl disulfide, methylallyl trisulfide, dially trisulfide).

Derivatives (FEMA): [1] garlic oil (2503).

10.4.25 Geranium (*Pelargonium graveolens* L'Her.; Geraniaceae)

P. graveolens grows in subtropical areas This specific species has great importance in the perfume industry, so it is cultivated on a large scale and its foliage is distilled for its scent. Essential oil (yield 0.1%) has a strong, rose-like odor with mint-like note. Concrete (0.2%) and absolute (0.13%) are also prepared. Most of the 200 tons of geranium products is isolated in China and Reunion Island. Major components are citronellol (30%), geraniol (12%), isomenthone (6%), and rose oxide (3%). Most uses of geranium oil are in fragrances, but it is also used in some natural rose-type flavors and as a minor component of natural raspberry, strawberry, blackcurrant, blueberry, and peach flavors. The minty note is a limiting factor in some of these uses.

Derivatives (FEMA): [1] geranium oil (2508); [2] concrete; [3] absolute.

10.4.26 Ginger (*Zingiber officinale* Roscoe; Zingiberaceae; FEMA 2520)

The plant is native to Asia, and is cultivated in tropical and subtropical countries. Brown ginger is produced from unpeeled rhizomes, whereas white ginger comes from skinned rhizomes. A total of 155 tons of ginger essential oil from rhizomes is mainly isolated in China and India (yield 0.25%–2%), and 150 tons of oleoresin is produced from seeds (yield 6%). Small quantities of the oil are also produced in Sri Lanka, Australia, and Jamaica. This oil has a warm, spicy odor. Its major components are zingiberene (35%), *ar*-curcumene (10%), and β-sesquiphellandrene (10%), whereas oleoresin contains 6- and 8-gingerols (up to 15%)—"hot" components. Terpeneless oil can be used to improve solubility in beverages, but this oil varies considerably by producer. Ginger oil can be added as a trace component to blackberry, loganberry, and raspberry flavors.

Derivatives (FEMA): [1] ginger oil (2522); [2] oleoresin (2523); [3] extract (2521).

10.4.27 Grapefruit (*Citrus paradisi* Macfad.; Rutaceae)

The grapefruit tree is cultivated in the south of United States, East Asia, Brazil, and Nigeria. Seven hundred tons of oil is produced in Brazil and the United States by expression of fresh peels of the fruit (yield 0.4%). It has a pleasant, citrus-like odor. Major components of the oil are limonene (88%–95%), but the most important are nootkatone (0.2%) and α-thioterpineol (0.2%). To improve the stability and solubility, the oil is very often concentrated. The main use of the oil is in grapefruit-flavored soft drinks and confectionery. The oil is not used much outside grapefruit flavors because of its high price and similarity to sweet orange oil, but sometimes can be used in orange, peach, and cola flavors. The grapefruit juice oil (essence) is produced in Florida as a by-product from concentrated juice oil production, with the yield 0.009%. The oil is used to add juicy character to grapefruit flavors and reconstituted this citrus juice.

Derivatives (FEMA): [1] grapefruit oil (2530); [2] concentrated grapefruit oil; [3] grapefruit concrete (2521); [4] grapefruit essence natural.

10.4.28 Juniper (*Juniperus communis* L. [FEMA 2602] and *J. osteosperma* Torr.; Cupressaceae)

The genus *Juniperus* includes up to 70 species. This small tree grows in on almost all dry areas of the world. Juniper berries are the part used; they have an aromatic odor and bitter taste. Essential oil is isolated from nonfermented berries (yield up to 2.6%). Oil isolated from fermented and subsequently distilled berries is used in gin production. Croatia is the main producer of juniper oil (12 tons/year). The main compounds are α-pinene (20%–40%), sabinene (3%–18%), myrcene (1%–6%), camphor (10%–18%), limonene (3%–8%), and bornyl acetate (12%–20%). Most of the oil is used in the production of gin flavors, but can also be used in traces in natural blueberry flavors.

Derivatives (FEMA): [1] juniper oil (2604); [2] extract (2603); [3] infusion.

10.4.29 Lavandin (*Lavandula hybrida* Rev.; Lamiaceae)

This plant grows in several varieties, because it is a hybrid between *L. latifolia* Vill. (spike lavender) and *L. officinalis* Chaix (true lavender) and is cultivated in southern France (750 tons annually). Essential oil (yield 1%–3%) is isolated from the flowering herb, which has camphoraceous, fresh, lavender-like odor. Oils from hybrids of lavandin have variable composition, but the main and the most important compounds are linalyl acetate (<40%) and linalool (<30%). Lavandin oil is used in fragrances; only small quantities can be used in tea and fruit flavors.

Derivatives (FEMA): [1] lavandin oil (2618); [2] lavandin absolute; [3] lavandin concrete.

10.4.30 Lavender (*Lavandula officinalis* Chaix; Lamiaceae; FEMA 2619)

The plant grows in Mediterranean areas and is cultivated in many European, North African, and American countries. The parts used are flowering tops and stalks.

Two hundred tons of lavender oil (yield 0.6%–1%), concrete (yield 1.5%–2%), and absolute (yield 50%–60% of concrete) is produced mainly in Bulgaria, France, and Russian republics. Major components are linalool (40%) and linalyl acetate (25%). The lavender products are mainly used in fragrances, for example, with combination with bergamot oil, in Earl Grey tea flavors. In small quantities, they can also be used as flavor component. Lavender concrete or absolute contain significant amount of coumarin—limited in food products.

Derivatives (FEMA): [1] lavender oil (2622); [2] lavender absolute (2620); [3] lavender concrete (2621).

10.4.31 Lavender Spike (*Lavandula latifolia* Vill.; Lamiaceae)

This herbaceous plant is morphologically very similar to *L. officinalis*. It grows in the mountain areas of the Mediterranean basin. Only 35 tons of this oil is still produced in Spain (yield 0.9%). Major components are linalool, linalyl acetate, and 1,8-cineole. This oil is much harsher than lavender or lavandin oils and is little used in fragrances or flavors.

Derivatives (FEMA): [1] lavender spike oil (3033).

10.4.32 Lemon (*Citrus limon* L.; Rutaceae)

This is an evergreen tree native to the Far East. A total of 370 tons of essential oil isolated by cold pressing of peels or peel pulp (yield 0.4%–4%) is produced annually, mainly in the United States, Italy, and Argentina. It has a characteristic odor and taste of outer lemon peel part. Major components are limonene (60%–90%), α- and β-pinene (<12%), γ-terpinene (<9%), and citral (geranial and neral, <3%). The citral level in U.S. lemon oils is usually low. A less valuable quality of the oil is obtained by steam distillation of peel. Essential oil (yield 0.16%–0.23%) from leaves and twigs (petitgrain oil) is produced mainly in Sicily, but in very small amounts. Its major components are limonene (30%), citral (25%), and neryl acetate (4%). This oil is used in lemon flavors. The distilled peel oil is often used to produce more stable and less sensitive to oxidation terpeneless oil. This oil has a sweet, almost rosy-fruit odor. Lemon oil is widely used in lemon and other natural flavors: pineapple, butterscotch, and banana flavors, and can be mixed with other citrus oils such as lime, orange, and grapefruit.

Derivatives (FEMA): [1] lemon oil (2625); [2] terpeneless lemon oil (2626); [3] extract (2623); [4] essence; [5] petitgrain lemon oil (2853); [6] tincture.

10.4.33 Lemongrass (*Cymbopogon flexuosus* and *C. citratus* Stapf.; Gramineae)

This grass grows in many tropical countries. The part used is the aerial part. Lemongrass has a strong, pungent, lemon-like odor. The essential oil (yield 0.2%–0.4%) from dried grass contains mainly citral (up to 85%). Other main compounds are geraniol, nerol, and citronellal but the most important are: isovaleric aldehyde, furfural, myrcene, limonene, 6-methyl-5-hepten-2-one, and various esters. Ten tons of lemongrass oil is produced annually from *C. flexuosus* mainly in India and China. Some oil is isolated from *C. citratus* in Guatemala and Brazil. Lemongrass oil is used as a source of natural citral for lemon flavors.

Derivatives (FEMA): [1] lemongrass oil (2624).

10.4.34 Lovage (*Levisticum officinale* Koch.; Apiaceae; FEMA 2649)

This is a perennial plant that grows wild in many European countries. The essential oil from green parts has less interesting odor than rhizome oil, so is manufactured in very low quantities. Less than 1 ton of oil from fresh or dried rhizome (yield 0.1%–1%) is produced annually, mainly in Benelux countries. The oil has a strong odor reminiscent of celery and angelica-like flavor. The most important components are phthalide lactones: 3-butylidene-4,5-dihydrophthalide (root: 62%, seed: 6%), *cis*- and *trans*-butylidenephthalide, and *cis*- and *trans*-ligustilide. The oil is used in seasoning blends and some natural meat and savor flavors.

Derivatives (FEMA): [1] lovage oil (2651); [2] lovage fluid extract (2650).

10.4.35 Mandarin (*Citrus reticulata* Blanco; Rutaceae)

This tree is native to China but was successfully introduced to Europe. The parts used are leaves, twigs, unripe fruits, fruits, and rind. The essential oil is isolated by cold pressing from rind of almost ripe fruits (yield 0.5%). One hundred twenty tons is produced mainly in Italy and China. Major components are limonene (65%–75%), γ-terpinene (16%–22%), and the most important: methyl *N*-methyl anthranilate (0.8%) and α-sinensal (0.05%). The oil is frequently concentrated (terpeneless) and widely used alone or in conjunction with orange oil in beverages, confectionery, and in many natural flavors (mango, peach, apricot). Small quantities of petitgrain mandarin oil are obtained by hydrodistillation from leaves, twigs, and unripe fruits (yield 0.2%) in Algeria, Italy, Brazil, Greece, and Spain. Major components are methyl-*N*-methyl anthranilate (30%; 98% in the terpeneless oil) and γ-terpinene (36%). This oil is important in strawberry, blueberry, mandarin, and orange flavors, whereas terpeneless oil is used for natural grape flavors.

Derivatives (FEMA): [1] mandarin oil (2657); [2] terpeneless mandarin oil; [3] petitgrain mandarin oil (2854); [4] tincture; [5] fluid extract.

10.4.36 Marjoram (*Origanum majorana* L., syn. *Majorana hortensis* Moench; Lamiaceae; FEMA 2662)

This herbaceous, annual or biennial plant is native to western Asia and North Africa, but is very famous in Europe, where dry flowering plants are used as a spice in many dishes. Thirty tons of oil is annually produced (yield 0.4%), mainly in Morocco. The major component is terpinen-4-ol (25%–35%) of nutmeg odor; this compound is frequently isolated from the oil. Other components are γ-terpinene (12%–20%), *cis*- and *trans*-sabinene hydrate (8%), and linalool (3%). The essential oil is used in seasoning blends. In addition, marjoram oil can help to replace nutmeg note in flavors.

10.4.37 Mustard Brown (*Brassica nigra* L. [FEMA 2760] and *B. juncea* L.; Brassicaceae)

This herbaceous 1-m-high plant is widespread throughout Europe, North Africa, and Asia. *B. nigra* is cultivated in Italy and Holland, whereas *B. juncea* is grown in northern India and southern Russia. The part used are seeds of irritating and sharp odor, which is the result of the presence of allyl isothiocyanate. Mustard is used in the food industry in several forms: whole seeds, ground seed meal: mustard cake or press cake, mustard flour, and prepared mustard. From the residue of press-cake (obtained after expressing the seed of the oil), the essential oil is isolated by steam distillation. This oil consists more than 90% of allyl isothiocyanate.

10.4.38 Mustard Yellow (*Brassica alba* L.; Brassicaceae; FEMA 2761)

This herbaceous plant is widespread at the same area as brown mustard. Taste of yellow mustard seeds is more warm and pungent than that of brown mustard. The seeds do not produce any volatiles when enzymatically treated. Seed mustard or sinapine from seed can be used as a spice.

10.4.39 Nutmeg (mace) (*Myristica fragrans* Houtt.; Myristicaceae; FEMA 2792)

This evergreen tree grows in many islands of the Indian Ocean (Java, Sumatra, Borneo, Penang, Banda). Only female trees bear nuts, which are dried to produce nutmeg, whereas dried aril yields the spice called mace, which possesses a flavor similar to that of nutmeg. The ratio of nutmeg to mace is 10:1. The most commercially important qualities of nutmeg and mace are Banda, Java, and Siauw. World production (mainly Indonesia and Sri Lanka) of nutmeg (yield 6.5%–8%) and mace (yield 10%–13%) oil is 300 tons/year. Mace oil has a fresher odor than nutmeg oil, which is isolated from free of fixed oil nutmeg. The *M. fragrans* oils contain mainly sabinene (20%) α- and β-pinene (30%), myristicin (10%), terpinen-4-ol (8%), and small amount of linalool, safrol, eugenol, and isoeugenol. Nutmeg oil is generally used in cola flavors and in seasoning blends.

Derivatives (FEMA): [1] nutmeg oil (2793); [2] nutmeg tincture; [3] nutmeg oleoresin; [4] fatty oil; [5] mace (2652); [6] mace oil (2653); [7] mace oleoresin (2654).

10.4.40 Onion (*Allium cepa* L.; Liliaceae)

Onion is a herbaceous plant native to Middle East, but is now widespread throughout the world. The consumption of this oldest vegetable bulb of strong, pungent, lasting odor is very large. Onion is used in a large number of recipes and preparations spanning almost the totality of the world's cultures. Onion bulbs are now available in fresh, frozen, canned, caramelized, pickled, powdered, chopped, dehydrated forms and can be used, in almost every type of food (cooked, fresh salads, spicy garnish). Three tons of onion bulb essential oil is produced each year mainly in Egypt (yield 0.02%). Major components of the oil are: dipropyl disulfide

(20%), dipropyl trisulfide (22%), propylpropenyl disulfide (9%), and methylpropyl trisulfide (17%), but the most important is dipropyl tetrasulfide (5%). The oil is mostly used in seasonings, but can be used to increase good effects in some natural flavors.

Derivatives (FEMA): [1] onion oil (2817); [2] oleoresin; [3] fluid concentrated water extract; [4] fluid extract.

10.4.41 Orange Bitter (*Citrus aurantium* L.; Rutaceae)

This is a tall tree (up to 10 m), native to the Far East, and also cultivated extensively in Mediterranean basin, Guinea, West Indies, Paraguay, and Brazil. The part used are leaves and twigs (petitgrain), flowers (neroli bigrade), and peel. A total of 150 tons of petitgrain oil is produced mainly in Paraguay (yield 0.2%). The main oil components are linalyl acetate (30%–40%) and linalool (up to 30%). Pyrazines (e.g., 2-methoxy-3-isobutyl pyrazine) play important role in petitgrain flavor creation. Oil is widely used in many fragrances and natural flavors. Two and a half tons of neroli oil is produced from bitter orange flowers (yield 0.2%). Major compounds of this oil are linalool (up to 50%), limonene (7%–18%), and linalyl acetate (<9%). Indole and methyl anthranilate, which constitute about 0.2% of the oil, are also very important as key odor components. Neroli oil is mainly used in fragrances and many fruit flavors. Additionally, a total of 30 tons of peel cold pressed essential oil is produced annually mainly in West India and Brazil (yield 0.4%). Major components are limonene (93%), decanal (0.2%), linalyl acetate (0.2%), and linalool (0.2%). The oil is generally used to modify sweet orange flavors and as the key ingredient in "Indian tonic" beverage flavors. Orange bitter oil also found a useful role in many other natural flavors.

Derivatives (FEMA): [1] petitgrain oil (2855); [2] bitter orange peel oil (2823); [3] terpeneless bitter orange oil; [4] sesquiterpeneless bitter orange oil; [5] tincture; [6] fluid extract; [7] orange flowers absolute (2818); [8] orange leaf absolute (2820); [9] neroli oil (2771).

10.4.42 Orange Sweet (*Citrus sinensis* L.; Rutaceae)

This is an evergreen tree of Oriental origin. The plant now has an enormous worldwide economic importance and is cultivated in many Mediterranean countries, and also in California and Florida. The parts used are leaves, flowers, ripe fruits, small, whole, unripe fruits, peels, and juice. The oils and various derivatives have a mild bitter, astringent flavor. More than 26,000 tons of the oil obtained by distillation from fresh peel or juice of the fruit (orange oil distilled) and from peels of partially ripened fruits (peel sweet oil) is produced each year (yield 0.3%–0.5%) mainly in Brazil, United States, Israel, and Italy. Major components are limonene (94%), myrcene (2%), octanal (0.4%), linalool (0.5%), and the most important, α-sinensal (0.02%) and β-sinensal (0.01%). The peel sweet oil is manufactured also as terpeneless oil, but both these oils have less fresh flavor than distillled orange oil. The other kind of orange oil is a by-product from concentrated orange juice production in the United States, Israel, and Brazil (yield 0.014%). It is produced together with a water phase distillate called orange esters (0.5%). The composition of the oil is similar to that of the peel oil, with significant fresh juice character. Orange oils are generally used in orange flavors and many other natural flavors. The production (hydrodistillation) of petitgrain oil from leaves, unripe and small fruit, twigs, and flowers is low, because of its limited application (perfumery).

Derivatives (FEMA): [1] orange oil distilled (2821); [2] orange terpeneless oil (2822); [3] peel sweet extract (2824); [4] peel sweet oil (2825); [5] orange peel sweet oil terpeneless (2826); [6] tincture; [7] infusion.

10.4.43 Origanum (*Origanum vulgare* L.; Lamiaceae)

This herbaceous plant is widespread throughout all continents. The flowering tops are the only part used. The oil is isolated from dried, flowering herb of various origanum species (yield <1%). These species differ considerably in both odor and flavor. *O. vulgare* is the most common herb. The oil is mainly used in seasonings.

Derivatives (FEMA): [1] origanum oil; [2] tincture; [3] infusion.

10.4.44 Paprika (*Capsicum frutescens* L., *C. annum* L. and Similar Varieties; Solanaceae; FEMA 2833)

C. annum is a large species, which includes many varieties such as chilli pepper, bell pepper, cayenne, pimento, and paprika. The parts used are fruits (dry and ground), which exhibit varying colors ranging from yellow, red to black. Oleoresin is the main product isolated from paprika. The seasoning is used in many cuisines to add color and flavor to a broad variety of dishes—to season and color rice, stews, and soups, goulash, and sausages. Paprika can range from sweet (mild, not hot) to spicy (hot).

Derivatives (FEMA): [1] paprika oleoresin (2834); [2] fluid extract; [3] tincture.

10.4.45 Parsley (*Apium petroselinum* L.; Apiaceae; FEMA 2835)

This is an annual or biennial herb that grows wild and is cultivated throughout the Mediterranean area, Hungary, Germany, France, Holland, and United States. Leaves, flowering tops, ripe seeds, and roots are the parts used. The essential oils from leaves and roots have a similar composition. Herb oil (2 tons/year) is produced from the whole plant (yield 0.06%–0.2%). The main component is myristicin. In contrast, the essential oil from seeds (yield 3%–6%) is either dominated by myristicin (60%–80%) or apiol (70%). Other components are 2,3,4,5-tetramethoxyallylbenzene (<20%), *p*-mentha-1,3,8-triene (<23%), and β-phellandrene (<14%). Five tons of seed oil is produced annually in Egypt and other countries (yield 1.5%–3.5%). Both herb and seed oil are used in seasoning blends.

Derivatives (FEMA): [1] parsley herb and seed oil (2836); [2] oleoresin (2837); [3] infusion.

10.4.46 Pepper, Black and White (*Piper nigrum* L.; Piperaceae; FEMA 2844 and 2850)

This perennial plant, which is native to southern India, is cultivated in India, Sunda Islands, Madagascar, and the Comoro Islands. The commercially available variants are Malabar and Lampong black pepper. The part used are the berries. Whole dried berries, consisting of the epicarp, mesocarp, and endocarp, are black pepper, whereas white pepper consist of berries harvested after removing the outer skin. Berries are separated from the stalk and soaked with water to remove the pericarp. Annual production of black pepper oil (yield 2%) is 8 tons, mainly from India, but 400 tons of oleoresin (yield 6%) is produced. Major components of the oil are β-caryophyllene (24%), limonene (20%), α- and β-pinene (16%), and in small amounts the most important ones, piperidine and piperine. The black pepper oleoresin consists of 40% of the pungent component piperine. White pepper seems to contain even more piperine than black pepper. Essential oil and oleoresin are also isolated from white pepper. Black and white pepper oil and oleoresin are used in seasoning blends. Black pepper retains both the odor and flavor characteristics of pepper, whereas white pepper has only the sharp, piquant flavor due to piperine.

Derivatives (FEMA): [1] black pepper oil (2845); [2] black pepper oleoresin (2846); [3] white pepper oil (2851); [4] white pepper oleoresin (2852).

10.4.47 Peppermint (*Menta piperita* L.; Lamiaceae; FEMA 2847)

M. piperita is a hybrid derived from three other *Mentha* species, and exists in two varieties. This plant is cultivated in central and southern Europe, North and South America, and Japan. Flowering tops and leaves are parts used. A total of 4500 tons of essential oil from flowering tops is produced annually from the United States (yield 0.3%–0.7%). Major components are menthol (34%–50%), menthone (12%–20%), menthyl acetate (4%–7%), 1,8-cineole (2%–5%), menthofuran (4%–9%), isomenthone (3%), and pulegone (2%–5%). The oil is used to give a peppermint flavor to a wide range of applications. It is also used in mint and herbal blends and in liquor flavors.

Derivatives (FEMA): [1] peppermint oil (2848); [2] fluid extract; [3] tincture; [4] infusion.

10.4.48 Rosemary (*Rosmarinus officinalis* L.; Lamiaceae; FEMA 2991)

Rosemary is an evergreen shrub that is native to Mediterranean regions and cultivated in Spain, France, Tunisia, Morocco, Croatia, and Italy. The flowering tops and leaves are the parts used. The yield of leaf essential oil is 0.5%–2.5%. The main components are α- (22%) and β-pinene (2%), camphene (11%), and camphor (10%–20%). From flowering tops, it is possible to isolate 0.5%–1.2% of the rosemary essential oil. The main constituents include α-pinene, camphene, 1,8-cineol, camphor, and bornyl acetate (10%–15%). Two hundred and fifty tons of this oil is produced annually. The main use of the oil is in seasoning blends.

Derivatives (FEMA): [1] rosemary leaf oil; [2] rosemary oil (2992); [3] rosemary oleoresin; [4] tincture; [5] fluid extract.

10.4.49 Sage (*Salvia officinalis* L.; Lamiaceae; FEMA 3000)

S. officinalis (dalmatin sage) is one from among 900 *Salvia* species. This evergreen, perennial herb grows wild and is cultivated in south Europe. Sage herb has a slightly peppery flavor and is used widely in cooking: for flavoring fatty meats, sausages, cheeses, and some drinks. Sage leaves have a warm, spicy flavor, and are raw materials for isolation of the essential oil (yield 0.5%–1.4%). Albania and Croatia are the main producers of herb oil (40 tons/year). Dalmatian sage oil has the best quality. Major components are α- and β-thujone (20%–45%), camphor (23%), 1,8-cineole (15%), and bornyl acetate (4%). The EU limits the content of thujones in food and beverages. Essential oil (Spanish sage oil) is also isolated from *S. lavandulaefolia*, at 5 tons annually (yield 0.8%). This herb is growing mainly in Spain, and has a camphoraceous, fresh odor. Main constituents are camphor (30%), 1,8-cineole (15%), α- and β-pinene (15%), linalool, and linalyl acetate. Both kinds of sage oils are used in seasoning blends and in liquor flavors.

Derivatives (FEMA): [1] sage oil (3001); [2] sage oleoresin (3002); [3] Spanish sage oil (3003); [4] infusion; [5] fluid extract; [6] tincture.

10.4.50 Sage Clary (*Salvia sclarea* L., Lamiaceae; FEMA 2320)

Sage clary is a biennial plant cultivated in the Mediterranean basin, central Europe, and Russia. Clary has a herb-like odor and wine-like taste. Essential oil is isolated from flowering tops and leaves (yield 0.7%–1.5%). Forty-five tons of oil is produced in Russia. Essential oil contains more than 250 compounds: aldehydes of C_6–C_9 carbon skeleton, linalyl acetate (56%–78%), linalool (7%–24%), germacrene D (2%–12%), and sclareol (<3%). Sclareol is the major component of the concrete and absolute (40%). The leaves are used as vegetables, whereas essential oil is used in fragrances, generally in Earl Grey tea flavors and at low levels in other flavors, and also as a flavor agent in wine, muscatel, and tobacco products.

Derivatives (FEMA): [1] clary sage oil (2321); [2] tincture.

10.4.51 Sassafras (*Sassafras albidum* Nutt. and *Ocotea pretiosa* Vell.; Lauraceae; FEMA 3011)

Sassafras is a genus of three species of deciduous trees, native to Eastern North America (*S. albidum*) and Eastern Asia (*S. randaiense* and *S. tzumu*). The parts used are the roots, bark, and leaves. *Sassafras* has a spicy odor, reminiscent of fennel flavor. The dried and ground leaves are used to make spice powder, used in the making of some meals. Essential oil is isolated from root (yield 1.8%) and bark (yield 6%–9%) by steam distillation. Four hundred and fifty tons of Brazilian sassafras oil, from *Ocotea pretiosa* (native to South America), is produced annually in Brazil. North American sassafras (from *Sassafras albidum*) oil is similar to the Brazilian oil. The main components of Brazilian and American oil are safrole (80%–90%) and camphor (1%–7%). The EU limits content of safrole in food and beverages. The oils are used exclusively in fragrances.

Derivatives (FEMA): [1] sassafras bark oil (3010); [2] sassafras leaf oil (safrole free).

10.4.52 Spearmint (*Mentha spicata* L. [FEMA 3030], *M. viridis* L. var. *crispa*, and *M. cardiaca* J. Gerard ex Baker; Lamiaceae)

Mentha species are herbaceous plants extensively cultivated in North America, and many European countries. Spearmint is derived from a range of species. The part used are flowering tops of warm, herbaceous odor. United States and China are the main producers of essential oil—1500 tons/year (yield 0.6%). About 80% of the U.S. oil is isolated from *M. spicata* (the rest is extracted from *M. cardiaca*), whereas Chinese oil is produced from *M. viridis* var. *crispa*. Major components are (–)-carvone (70%) and limonene (10%). 1,8-Cineole, myrcene, carvyl acetate, dihydrocarvyl acetate, and dihydrocarveol constituted about 10%. The raw oil is usually rectified to remove the harsh sulfurous low boiling fraction yield terpeneless oil. The essential oil is mainly used as a mint flavor in chewing gum and oral hygiene products.

Derivatives (FEMA): [1] spearmint oil (3032); [2] terpeneless spearmint oil (3031).

10.4.53 Star Anise (*Illicium verum* Hook.; Illiciaceae; FEMA 2095)

This tall evergreen tree is native to China and Indochina. The oil is isolated from dried, ripe seeds. A total of 410 tons of this oil is produced annually, mainly in China (yield 8%). Major components are anethole (85%–90%), limonene (<8%), and anisaldehyde (<1%). The oil is used in liquor, root beer, and blackberry flavors.

Derivatives (FEMA): [1] star anise oil (2096).

10.4.54 Tarragon (*Artemisia dracunculus* L.; Asteraceae; FEMA 3043)

This perennial herb grows in wide areas of Europe across central and eastern Asia to India, North America, and Mexico. Parts used are flowering tops and leaves. Tarragon has a sweet, spicy, basil-like, anise-like flavor. It is a famous herb used in cooking, suitable for chicken, lasagna, fish, and egg dishes. Ten tons of essential oil is produced annually in Italy, France, and Hungary (yield 0.3%–1.4%). Major components of the oil are methyl chavicol (68%–80%), (*Z*)-β- and (*E*)-β-ocimene (15%–18%), and limonene (2%–5%). The material is mainly used in seasoning blends, but can also give an interesting note in root beer flavors. Tarragon is used to flavor a popular carbonated soft drink in Ukraine and Russia.

Derivatives (FEMA): [1] tarragon essential oil (2412); [2] oleoresin; [3] infusion; [4] tincture.

10.4.55 Thyme (*Thymus vulgaris* L. and *T. zygis* var. *gracilis* Boiss; Lamiaceae; FEMA 3063)

Thyme is a shrub native and cultivated in Mediterranean basin, eastern and central Europe, and North America. Common thyme, *T. vulgaris*, and some other species are used in cooking: to flavor meats, soups, and stews. Steam distillation from leaves, stems, and mainly flowering tops of *T. vulgaris* and *T. zygis* yield the thyme essential oil (about 0.5%–1.2%). The main oil producer (25 tons/year) is Spain. Two kinds of oil are known: red (crude of strong, aromatic flavor) and white (redistilled of milder odor). The major components of the oil are thymol (37%–55%), *p*-cymene (14%–28%), γ-terpienene (4%–11%), linalool (3%–7%), carvacrol (0.5%–6%), and myrcene (1%–3%). The oil is used as an agent in seasoning blends and in traces in many flavors.

Derivatives (FEMA): [1] red thyme oil (3064); [2] white thyme oil (3065); [3] oleoresin; [4] fluid extract; [5] tincture; [6] absolute.

10.4.56 Turmeric (*Curcuma longa* L.; Zingiberaceae; FEMA 3085)

C. longa species is the one used as a flavor agent. This perennial herb is native to south Asia. The parts used as spice are rhizomes of fresh, sweet orange-like, ginger-like, bitter flavor. It has become the key ingredient for many Indian, Persian, Thai, and Malaya dishes. Very little amount of essential oil is produced from rhizome (yield 1.3%–5.5%), but 100 tons of oleoresin (yield 30%) is produced each year mainly in India. It contains several compounds, and the main components are α- and β-turmerone (30%–60%), *ar*-turmerone (about 20%), and the most important, borneol (<0.5%). Oil and oleoresin are used to provide color in blends, in cheeses, yogurt, dry mixes, salad dressings, winter butter, and margarine. Turmeric is also used to give a yellow color to prepared mustards, canned chicken broths, and other foods (as a much cheaper replacement for saffron). Turmeric is also used to give boiled white rice a golden color.

Derivatives (FEMA): [1] turmeric oil (3086); [2] oleoresin (3087); [3] fluid extract; [4] tincture.

10.4.57 Valerian (*Valeriana officinalis* L.; Valerianaeae)

This perennial herb is native to Europe and Asia. The parts used are rhizomes and roots. Very little essential oil is still produced (yield 0.4%–0.6%) in Belgium, Croatia, and France. The main components are bornyl acetate (32%–44%), camphene (16%–25%), α- and β-pinene (6%–12%). However, the most important flavor component is isovaleric acid (1%–4%). The oil is used in many fruit flavors but at low levels.

Derivatives (FEMA): [1] valerian root oil (3100); [2] valerian root extract (3099); [3] fluid extract; [4] tincture.

REFERENCES

Ashurst, P.R. 1999. *Food Flavorings*, 3rd edn. Gaithersburg, Maryland, US: Aspen Publishers.

Burdock G.A. 2002. *Handbook of Flavor Ingredients*, 4th edn. Boca Raton, US: CRC Press.

Cseke, L., A. Kirakosyan, P. Kaufman, S. Warber, J. Duke, and H. Brielmann. 2006. *Natural Products from Plants*. Boca Raton, US: Taylor & Francis Group.

Lawrence, B.M. 1985. A review of world production of essential oils. *Perf. Flav.* 10(5): 1–20.

Lawrence, B.M. 1985–2007. Progress in essential oil. *Perf. Flav.*, issues of volumes 10–32.

Raghavan, S. 2007. *Handbook of Spices, Seasonings, and Flavorings*, 2nd edn. Boca Raton, New York, US: Taylor & Francis Group.

Reineccius, G. 1994. *Source Book of Flavors*, 2nd edn. London, GB: Chapman & Hall.

Rovira, D.D. 2008. *Dictionary of Flavors*, 2nd edn. Iowa, US: Wiley-Blackwell.

Rowe, D.J. 2005. *Chemistry and Technology of Flavors and Fragrances*, Oxford, UK: Blackwell Publishing Ltd.

Wright, J. 2004. *Flavor Creation*, Carol Stream, Illinois, US: Allured Publishing Corporation.

11 Functional (Nonflavor) Properties of Flavor Compounds

Alfreda Wei and Takayuki Shibamoto

CONTENTS

11.1 FLAVOR CHEMICALS FORMED BY MAILLARD REACTIONS: HETEROCYCLIC FLAVOR COMPOUNDS

11.1.1 Introduction

Flavor compounds offer a secondary function to foods, after nutrition. They have received much attention as the chemicals that play an important role in the palatability of foods. Cooking has been practiced since ancient times to increase food palatability. It has been well known that heat treatment produces preferable toasted and roasted odors and attractive colors.

Many volatile flavor chemicals have been isolated and identified in cooked foods. Among the flavor chemicals identified in cooked foods, numerous heterocyclic compounds—which comprise one-fourth of the volatile compounds identified in foods (Fernandez et al. 2002)—have been reported as the chemicals responsible for roasted or toasted flavors (Shibamoto 1980, 1983). There are also several comprehensive

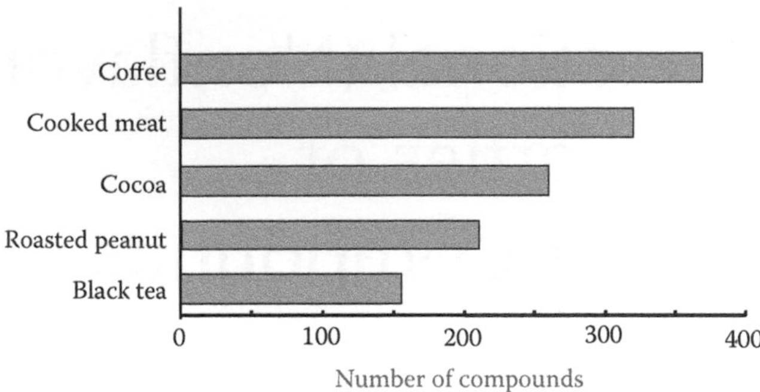

FIGURE 11.1 Approximate number of heterocyclic compounds found in typical foods.

reviews on heterocyclics in flavor chemistry (Flament 1975; Vernin 1979). Figure 11.1 shows the approximate number of volatile heterocyclic compounds found in typical heat-treated foods and beverages. As this figure shows, roasted coffee contains nearly 400 volatile heterocyclic compounds. The approximate numbers of each type of heterocyclic compound found in coffee are shown in Figure 11.2. Among them, furans were found in the greatest number (142), followed by pyrazines (99) and pyrroles (80) (Flament 2002).

Antioxidants have received much attention among food scientists as inhibitors of lipid peroxidation. Lipid peroxidation and DNA damage caused by reactive oxygen species are associated with various diseases, including cancer, cardiovascular diseases, cataracts, atherosclerosis, diabetes, malaria, arthritis, and aging (Huang et al. 1999; Beckman and Ames 1998). Synthetic antioxidants, such as BHA and BHT,

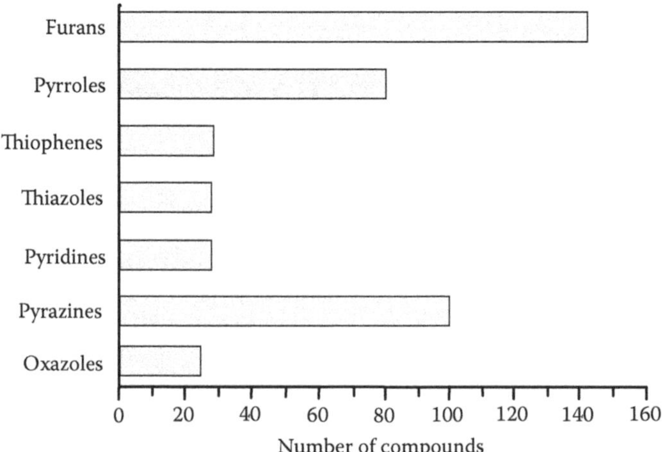

FIGURE 11.2 Approximate numbers of each type of heterocyclic compounds found in roasted coffee.

have been used to maintain the quality of foods (Henderson et al. 1999). However, BHA and BHT have been reported to demonstrate carcinogenic effects (Hocman 1988). Therefore, in addition to natural antioxidants found in plants (such as vitamins, polyphenols, and flavonoids), volatile chemicals including heterocyclic flavor chemicals found in cooked foods and beverages have begun to receive attention as nontoxic and safe antioxidants.

11.1.2 Antioxidant Activities of Maillard Reaction Products

In 1912, the French chemist Maillard reported that the reaction between an amino acid and sugar produced various materials via heat treatment. In 1953, the

TABLE 11.1
Typical Antioxidant Studies on Maillard Reaction Products (MRPs)

Model System	Nature of Sample	Method	Activity	Reference
Porcine plasma protein/glucose, fructose, galactose	Aqueous reaction mixture	DPPH	Coincidental with browning development	Benjakul et al. 2005
Lysine/glucose, fructose, ribose	High-m.w. MRPs	DPPH	Protection against both H_2O_2- and AAPH-induced intracellular oxidation	Kitts and Hu 2005
		ORAC		
Glycine, lysine/ glucose, ribose	Fractions from aqueous solutions	ORAC	High-m.w. MRPs did not always have higher activities than low-m.w. MRPs	Chen and Kitts 2008
Glycine, diglycine, triglycine/glucose	Aqueous reaction mixture	ABTS	Diglycine/glucose produced most potent antioxidants	Kim and Lee 2009
		FRAP		
Asparagines, glycine, arginine/ glucose	Aqueous reaction mixture, high- and low-m.w. MRPs	ABTS	High-m.w. MRPs > original MRPs > low-m.w. MRPs	Zhang et al. 2008
Lysine/lactose	Aqueous reaction mixture, galactosylisomaltol, pyrraline	RAE	High-m.w. MRPs showed highest as well as two chemicals	Monti et al. 1999
Lysine/glucose, fructose, ribose	Aqueous reaction mixture, high- and low-m.w. MRPs	RAE	High-m.w. MRPs exhibited higher activity than low-m.w. MRPs	Hao and Kitts 2004

Note: m.w., molecular weight; RAE, relative antioxidative efficiency.

comprehensive review on the Maillard reaction was published (Hodge 1953). The Maillard reaction became defined as the nonenzymatic reaction between reducing sugars and amino acids or proteins, which produces a complex series of compounds (Maillard reaction products or MRPs). Since then, many studies have been performed using the so-called Maillard reaction systems to investigate subjects related to cooked foods and beverages (including flavors, tastes, colors, and textures), as well as biological properties (mutagenicity, carcinogenicity, and antioxidant) (Lee and Shibamoto 2002a).

Early studies on MRPs focused on melanoidin polymers (Obretenov and Vernin, 1998), including investigations on the antioxidant activity of melanoidins (Kim and Lee 2009). MRPs have been shown to inhibit oxidation in model systems (Mastrocola and Munari 2000; Mastrocola et al. 2000) as well as in storage experiments with food products (Nicoli et al. 1997). Table 11.1 summarizes recent studies on the antioxidant activities of the Maillard reaction mixtures obtained from amino acid/sugar model systems. It is obvious that the high-molecular weight fractions of Maillard reaction mixtures exhibited greater antioxidant activity than the low-molecular weight fractions. The principle antioxidant component in high-molecular weight MRPs is melanoidins. Recently, investigations of the principle antioxidants in MRPs have been performed. Consequently, volatile flavor chemicals formed in the Maillard reaction have come to receive much attention as possible antioxidants (Lee and Shibamoto 2002a).

Figure 11.3 shows the antioxidant activities of dichloromethane extracts obtained from Maillard model systems consisting of glucose and eight amino acids tested by

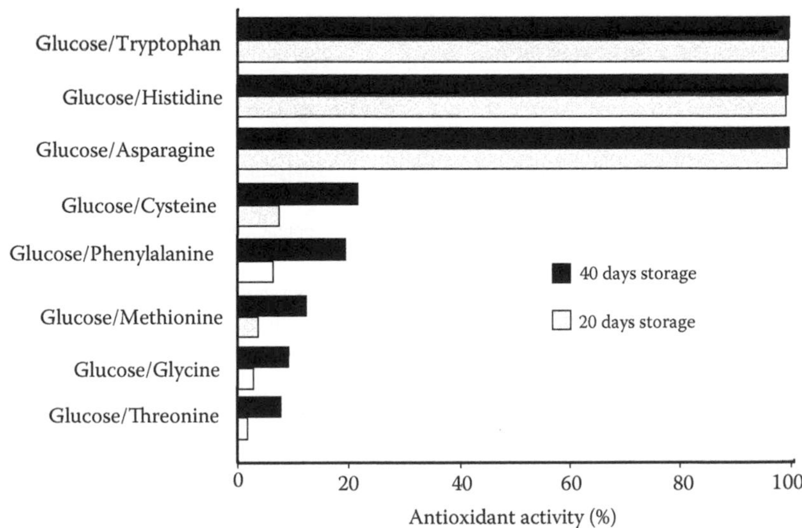

FIGURE 11.3 Antioxidant activities of dichloromethane extracts obtained from Maillard model systems consisting of glucose and eight amino acids, as tested by aldehyde/carboxylic acid assay.

the aldehyde/carboxylic acid assay (Osada and Shibamoto 2006). This figure shows the results at the level of 10 μg/ml. The extracts from the model systems of glucose/asparagine, histidine, or tryptophan inhibited hexanal oxidation by 100% at the level of 10 μg/ml of extract over 40 days. The extracts from all eight-model systems inhibited hexanal oxidation by 100% at levels greater than 50 μg/ml over 40 days. Seven column chromatographic fractions from the extract of a glucose/asparagine model system inhibited hexanal oxidation by 100% at levels greater than 10 μg/ml over 40 days.

11.1.3 Aldehyde/Carboxylic Acid Assay for Antioxidant Activity

The aldehyde/carboxylic acid assay is a simple and convenient test for evaluating the effects of antioxidants against slow oxidation phenomena occurring over prolonged periods, such as during the shelf-life of foods (Moon and Shibamoto 2009). This assay is based on the conversion mechanism of alkylaldehyde to alkylcarboxylic acid in the presence of reactive radicals (Nonhebel et al. 1979). This conversion occurs stoichiometrically, and the reduction of aldehyde or formation of carboxylic acid in a dichloromethane solution is easily monitored by gas chromatography (Macku and Shibamoto 1991). A 3% hexanal in dichloromethane solution is most commonly used and oxidation of hexanal is induced by heat, O_2, or H_2O_2. One drawback of this method is that the test sample must be lipid soluble.

11.1.4 Antioxidant Activities of Volatile Heterocyclic Compounds

Volatile compounds, in particular heterocyclic flavor chemicals, obtained from a sugar/amino acid model system have been reported to inhibit the oxidation of lipids (Elizalde et al. 1992; Shaker et al. 1995). Also, heterocyclic flavor chemicals (formed by the Maillard reaction) found in brewed coffee exhibited antioxidative activity (Yanagimoto et al. 2002). These studies clearly indicate that some heterocyclic flavor chemicals possess antioxidative activities (El-Massary et al. 2003).

As noted above, there are many reports on the antioxidant activity of volatile heterocyclic compounds. Table 11.2 summarizes the antioxidant activity of representative heterocyclic compounds. Most antioxidant activities were tested using the aldehyde/carboxylic acid (Ald/Carb. acid) assay. Details of the assays shown in Table 11.2 have been reported (Moon and Shibamoto 2009). Only antioxidant values at typical concentrations are shown in Table 11.2. However, most compounds showed dose-dependent activities (Figure 11.4). All compounds except thiazole exhibited increasing activities with increasing doses. Among unsubstituted heterocyclic compounds, pyrrole exhibited the greatest antioxidant activity followed by furan. Its activity was almost 10 times greater than those of others. It is interesting that antioxidant activity is changed by a substituent on a heterocyclic ring. Figure 11.5 shows relative antioxidant activity for alkyl- and acetyl-substituted heterocyclic compounds. Antioxidant activity increased with the addition of either alkyl- or

TABLE 11.2
Antioxidant Activity of Heterocyclic Compounds Tested by Various Antioxidant Assays

Compound	Activity (%)	Assay	Amount Tested	References
Furan	11.1	Peroxide value	1 M	Cho and Kim 1994
	4.8	Carbonyl value	1 M	Cho and Kim 1994
	84.5	Ald/Carb. acid	500 μg/ml	Yanagimoto et al. 2002; Fuster et al. 2000
	6.1	Ald/Carb. acid	50 μg/ml	Yanagimoto et al. 2002
Furfural	3.9	Peroxide value	1 M	Cho and Kim 1994
	6.4	Carbonyl value	1 M	Cho and Kim 1994
	50.3	Ald/Carb. acid	500 μg/ml	Yanagimoto et al. 2002
	5.0	Ald/Carb. acid	50 μg/ml	Yanagimoto et al. 2002
DMHF [2,5-dimethyl-4-hydroxy-3(2*H*)-furanone]	21.7	Peroxide value	1 M	Cho and Kim 1994
	18.5	Ald/Carb. acid	500 μg/ml	Fuster et al. 2000
	16.4	Carbonyl value	1 M	Cho and Kim 1994
	94.5	Ald/Carb. acid	100 μg/ml	Eiserich et al. 1992
Oxazole	90.0	Ald/Carb. acid	50 μg/ml	Eiserich et al. 1992
2,4,5-trimethyloxazole				
	–3.7	Peroxide value	1 M	Cho and Kim 1994
	–3.2	Carbonyl value	1 M	Cho and Kim 1994
	90.9	Ald/Carb. acid	100 μg/ml	Eiserich et al. 1992
Thiazole	10.1	Ald/Carb. acid	500 μg/ml	Yanagimoto et al. 2002; Fuster et al. 2000
	14.5	Ald/Carb. acid	50 μg/ml	Yanagimoto et al. 2002
2,4,5-Trimethylthiazole	–1.0	Peroxide value	1 M	Cho and Kim 1994
	–7.0	Carbonyl value	1 M	Cho and Kim 1994
	55.1	Ald/Carb. acid	100 mg/ml	Eiserich et al. 1992
Pyrazine	19.5	Ald/Carb. acid	500 μg/ml	Yanagimoto et al. 2002
	12.3	Ald/Carb. acid	50 μg/ml	Yanagimoto et al. 2002
2,3,5-Trimethylpyrazine	14.1	Peroxide value	1 M	Cho and Kim 1994
	18.2	Carbonyl value	1 M	Cho and Kim 1994
	10.2	Ald/Carb. acid	100 μg/ml	Eiserich et al. 1992
Pyrrole	54.5	Peroxide value	1 M	Cho and Kim 1994

(*continued*)

TABLE 11.2 (Continued)
Antioxidant Activity of Heterocyclic Compounds Tested by Various Antioxidant Assays

Compound	Activity (%)	Assay	Amount Tested	References
	31.6	Carbonyl value	1 M	Cho and Kim 1994
	100.0	Ald/Carb. acid	50 μg/ml	Yanagimoto et al. 2002; Fuster et al. 2000
	6.5	Ald/Carb. acid	5 μg/ml	Yanagimoto et al. 2002
1-Methylpyrrole	100.0	Ald/Carb. acid	50 μg/ml	Yanagimoto et al. 2002
	5.6	Ald/Carb. acid	5 μg/ml	Yanagimoto et al. 2002
Pyrrole-2-carboxyaldehyde	100.0	Ald/Carb. acid	50 μg/ml	Yanagimoto et al. 2002
	89.2	Ald/Carb. acid	5 μg/ml	Yanagimoto et al. 2002
Imidazole	54.5	Peroxide value	1 M	Cho and Kim 1994
	50.3	Carbonyl value	1 M	Cho and Kim 1994
	37.9	Ald/Carb. acid	2 μg/ml	Shaker et al. 1995
Thiophene	46.3	Ald/Carb. acid	1 mM	Eiserich et al. 1992
	63.2	Ald/Carb. acid	500 μg/ml	Yanagimoto et al. 2002
	11.2	Ald/Carb. acid	50 μg/ml	Yanagimoto et al. 2002
	32.5	Ald/Carb. acid	1 mM	Eiserich and Shibamoto 1994
Pyridine	41.1	Ald/Carb. acid	100 μg/ml	Macku and Shibamoto 1991

acetyl substituents, as in the case of pyrrole. Antioxidant activity was increased by an alkyl group but was decreased by an acetyl group. This phenomenon may be attributable to the nature of electron donating (alkyl group) or withdrawing (acetyl group). The scavenging of a hydroxyl radical occurred at the more electron dense ring carbon atom of a heterocyclic compound.

When unsubstituted heterocyclic compounds were reacted with hydrogen peroxide (hydroxyl radical), all generated oxidized products (Yanagimoto et al. 2002). For example, the reaction between 1-methylpyrrole and hydrogen peroxide in a dichloromethane solution produced 1,5-dihydro-1-methyl-2*H*-pyrrole-2-one and 1-methyl-2,5-pyrrolidinedione as major products. The proposed hydroxyl radical scavenging mechanism by 1-methylpyrrole is shown in Figure 11.6. According to this mechanism,

Compound		Unsubstituted	Alkyl-	Acetyl-
Pyrrole	 N H	++++	+++++	+++++
Furan	 O	+++	+++++	++
Thiophene	S	+++	+++++	+
Imidazole	 N N H	++	–	–
Thiazole	 N S	++	+++	–

FIGURE 11.4 Dose-dependent antioxidant activities of unsubstituted heterocyclic compounds, as measured by aldehyde/carboxylic acid assay.

1 mole of 1-methylpyrrole traps 2 moles of hydrogen radicals. These reports indicate that heterocyclic flavor chemicals possess strong antioxidant activities.

11.1.5 Summary

The antioxidant activities of flavor chemicals are not as strong as the known antioxidants—α-tocopherol, flavonoids, and BHT. However, since tremendous numbers of these flavor chemicals are present in foods and natural plants, their combined activity might be comparable to those of known antioxidants.

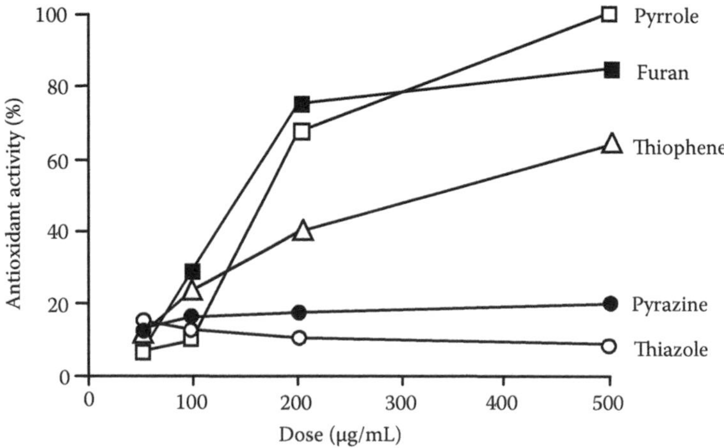

FIGURE 11.5 Relative antioxidant activity of alkyl- and acetyl-substituted heterocyclic compounds.

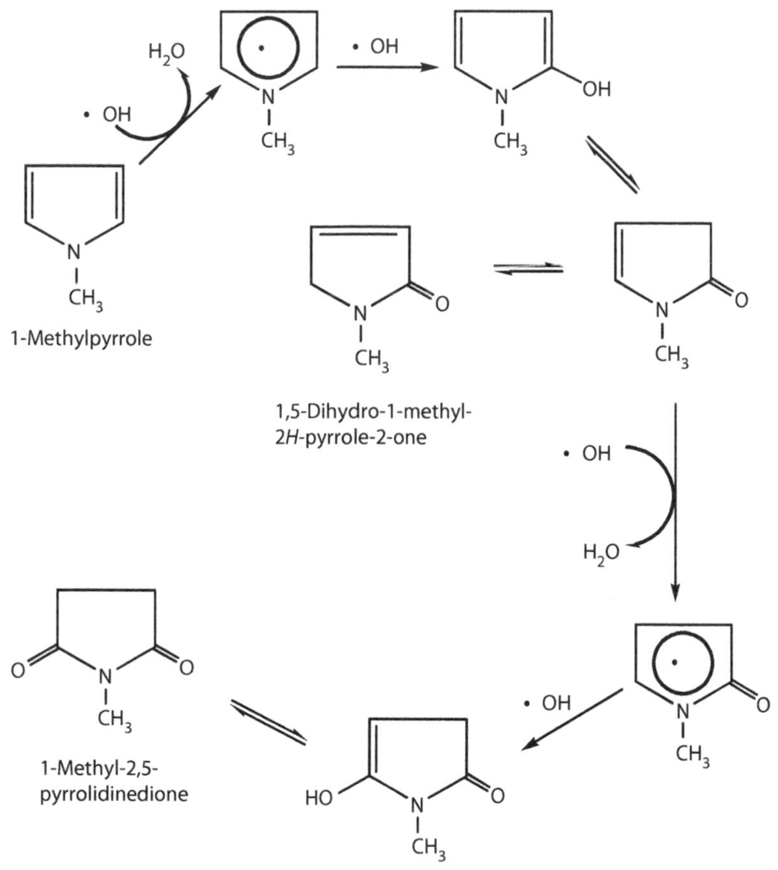

FIGURE 11.6 Proposed hydroxyl radical trapping mechanisms by 1-methylpyrrole.

11.2 ESSENTIAL OILS AND THEIR FLAVOR COMPONENTS

11.2.1 Introduction

Traditionally used as flavoring agents in foods, plant essential oils have received renewed scrutiny for the potential bioactive functions they possess. Since ancient times, numerous civilizations have utilized the therapeutic properties of essential oils in folk medicine and aromatherapy. More recently, demand for essential oils is based on consumers seeking low cost, natural, and low toxicity alternatives to prescription medications. Whereas some essential oils possess beneficial physiological effects at certain concentrations, others may not, and a more thorough understanding of their safety and efficacy can only be obtained by evaluating their bioactivities in various systems.

Although the volatile compositions of essential oils are complex and varied in nature, most constituents present are monoterpenes, sesquiterpenes, or oxygenated derivatives of these compounds. The biosynthetic pathways of these compounds have been described (Dewick 2001). The functional properties exhibited by essential oils is influenced by many factors, including climate, growing region, harvest times, extraction techniques (Arctander 1960), and the part of the plant from which the essential oil was extracted from. Among the functional properties demonstrated by essential oils, herbal and spice essential oils have been valued in food systems for their antioxidant and antimicrobial activities.

Interest in essential oils, their volatile constituents, and their bioactive properties has promoted numerous studies that investigate their potential roles in health and disease prevention. Among the diverse functional activities exhibited by essential oils, an overview of their antioxidant, antiinflammatory, antimicrobial, antiviral, and anticarcinogenic properties is included.

11.2.2 Antioxidant and Antiinflammatory Activities

In addition to heterocyclic flavor chemicals generated during the Maillard reaction, essential oils have also been reported to exhibit antioxidant activities. To date, numerous assays have been developed to study the antioxidant capacities of test compounds (Huang et al. 2005; Moon and Shibamoto 2009). The phenolic constituents present in essential oils can behave as antioxidants by serving as hydrogen donors, free radical scavengers, reducing agents, or singlet oxygen quenchers (Rice-Evans et al. 1997). An accurate assessment of the potential antioxidant capacities of a test compound would require ascertaining its actions in these assay systems and in in vivo systems.

Essential oils have been considered potential substitutes for synthetic antioxidants in order to prevent lipid oxidation and extend the shelf life of foods. Using the aldehyde/carboxylic acid assay, volatile extracts of thyme (*Thymus vulgaris*) (at 10 μg/ml) and basil (*Ocimum basilicum*) (at 50 μg/ml) exhibited inhibitory activities comparable to BHT (Lee and Shibamoto 2002b). Parsley (*Petroselinum sativum* H.) seed and jasmine (*Jasminum officinale* L.) essential oils showed equal effectiveness as α-tocopherol in inhibiting hexanal oxidation at 500 μg/ml after 40 days (Wei and Shibamoto 2007a). In contrast, essential oils of ginger (*Zingiber officinale* R.), chamomile (*Anthemis nobilis* L.), and peppermint (*Mentha piperita* L.) showed low inhibitory activities against hexanal oxidation at 500 μg/ml (Wei and Shibamoto 2007a).

The association between lipid peroxidation of biological moieties and development of diseases such as aging (Beckman and Ames 1998) may also be influenced by essential oils. The inclusion of essential oils as fragrances in cosmetic products was popular among ancient Egyptians, Greeks, and Romans. More recently, the potential application of essential oils as antioxidants against oxidation of skin lipids was demonstrated. Parsley (*P. sativum* H.) seed essential oil at 500 μg/ml was shown to inhibit ultraviolet radiation-induced oxidation (UVR, $\lambda = 300$ nm) of squalene by 67% (Wei and Shibamoto 2007a). In comparison, α-tocopherol inhibited squalene oxidation by 76% at the same concentration. Mixtures of essential oils (thyme; *T. vulgaris*) or clove (*Eugenia caryophyllata*) leaf with rose (*Rosa damascena*), cinnamon (*Cinnamomum zeylanicum*) leaf, or parsley (*P. sativum* H.) seed have also been reported to inhibit UVR-induced oxidation of squalene (Wei and Shibamoto 2007b).

The presence of reactive oxygen species at sites of tissue injury has also been suggested as an agent for inducing inflammatory responses. Conditions such as stroke (Muir et al. 2007) and hypertension (Vaziri and Rodriguez-Iturbe 2006) have been associated with inflammation. Animal models and biochemical assays based on the complex cascade of reactions that occurs during inflammation are often used to test for antiinflammatory effects. Black cumin (*Nigella sativa* L.) seed essential oil demonstrated antiinflammatory properties in carrageenan-induced paw edema tests and analgesic effects in the acetic acid-induced writhing test, formalin test, and light tail flick test in rats (Hajhashemi et al. 2004). The antiinflammatory property of cinnamon (*Cinnamomum osmophloeum*) leaf essential oil was demonstrated based on its ability to inhibit inflammatory cytokines (Chao et al. 2005). In the presence of this oil, expression of prointerleukin-1-β protein stimulated by lipopolysaccharide-treated J774A.1 murine macrophage cells was inhibited. Secretion of lipopolysaccharide-induced tumor necrosis factor-α was slightly reduced in J774A.a macrophage cells treated with this oil. A dose-dependent decrease in interleukin-1-β protein secretion was also observed with increasing concentrations of essential oil. Cinnamaldehyde, a major constituent from this oil, was also found to inhibit interleukin-1-β protein and tumor necrosis factor-α secretion at low concentrations (Chao et al. 2008).

11.2.3 Antimicrobial and Antiviral Activities

In human medicine, the growing prevalences of antibiotic-resistant pathogens and drug-resistant viral strains have stimulated research into the use of essential oils as potential alternative antimicrobial or antiviral agents (Filipowicz et al. 2003; Ebrahimi et al. 2008; Schnitzler et al. 2007; Astani et al. 2010). The essential oil of juniper (*Juniperus communis* L.) berry was found to be potent against antibiotic-sensitive strains of *Candida albicans*, antibiotic-resistant *Acinetobacter baumanii*, and *Staphylococcus aureus* (both the reference strain and the antibiotic resistant strain) using a modified disk diffusion method. Furthermore, the chirality of the enantiomers of α-pinene (a constituent in the juniper berry essential oil) was found to influence the potencies against the bacterial and fungal strains differently. More potent activities were observed by (–)-α-pinene against *C. albicans* and *A. baumanii*, compared to its enantiomer and to the other constituents (both enantiomers of β-pinene and limonene, (+)-sabinene, *p*-cymene, and terpinen-4-ol) tested. However, (+)-α-pinene showed potency against *S. aureus*, whereas (–)-α-pinene did not (Filipowicz et al. 2003).

In a recent report, essential oils from eucalyptus, tea tree, and thyme, and their major constituents (α-terpinene, γ-terpinene, α-pinene, *p*-cymene, terpinen-4-ol, α-terpineol, thymol, citral, and 1,8-cineole) were evaluated for antiviral potentials against herpes simplex virus type 1 in vitro. Viral multiplication was suppressed by >96% for all essential oils and >80% for all monoterpenes, except for the less active 1,8-cineole. Plaque formation was significantly reduced when viral cells were pretreated with essential oils or the monoterpenes for 1 h before infection. It was suggested that the mechanism by which the tested compounds exhibited antiviral activities involved disrupting the viral structures needed for adsorbing or entering into host cells (Astani et al. 2010).

Global interest to seek safe alternatives to additives used for controlling foodborne pathogens and costly food poisoning outbreaks have renewed efforts to find natural ingredients with antimicrobial properties. Many kinds of spice essential oils have been reported to demonstrate both antibacterial and antifungal actions (Smith-Palmer et al. 1998; Kalemba and Kunicka 2003; Burt 2004). Because of the diversity in microorganisms (gram negative and positive bacteria and fungi) that cause food spoilage, the bioactive components present in essential oils may have differing functions against each microorganism. As a consequence, the overall antimicrobial activity of the essential oil is influenced. Enhancement of antimicrobial activities may be accomplished by mixing essential oils for synergistic or additive effects. Ultimately, the impact that essential oils would have when added to a food system in terms of altering the composition of the food and its organoleptic properties must also be considered.

The antibacterial activities of lemon balm (*Melissa officinalis*), marjoram (*Origanum majorana*), oregano (*Origanum vulgare*), and thyme (*T. vulgaris*) essential oils against spoilage bacteria (*Enterobacter* spp., *Listeria* spp., *Lactobacillus* spp., and *Pseudomonas* spp.) were investigated by Gutierrez et al. (2009) using the agar dilution method, the agar well diffusion test, and the absorbance-based microtiter plate assay. Their assessments of the minimum inhibitory concentrations showed that oregano and thyme essential oils possessed the greatest antibacterial activities. Combinations of oregano/thyme, oregano/lemon balm, and thyme/lemon balm essential oils were shown to be effective against *Listeria* strains by the checkerboard method, whereas additive effects against *Enterobacter cloacae*, *Pseudomonas fluorescens*, and *Listeria innocua* occurred from the oregano/thyme essential oil combination.

11.2.4 Anticarcinogenic Activities

Essential oils and their volatile constituents also possess the abilities to inhibit cancer formation and/or progression, and inhibit tumor cell expression. The mechanisms by which these occur have been reviewed (Edris 2007; Crowell 1999). Oregano oil was examined for its abilities to inhibit the replication of four cell lines, including two derived from human carcinomas (Sivropoulou et al. 1996). Cell death for all cell lines occurred for all dilutions of oregano essential oil up to 1/1000. However, no effects were observed in the human carcinoma cell lines when the oil was diluted 1/5000.

One enzyme commonly used as a biomarker in carcinogenicity studies is glutathione-*S*-transferase, which metabolizes carcinogens to less toxic forms so that they can be removed from the body. Screening of potential anticarcinogenic compounds, such as essential oils, can be accomplished by studying the abilities of the test compounds to induce increased glutathione-*S*-transferase levels (Lam and Zheng 1991). Basil, cumin, and poppy essential oils were found to inhibit 3,4-benzo(*a*)pyrene-induced squamous cell carcinoma based on stimulation of glutathione-*S*-transferase in Swiss mice (Aruna and Sivaramakrishnan 1996).

Increased diagnoses of human nonmelanoma skin cancers as a consequence of ozone depletion and UVR have raised awareness in the search for chemopreventive agents that inhibit changes in damaged skin and skin cancers. Sandalwood essential oil and one of its major constituents, alpha-santalol, have been shown to inhibit different stages of skin tumor development in mice (Dwivedi and Abu-Ghazaleh 1997; Dwivedi and Zhang 1999; Dwivedi et al. 2003). Topical applications of α-santalol (5% w/v) to hairless SKH-1 mice were shown to protect against ultraviolet B (UVB) radiation-induced (λ = 290–320 nm) tumor initiation (12-*O*-tetradecanoylphorbol-13-acetate was used for tumor promotion), UVB-induced tumor promotion (7,12-dimethylbenzanthracene was used for tumor initiation), and UVB-induced complete carcinogenesis (Dwivedi et al. 2006). Significant reductions in tumor incidences and multiplicity were observed, suggesting that α-santalol may be a potential chemopreventive agent for UVR-induced skin cancers.

11.2.5 Summary

Interest in finding natural alternative agents to synthetic chemicals and drugs has stimulated extensive analyses into the functional properties of flavor compounds. The safety and efficacy of these compounds will also have to be considered in order to accurately define their potential roles as preventive agents. By doing so, a better understanding of the mechanisms by which flavor chemicals exert their pharmacological activities can be obtained in order for us to most effectively use them.

REFERENCES

Arctander, S. 1960. *Perfume and Flavor Materials of Natural Origin*. Denmark: Det Hoffensbergske Etablissement.

Aruna, K., and V.M. Sivaramakrishnan. 1996. Anticarcinogenic effects of the essential oils from cumin, poppy and basil. *Phytother. Res.* 10: 577–580.

Astani, A., J. Reichling, and P. Schnitzler. 2010. Comparative study on the antiviral activity of selected monoterpenes derived from essential oils. *Phytother. Res.* 24: 673–679.

Beckman, K.B., and B.N. Ames. 1998. The free radical theory of aging matures. *Physiol. Rev.* 78: 547–581.

Benjakul, S.,W. Lertittikul, and F. Bauer. 2005. Antioxidant activity of Maillard reaction products from a porcine plasma protein-sugar model system. *Food Chem.* 93: 189–196.

Burt, S. 2004. Essential oils: Their antibacterial properties and potential applications in foods—A review. *Int. J. Food Microbiol.* 94: 223–253.

Chao, L.K., K.F. Hua, H.Y. Hsu, S.S. Cheng, I.F. Lin, S.T. Chen, and S.T. Chang. 2008. Cinnamaldehyde inhibits pro-inflammatory cytokines secretion from monocytes/macrophages through suppression of intracellular signaling. *Food Chem. Toxicol.* 46: 220–231.

Chao, L.K., K.F Hua, H.Y Hsu, S.S. Cheng, J.Y. Liu, and S.T. Chang. 2005. Study on the anti-inflammatory activity of essential oil from leaves of *Cinnamomum osmophloeum*. *J. Agric. Food Chem.* 53: 7274–7278.

Chen, X.-M., and D.D. Kitts. 2008. Antioxidant activity and chemical properties of crude and fractionated Maillard reaction products derived from four sugar-amino acid Maillard reaction model systems. *Ann. N.Y. Acad. Sci.* 1126: 220–224.

Cho, Y.-H., and D.H. Kim. 1994. Antioxidant activity of low-molecular weight compounds identified as Maillard reaction products on refined soybean oil. *Food Biotechnol.* 3: 148–151.

Crowell, P.L. 1999. Prevention and therapy of cancer by dietary monoterpenes. *J. Nutr.* 129: 775S–778S.

Dewick, P.M. 2001. *Medicinal Natural Products: A Biosynthetic Approach*, 2nd ed. Chichester, England: John Wiley & Sons.

Dwivedi, C. and A. Abu-Ghazaleh. 1997. Chemopreventive effects of sandalwood oil on skin papillomas in mice. *Eur. J. Cancer Prev.* 6: 399–401.

Dwivedi, C. and Y. Zhang. 1999. Sandalwood oil prevents skin tumor development in CD-1 mice. *Eur. J. Cancer Prev.* 8: 449–455.

Dwivedi, C., X. Guan, W.L. Harmsen, A.L. Voss, D.E. Goetz-Parten, E.M. Koopman, K.M. Johnson, H.B. Valluri, and D.P. Matthees. 2003. Chemopreventive effects of a-santalol on skin tumor development in CD-1 and SENCAR mice. *Cancer Epidemiol. Biomarkers Prev.* 12: 151–156.

Dwivedi, C., H.B. Valluri, and R. Agarwal. 2006. Chemopreventive effects of alpha-santalol on ultraviolet B radiation-induced skin tumor development in SKH-1 hairless mice. *Carcinogenesis* 27: 1917–1922.

Ebrahimi, S.N., J. Hadian, M.H. Mirjalili, A. Sonboli, and M. Yousefzadi. 2008. Essential oil composition and antibacterial activity of *Thymus caramanicus* at different phenological stages. *Food Chem.* 110: 927–931.

Edris, A.E. 2007. Pharmaceutical and therapeutic potentials of essential oils and their individual volatile constituents: A review. *Phytother. Res.* 21: 308–323.

Eiserich, J.P., and T. Shibamoto. 1994. Antioxidant activity of volatile heterocyclic compounds. *J. Agric. Food Chem.* 42: 1060–1063.

Eiserich, J.P., C. Macku, and T. Shibamoto. 1992. Volatile antioxidants formed from an L-cysteine/D-glucose Maillard model system. *J. Agric. Food Chem.* 40: 1982–1988.

Elizalde, B.E., F. Bressa, and M.D. Rosa. 1992. Antioxidative action of Maillard reaction volatiles—Influence of Maillard solution browning level. *J. Am. Oil Chem. Soc.* 69: 331–334.

El-Massary, K., A. Farouk, and A. El-Ghorab. 2003. Volatile constituents of glutathione–ribose model system and its antioxidant activity. *Amino Acids* 24: 171–177.

Fernandez, S., S. Kerverdo, E. Dunach, and L. Liizzani-Cuvelier. 2002. Heterocycles in flavour chemistry. *Actual. Chim.* 4: 4–14.

Filipowicz, N., M. Kamiński, J. Kurlenda, M. Asztemborska, and J.R. Ochocka. 2003. Antibacterial and antifungal activity of juniper berry oil and its selected components. *Phytother. Res.* 17: 227–231.

Flament, I. 1975. Heterocyclics in flavor chemistry. Five and six-membered rings containing oxygen, nitrogen and sulfur atoms; monocyclic condensed bicyclic components. In *Aroma Research. Proceedings of International Symposium*, ed. H. Maarse and P.J. Groenen, 221–237. Wageningen, Netherlands: Cent. Agric. Publ. Doc.

Flament, I. 2002. *Coffee Flavor Chemistry*. New York: John Wiley & Sons, Ltd.

Fuster, M.D., A.E. Mitchell, H. Ochi, and T. Shibamoto. 2000. Antioxidative activities of heterocyclic compounds formed in brewed coffee. *J. Agric. Food Chem.* 48: 5600–5603.

Gutierrez, J., C. Barry-Ryan, and P. Bourke. 2009. Antimicrobial activity of plant essential oils using food model media: Efficacy, synergistic potential and interactions with food components. *Food Microbiol.* 26: 142–150.

Hajhashemi, V., A. Ghannadi, and H. Jafarabadi. 2004. Black cumin seed essential oil, as a potent analgesic and anti-inflammatory drug. *Phytother. Res.* 18: 195–199.

Hao, J., and D.D. Kitts. 2004. Antioxidant activity of sugar-lysine Maillard reaction products in cell free and cell culture systems. *Arch. Biochem. Biophys.* 429: 154–163.

Henderson, D.E., A.M. Slickman, and S.K. Henderson. 1999. Quantitative HPLC determination of the antioxidant activity of capsaicin on the formation of lipid hydroperoxides of linoleic acid: A comparative study against BHT and melatonin. *J. Agric. Food Chem.* 47: 2563–2570.

Hocman, G. 1988. Chemoprevention of cancer—Phenolic antioxidants (BHT, BHA). *Int. J. Biochem.* 20: 639–651.

Hodge, J.E. 1953. Dehydrated foods—Chemistry of browning reactions in model systems. *J. Agric. Food Chem.* 1: 928–943.

Huang, D.J., B.X. Ou, and R.L. Prior. 2005. The chemistry behind antioxidant capacity assays. *J. Agric. Food Chem.* 53: 1841–1856.

Huang, Y.L., J.Y. Sheu, and T.H. Lin. 1999. Association between oxidative stress and changes of trace elements in patients with breast cancer. *Clin. Biochem.* 32: 1069–1072.

Kalemba, D., and A. Kunicka. 2003. Antibacterial and antifungal properties of essential oils. *Curr. Med. Chem.* 10: 813–829.

Kim, J.-S., and Y.S. Lee. 2009. Antioxidant activity of melanoidins from different sugar/amino acid model systems: Influence of the enantiomer type. *Food Sci. Technol. Int.* 15: 291–297.

Kitts, D.D., and C. Hu. 2005. Biological and chemical assessment of antioxidant activity of sugar-lysine model Maillard reaction products. *Ann. N.Y. Acad. Sci.* 1043: 501–512.

Lam, L.K.T., and B.-L. Zheng. 1991. Effects of essential oils on glutathione-*S*-transferase activity in mice. *J. Agric. Food Chem.* 39: 660–662.

Lee, K.-G., and T. Shibamoto. 2002a. Toxicology and antioxidant activities of non-enzymatic browning reaction products: Review. *Food Rev. Int.* 18: 151–175.

Lee, K.-G., and T. Shibamoto. 2002b. Determination of antioxidant potential of volatile extracts isolated from various herbs and spices. *J. Agric. Food Chem.* 50: 4947–4952.

Macku, C., and T. Shibamoto. 1991. Volatile antioxidants produced from heated corn oil/glycine model system. *J. Agric. Food Chem.* 39: 1990–1993.

Mastrocola, D., and M. Munari. 2000. Progress of the Maillard reaction and antioxidant action of Maillard reaction products in preheated model systems during storage. *J. Agric. Food Chem.* 48: 3555–3559.

Mastrocola, D., M. Munari, M. Cioroi, and C.R. Lerici. 2000. Interaction between Maillard reaction products and lipid oxidation in starch-based model systems. *J. Sci. Food Agric.* 80: 684–690.

Monti, S.M., A. Ritieni, G. Graziani, G. Randazzo, L. Mannina, A.L. Segre, and V. Fogliano. 1999. LC/MS analysis and antioxidative efficiency of Maillard reaction products from lactose–lysine model system. *J. Agric. Food Chem.* 47: 1506–1513.

Moon, J.-K., and T. Shibamoto. 2009. Antioxidant assays for plant and food components. *J. Agric. Food Chem.* 57: 1655–1666.

Muir, K.W., P. Tyrrell, N. Sattar, and E. Warburton. 2007. Inflammation and ischaemic stroke. *Curr. Opin. Neurol.* 20: 334–342.

Nicoli, M.C., M. Anese, M.T. Parpinel, S. Franceschi, and C.R. Lerici. 1997. Loss and/or formation of antioxidants during food processing and storage. *Cancer Lett.* 114: 71–74.

Nonhebel, D.C., J.M. Tedder, and J.C. Walton. 1979. *Radicals*, p. 157. London: Cambridge University Press.

Obretenov, T., and G. Vernin. 1998. Melanoidins in the Maillard reaction. *Dev. Food Sci.* 40: 455–482.

Osada, Y., and T. Shibamoto. 2006. Antioxidative activity of volatile extracts from Maillard model systems. *Food Chem.* 98: 522–528.

Rice-Evans, C.A., N.T. Miller, and G. Paganga. 1997. Antioxidant properties of phenolic compounds. *Trends Plant Sci.* 4: 304–309.

Schnitzler, P., C. Koch, and J. Reichling. 2007. Susceptibility of drug-resistant clinical herpes simplex virus type 1 strains to essential oils of ginger, thyme, hyssop, and sandalwood. *Antimicrob. Agents Chemother.* 51: 1859–1862.

Shaker, E.S., M.A. Ghazy, and T. Shibamoto. 1995. Antioxidant activity of volatile browning reaction products and related compounds in a hexanal/hexanoic acid system. *J. Agric. Food Chem.* 43: 1017–1022.

Shibamoto, T. 1980. Heterocyclic compounds found in cooked meats. *J. Agric. Food Chem.* 28: 237–243.

Shibamoto, T. 1983. Heterocyclic compounds in browning and browning/nitrite model systems: Occurrence, formation mechanisms, flavor characteristics and mutagenic activity. In *Instrumental Analysis of Foods*, ed. I.G. Charalambous and G. Inglett, 229–278. New York: Academic Press.

Sivropoulou, A., E. Papanikolaou, C. Nikolaou, S. Kokkini, T. Lanaras, and M. Arsenakis. 1996. Antimicrobial and cytotoxic activities of *Origanum* essential oils. *J. Agric. Food Chem.* 44: 1202–1205.

Smith-Palmer, A., J. Stewart, and L. Fyfe. 1998. Antimicrobial properties of plant essential oils and essences against five important food-borne pathogens. *Lett. Appl. Microbiol.* 26: 118–122.

Vaziri, N.D., and B. Rodriguez-Iturbe. 2006. Mechanisms of disease: Oxidative stress and inflammation in the pathogenesis of hypertension. *Nat. Clin. Pract. Nephrol.* 2: 582–593.

Vernin, G. 1979. Heterocycles in food aromas. I. Structure and organoleptic properties. *Parfums, Cosmet., Aromes* 27: 77–86.

Wei, A., and T. Shibamoto. 2007a. Antioxidant activities and volatile constituents of various essential oils. *J. Agric. Food Chem.* 55: 1737–1742.

Wei, A. and T. Shibamoto. 2007b. Antioxidant activities of essential oil mixtures toward skin lipid squalene oxidized by UV irradiation. *Cutan. Ocul. Toxicol.* 26: 227–233.

Yanagimoto, K., K.-G. Lee, H. Ochi, and T. Shibamoto. 2002. Antioxidative activity of heterocyclic compounds found in coffee volatiles produced by Maillard reaction. *J. Agric. Food Chem.* 50: 5480–5484.

Zhang, L., Q. Li, Z. Yin, and H. Jing. 2008. Physiochemical properties and antioxidant activity of three glucose-amino acid model Maillard reaction products. *Zhongguo Shipin Xuebao* 8:12–22.

12 Flavors from Cheeses

Henry-Eric Spinnler

CONTENTS

12.1 INTRODUCTION

In cheese, flavors are mainly issued from the catabolism of lactose, citrate, proteins, and lipids by the microorganisms that are growing first in the milk and then in or over the cheese curd. Some of the volatile compounds found in cheeses are issued from the ones found in milk, but their role in cheese flavor has, until now, never been confirmed (Martin et al. 2002; Cornu et al. 2005, 2009). However, they are very good elements of traceability and it is possible to know, using gas chromatography/mass spectrometry and databanks made from the analysis of the forage (Walker et al. 2004), if the milk used for cheese making, comes from a specific region or another.

Therefore, the role of the microbial ecosystem that develops inside and at the surface of the cheese has a major role on the construction of the cheese flavor typicality. The specific role of microorganisms present inside or at the surface of the cheese depends on the cheese technology used to make the cheese. For example, the size of a pressed cooked Swiss-type cheese, leading to a small surface/volume ratio, will give a major role to the anaerobic bacteria developing inside the cheese. On the other hand, in washed rind cheeses such as Munster, or mold ripened cheeses (e.g., Camembert), where the surface/volume ratio is high, the surface microflora will have a major role.

The different steps of the technological diagram in use—(1) heating of the milk, (2) curdling, (3) treatment of the curd (cutting, stirring, heating, draining), and (4) ripening—will determine the microbial ecosystems. The diversity of their metabolism and therefore the production of a large variety of flavor compounds, will lead to the large variety of organoleptic properties of the cheeses.

Heat treatment of the milk has an impact on milk flavors, which could be important in fresh cheeses. A heat treatment at a temperature of 100°C can generate a series of γ- and δ-lactones from the hydroxy acids present in the milk triglycerides (Alewijn et al. 2007). This will give the typical flavor of warm milk; the compounds involved have coconut flavors or peach, apricot flavors (Dufossé et al. 1994).

At higher temperatures, sulfur compounds can be produced from cystein degradation giving rotten eggs or cabbage flavors (De Wit and Nieuwenhuijse 2008). These compounds are very volatile for some (H_2S, COS, CH_3SH) and will disappear quickly, while others can participate in fresh cheese flavors [dimethyl sulfide (DMS), dimethyl disulfide (DMDS)] and can be positive at low concentrations.

In this chapter, we describe the major compounds involved in different cheese varieties. For each type of cheese, we focus on the two biochemical factors involved in their occurrence: milk component and microbial activities transforming the milk component.

12.2 FLAVORS FROM YOGURTS, FRESH CHEESES, AND MOZZARELLA

In these products, the role of the lactic acid bacteria (LAB) is essential. The starter LAB (mainly *Streptococcus thermophilus*, *Lactococcus lactis* ssp. *lactis*, *L. lactis* ssp. *lactis* biovar. *diacetylactis*, *L. lactis* ssp. *cremoris*, *Lactobacillus* sp., and *Leuconostoc* spp.) by reducing the pH, will play a role in milk clotting and generate the flavors of fresh products. For example, the acidification will permit the short-chain fatty acid to become volatile as soon as the pH is below their pK_a (close to pH = 5.0 and depending on the acid considered). LAB are also important in most cheeses, via the volatile compounds they are able to produce, but they have a major role in fresh products such as butter, cottage cheese, fromage frais, mozzarella, or yogurts.

Strains of *L. lactis* ssp. *lactis* biovar. *diacetylactis* and *Leuconostoc* are able to produce diacetyl and acetoin from the milk citrate. LAB have also a large diversity of enzymatic systems able to hydrolyze the proteins in peptides and amino acids. They are able to realize the amino acid breakdown through transamination (see the Chapter 6). It has been shown that the speed of flavor production during cheese ripening is not limited by the amount of amino acids but by the amino acid breakdown. Yvon et al. (2000) have clearly shown that the transamination step providing α-keto acids as direct precursors of aldehydes was a limiting step. This pathway permits the conversion of amino acids into aldehydes, with malty flavors, or into alcohols and esters with fruity flavors. Other aldehydes such as hexanal and sulfur compounds such as DMS are also significant in the flavors of these products.

In yogurt, numerous efforts have been made to quantify the important flavor compounds. The work of Ott et al. (1997, 1999) probably offers a more complete treatment of the subject (Table 12.1). Five compounds seem to be significant: acetaldehyde, DMS, diacetyl, 2,3-pentanedione, and benzothiazole. However, in this product, these authors have shown the particular sensitivity of the panel to small change in acidity on flavor perception (Ott et al. 2000). On flavored yogurts, change in flavors was noticeable when the texture was modified. The change in texture physically changes the flavor release in the mouth, but it may also interact with the flavor perception (Saint-Eve et al. 2006).

TABLE 12.1
Comparison of Absolute Measurement of Flavor Compounds in Milk and Yogurt with Their Olfactive Impact

	Milk[a]		Yogurt		
Compound	**(mg/kg)**	**SNIF[b]**	**(mg/kg)**	**Mix lac+[c] (mg/kg)**	**SNIF**
Acetaldehyde	nd	0	1.8–16.8	16.57	7006
Acetone	2.7	nd	2.3–4.2	4.03	nd
Dimethyl sulfide	0.027	5269	0.013–0.048	0.022	3969
Ethanol	1.3	nd	1.2–5.1	1.95	nd
2-Butanone	0.20	nd	0.11–0.69	0.31	nd
2-Pentanone	0.060	nd	0.024–0.066	0.054	nd
2,3-Butanedione	nd	3069	0.31–3.62	1.35	8561
2,3-Pentanedione	nd	nd	0.02–0.27	0.13	3549
2-Heptanone	0.14	nd	0.08–0.16	0.13	nd
Benzaldehyde	0.030	nd	0.027–0.128	–	nd
Benzothiazole	0.38	1779	0.13–1.10	0.47	3486

Source: Ott, A., Germond, J.E., Baumgartner, M., Chaintreau, A., *J. Agric. Food Chem.* 47, 2379–2385, 1999. With permission.

Note: nd, not detected.

[a] Milk fortified with 2.5% skimmed milk powder and heat-treated at 98°C for 15 min.

[b] Area of the GC–olfactometry peaks.

[c] Same yogurt with indicated strains used for determination of key aroma compound by sniffing.

12.3 RIPENED SOFT CHEESES

Because of its low pH at the start of draining, the curd made to manufacture ripened soft cheeses is usually fragile and demineralized. It only permits draining by gravity. Therefore, the curd of soft cheeses is characterized by a high humidity, a low level of mineralization, and usually a low pH at the start of ripening. These factors will lead to small cheeses (which means high surface/volume ratio), comprising of a curd having a low buffer capacity and therefore allows quick changes in cheese biochemical composition, due to the surface microflora activity. Therefore, they will ripen quickly with a shorter shelf life than pressed cheeses. In soft cheeses, microbial growth at the surface is stimulated by oxygen and can be separated into two phases. The first one is attributable to acidophilic yeasts and filamentous fungi; the second one starts when the pH exceeds 5.8, permitting the growth of ripening bacteria.

12.4 MOLD RIPENED SOFT CHEESES

The succession of flora is observed in Camembert or Brie cheeses. During the first phase, the yeast and fungi will develop and will produce flavors evoking fermented apples. This typical flavor is mainly due to the production of esters and aldehydes from amino acids during their growth phase. It is yeasts such as *Kluyveromyces lactis* and *Debaryomyces hansenii* that grow first after draining. They are very efficient for the breakdown of amino acids, present in the milk, or produced by LAB. Their breakdown is carried out using mainly the Ehrlich–Neubauer pathway to produce branched chain aldehydes, alcohols, and esters. *Debaryomyces* produce more aldehydes and is associated with malt flavors although *Kluyveromyces* produces more esters and alcohols (Martin et al. 2001).

After these yeasts, other fungi such *Geotrichum candidum* are also growing. Because its activity lasts until the end of the ripening (Bonaiti 2004), *G. candidum* is involved in the typicality of the mold ripened cheeses (Molimard et al. 1997b), in contrast to other yeasts such as *Kluyveromyces*, whose activity quickly slows down when other ripening microorganisms are growing. *G. candidum* is a hemiascomycete yeast that is also very efficient on amino acid breakdown and able to produce sulfur compounds from methionine (Berger et al. 1999a). From leucine, it produces isovaleric acid. Isovaleric acid, by its significant cheese odor, its concentration level in Camembert, and its olfactive thresholds, very likely contributes to the Camembert cheese flavor.

When no more sugars or lactate are available, *Penicillium camemberti* is able to produce enzymes having a strong impact on cheese flavors, such as the phenylalanine ammonia lyase, which is able to produce styrene taint from phenylalanine (Spinnler 2003; Pagot et al. 2007).

G. candidum and *P. camemberti* are very efficient on fat breakdown. Fat has a role in texture; it acts as a lubricant, which gives the cheese a soft sensation in the mouth. The monoglycerides are very efficient emulsifiers and may reduce the size of fat globules in the cheese, which may help to induce a smoother mouthfeel and may also change the flavor release (Miettinen et al. 2002). *P. camemberti* produces large quantities of an extracellular alkaline lipase (pH optimum: 9.0). At pH 6.0, this enzyme retains 50% of its maximal activity and remains very active at the temperature of ripening (Lamberet and Lenoir 1976). It is the main lipolytic agent in Camembert cheese. It is more active on triglycerides composed of low-molecular weight fatty acids. Other acids, such as C1 to C4 or branched chain C4, C5 acids, result from the action of microorganisms on amino acids.

Lipolysis is not homogeneous throughout the cheese and mainly occurs under the rind (Hassouna and Guizzani 1995). The association of this phenomenon with proteolysis and particularly the relatively high pH give the very typical texture of the soft part of the cheese under a Camembert rind after a long ripening.

Short and intermediate chain, even numbered fatty acids (4 to 12 carbons), have high perception thresholds, but their concentration in cheese leads to a sensation and each has a characteristic note (Molimard and Spinnler 1996). However, the role of small free fatty acids with an even number of carbons from (C2 to C12) is questionable in this type of cheeses because during the ripening stage, the lipolysis increases their levels but at the same time the pH rises because of the lactate uptake by *P. camemberti*. These fatty acids have quite high olfactive thresholds from a few mg kg^{-1} in water to several hundred mg kg^{-1} in fat. It is in fact the undissociated form of the acids that is aromatic. This form is found in the fat phase of cheese, whereas the aqueous phase contains both forms, undissociated and ionized. Low pH reduces ionization and increases volatility of the acids in water. In Camembert, the level of lipolysis is smaller than in blue cheeses, but with the concentrations measured, it is reasonable to think that the short- and medium-chain free fatty acids may directly have a role on the mold ripened cheese flavors (Molimard and Spinnler 1996). Octanoic, 4-methyl-octanoic, and especially 4-ethyloctanoic acids, have odorous notes that make them particularly important in mold ripened goat cheeses (Molimard et al. 1997a). In these cheeses, branched chain fatty acids have much lower thresholds than the linear fatty acids. 4-Ethyloctanoic acid has an olfactive threshold about 500 times smaller than that of decanoic acid, which is linear with the same number of carbons. These fatty acids play a major role in goat cheese typicality. Young cheeses that are unlipolyzed are much less goaty than the more ripened ones. These branched chain fatty acids are also present in ewes' milk cheeses (Ha and Lindsay 1991) but not in cows' milk cheeses. According to their concentration and perception thresholds, volatile fatty acids can contribute to the aroma of the cheese or, for some, even give a rancidity defect. Lipolysis can also be responsible for the lactones found in Camembert (γ-decalactone, δ-decalactone, γ-dodecalactone, and δ-dodecalactone). In particular, when raw milk cheese is considered, heat treatment cannot be responsible of their occurrence in the cheese.

Fatty acids are sources of methyl ketones (2 pentanone, 2-heptanone, 2-nonanone mainly) with pungent cheesy fruity flavors. *P. camemberti*, by β-oxidation, is able to generate these important compounds from FFA, whereas intrachain oxidation of unsaturated fatty acids is the source of 1-octene-3-ol. This last compound is produced when the preferred substrates for *P. camemberti* lactate or sugars are exhausted (Spinnler et al. 1992).

If some sulfur compounds are also characteristic of the complex flavors of long ripened mold ripened cheeses, polysulfides (e.g., DMDS), methyl thioesters (e.g., methyl thioisovalerate), or thioacetal (e.g., 2,4-dithiapentane or 2,3,5-trithiahexane) (Molimard and Spinnler 1996), most of these compounds are present in much higher quantities in smear cheeses (see below).

Terpene alcohols such as 2-methylisoborneol (2-MIB; which has a musty flavor) are produced by *P. camemberti*. 2-MIB has a musty flavor but a very low detection threshold (0.1 μg kg^{-1}), which accounts for its role in soft and mold ripened cheeses (Karahadian et al. 1985a, 1985b).

12.5 WASHED RIND SOFT CHEESES

In washed rind cheeses, the wash of the rind will provoke the disappearance of filamentous fungi such as *Penicillium* and provokes the development of the ripening bacteria that will produce the smear.

This smear is composed of a large diversity of yeasts and bacteria (Irlinger and Bergère 1999). In Livarot cheese, up to 82 different strains of bacteria and yeasts have been detected. Among these, the coryneform bacteria and coagulase-deficient staphylococci (Irlinger et al. 1997) are the more frequent.

Although the new results on washed rind cheeses have shown that *Brevibacterium linens* and *Brevibacterium aurantiacum* are not the more abundant species among surface bacteria, they are probably the bacteria that are more studied for their flavor capacity to produce typical flavors in washed rind cheeses. The genome of *B. linens* has been sequenced (Ganesan et al. 2004). Its flavoring capabilities are mainly caused by its specific amino acid catabolism. *B. linens* produces extracellular proteolytic enzymes and an active methionine γ-lyase. Rattray and Fox (1999) reviewed in detail the properties of the proteolytic systems of *B. linens*.

In this chapter, we focus on the metabolism of sulfur amino acids, which is the source of the main flavor compounds produced by this bacteria and has been intensively studied (Figure 12.1) (Forquin 2010). *B. linens* is able to produce from the methionine a whole series of sulfur compounds whose starting point is the methane thiol (MTL). This very volatile and very reactive compound is the source of numerous other compounds:

- By oxidation of different polysulfides such as dimethyl disulfide (DMDS), dimethyl trisulfide (DMTS), and dimethyl quadrisulfide (DMQS) (Ferchichi et al. 1985; Dias and Weimer 1998; Arfi et al. 2003).
- By reaction with different acyl CoA present in the bacteria, it will also generate a series of *S*-methyl thioester methyl thioacetate, methyl thiobutyrate, methyl thioisovalerate, and methyl thiohexanoate (Berger et al. 1999b; Sourabié 2009).

These compounds are the key flavor compounds of the washed rind cheeses, and their olfactive impact have been studied by Martin et al. (2004). Since the discovery of these productions by *B. linens*, it has been shown that other microorganisms very common in smear cheeses, such as *G. candidum* (Berger et al. 1999a), yeasts

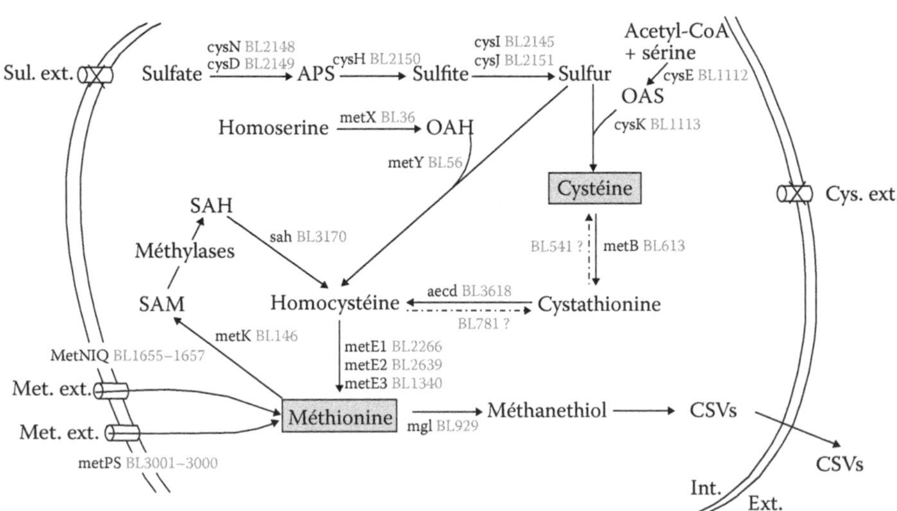

FIGURE 12.1 Sulfur metabolism by *B. aurantiacum*. Numbers in gray indicate probable implicated enzymes from gene annotation in *B. aurantiacum* genome. (From Forquin, M.P., Étude de *Brevibacterium aurantiacum*, une bactérie d'affinage de fromage: de son métabolisme du soufre à son interaction avec *Kluyveromyces lactis*, PhD thesis, AgroParisTech, 6th July 2010. With permission.)

(Spinnler et al. 2001) or *Proteus vulgaris* (Deetae et al. 2009), were also able to produce these compounds.

New sulfur compounds are still to be researched in washed rind cheeses; the presence of fat, for example, makes it difficult to conduct research on thiols, which have very low olfactive thresholds. Recently, a new thiol the ethyl 3-mercaptopropionate has been detected in Munster cheese (Sourabié et al. 2008).

12.6 BLUE CHEESES

Blue cheeses undergo a more intense lipolysis, reaching up to 50 mEq/100 g of fat in a Danish Blue, whereas the rate is much lower in mold surface ripened cheeses. The fatty acids produced from the lipolysis are the source of a whole series of methyl ketones and secondary alcohols.

A homologous series of methyl ketones with an odd number of carbon atoms, from C3 to C15, are some of the most important aroma compounds in Blue cheeses and surface mold ripened cheese (Gallois and Langlois 1990). It is probable that longer methyl ketones are not as important as the intermediate ones because of their lipophilicity, which probably limits their volatility in a fatty matrix such as that in cheese.

From fat are also issued γ-decalactone, δ-decalactone, γ-dodecalactone, and δ-dodecalactone. They have also been identified in Blue cheeses (Gallois and Langlois 1990). Recently, the effect of the application of high pressure (600 MPa) on milk used to make Blue cheese has been tested (Voigt et al. 2010); a general decrease in microflora was observed and in particular, mold development with a significant reduction of proteolysis was noted, but without any effect on the sensory properties.

The increasing demand for Blue cheese aroma compounds from the food industry gave rise to much work on the production of methyl ketones by mold, especially *P. roqueforti* (Creuly et al. 1992). The production of heptan-2-one continues to attract attention because it is preponderant in Blue-type cheeses. New flavorings with high aromatic power appeared on the market to flavor sauces, crackers, etc.

12.7 PRESSED CHEESES (CHEDDAR, HOLLAND TYPE, SWISS TYPE, AND ITALIAN HARD CHEESES)

In pressed cheeses, because of the low surface/volume ratio, it is the anaerobic flora that is probably more important (even if the surface flora when a rind is made should not be neglected). During the ripening, the role of the nonstarter LAB, *Lactobacillus* or *Propionibacterium*, is documented (Beresford and Cogan 2000; Thierry and Maillard 2002).

In Emmenthal cheeses, the ripening in a warm room (23°C) for 3 weeks will provoke the development of propionic bacteria that will open the cheese with many eyes and will give a mild dairy flavor to these cheeses. It is mainly the breakdown of the amino acid that provides the main flavoring characteristics of pressed cooked cheeses (Thierry and Maillard 2002). Recently, however, Richoux et al. (2008) have demonstrated that limitation in ethanol in the cheese would provide lower amount of esters than what is possible when a model curd is enriched in ethanol. It is probable that by playing on this factor, it would be possible to enhance fruity notes in this type of cheese. From the data issued from *Propionibacterium freudenreichii* genome (Falentin et al. 2010), Dherbécourt et al. (2010) showed evidence for an esterase activity of *Pro. freudenreichii* and validated its importance in cheese in making overexpressed mutants. The lipolysis (less than 2%) level is low in these cheeses, and a too high level can be responsible for flavor defects. The heating temperature of the curd has an important impact on the development of flavor compounds as reported by Sheehan et al. (2007).

In Italian hard cheeses on the contrary, the level of lipolysis is high and has an impact on the free fatty level, which has a role on the flavors of Grana Padano or Parmiggiano Reggiano (Moio and Adeo 1998; Qian and Reineccius 2002). In Parmiggiano Reggiano, these authors have shown that butanoic, hexanoic, octanoic, and decanoic acids are the major free fatty acids contributing to the cheesy, lipolyzed aroma of this product. In this technology, the use of lamb rennet, the large occurrence of lipolytic ripening flora and, more recently, of lipases are at the origin of the sharp flavor of these cheeses.

Ethyl butanoate, ethyl hexanoate, ethyl octanoate, ethyl propanoate, ethyl pentanoate, ethyl heptanoate, and ethyl decanoate are the major esters contributing to fruity aroma (Qian and Reineccius 2003a, 2003b). Several aldehydes and ketones such as 2-methylbutanal, 3-methylbutanal, 2,4-hexadienal, 2-butenal, pentanal, hexanal, heptanal, diacetyl, phenylacetaldehyde, DMTS, and methional were identified to be odor-active. It was found that 2,3-dimethylpyrazine, 2,6-dimethyl-pyrazine, 2,5-dimethyl-3-ethylpyrazine, trimethyl-pyrazine, 5-ethyl-2-methylpyridine, 2,3-dimethyl-5-ethylpyrazine, 2,3,5-trimethyl-6-ethyl-pyrazine, furfural, and 2-furanmethanol contribute to the nutty, roasted aroma in this cheese.

In Gouda, Van Leuven et al. (2008) analyzed the changes in perception of Gouda made with raw milk and with pasteurized milk. They have clearly shown that young Gouda (6 weeks ripening), made with raw milk, are significantly more fruity and nutty than the one made with pasteurized milk. The level of methyl ketone is also significantly lower in pasteurized milk; the reverse is observed for the lactone series, except for γ-nonalactone that only appears later. However, it is only after a period of 4 months that the ester level is high enough to measure and are increasing during Gouda ripening up to the end of the ripening time tested (10 months). With ripening time, branched chain aldehydes, 2-phenyl ethanol, and DMDS are also significantly increasing, but lactones concentrations are decreasing.

The more studied of the cheeses is probably Cheddar. However, Zehentbauer and Reineccius (2002) made a complete summary of the key odorants in Cheddar by GC olfactometry (Table 12.2). The analysis was based on their mass spectra, retention times on at least two different columns, and odor character at the sniffing port. This method has found the highest FD factors for 2,3-butandione, (*Z*)-4-heptenal, methional, 2-acetyl-1-pyrroline, DMTS, 1-octen-3-one, (*Z*)-1,5-octadien-3-one, furaneol, (*Z*)-2-nonenal, and (*E*)-2-nonenal, suggesting that these are the most potent odorants in Cheddar cheese. Three additional compounds remain unknown.

TABLE 12.2
FD Factors Found for Odorants (Neutral and Basic) Mild Cheddar Cheese Using AEDA Methodology [Both Ether and Dichloromethane (DCM) Solvents]

	RI[b] on Capillary		FD Factor on DB-5 Column		
Odor Character[a]	**DB-5**	**DB-FFAP**	**Diethyl Ether**	**DCM**	**Compound[c]**
Potato	901	1449	64 (32[d], 2[e])	32	Methional
Fruity	1494	2179	64 (128[d])	32	δ-Decalactone
Soapy	1656	2394	64 (64[d])	32	(*Z*)-6-Dodecen-γ-lactone
Buttery	595	976	16 (16[d,e])	4	2,3-Butandione
Fecal	1398	–	16 (64[d])	16	Skatol
Metallic	1573	–	8	8	Unknown
Malty	646	936	8 (NF[f])	2	2/3-Methylbutanal
Mushroom	974	1295	8 (4[d],2[e])	8	1-Octen-3-one[g]
Green	1145	1499	8 (4[d])	8	(*Z*)-2-Nonenal[g]
Green-fatly	1155	1527	8 (8[d])	8	(*E*)-2-Nonenal[g]
Green	800	1079	4 (1[e])	4	Hexanal
Green	999	1280	4 (NF)	1	Octanal
Sulfurous, roasted	1079	–	4	2	Unknown
Roasted	1101	1750	4 (NF)	1	2-Acetyl-2-thiazoline[g]
Soapy	1292	–	4	4	Unknown
Fruity	1386	1805	4 (32[d])	4	β-Damascenone[g]
Roasted	918	1327	2 (NF)	1	2-Acetyl-1-pyrroline[g]
Green	1108	1355	2 (16[d])	–	Nonanal
Cucumber	1150	1576	2 (NF)	–	(*E*,*Z*)-2,6-Nonadienal[g]
Soapy	1209	1693	2 (NF)	–	(*E*,*E*)-2,4-Nonadienal[g]

Source: Zehentbauer, G., Reineccius, G.A., *Flav. Fragr. J.* 17, 300–305, 2002. With permission.
[a] Odor character at sniffing port.
[b] RI, retention index.
[c] Compound identified by comparison with reference compound based on: odor character, retention index on two different columns and mass spectrum.
[d] FD factor 8 from Milo and Reinecius using AEDA method.
[e] FD factor from Milo and Reinecius using headspace method.
[f] Not found by Milo and Reineccius.
[g] A mass spectrum could not be obtained; identification was performed on the basis of remaining criteria.

Aldehydes and ketones are usually transitory compounds in cheese since they are transformed into alcohols (in anaerobiosis) or into corresponding acids, but it seems that they are very important in all pressed cheeses.

In semihard cheeses made with goat milk, Sheehan et al. (2009) have shown that an increased bovine milk content resulted in decreased levels of butanoic to octanoic acids, ethyl and methyl esters of acetic to octanoic acids, and acyclic, mono and bicyclic terpenes and waxy/goaty flavors to a degree dependent on the proportion of each milk type.

12.8 CONCLUSIONS

Because of the complexity of milk composition, the diversity of microbial flora, and the diversity of technologies, the flavors of dairy products keep their part of the mystery. A key for a better control of fermented dairy products is probably the microflora: the one added as starter, but also the contaminations of the milk and curd during cheese manufacturing, regardless of the hygiene and cleaning procedures applied in the dairy plants.

Another point is the interaction between the aroma perception and the other perception such as taste and texture. Looking for a better understanding of salt effect in cheese, Saint-Eve et al. (2009) recently reported evidence for the differences in aroma release and olfactory perception between different model cheeses when the salt and fat contents varied. Basically, the new knowledge in the field of perception is probably an important window open to innovation in this field.

Fermented products are generally very traditional foods. It is interesting to see how complex they are and the level of science necessary to understand and control their production at an industrial level.

REFERENCES

Alewijn, M., B.A. Smit, E.L. Sliwinski, and J.T.M. Wouters. 2007. The formation mechanism of lactones in Gouda cheese. *Int. Dairy J.* 17: 59–66.

Arfi, K., R. Tâche, H.E. Spinnler, and P. Bonnarme. 2003. Dual influence of the carbon source and L-methionine on the synthesis of sulfur compounds in the cheese-ripening yeast *Geotrichum candidum. Appl. Microbiol. Biotechnol.* 61: 359–365.

Berger, C., J.A. Khan, P. Molimard, N. Martin, and H.E. Spinnler. 1999a. Production of sulfur flavors by 10 strains of *G. candidum. Appl. Environ. Microbiol.* 65: 5510–5514.

Beresford, T., and T.M. Cogan. 2000. *Role of Lactobacilli in flavour development of Cheddar cheese.* Project report 2000, DPRC no. 34. Fermoy, Cork, Ireland: Teagasc Dairy Products Research Centre Moorepark.

Berger, C., N. Martin, S. Collin, L. Gijs, J.A. Khan, G. Piraprez, H.E. Spinnler, and E.N. Vulfson. 1999b. Combinatorial approach to flavor analysis. 2. Olfactory investigation of a library of *S*-methyl thioesters and sensory evaluation of selected components. *J. Agric. Food Chem.* 47: 3274–3279.

Bonaiti, C., 2004. Approches dynamiques des fonctions et des interactions microbiennes dans un écosystème reconstitute par une méthode d'omission: Exemple de l'affinage du fromage de Livarot, 200 pp. Thèse de Doctorat de l'Institut National Agronomique Paris-Grignon.

Cornu, A., N. Kondjoyan, B. Martin, I. Verdier-Metz, P. Pradel, J.L. Berdagué, J.-B. Coulon, et al. 2005. Terpene profiles in Cantal and Saint-Nectaire-type cheese made from raw or pasteurised milk. *J. Sci. Food Agric.* 85: 2040–2046.

Cornu, A., N. Rabiau, N. Kondjoyan, I. Verdier-Metz, P. Pradel, P. Tournayre, J.L. Berdagué, and B. Martin. 2009. Odour-active compound profiles in Cantal-type cheese: Effect of cow diet, milk pasteurization and cheese ripening. *Int. Dairy J.* 19: 588–594.

Creuly, C., C. Larroche, and J.B. Gros. 1992. Bioconversion of fatty acids into methyl ketones by spores of *Penicillium roqueforti* in a water–organic solvent, two-phase system. *Enzym. Microb. Technol.* 14: 669–678.

De Wit, R., and H. Nieuwenhuijse. 2008. Kinetic modelling of the formation of sulphur-containing flavour components during heat-treatment of milk. *Int. Dairy J.* 18: 539–547.

Deetae, P., J. Mounier, P. Bonnarme, H.E. Spinnler, F. Irlinger, and S. Helinck. 2009. Effects of *Proteus vulgaris* growth on the establishment of a cheese microbial community and on the production of volatile aroma compounds in a model cheese. *J. Appl. Microbiol.* 107: 1404–1413.

Dherbécourt, J., H. Falentin, J. Jardin, M.-B. Maillard, F. Baglinière, F. Barloy-Hubler, A. Thiery, et al. 2010. Identification of a secreted lipolytic esterase in *Propionibacterium freudenreichii*, a ripening process bacterium involved in Emmental cheese lipolysis. *Appl. Environ. Microbiol.* 76: 1181–1188.

Dias, B., and B. Weimer. 1998. Conversion of methionine to thiols by lactococci, lactobacilli, and brevibacteria. *Appl. Environ. Microbiol.* 64: 3320–3326.

Dufossé, L., A. Latrasse, and H.E. Spinnler. 1994. Importance des lactones dans les arômes alimentaires: Structures, distribution, propriétés sensorielles et biosynthèse. *Sci. Aliments* 14: 17–50.

Falentin, H., S.M. Deutsch, G. Jan, V. Loux, V. Thierry, S. Parayre, M.-B. Maillard, et al. 2010. The complete genome of *Propionibacterium freudenreichii* CIRM-BIA1T, a hardy actinobacterium with food and probiotic applications. *PLoS ONE* 5(7): e11748.

Ferchichi, M., D. Hemme, M. Nardi, and N. Pamboukdjian. 1985. Production of methanethiol from methionine by *Brevibacterium linens* CNRZ 918. *J. Gen. Microbiol.* 131: 715–723.

Forquin, M.P. 2010. Étude de *Brevibacterium aurantiacum*, une bactérie d'affinage de fromage: de son métabolisme du soufre à son interaction avec *Kluyveromyces lactis*. PhD thesis, AgroParisTech, 6 July 2010.

Gallois, A., and D. Langlois. 1990. New results in the volatile odorous compounds of French cheese. *Lait* 70: 89–106.

Ganesan, B., K. Seefeldt, and B.C. Weimer. 2004. Fatty acid production from amino acids and α-keto acids by *Brevibacterium linens* BL2. *Appl. Environ. Microbiol.* 70: 6385–6393.

Ha, J.K., and R.C. Lindsay. 1991. Contribution of cow, sheep and goat milks to characterizing branched chain fatty acids and phenolic flavors to varietal cheeses. *J. Dairy Sci.* 74: 3267–3274.

Hassouna, M., and N. Guizzani. 1995. Evolution de la flore microbienne et des caractéristiques physico-chimiques au cours de la maturation du fromage Tunisien de type camembert fabriqué avec du lait pasteurisé. *Microbiol. Hyg. Alim.* 18: 14–23.

Irlinger, F., and J.L. Bergère. 1999. Use of conventional biochemical tests and analyses of ribotype patterns for classification of micrococci isolated from dairy products *J. Dairy Res.* 66: 91–103.

Irlinger, F., A. Morvan, N. El Sohl, and J.L. Bergère. 1997. Taxonomic characterization of coagulase-negative *Staphylococci* in ripening flora from traditional French cheeses. *Syst. Appl. Microbiol.* 20: 319–328.

Karahadian, C., D.B. Josephson, and R.C. Lindsay. 1985a. Volatile compounds from *Penicillium* sp. contributing musty–earthy notes to Brie and Camembert cheese flavors. *J. Agric. Food Chem.* 33: 339–343.

Karahadian, C., D.B. Josephson, and R.C. Lindsay. 1985b. Contribution of *Penicillium* sp. to the flavors of Brie and Camembert cheese. *J. Dairy Sci.* 68: 1865–1877.

Lamberet, G., and J. Lenoir. 1976. Les caractères du système lipolytique de l'espèce *Penicillium caseicolum*: Purification et propriété de la lipase majeure. *Lait* 56: 622–644.

Martin, N., C. Berger, C. Le Du, and H.E. Spinnler. 2001. Aroma compound production in cheese curd by coculturing with selected yeast and bacteria. *J. Dairy Sci.* 84: 2125–2135.

Martin, N., V. Neelz, and H.E. Spinnler. 2004. Suprathreshold intensity and odour quality of sulphides and thioesters. *Food Qual. Prefer.* 15: 247–257.

Martin, B., I. Verdier-Metz, A. Cornu, P. Pradel, S. Hulin, S. Buchin, D. Dupont, J.L. Lamaison, A.P. Carnat, J.L. Berdagué, and J.B. Coulon. 2002. Do terpenes influence the flavour of cheeses? II. Cantal cheese. *Caseus Int.* 3: 25–27.

Miettinen, S.M., H. Tuorila, V. Piironen, K. Vehkalahti, and L. Hyvönen. 2002. Effect of emulsion characteristics on the release of aroma as detected by sensory evaluation, static headspace gas chromatography, and electronic nose. *J. Agric. Food Chem.* 50: 4232–4239.

Milo, C., and G. A. Reineccius. 1997. Identification and quantification of potent odorants in regular-fat and low-fat mild cheddar cheese. *J. Agric. Food Chem.* 45: 3590–3594.

Moio, L., and F. Addeo. 1998. Grana Padano cheese aroma. *J. Dairy Res.* 65: 317–333.

Molimard, P., J.L. Le-Quere, and H.E. Spinnler. 1997a. Les lipides et la flaveur des produits laitiers *Oléagin. Corps Gras Lipides* 4: 301–311.

Molimard, P., I. Lesschaeve, S. Issanchou, M. Brousse, and H.E. Spinnler. 1997b. Effect of the association of surface flora on the sensory properties of mould ripened cheese. *Lait* 77: 181–187.

Molimard, P., and H.E. Spinnler. 1996. Review: Compounds involved in the flavor surface mold ripened cheeses: Origins and properties. *J. Dairy Sci.* 79: 169–184.

Ott, A., L.B. Fay, and A. Chaintreau. 1997. Determination and origin of the aroma impact compounds of yogurt flavor. *J. Agric. Food Chem.* 45: 850–858.

Ott, A., J.E. Germond, M. Baumgartner, and A. Chaintreau. 1999. Aroma comparisons of traditional and mild yogurt. Headspace–GC quantitation of volatiles and origin of *R*-diketones. *J. Agric. Food Chem.* 47: 2379–2385.

Ott, A., A. Hugi, M. Baumgartner, and A. Chaintreau. 2000. Sensory investigation of yogurt flavor perception: Mutual influence of volatiles and acidity. *J. Agric. Food Chem.* 48: 441–450.

Pagot, Y., J.M. Belin, F. Husson, and H.E. Spinnler. 2007. Metabolism of phenylalanine and biosynthesis of styrene in *Penicillium camemberti*. *J. Dairy Res.* 74: 180–185.

Qian, M., and G. Reineccius. 2002. Importance of free fatty acids in Parmesan Cheese. Conference Information: Symposium on Heteroatomic Aroma Compounds, 2000, Washington, D.C. *Heteroatomic Aroma Compounds*. Book Series: ACS Symposium Series vol. 826: 251–263.

Qian, M., and G. Reineccius. 2003a. Quantification of aroma compounds in Parmigiano Reggiano cheese by a dynamic headspace gas chromatography–mass spectrometry technique and calculation of odor activity value. *J. Dairy Sci.* 86: 770–776.

Qian, M., and G. Reineccius. 2003b. Static headspace and aroma extract dilution analysis of Parmigiano Reggiano cheese. *J. Food Sci.* 68: 794–798.

Rattray, F.P., and P.F. Fox. 1999. Aspect of enzymology and biochemical properties of *Brevibacterium linens* relevant to cheese ripening: A review. *J. Dairy Sci.* 82: 891–909.

Richoux, R., M.-B. Maillard, J.R. Kerjean, S. Lortal, and A. Thierry. 2008. Enhancement of ethyl ester and flavour formation in Swiss cheese by ethanol addition. *Int. Dairy J.* 18: 1140–1145.

Saint-Eve, A., N. Martin, H. Guillemin, E. Semon, E. Guichard, and I. Souchon. 2006. Flavored yogurt complex viscosity influences real-time aroma release in the mouth and sensory properties. *J. Agric. Food Chem.* 54: 7794–7803.

Saint-Eve, A., C. Lauverjat, C. Magnan, I. Déléris, and I. Souchon. 2009. Reducing salt and fat content: Impact of composition, texture and cognitive interactions on the perception of flavoured model cheeses. *Food Chem.* 116: 167–175.

Sheehan, J.J., M.A. Fenelon, M.G. Wilkinson, and P.L.H. McSweeney. 2007. Effect of cook temperature on starter and non-starter lactic acid bacteria viability, cheese composition and ripening indices of a semi-hard cheese manufactured using thermophilic cultures. *Int. Dairy J.* 17: 704–716.

Sheehan, J.J., A.D. Patel, M.A. Drake, and P.L.H. McSweeney. 2009. Effect of partial or total substitution of bovine for caprine milk on the compositional, volatile, non-volatile and sensory characteristics of semi-hard cheeses. *Int. Dairy J.* 19: 498–509.

Sourabié, A.M. 2009. Étude de la production de composés soufrés volatils par voie biotechnologique. PhD thesis, AgroParisTech, 30th November 2009.

Sourabié, A.M., H.E. Spinnler, P. Bonnarme, A. Saint-Eve, and S. Landaud. 2008. Identification of a powerful aroma compound in Munster and Camembert cheeses: Ethyl 3-mercaptopropionate. *J. Agric. Food Chem.* 56: 4674–4680.

Spinnler, H.E. 2003. Off flavours due to interactions between food components. In *Taint and Off-flavours in Food*, ed. B. Baigrie, 176–186. Cambridge, UK: Woodhead Publishing Ltd.

Spinnler, H.E., C. Berger, C. Lapadatescu, and P. Bonnarme. 2001. Production of sulfur compounds by several yeasts of technological interest for cheese ripening. *Int. Dairy J.* 11: 245–252.

Spinnler, H.E., O. Grosjean, and I. Bouvier. 1992. Effect of culture parameters on the production of styrene (vinyl benzene) and 1-octene-3-ol by *Penicillium caseicolum*. *J. Dairy Res.* 59: 533–541.

Thierry, A., and M.-B. Maillard. 2002. Production of cheese flavour compounds derived from amino acid catabolism by *Propionibacterium freudenreichii*. *Lait* 82: 17–32.

Van Leuven, I., T. Van Caelenberg, and P. Dirinck. 2008. Aroma characterisation of Gouda-type cheeses. *Int. Dairy J.* 18: 790–800.

Voigt, D.D., F. Chevalier, M.C. Qian, and A.L. Kelly. 2010. Effect of high-pressure treatment on microbiology, proteolysis, lipolysis and levels of flavour compounds in mature blue-veined cheese. *Innov. Food Sci. Emerg. Technol.* 11: 68–77.

Walker, G.P., F.R. Dunshea, and P.T. Doyle. 2004. Effects of nutrition and management on the prediction and composition of milk fat and protein: A review. *Aust. J. Agric. Res.* 55: 1009–1028.

Yvon, M., E. Chambellon, A. Bolotin, and F. Roudot-Algaron. 2000. Characterisation and role of the branched chain aminotransferase (BcatT) isolated from *Lactococcus lactis* subsp. *cremoris* NCDO 763. *Appl. Environ. Microbiol.* 66: 571–577.

Zehentbauer, G., and G. Reineccius. 2002. Determination of key aroma components of Cheddar cheese using dynamic headspace dilution assay. *Flav. Fragr. J.* 17: 300–305.

13 Red Meat Flavor

Jennie M. Hodgen and Chris R. Calkins

CONTENTS

13.1 INTRODUCTION

Meat flavor is a complex trait. It is affected by small changes in water-soluble compounds, lipids, cooking procedures, aging periods, packaging systems, and storage conditions. Different species' flavors are influenced by numerous factors including the animals' digestive system, breed, animal age and weight, gender, diets, and production systems. These differences, especially the lipid portion, help differentiate between species and create opportunities to enhance or degrade the flavor acceptability and intensity. This chapter will focus on what creates flavor differences within the different meat species and the role of postharvest factors on the flavor development of meat.

13.2 BEEF

Unlike other foods, an official lexicon has not been developed for many of the meat species. Research is under way (Adhikari and Miller 2010; Maughan and Martini 2010; Philip et al. 2010), however, and a beef lexicon should be published soon to standardize beef flavor descriptors and scores. Forty-one aroma and flavor attributes were identified, defined, and referenced for beef with the "Major Notes" including beef flavor, brown/roasted, bloody/serumy, fat-like, metallic, liver-like, green-hay-like, and all five standard tastes (Philip et al. 2010). Since this represents a new research segment, flavor descriptions for meat have only been used minimally with flavor desirability and/or flavor intensity used much more frequently to assess differences among samples.

13.2.1 Breed

The genotype of a beef animal may play a small role in some flavor intensity differences seen in meat (Mandell et al. 1997). Unaged steaks from *Bos taurus* have statistically higher scores than those from *Bos indicus* for flavor desirability, but differences were less than 0.5 on an 8-point scale, and most differences in flavor disappeared after 16 days postmortem (Winer et al. 1981). Meat from Simmental steers was found by a trained panel to have more beef flavor intensity than meat from Red Angus (Laborde et al. 2001) but, with differences of less than 1 on a 15-point scale, distinct differences may be difficult to detect by an average consumer. There is speculation that any flavor differences due to breed are attributable to differences in fat deposition and fatty acid profiles, which may be observed in English breeds compared to Continental breeds when the animals come from different production systems or are not finished to the same physiological endpoint. In fact, heritability of flavor intensity is not different from zero (Splan et al. 1998), whereas beef flavor has been shown to be hugely impacted by environmental effects (Nephawe et al. 2004). The majority of research done on breed effect has focused on the *longissimus dorsi* muscle, so little is known about the breed effect on the flavor of other beef muscles.

13.2.2 Animal Age

As animals age, there is an increase in myoglobin within the muscle that typically results in meat from older animals having a higher flavor intensity than younger animals. Early work in steers and heifers aged 300–700 days showed that animal age is usually positively correlated to more acceptable flavor scores, but in intact males, the increase in animal age decreases the acceptance of the flavor even though the beef flavor intensity increases (Field et al. 1966). Other work has shown a linear decrease in flavor acceptability as carcass maturity increases (Smith et al. 1983). Another study found that heifers older than 336 days had decreased flavor acceptability (Jacobson and Fenton 1956), whereas a study comparing bulls and steers found that adding 100 days to the slaughter age increased the percentage of undesirable flavor, exceeding 20% for both groups (Reagan et al. 1971). Beef animals fed concentrate diets from a young age—termed calf-fed—have steaks with fewer panelists finding undesirable flavor and slightly more steaks from calf-fed animals having desirable flavor than steaks from yearling-fed beef animals (Brewer et al. 2007). Other studies have clearly shown that different muscles are impacted differently by animal age (Stelzleni et al. 2007; Nelson et al. 1930). In general, as beef animals get past 30 months of age, flavor intensity increases and acceptability decreases. This can be modified by the level of nutrition for the animal and by the sex of the animal.

13.2.3 Gender

Steers and heifers do not have as intense a beef flavor compared to cows and intact males, but they do have fewer or less intense off-flavors. Like other factors, the influence of sex on flavor depends on numerous conditions including production system, diet, and animal age. In a study of animals ranging from 300 to 699 days of age, no differences in flavor desirability were found between steers, heifers, and bulls until 600 days of age when steers and heifers were more acceptable in flavor than bulls even though there were similar flavor scores for Hereford bulls, heifers, and steers (Field et al. 1966). Meat from intact males had an increase in livery taste and odor and bloody taste that is related to increased levels of 2-propanone (Gorraiz et al. 2002). Muscles from USDA Select steers have been shown to have less intense beef flavor than cow muscles (Stelzleni 2006), whereas differences in flavor between steers and heifers have not been observed (Wilson et al. 1976).

13.2.4 Diet

Most of the research on cattle diets has compared meat from grass-finished animals to meat from grain-finished animals. Acceptability of meat from cattle on different production systems depends greatly on consumer preference (Sitz et al. 2005; Killinger et al. 2004); meat from grass-finished animals was not found to lack flavor but was characterized as having an off-flavor (Brown et al. 1979). Most differences in flavor due to diet can be explained to some extent by the level of energy intake, days on feed (corresponding to growth rate), age of the animal, fat deposition and composition, and carcass weight. In the United States, it is usually shown that meat from cattle fed grain-based diets (namely, corn) has more desirable flavor than beef from grass-finished animals. Although quality of grasses may affect results, orchard grass, ryegrass, forage sorghum, bluegrass, fescue, arrow leaf clover, Bermuda, sudan, millet, and combinations of these grasses typically cause a decrease in desirability of beef flavor (Melton 1990). High-quality alfalfa, Timothy and alfalfa, barley, and 50:50 corn/barley diets produce meat similar in flavor to beef from animals finished on a corn diet (Melton 1983; Miller et al. 1996).

The majority of differences in flavor desirability due to diet can be traced to changes in lipid deposition, lipid composition, and the resulting aroma and taste changes. Fishy and bloody notes in beef from grass-finished animals have been associated with an increase in 18:1 *trans* isomers, especially conjugated linoleic acid (CLA) *cis*-9, *trans*-11 (Nuernberg et al. 2005). Decreases in palmitoleic acid, vaccenic acid, and eicostarienoic acid with increases in eicosadienoic acid and sodium have been related to an increase in the perception of liver-like off-flavor (Jenschke et al. 2007). Since oleic and linoleic acid are higher in meat from grain-finished cattle and α-linolenic is higher in beef from grass-finished cattle, compounds derived from these acids are more prevalent in the meat of cattle raised with the specific feedstuff (Larick et al. 1987). In many cases, it is not the presence or absence of a compound that affects the beef flavor but the concentration of the compound or the combination of concentrations of a variety of compounds. For instance, grain-finished beef is higher in δ-tetradecalactone and δ-hexadecalactone (Larick et al. 1987), whereas levels of pentanal, toluene, 1-ethyl-2-methylbenzene, and an unknown compound accounted for 51% of the variation in flavor intensity between meat from grass- and grain-finished animals (Melton 1990). Animals on a concentrate diet have meat with more aldehydes (Descalzo et al. 2005), whereas phenolic compounds (Vasta et al. 2006) and terpenoid-type compounds (Larick et al. 1987) are more common in meat from grass-finished animals.

13.3 PORK

The deposition of flavor compounds in pork occurs differently than in beef or lamb because pork has a monogastric digestive system where fatty acids are not biohydrogenated as in ruminants. Compounds in the diet therefore are more likely to appear in the fat or lean tissue of pork similar to the dietary compounds consumed.

13.3.1 Breed

Most studies of swine breeds used different breed sires on a given breed of females and show no breed effect on pork flavor intensity and/or desirability. In large studies where production system, gender, and animal age were held constant, panelists characterized meat from Hampshires as having more intense flavor than Durocs (Jeremiah et al. 1999). Both of those breeds have been reported to have meat with more intense pork flavor than meat from Landraces and Yorkshires. This increase in flavor intensity was not correlated to overall acceptability, presumably because of the relative low variation among animals. Nevertheless, Durocs and Hampshires had a slight advantage in flavor desirability and ultimately overall palatability when all other factors were held constant (Jeremiah et al. 1999). Compared to traditional breeds—Berkshire and Tamworth—the more modern breeds have been found statistically to have more abnormal pork flavor and lower flavor like ratings, yet overall acceptability is minimally affected since these differences were small. Other factors such as muscle or diet typically have a bigger overall impact on pork flavor than breed (Wood et al. 2004). The effect of breed on flavor may be overshadowed by the wide variation in tenderness and juiciness effects among animals and breeds. For instance, flavor differences are usually not seen between negative and positive halothane gene-carrying animals that are predisposed for pale, soft, and exudative meat even though overall acceptability is reduced and significantly inferior due to tenderness and juiciness issues (Brewer et al. 2002; Ellis et al. 1996, 1998).

13.3.2 Gender

Gender differences in flavor or off-flavor most often appear when comparing barrows and/or gilts to fully mature sows and boars. Gilts have a tendency to be leaner than barrows so meat from barrows will be rated slightly higher for flavor intensity than gilts if a common fat level is not met (Martel et al. 1988; Jeremiah et al. 1992). In a few studies where boars were harvested at a young age, similar to barrows or gilts, no difference in flavor intensity or desirability were detected. In most cases, however, off-flavors in boar muscle resulted in lower flavor desirability. Boar taint has been attributed to skatole and 16-androstene compounds (Pearson et al. 1995). Both compounds are lipophilic so are preferentially deposited in the fat although skatole is hydrophilic so it can also be deposited in lean tissue. Boar taint is generally characterized by a strong odor when cooked, and some have suggested that it is not always detected when the meat is prepared in a separate area such that consumers are not subjected to the cooking aroma. At 100 kg harvest weight, boars on an adequate-protein diet had more off-flavor than barrows or gilts on a high- or adequate-protein diet and boars on a high-protein diet. By the time swine reached 110 kg regardless of protein level in the diet, boars had more off-flavors (Nold et al. 1997).

Gender can play a role in the acceptability of pork flavor. Barrows tend to have slightly more acceptable flavor scores than gilts although the differences are slight and probably not meaningful to the average consumer. Boar taint has been widely studied and can be very unpleasant for a person who is very sensitive to the compounds. A small portion of the human population is anosmic to androstenone and this lack of sensitivity to the compounds means boar taint is not a factor in determination of flavor desirability for them.

13.3.3 Diet

With the push to increase the amount of *n*–3 polyunsaturated fatty acids (PUFA) in the human diet, the pork industry has investigated changes in swine diets to alter the fatty acid composition of pork. Feeding crushed linseed for short durations (<60 days) can increase the PUFA level with minimal changes to pork odor and taste although overall flavor decreases as does shelf-life. This short-term feeding can also reduce the amount of skatole in the backfat (Kouba et al. 2003). Feeding a CLA supplement before harvest tended to lead to a decrease in flavor acceptability and/or increase of off-flavors (D'Souz and Mullan 2002). Many of the off-flavor issues are a result of the increase in aldehydes, indicating oxidation (Larick et al. 1992). Feeding high levels (up to 20%) of monounsaturated fat to swine yielded no change in flavor intensity (St. John et al. 1987) although off-flavors may be more prevalent with canola oil-supplemented monounsaturated diets than normal, safflower oil, or sunflower oil swine diets (Miller et al. 1990). Dietary crude protein and lysine levels have no effect on longissimus muscle flavor desirability or off-flavor (Wood et al. 2004).

In today's commercial swine herd, pork flavor is predominantly impacted by animal diet because of the homogeneity among animals. However, since much of pork is sold unfrozen or used in further processing, diet manipulations may subsequently affect flavor later in the shelf life of the product as a consequence of oxidation.

13.4 LAMB

Since the production systems vary from country to country, a person's history of lamb consumption plays a critical role in an individual's personal acceptability of lamb meat flavor. Differences in familiarity with lamb flavor can result in wide variation in acceptability.

Flavor development in lamb is similar to beef so many flavor components are developed preharvest depending on animal age, gender, and diet. The lipid portion of lamb contributes 4-hydroxydodec-*cis*, 6-enoic acid lactone (*cis*-γ-dodec-6-enolactone), which provides a sweet aroma and flavor, whereas *trans-trans*-deca-2,4-dienal imparts an oily odor and taste when cooked (Park 1975). It is thought that the sweet component is responsible for lower hedonic ratings of sheep meat flavor acceptability.

13.4.1 Breed

Although other palatability traits may be different, sheep breed does not have a large effect on flavor except when comparing breeds raised for wool versus meat breeds (Duckett and Kuber 2001; Webb et al. 1994; Griffin et al. 1992; Sink and Caporaso 1977). However, breed can have an effect on flavor when fatty acid concentrations are different. For example, Soay sheep finished on lowlands grass can have similar flavor characteristics to concentrate fed Suffolk but differ from Welsh Mountain or Suffolk finished on grass. Soay sheep tend to yield meat perceived to have higher intensity of livery flavor thought to be caused by higher myoglobin levels (Fisher et al. 2000).

13.4.2 Animal Age and Slaughter Weight

Mutton and lamb have distinct flavor differences, with the meat from younger animals typically preferred because of less intense flavor and perception of off-flavor. The intermediate chain fatty acids of 4-methyl C9 and C10 fatty acids as well as caprylic acid are found in higher levels in older animals and are believed to be associated with the characteristic odor (Wong et al. 1975; Cramer 1983). The compounds are found in lower concentrations in lamb and help us differentiate among other species. Mutton odor has also been linked to skatole (3-methylindole) and, to a lesser extent, methylphenol compared with younger lamb meat (Young et al. 2003) as well as other heterocyclic sulfur-containing compounds.

In studies where flavor profiles were conducted, desirable lamb meat aromatics were more intense in meat from lambs less than 9 months of age compared with those that were between 9 and 15 months of age. The undesirable, unidentifiable off-odors as well as an inappropriate bitter taste were more pronounced in meat from animals that were between 12 and 15 months than those in the younger groups. Some aromatics were more intense in younger animals such as fatty/rancid (Jeremiah et al. 1998). It has been hypothesized that lamb flavor may have a quadratic relationship to animal age and physiological maturity so testing at different ages of animals may lead to different results.

Because of the variety in production systems, differences in slaughter weights exist with animals of the same age. When profiling the flavor of lamb meat at different slaughter weights, meat from light weight lambs (31.8 kg–40.4 kg) have higher aromatics of chemical/sour and metallic, but a much lower intensity of fishy and rancid/fatty odors. Meat from animals weighing between 40.5 kg and 67.7 kg had a higher incidence of bloody aromatics. The heavier animals (58.9 kg and 76.8 kg) had meat that had a more appropriate browned aftertaste. Whereas the lightest animals had meat with the highest metallic and wooly aftertaste and lowest livery aftertaste, the heavier (58.9 kg and 76.8 kg) animals' meat had the highest incidence of undesirable, off-flavored aftertaste (Jeremiah et al. 1998). In studies with much lighter animals, the lighter lambs were found to have the best flavor (Martinez-Cerezo et al. 2005) or less intense flavor (Teixeira et al. 2005) although this may have been confounded by tenderness differences between the weight groups. Thus there are both positive and negative effects on flavor due to increased slaughter weight.

13.4.3 Gender

Research regarding flavor differences among genders has shown mixed results as studies have often confounded slaughter weight and production systems. Meat from ewes has less wooly aromatics—which have been associated with sulfur compounds in adipose tissue used to sustain wool growth—than rams, whereas wethers had a more acceptable fatty/tallow aromatic and aftertaste than rams. Rams' meat had less chemical/sour aftertaste, which demonstrated that ewe and ram's meat was more well-balanced in flavor than wether's meat (Jeremiah et al. 1998). However, other researchers have indicated no difference in meat flavor from wethers or ewes (Summers et al. 1978) or between rams and wethers (Batcher et al. 1969), whereas some have ranked wethers more acceptable than ewes and ewes more acceptable than rams (Sink and Caporaso 1977).

13.4.4 Diet

Most studies find that diet has the biggest influence on the flavor of lamb. Like other species, lambs finished on corn-based diets have been found to have meat with a much milder taste than animals finished with forages. Adding corn to grass-based diets lead to less intense grassy and lamby flavor (Bailey et al. 1994). The total concentrations of volatile organic compounds are much less in lamb meat from corn and corn-based diets than other diets especially those on ryegrass. While caprylic, C9, and C10 fatty acids contribute to mutton flavor, differences in diet can be seen from free fatty acids of C6–C9 with decanoic acid being lower in lamb meat from corn diets. The grassy flavors of lamb were associated with intermediate chain fatty acids, aldehydes, ketones, and diterpenoids, whereas higher intensity of lamby flavor scores in meat from sheep fed different diets when aldehydes, ketones, and bovolide, were noted (Bailey et al. 1994). Another study showed 3-methylindole to be higher in meat from pasture-fed animals than those finished on corn (Young et al. 2003). These differences may lead to the perception of more pronounced livery flavor, less fatty flavor, and less "typical" lamb flavor in meat from grass-finished sheep (Priolo et al. 2002). The diets with higher concentrations of long chain fatty acids can also impart off-flavor because of increased development of oxidative products (Fisher et al. 2000).

13.4.5 Other Factors

Lipid-protected supplements in the diet decrease *cis*-γ-dodec-6-enolactone concentrations and result in more desirable meat flavor (Park et al. 1976). It has also been shown that colder frozen storage (–40°C) maintains lamb flavor and odor while retarding off-flavors probably due to a reduction in lipid oxidation (Carlucci et al. 1999).

Ultimately, lamb flavor acceptability is very dependent on regional location and familiarity of the consumer with meat raised under certain production management styles.

13.5 PROCESSING EFFECTS ON MEAT FLAVOR

Hornstein and Crowe (1960) established that the meaty flavor of the lean portion of beef and pork is quite similar as detected by the human nose and chromatographically. Their hypothesis stated that the lean meat flavor origin came from low-molecular weight, water-soluble compounds that provide an interaction with amino acids, low-molecular weight carbohydrates, and polypeptides to help give the cooked meat flavor. This is reinforced with each study thereafter that finds no single compound or class of compounds is solely responsible for the flavor of cooked meat as hundreds of compounds influence cooked meat flavor.

More research has been conducted with beef regarding postharvest practices that affect flavor in intact muscle than with other species, but many believe similar characteristics are present for other species as well.

13.5.1 Muscle

Most of the meat research regarding specific muscles focuses on tenderness since there is 3–4 times the variation in tenderness between muscles compared to flavor (Shackelford et al. 1995; Wulf and Page 2000). Little difference in flavor intensity is detected between different beef muscles although there is a trend in rankings (Table 13.1) that those with more heme iron content are slightly more intense (Calkins and Hodgen 2007). Although in beef there are minimal differences in flavor intensity, differences are seen in off-flavor and flavor desirability, which is important since flavor desirability is always highly correlated to overall palatability. In chuck and round muscles typically perceived to have more off-flavors than middle meats, *M. infraspinatus* (top blade/flat iron) has the least off-flavors and lowest frequency of sour notes. When a muscle from the chuck or round is perceived to have an off-flavor, other muscles from that animal have an increased chance of also having an undesirable flavor (Meisinger et al. 2006). The Canadian Cattlemen's Association listed seven beef muscles as meeting 95% flavor acceptability by consumers: *M. teres major* (petite tender), *M. psoas major* (tenderloin), *M. longissimus thoracis* (rib roll), *M. longissimus lumborum* (strip loin), *M. ilio psoas*, *M. spinalis dorsi*, and *M. subscapularis* (Jeremiah et al. 2003).

TABLE 13.1
Ranking of Muscles[a] for Flavor Intensity from Different Studies

Rank[b]	1	2	3	4	5	6	7	8	9	10	11
1	LD^{c}	IF^{c}	DI	BF^{c}	TB^{c}	CP	LD^{c}	LD^{c}	SV^{c}	SV^{c}	LD
2	QF^{c,d}	SV^{c}	IF	PM^{c,d}	SM^{c}	SP	BF^{d}	SM^{c}	VI^{c,d}	MD^{c,d}	GM
3	ST^{d,e}	CP^{c}	BF	GM^{c,d}	LD^{c}	TB	SM^{d,e}	BF^{c}	CO^{c,d,e}	GR^{c,d,e}	SM
4	GM^{d,c,f}	LD^{c}	SV	SM^{c,d,e}		SV	TB^{d,e,f}	GM^{c}	MD^{c,d,e,f}	PP^{c,d,e}	
5	SM^{e,f,g}	PP^{c,d}	IP	TB^{c,d,e,f}		RB	RF^{e,f,g}		IF^{c,d,e,f}	TB^{c,d,e}	
6	TB^{e,f,g}	TB^{c,d}	PM	RF^{d,e,f,g}		SL	ST^{f,g}		TB^{d,e,f,g}	IF^{c,d,e}	
7	BF^{f,g,h}	SP^{d,e}	TM	LD^{d,e,f,g}		LT	SP^{f,g}		RB^{d,e,f,g}	BF^{c,d,e,f}	
8	SP^{f,g,h}	RB^{e}	TB	SV^{d,e,f,g}		SS	GM^{f,g}		VM^{d,e,f,g}	CO^{c,d,e,f}	
9	PM^{g,h}	BB^{e}	SP	IF^{e,f,g}			AD^{g}		SS^{e,f,g,h}	VI^{c,d,e,f}	
10	IF^{h}		AD	ST^{e,f,g}			IF^{g}		LT^{e,f,g,h}	SL^{c,d,e,f,g}	
11			TF	PP^{f,g}			PM^{h}		SP^{e,f,g,h}	LT^{c,d,e,f,g}	
12			GM	SP^{g}					ST^{e,f,g,h}	VL^{c,d,e,f,g,h}	
13			GR						PP^{e,f,g,h}	SS^{d,e,f,g,h}	
14			RA						VL^{e,f,g,h}	RB^{d,e,f,g,h}	
15			PP						SF^{e,f,g,h}	SM^{d,e,f,g,h}	
16			SD						BF^{e,f,g,h}	SP^{e,f,g,h}	
17			OA						GR^{f,g,h}	VM^{e,f,g,h}	
18			SS						RF^{f,g,h}	SF^{e,f,g,h}	
19			IC						BT^{g,h}	RF^{e,f,g,h}	
20			VL						SL^{h}	AD^{f,g,h}	
21			TR						SM^{h,i}	BT^{g,h}	
22			SM						AD^{i}	ST^{h}	
23			RF								
24			LD								
25			ST								

Source: Calkins, C.R. and Hodgen, J.M., *Meat Sci.*, 77, 63–80, 2007.

Note: 1 Shackelford et al. 1995; 2 Paterson and Parish 1986; 3 Jeremiah et al. 2003; 4 Carmack et al. 1995; 5 Jeremiah et al. 1985; 6 Molina et al. 2005; 7 Rhee et al. 2004; 8 Wheeler et al. 2000; 9 Brickler 2000, dry cookery; 10 Brickler 2000, wet cookery; 11 Wulf et al. 2000.

[a] AD = *M. adductor*; BB = *M. biceps brachii*; BF = *M. biceps femoris*; BT = *M. brachiocephalicus*; CP = *M. complexus*; DI = *Diaphragm*; GM = *M. gluteus medius*; GR = *M. gracillis*; IC = *intercostal muscles*; IF = *M. infraspinatus*; IP = *M. ilio psoas*; LD = *M. longissmus dorsi*; LT = *M. latissimus dorsi*; MD = *M. multifidus dorsi*; OA = *M. obliquus abdominus internus*; PM = *M. psoas major*; PP = *M. pectoralis profundi*; QF = *M. quadriceps femoris*; RA = *M. rectus abdominis*; RB = *M. rhomboideus*; RF = *M. rectus femoris*; SD = *M. spinalis dorsi*; SF = *M. superficial pectoral*; SL = *M. splenius*; SM = *M. semimembranosus*; SP = *M. supraspinatus*; SS = *M. subscapularis*; ST = *M. semitendinosus*; SV = *M. serratus ventralis*; TB = *M. triceps brachii*; TF = *M. tensor faciae latae*; TM = *M. teres major*; TR = *M. trapezius*; VI = *M. vastus intermedius*; VL = *M. vastus lateralis*; VM = *M. vastus medialis*.

[b] Samples are ordered from the most intense beef flavor intensity to the least (bland).

[c–i] Means within column without common superscript differ.

Pork and lamb are fabricated into fewer pieces than beef because of the relative size, so less work has been done comparing the flavor of individual muscles in those species. The pork *M. psoas major* had higher (more desirable) sensory scores than the *M. longissimus* muscle for pork flavor, flavor liking, and overall liking (Wood et al. 2004), whereas the *M. longissimus* muscle had less off-flavors than the *M. semitendinosus* partially due to differences in fat content (Nold et al. 1997). As more interest in alternative merchandising of the cuts from those species grows, there will be a more concerted effort to evaluate tenderness and flavor differences among muscles.

13.5.2 Aging

As with tenderness of meat, aging meat has a profound effect on the flavor. Since aging affects sugars, organic acids, peptides, free amino acids, ATP metabolites, and enzyme activity, different flavors can be impacted through the Maillard reaction during cooking.

Desirable beef flavors such as beefy, brothy, browned/caramelized, and sweet gradually start to decline after 4–10 days postmortem, whereas bitter, sour, painty, and cardboardy slowly become more pronounced (Spanier et al. 1997; Monson et al. 2005). Meat from beef breeds require less aging time to reach optimum flavor and palatability scores than dairy or dual-purpose breeds (Monson et al. 2005). In beef, optimum flavor can be reached before the development of off-flavors (Spanier et al. 1997; Monson et al. 2005). By 21 days of aging, overall aroma and liver aroma have increased significantly in strip steaks as do overall and livery taste. The aging period allows products to break down and interact with other compounds to produce new flavor compounds (Campo et al. 1999). Acid flavor can start developing after 10 days of aging (Spanier et al. 1997).

There are conflicting results about the effect of dry aging versus wet aging in beef. In general, most consumers do not detect differences until the intramuscular fat content gets high and then wet aging is preferred (Sitz et al. 2005), although trained panels have found higher flavor intensity, dry-aged flavor, and brown, roasted aromas in dry aged beef compared to wet aged beef.

While more panelists detect sour notes by 14 days of aging in pork, fewer people detect an off-flavor after 14 days of aging (Jaurez et al. 2009). About 8 days postmortem, aging in appropriate storage conditions appears to maximize pork flavor development. Diet, gender, and breed differences in pork impact flavor more than aging does.

Because of the shorter chain fatty acids in lamb, extended aging does not have as beneficial an effect as seen in beef because of the development of undesirable flavor components. Four days of aging (when disregarding muscle) seems to yield greater overall flavor acceptability although benefits to tenderness make 4–16 days more acceptable for overall flavor perception. The *semitendinosus* and *semimembranosus* muscles show minimal flavor benefits due to aging, although in general the *semitendinosus* does have slightly higher flavor scores with shorter aging. The *gluteo biceps* seems to receive higher scores with intermediate aging (4–16 days; Martinez-Cerezo et al. 2005).

13.5.3 Cooking Method

Grilling is the predominant cooking method of top loin steaks in U.S. households, although flavor desirability was high for all cooking methods in all cities studied (Lorenzen et al. 1999).

Many factors influence desirability and intensity of flavor for top sirloin steak. When pan-frying and broiling, sensory ratings were higher with a lower degree of doneness, whereas a higher degree of doneness was preferred when stir-frying and simmering/stewing top sirloin steaks. USDA grade only affected desirability scores when indoor grilling was used. Flavor was more intense for top sirloin when it was grilled outdoors, broiled, or pan-fried at lower degrees of doneness. Overall, indoor grilling of sirloin steak resulted in the highest flavor ratings (Savell et al. 1999).

Top round beef steaks generally had more desirable flavor when simmered or stewed, whereas broiling and outdoor grilling were not as acceptable. This muscle has higher amounts of connective tissue so moist cooking methods that solubilize connective tissue and enhance tenderness probably help improve flavor ratings. There were no differences in flavor desirability when round steaks were cooked to well done or medium degree of doneness by braising and stir frying. There is probably little meaningful difference in flavor intensity due to cooking method for top round steaks (Neely et al. 1999).

13.5.4 Degree of Doneness

Overall satisfaction is more dependent on flavor than tenderness when top loin steaks are cooked to a higher degree of doneness. Generally, ratings were higher for flavor desirability when top loin steaks were cooked to medium rare or less, but medium and well done steaks received higher flavor scores than medium well done steaks (Lorenzen et al. 1999).

In an in-home study, top sirloin steaks were cooked to well done or above approximately 30% of the time (Savell et al. 1999). Outdoor grilling and broiling (the most common cooking method in the study) generated the lowest flavor desirability scores when the meat was cooked to well done or more. In beef studies where a single cooking method was used, top sirloin steaks received the highest scores when cooked to lower end point temperatures (Luchak et al. 1998).

Top round steaks received higher ratings when cooked to medium or well done compared to medium well (Neely et al. 1999).

Some studies have found that pork flavor desirability improves and abnormal flavors decrease for both pork loin steaks and roasts as degree of doneness increases (Wood et al. 1995; Heymann et al. 1990), but the effect on tenderness and juiciness lead to the recommendation to cook to lower temperatures to maximize overall palatability. More recently, pork loin chops from commercial lines of swine were not found to differ in flavor characteristics when trained panelists focused on lean flavor at varying degrees of doneness (Moeller et al. 2010a), or by consumers who rated flavor like and flavor level (Moeller et al. 2010b).

Numerous preharvest and postharvest factors affect the flavor of red meat. Small changes to any factor can give a consumer a different perception of the flavor of the product. Like with other food products, flavor preference of red meat is closely linked to past personal experiences.

REFERENCES

Adhikari, K., and R. Miller. 2010. Beef lexicon. Reciprocation Sessions. American Meat Science Association 63rd Reciprocal Meat Conference, June 21, 2010, Texas Tech University, Lubbock, TX.

Bailey, M.E., J. Suzuki, L.N. Fernando, H.A. Swartz, and R.W. Purchas. 1994. Influence of finishing diets on lamb flavor. In *Lipids in Food Flavors*, ed. C. Ho et al., 170–185. ACS Symposium Series. Washington, D.C.: American Chemical Society.

Batcher, O.M., A.W. Brant, and M.S. Kunze. 1969. Sensory evaluation of lamb and yearling mutton flavors. *J. Food Sci.* 34: 272–276.

Brewer, P.S., J.M. James, C.R. Calkins, R.M. Rasby, T.J. Klopfenstein, and R.V. Anderson. 2007. Carcass traits and *M. longissimus lumborum* palatability attributes of calf- and yearling-finished steers. *J. Anim. Sci.* 85: 1239–1246.

Brewer, M.S., J. Jensen, A.A. Sosnicki, B. Fields, E. Wilson, and F.K. McKeith. 2002. The effect of pig genetics on palatability, color, and physical characteristics of fresh pork loin chops. *Meat Sci.* 61: 49–256.

Brickler, J.E. 2000. The effect of carcass weight, yield grade, quality grade, and cooking method on physical attributes, Warner-Bratzler Shear Force, and sensory panel characteristics on muscles of the beef chuck and round. Masters Thesis. University of Florida, Gainesville, FL.

Brown, H.G., S.L. Melton, M.J. Riemann, and W.R. Backus. 1979. Effects of energy intake and food source on chemical changes and flavor of ground beef during frozen storage. *J. Anim. Sci.* 48: 338–347.

Calkins, C.R., and J.M. Hodgen. 2007. A fresh look at beef flavor. *Meat Sci.* 77: 63–80.

Campo, M.M., C. Sanudo, B. Panea, P. Alberti, and P. Santolaria. 1999. Breed type and ageing time effects on sensory characteristics of beef strip loin steaks. *Meat Sci.* 51: 383–390.

Carlucci, A., F. Napolitano, A. Girolami, and E. Monteleone. 1999. Methodological approach to evaluate the effects of age at slaughter and storage temperature and time on sensory profile of lamb meat. *Meat Sci.* 52: 391–395.

Carmack, C.F., C.L. Kastner, M.E. Dikeman, J.R. Schwenke, and C.M. Garcia Zepeda. 1995. Sensory evaluation of beef-flavor-intensity, tenderness, and juiciness among major muscles. *Meat Sci.* 39: 143–147.

Cramer, D.A. 1983. Chemical compounds implicated in lamb flavor. *Food Technol.* 37: 249–257.

Descalzo, A.M., E.M. Insani, A. Biolatte, A.M. Sancho, P.T. Garcia, N.A. Pensel, and J.A. Josifovich. 2005. Influence of pasture or grain-based diets supplemented with vitamin E on antioxidant/oxidative balance of Argentine beef. *Meat Sci.* 70: 35–44.

D'Souz, D.N., and B.P. Mullan. 2002. The effect of genotype, sex and management strategy on the eating quality of pork. *Meat Sci.* 60: 95–101.

Duckett, S.K., and P.S. Kuber. 2001. Genetic and nutritional effects on lamb flavor. *J. Anim. Sci.* 79: 249–254.

Ellis, M., M.S. Brewer, D.S. Sutton, H.-Y. Lan, R.C. Johnson, and F.K. McKeith. 1998. Aging and cooking effects on sensory traits of pork from pigs of different breed lines. *J. Muscle Foods* 9: 281–291.

Ellis, M., A.J. Webb, P.J. Avery, and I. Brown. 1996. The influence of terminal sire genotype, sex, slaughter weight, feeding regime, and slaughter house on growth performance, and carcass and meat quality in pigs on the organoleptic properties of fresh pork. *J. Anim. Sci.* 74: 521–530.

Field, R.A., G.E. Nelms, and C.O. Schoonover. 1966. Effects of age, marbling, sex on palatability of steers. *J. Anim. Sci.* 25: 360–366.

Fisher, A.V., M. Enser, R.I. Richardson, J.D. Wood, G.R. Nute, E. Kurt, L.A. Sinclair, and R.G. Wilkinson. 2000. Fatty acid composition and eating quality of lamb types derived from four diverse breed × production systems. *Meat Sci.* 55: 141–147.

Gorraiz, C., M.J. Beriain, and K. Insausti. 2002. Effect of aging time on volatile compounds, odor, and flavor of cooked beef from Pirenaica and Friesian bulls and heifers. *J. Food Sci.* 67: 916–922.

Griffin, C.L., M.W. Orcutt, R.R. Riley, G.C. Smith, J.W. Savell, and M. Shelton. 1992. Evaluation of palatability of lamb, mutton, and chevon by sensory panels of various cultural backgrounds. *Small Ruminant Res.* 8: 67–74.

Heymann, H., H.B. Hedrick, M.A. Karrasch, M.K. Eggeman, and M.K. Ellersieck. 1990. Sensory and chemical characteristics of fresh pork roasts cooked to different end point temperatures. *J. Food Sci.* 55: 613–617.

Hornstein, I., and P.F. Crowe. 1960. Flavor studies on beef and pork. *J. Agric. Food Chem.* 8: 494–498.

Jacobson, M., and F. Fenton. 1956. Effects of three levels of nutrition and age of animals on quality of beef. I. Palatability, cooking data, moisture, fat, and nitrogen. *Food Res.* 21: 415–426.

Jaurez, M., W.R. Caine, I.L. Larsen, W.M. Robertson, M.E.R. Dugan, and J.L. Aalhus. 2009. Enhancing pork loin quality attributes through genotype, chilling method, and ageing time. *Meat Sci.* 83: 447–453.

Jenschke, B.E., J.M. Hodgen, J.L. Meisinger, A.E. Hamling, D.A. Moss, M. Lundesjo Ahnstrom, K.M. Eskridge, and C.R. Calkins. 2007. Unsaturated fatty acids and sodium affect the liver-like off flavor in cooked beef. *J. Anim. Sci.* 85: 3072–3078.

Jeremiah, L.E., L.L. Gibson, J.L. Aalhus, and M.E.R. Dugan. 2003. Assessment of palatability attributes of the major beef muscles. *Meat Sci.* 65: 949–958.

Jeremiah, L.E., J.P. Gibson, L.L. Gibson, R.O. Ball, C. Aker, and A. Fortin. 1999. The influence of breed, gender, and PSS (Halothane) genotype on meat quality, cooking loss, and palatability of pork. *Food Res. Int.* 32: 59–71.

Jeremiah, L.E., S.D.M. Jones, G. Kruger, A.K. Tong, and R. Gibson. 1992. The effects of gender and blast chilling time and temperature on cooking properties and palatability of pork longissimus muscle. *Can. J. Anim. Sci.* 72: 501–506.

Jeremiah, L.E., A.H. Martin, and A.C. Murray. 1985. The effects of various post-mortem treatments on certain physical and sensory properties of three different bovine muscles. *Meat Sci.* 12: 155–176.

Jeremiah, L.E., A.K. Tong, and L.L. Gibson. 1998. The influence of lamb chronological age, slaughter weight, and gender. Flavor and texture profiles. *Food Res. Int.* 31: 227–242.

Killinger, K.M., C.R. Calkins, W.J. Umberger, D.M. Feuz, and K.M. Eskridge. 2004. A comparison of consumer sensory acceptance and value of domestic beef steaks and steaks from a branded, Argentine beef program. *J. Anim. Sci.* 82: 3302–3307.

Kouba, M., M. Enser, F.M. Whittington, G.R. Nute, and J.D. Wood. 2003. Effect of high-linolenic acid diet on lipogenic enzyme activities, fatty acid composition, and meat quality in the growing pig. *J. Anim. Sci.* 81: 1967–1979.

Laborde, F.L., I.B. Mandell, J.J. Tosh, J.W. Wilton, and J.G. Buchanan-Smith. 2001. Breed effects on growth performance, carcass characteristics, fatty acid composition, and palatability attributes in finishing steers. *J. Anim. Sci.* 79: 355–365.

Larick, D.K., H.B. Hedrick, M.E. Bailey, J.E. Williams, D.L. Hancock, G.B. Garner, and R.E. Morrow. 1987. Flavor constituents of beef as influenced by forage- and grain-feeding. *J. Food Sci.* 52: 245–251.

Larick, D.K., B.E. Turner, W.D. Schoenherr, M.T. Coffey, and D.H. Pilkington. 1992. Volatile compound content and fatty acid composition of pork as influenced by linoleic acid content of the diet. *J. Anim. Sci.* 70: 1397–1403.

Lorenzen, C.L., T.R. Neely, R.K. Miller, J.D. Tatum, J.W. Wise, J.F. Taylor, M.J. Buyck, J.O. Reagan, and J.W. Savell. 1999. Beef customer satisfaction: Cooking method and degree of doneness effects on the top loin steak. *J. Anim. Sci.* 77: 637–644.

Luchak, G.L., R.K. Miller, K.E. Belk, D.S. Hale, S.A. Michaelson, D.D. Johnson, R.L. West, F.W. Leak, H.R. Cross, and J.W. Savell. 1998. Determination of sensory, chemical, and cooking characteristics of retail beef cuts differing in intramuscular fat and external fat. *Meat Sci.* 50: 55–72.

Mandell, I.B., E.A. Gullett, J.W. Wilton, R.A. Kemp, and O.B. Allen. 1997. Effects of gender and breed on carcass traits, chemical composition, and palatability attributes in Hereford and Simmental bulls and steers. *Livest. Prod. Sci.* 49: 235–248.

Martel, J., F. Minvielle, and L.M. Poste. 1988. Effects of crossbreeding and sex on carcass composition, cooking properties and sensory characteristics of pork. *J. Anim. Sci.* 66: 41–46.

Martinez-Cerezo, S., C. Sanudo, B. Panea, and J.L. Olleta. 2005. Breed, slaughter weight and ageing time effect on consumer appraisal of three muscles of lamb. *Meat Sci.* 69: 797–805.

Maughan, C., and S. Martini. 2010. Development of a flavor lexicon for meat samples. American Meat Science Association 63rd Reciprocal Meat Conference, Texas Tech University, Lubbock, TX, p. 41.

Melton, S.L. 1983. Effect of forage feeding on beef flavor. *Food Technol.* 37: 239–248.

Melton, S.L. 1990. Effects of feed on flavor of red meat: A review. *J. Anim. Sci.* 68: 4421–4435.

Meisinger, J.L., J.M. Hodgen, and C.R. Calkins. 2006. Flavor relationships among muscles from the chuck and round. *J. Anim. Sci.* 84: 2826–2833.

Miller, M.F., L.C. Rockwell, D.K. Lun, and G.E. Carstens. 1996. Determination of the flavor attributes of cooked beef from cross-bred Angus steers fed corn- or barley-based diets. *Meat Sci.* 44: 235–243.

Miller, M.F., S.D. Shackelford, K.D. Hayden, and J.O. Reagan. 1990. Determination of the alteration in fatty acid profiles, sensory characteristics and carcass traits of swine fed elevated levels of monounsaturated fats in the diet. *J. Anim. Sci.* 68: 1624–1631.

Moeller, S.J., R.K. Miller, T.L. Aldredge, K.E. Logan, K.K. Edwards, H.N. Zerby, M. Boggess, J.M. Box-Steffensmeier, and C.A. Stahl. 2010a. Trained sensory perception of pork eating quality as affected by fresh and cooked pork quality attributes and end-point cooked temperature. *Meat Sci.* 85: 96–103.

Moeller, S.J., R. K. Miller, K.K. Edwards, H.N. Zerby, K.E. Logan, T.L. Aldredge, C.A. Stahl, M. Boggess, and J.M. Box-Steffensmeier. 2010b. Consumer perceptions of pork eating quality as affected by pork quality attributes and end-point cooked temperature. *Meat Sci.* 84: 14–22.

Molina, M.E., B.L. Gwartney, R.L. West, and D.D. Johnson. 2005. Enhancing palatability traits in beef chuck muscles. *Meat Sci.* 71: 52–61.

Monson, F., I. Sierra, and C. Sanudo. 2005. Influence of breed and ageing time on the sensory meat quality and consumer acceptability in intensively reared beef. *Meat Sci.* 71: 471–479.

Neely, T.R., C.L. Lorenzen, R.K. Miller, J.D. Tatum, J.W. Wise, J.F. Taylor, M.J. Buyck, J.O. Reagan, and J.W. Savell. 1999. Beef customer satisfaction: Cooking method and degree of doneness effects on the top round steak. *J. Anim. Sci.* 77: 653–660.

Nelson, P.M., B. Lowe, and H.D. Helser. 1930. Influence of the animal's age upon quality and palatability of beef. *Iowa Agric. Exp. Stn. Bull.* 272.

Nephawe, K.A., L.V. Cundiff, M.E. Dikeman, J.D. Crouse, and L.D. Van Vleck. 2004. Genetic relationships between sex-specific traits in beef cattle: Mature weight, weight adjusted for body condition score, height and body condition score of cows, and carcass traits of their steer relatives. *J. Anim. Sci.* 82: 647–653.

Nold, R.A., J.R. Romans, W.J. Costello, J.A. Henson, and G.W. Libal. 1997. Sensory characteristics and carcass traits of boars, barrows, and gilts fed high- or adequate-protein diets and slaughtered at 100 or 110 kilograms. *J. Anim. Sci.* 75: 2641–2651.

Nuernberg, K., J.D. Wood, N.D. Scollan, R.I. Richardson, G.R. Nute, G. Nuernberg, D. Dannenberger, J. Voigt, and K. Ender. 2005. Effect of a grass-based and a concentrate feeding system on meat quality characteristics and fatty acid composition of longissimus muscle in different cattle breeds. *Livestock Prod. Sci.* 94: 137–147.

Park, R.J. 1975. The effect of meat flavor of period of feeding a protected lipid supplement to lambs. *J. Food Sci.* 40: 1217–1221.

Park, R.J., A.L. Ford, and D. Ratcliff. 1976. The influence of two kinds of protected lipid supplement on the flavor of lamb. *J. Food Sci.* 41: 633–635.

Paterson, B.C., and F.C. Parrish Jr. 1986. A sensory panel and chemical analysis of certain beef chuck muscles. *J. Food Sci.* 51: 876–879, 896.

Pearson, A.M., R.L. Dickson, T.W. Hill, and D.W. Holtan. 1995. Boar taint: Androstenones or skatole? A review. *Proceedings of the 41st International Congress of Meat and Technology*, San Antonio, TX, pp. 78–79.

Philip, C., N. Bhumiratana, L.V. Araujo, K. Adhikari, E. Chambers IV, and R.K. Miller. 2010. Development of a lexicon for the description of beef aroma and flavor. American Meat Science Association 63rd Reciprocal Meat Conference, Texas Tech University, Lubbock, TX, p. 68.

Priolo, A., D. Micol, J. Agabriel, S. Prache, and E. Dransfield. 2002. Effect of grass or concentrate feeding systems on lamb carcass and meat quality. *Meat Sci.* 62: 179–185.

Reagan, J.O., Z.L. Carpenter, G.C. Smith, and G.T. King. 1971. Comparison of palatability traits of beef produced by young bulls and steers. *J. Anim. Sci.* 32: 641–646.

Rhee, M.S., T.L. Wheeler, S.D. Shackelford, and M. Koohmaraie. 2004. Variation in palatability and biochemical traits within and among eleven beef muscles. *J. Anim. Sci.* 82: 534–550.

Savell, J.W., C.L Lorenzen, T.R. Neely, R.K. Miller, J.D. Tatum, J.W. Wise, J.F. Taylor, M.J. Ruyck, and J.O. Reagan. 1999. Beef customer satisfaction: Cooking method and degree of doneness effects on the top sirloin steak. *J. Anim. Sci.* 77: 645–652.

Shackelford, S.D., T.L. Wheeler, and M. Koohmaraie. 1995. Relationship between shear force and trained sensory panel tenderness ratings of 10 major muscles from Bos indicus and Bos taurus cattle. *J. Anim. Sci.* 73: 3333–3340.

Sink, J.D., and F. Caporaso. 1977. Lamb and mutton flavour: Contributing factors and chemical aspects. *Meat Sci.* 1: 119–127.

Sitz, B.M., C.R. Calkins, D.M. Feuz, W.J. Umberger, and K.M. Eskridge. 2005. Consumer sensory acceptance and value of domestic, Canadian, and Australian grass-fed beef steaks. *J. Anim. Sci.* 83: 2863–2868.

Smith, G.C., J.W. Savell, H.R. Cross, and Z. Carpenter. 1983. The relationship of USDA quality grade to beef flavor. *Food Technol.* 37: 233–238.

Spanier, A.M., M. Flores, K.W. McMillin, and T. D. Bidner. 1997. The effect of postmortem aging on meat flavor quality in Brangus beef. Correlation of treatments, sensory, instrumental and chemical descriptors. *Food Chem.* 59: 531–538.

Splan, R.K., L.V. Cundiff, and L.D. Van Vleck. 1998. Genetic parameters for sex-specific traits in beef cattle. *J. Anim. Sci.* 76: 2272–2278.

St. John, L.C., C.R. Young, D.A. Knabe, L.D. Thompson, G.T. Schelling, S.M. Grundy, and S.B. Smith. 1987. Fatty acid profiles and sensory and carcass traits of tissues from steers and swine fed an elevated monounsaturated fat diet. *J. Anim. Sci.* 64: 1441–1447.

Stelzleni, A. 2006. Feeding and aging effects on carcass composition, fatty acid profiles and sensory attributes of muscles from cull cow carcasses. PhD dissertation, University of Florida, Gainesville, FL.

Stelzleni, A.M., L.L. Patten, D.D. Johnson, C.R. Calkins, and B.L. Gwartney. 2007. Benchmarking carcass characteristics and muscles from commercially identified beef and cull dairy cull cow carcasses for Warner–Bratzler shear force and sensory attributes. *J. Anim. Sci.* 85: 2631–2638.

Summers, R.L., J.D. Kemp, D.G. Ely, and J.D. Fox. 1978. Effects of weaning, feeding systems, and sex of lamb on lamb carcass characteristics and palatability. *J. Anim. Sci.* 47: 622–629.

Teixeira, A., S. Batista, R. Delfa, and V. Cadavez. 2005. Lamb meat quality of two breeds with protected origin designation. Influence of breed, sex, and live weight. *Meat Sci.* 71: 530–536.

Vasta, V., and Priolo, A. 2006. Ruminant fat volatiles as affected by diet. A review. *Meat Sci.* 73: 218–228.

Webb, E.C., M.J.C. Bosman, and N.H. Casey. 1994. Dietary influences on subcutaneous fatty acid profiles and sensory characteristics of Dorper and SA mutton Merino wethers. *South African J. Food Sci. Nutr.* 6: 45–50.

Wheeler, T.L., S.D. Shackelford, and M. Koohmaraie. 2000. Relationship of beef longissimus tenderness classes to tenderness of gluteus medius, semimembranosus, and biceps femoris. *J. Anim. Sci.* 78: 2856–2861.

Wilson, L.L., J.R. MCCurley, J.H. Ziegler, and J.L. Watson. 1976. Genetic parameters of live and carcass characters from progeny of polled Hereford sires and Angus–Holstein cows. *J. Anim. Sci.* 43: 569–576.

Winer, L.K., P.J. David, C.M. Bailey, M. Reid, T.P. Ringkob, and M. Stevenson. 1981. Palatability characteristics of the longissimus muscle of young bulls representing divergent beef breeds and crosses. *J. Anim. Sci.* 53: 387–394.

Wong, E., C.B. Johnson, and L.N. Nixon. 1975. The contribution of 4-methyloctanic (hircinoic) acid to mutton and goat meat flavor. *N. Z. J. Agric. Res.* 18: 261–266.

Wood, J.D., G.R. Nute, G.A.J. Fursey, and A. Cuthbertson. 1995. The effect of cooking conditions on the eating quality of pork. *Meat Sci.* 40: 127–135.

Wood, J.D., G.R. Nute, R.I. Richardson, F.M. Whittington, O. Southwood, G. Plastow, R. Mansbridge, N. da Costa, and K.C. Chang. 2004. Effects of breed, diet and muscle on fat deposition and eating quality in pigs. *Meat Sci.* 67: 651–667.

Wulf, D.M. and J.K. Page. 2000. Using measurements of muscle color, pH, and electrical impedance to augment the current USDA beef quality grading standards and improve the accuracy and precision of sorting carcasses into palatability groups. *J. Anim. Sci.* 78: 2595–2607.

Young, O.A., G.A. Lane, A. Priolo, and K. Fraser. 2003. Pastoral and species flavour in lambs raised on pasture, lucerne or maize. *J. Sci. Food Agric.* 83: 93–104.

14 Flavor of Wine

Vicente Ferreira and Felipe San Juan

CONTENTS

14.1 GENERAL CONSIDERATIONS

Wines are highly valued beverages produced from grape juice, being grapes quite complex fruits from the chemical point of view. During wine production, a series of physical, chemical, and biochemical processes take place, many of them having a deep influence on the final wine chemical composition. Different enzymatic, oxidative, hydrolytic, and chemical-dissolution processes predominate before fermentation. During fermentation, major changes are due to the metabolism of yeast that secretes to the medium large amounts of secondary metabolites, many of which have relevant sensory properties. After the alcoholic fermentation, biological changes are slower yet very important, since dead yeast cells still display enzymatic activities and lactic bacteria are active. Later on, changes still are slower, but very relevant from the sensory point of view. In this last step of aging, most changes are mainly of chemical nature: acid hydrolytic processes, chemical rearrangements, oxidations, reductions, polymerizations, adduct formation, or extractions from wood, to name some of the most relevant. This large series of processes, consequence of the different technological options, introduce a huge chemical complexity and also can introduce a relatively large chemical diversity. The chemical complexity manifests not only on the large number of different molecules present in the medium, but on the complex and for the most unknown interactions that these molecules can exert between them. Similarly, the chemical diversity will manifest on the relatively large ranges at which some compounds will be found, depending on the wine type, origin, age, or wine making method.

When it comes to flavor, wine flavor is a complex percept that can be divided into two major items: wine (orthonasal) aroma and wine (in mouth) flavor. The first one is obviously majorly caused by aroma-active volatile molecules, although major compounds in the matrix play an important but not well-understood role on the release of these molecules (Dufour and Bayonove 1999a, 1999b; Dufour and Sauvaitre 2000; Robinson et al. 2009; Tsachaki et al. 2009). Recent experimental work has revealed that differences introduced by the retentive power of the matrix can have a deep sensory effect on wine orthonasal aroma properties (Saenz-Navajas et al. 2010a). Such unexpected effect was attributed to the fact that different nonvolatile compositions can interact differentially toward different chemical structures or functionalities. This in turn means that, for a nominally equivalent volatile composition, substantially different headspace volatile profiles leading to different aroma properties can be obtained (Saenz-Navajas et al. 2010b). Wine in mouth flavor is yet more complex, because of the additional chemical and perceptual complexity introduced by tastants and astringent compounds. It is not only that a relatively large number of active-tastants can be found on wine (Hufnagel and Hofmann 2008a, 2008b; Saenz-Navajas et al. 2010c), but that strong and not well-understood perceptual interactions between tastants, astringent, and aroma compounds do exist, as has been recently highlighted (Saenz-Navajas et al. 2010b). Therefore, it can be said that within wine flavor, wine aroma-active molecules are:

1. Directly responsible for wine (orthonasal) aroma, albeit the nonvolatile matrix plays an indirect role as was aforementioned.
2. Directly responsible for wine retronasal aroma, although in this case not only the nonvolatile matrix plays an indirect role, but tastants and astringent molecules can exert a relevant effect through perceptual interactions.
3. Indirectly involved in taste and astringency and in the general "in mouth" sensory perception, most likely via perceptual interactions. As recently discussed, such interactions are particularly effective in white wines and seem to be of just a secondary importance in red wines (Saenz-Navajas et al. 2010b).

This subchapter deals exclusively with aroma-active and volatile molecules, responsible for the aforementioned flavor perceptions. In this context, it is noteworthy that only a small fraction of volatile molecules are present at levels high enough to trigger an olfactory signal, and because of that, the study of wine flavor is not the same as the study of the wine volatile profile. To study wine flavor, aroma relevant molecules must be first screened from the volatile profile by using Gas chromatography–olfactometry (GC-O).

14.2 WINE AROMA COMPOUNDS DETECTED BY GAS CHROMATOGRAPHY–OLFACTOMETRY

GC-O is the most useful screening technique for the study of wine aroma. Readers interested in the specific aspects of the application of this technique to the study of wine aroma are referred to previous reviews (Ferreira et al. 2001; Ferreira and Cacho 2009), whereas here our focus is on the detected aroma compounds. Table 14.1 gives a summary of most of the aromatic zones detected by GC-O in the different

TABLE 14.1
Odor Regions Detected in Wine Extracts by Different Researchers in Effluent of Gas Chromatography–Olfactometer

R.I. pol[a]	R.I. npol[b]	Odor Description	Identity	Observations	Reference
750	<600	Apple, fruit	Acetaldehyde		
885	<700	Fruity, solvent	Ethyl acetate		
900	730	Ethereal, green	1,1-Diethoxyethane	Found in oxidized wines	Escudero et al. 2000a
900	757	Green, aldehyde	2,4,5-Trimethyldioxolane	Found in oxidized wines	Escudero et al. 2000a
910	663	Bread crust, aldehyde	2-Methylbutanal		Ferreira et al. 2009
935	713	Sweet, fruity, alcoholic	Isopropyl acetate		Ferreira et al. 2009
937	755	Sweet, solvent	Ethyl propanoate		Komes et al. 2006
953	814	Fruity, pineapple	Ethyl 2-methylpropanoate		
969	<700	Cream	2,3-Butanodione		
974	716	Fruit	Propyl acetate		Escudero et al. 2007
984		Plastic, glue	Unknown		Ferreira et al. 2009
1012	906	Orange peel, citric, alcoholic	Unknown	Coelution with isobutyl acetate in polar column	Ferreira et al. 2009
1013	771	Fruity	Isobutyl acetate	Highest in young wines	Ferreira et al. 2009
1043	812	Acid fruit	Ethyl butyrate		
1052	910	Cabbage	Dimethyl sulfide		Guth 1997; Kotseridis and Baumes 2000
1057	695	Cream, butter	2,3-Pentanodione		Ferreira et al. 2009
1061	860	Sweet fruit	Ethyl 2-methylbutyrate	Highest in aged wines	
1075	864	Berry, blackberry	Ethyl isovalerate	Highest in aged wines	
1079	800	Green	Hexanal	Found in muscadine grape juice	Chisholm et al. 1995
1095		Green, grass	2,5,-Dimethyl-1,4-dioxane		Baek et al. 1997; Lopez et al. 1999
1096	<700	Fusel, alcohol	Isobutanol		

(continued)

TABLE 14.1 (Continued)
Odor Regions Detected in Wine Extracts by Different Researchers in Effluent of Gas Chromatography–Olfactometer

R.I. pol[a]	R.I. npol[b]	Odor Description	Identity	Observations	Reference
1112	904	Hop, rubber	3-Methyl-2-buten-thiol	Found in Sauternes	Bailly et al. 2006
1121		Fish	Unknown		Escudero et al. 2007
1128	886	Banana	Isoamyl acetate	Highest in young wines	
1142	941	Fruit, strawberry	Ethyl 2-methylpentanoate	Highest in aged wines	Campo et al. 2006
1150	908	Fruity, ester, mint,	Ethyl pentanoate		Aznar et al. 2001
1185	960	Fruity, ester	Ethyl 3-methylpentanoate	Highest in aged wines	Campo et al. 2006
1198	969	Fruity, ester	Ethyl 4-methylpentanoate	Highest in aged wines	Campo et al. 2006
1202	857	Grass	*E*)-2-Hexenal	Highest in some oxidized wines	Chisholm et al. 1995
1219	753	Alcohol, harsh	Isoamyl alcohol	Highest intensity in olfactometry	
1230		Mushroom, grass	Unknown		Escudero et al. 2007
1238	1005	Green, apple	Ethyl hexanoate		
1286	1008	Banana	Hexyl acetate		Campo et al. 2005
1290	711	Wet, flowery	Acetoin		Ferreira et al. 2002
1292	952	Solvent	Furfuryl ethyl ether		Ferreira et al. 2009
1293	1004	Solvent, lemon, bitter	Octanal		
1302	976	Earthy, mushroom	1-Octen-3-one	Found in wines made of rotten grapes	Baek et al. 1997
1310	977	Onion, sunflower seeds	2-Methyl-3-furanthiol		Bouchilloux et al. 1998a, 1998b; Kotseridis and Baumes 2000
1310		Grass	Unknown		Ferreira et al. 2009
1337	1112	Roses	(*Z*)-Rose-oxide	Key odor of Gewurztraminer	Guth 1997
1359	888	Green, grass	1-Hexanol		
1360	970	Cabbage	Dimethyl trisulfide	Only detected in headspace	Guth 1997

1374	940	Patchuli, green, bitter	4-Mercapto-4-methylpentan-2-one	Key odor of Sauvignon Blanc and Schreube	Guth 1997
1378	1064	Geranium-like	Unknown	Found in Carmenere	San Juan (unpublished)
1386		Pleasant	Unknown		Lopez et al. 1999
1389	870	Freshly cut grass	(*Z*)-3-Hexenol		
1393		Incense			Lopez et al. 1999
1408		Wet, green, grass	2-Butoxyethanol		Escudero et al. 2000a
1422		Mushroom	1-Nonen-3-one		Cullere et al. 2004
1425	1130	Fruity, sweet	Ethyl cyclohexanoate	Highest in fortified aged wines	Campo et al. 2006
1430	987	Strawberry	Ethyl 2-hydroxy-3-methylbutyrate	Highest in fortified aged wines	Campo et al. 2006
1433	985	Toasted, mushroom	1-Octen-3-ol		Escudero et al. 2000a
1434	1040	Earthy, cork	3,5-Dimethyl-2-methoxypyrazine	Responsible for cork taint	Simpson et al. 2004; Ferreira et al. 2009
1436	907	Coffee, toasted	Furfurylthiol		Tominaga et al. 2000
1439	1201	Sweet, fruit soap	Ethyl octanoate		
1445	1093	Pepper, earthy, green	3-Isopropyl-2-methoxypyrazine		Ferreira et al. 2009
1450	985	Potato, cooked vegetables	3-Methyl-thiopropanal	Highest in some oxidized wines	Escudero et al. 2000a
1455	<700	Vinegar	Acetic acid		
1459		Citrus zest	4-Mercapt-4-methylpentan-2-ol	Found in Sauvignon Blanc	Bouchilloux et al. 1998a, 1998b
1474		Food, seed	Unknown		Lopez et al. 1999
1474	828	Incense, flowery	Furfural		Aznar et al. 2001
1495		Caramel, bakery	Unknown	Found in aged champagne	Escudero et al. 2000a; Escudero et al. 2000b
1507	1147	Chlorine, green, metallic	(*Z*)-2-Nonenal	Found in aged red wines	Ferreira et al. 2009
1509	1172	Pepper, earthy	3-*sec*-Butyl-2-methoxypyrazine		Escudero et al. 2007
1517	1209	Green	Decanal		Kotseridis and Baumes 2000
1520		Plastic, gas oil	2-Methyltetrahydrotiophenone		Aznar et al. 2001

(continued)

TABLE 14.1 (Continued)
Odor Regions Detected in Wine Extracts by Different Researchers in Effluent of Gas Chromatography–Olfactometer

R.I. pol[a]	R.I. npol[b]	Odor Description	Identity	Observations	Reference
1524	935	Burnt, marshmallow, muscadine	Ethyl 3-hydroxybutyrate		Baek et al. 1997
1528		Fruity, toasted	Unknown		Lopez et al. 1999
1532	1158	Wet, rancid	(*E*)-2-nonenal	Found in oxidized wines	Escudero et al. 2000a
1535	1181	Green, pepper (capsicum)	3-Isobutyl-2-methoxypirazine		Escudero et al. 2007
1539	962	Bitter, cherry	Benzaldehyde	Found in oxidized wines	Escudero et al. 2000a
1542		Toasty, caramel	Unknown *m*/*z* 144		Cutzach et al. 1997
1543	1315	Pleasant, complex	Vitispirane		Lopez et al. 1999
1545	1060	Strawberry	Ethyl 2-hydroxy-4-methylpentanoate	Highest in aged wines (Madeira and Sherry)	Campo et al. 2006
1551	1106	Flowery	Linalool		
1566		Lavender	Unknown		Lopez et al. 1999
1588	796	Acid, cheese	Isobutyric acid		Cullere et al. 2004
1572		Meat	Unknown		Lopez et al. 1999
1582	1156	Cucumber	(*E*,*Z*)-Nona-2,6-dienal		Baek et al. 1997; Kotseridis and Baumes 2000; Aznar et al. 2001
1593	1158	Earth, humidity	2-Methylisoborneol		Ferreira et al. 2009
1620		Cooked leeks	3-Mercapto-3-methylbutanol	Found in Sauvignon blanc	Bouchilloux et al. 1998a, 1998b
1620	1103	Orange	Methyl benzoate		Aznar et al. 2001
1622	1021	Toasted, burnt	2-Acetylpirazine	Found in aged wines	Cullere et al. 2004
1623	796	Cheese	Butyric acid		
1624	1062	Woody, oily	Ethyl furoate		Lopez et al. 1999
1634	1399	Pleasant, soap	Ethyl decanoate		Lopez et al. 1999

1635	1047	Honey	Phenylaceteldehyde	Highest in some oxidized wines	Baek et al. 1997; Kotseridis and Baumes 2000
1654	1449	Flower, fruity	Isoamyl octanoate		Lopez et al. 1999
1655	878	Blue cheese	Isovaleric acid		
1662	1179	Ripe fruit	Ethyl benzoate		Lopez et al. 1999
1675	1170	Onion, barbecue, roasty	2-Methyl-(3-methyldithio)-furane		Cullere et al. 2008; Ferreira et al. 2009
1676	1153	Honey	2,6,6-Trimehtylcyclohexen-2-ene-1,4-dione	Found in Douro fortified wines	Rogerson et al. 2001
1678		Fruity, violets, fruit candy	Unknown		Lopez et al. 1999
1684	1199	Pleasant, sweet	α-Terpineol		Lopez et al. 1999
1686		Fruit, candy	Unknown		Lopez et al. 1999
1708		Acid, root, grass	Unknown		Lopez et al. 1999
1709	1158	Anise, sweet, medicinal	1,3-Dimethoxibenzene	Found in port wines	Rogerson et al. 2002
1723	994	Cooked vegetables, onion, garlic	3-Methylthiopropanol (methionol)		
1725		Grapefruit, banana	3-Mercaptohexyl acetate		Kotseridis and Baumes 2000
1740	1181	Honey, sweet, licorice	3-Methyl-2,4-nonadione		Pons et al. 2008; Ferreira et al. 2009
1748		Tree, fruit	Unknown		Lopez et al. 1999
1753		Kerosene	1,1,6-Trimethyl-1,2-dihydronaphthalene	Found in oxidized wines	Silva Ferreira et al. 2003
1758		Coconut	Unknown		Lopez et al. 1999
1763	1227	Citric	Citronellol		Guth 1997
1771	1167	Cucumber, melon	(*E*,*Z*)-2,6-Nonadien-1-ol	Found in muscadine grape juice	Baek et al. 1997
1772		Flower, licorice	Unknown		Lopez et al. 1999
1775		Flowery	Unknown		Lopez et al. 1999
1782	1278	Sweet, honey	Ethyl phenylacetate		Tat et al. 2007a
1793		Vegetables	Unknown		Lopez et al. 1999

(continued)

TABLE 14.1 (Continued)
Odor Regions Detected in Wine Extracts by Different Researchers in Effluent of Gas Chromatography–Olfactometer

R.I. pol[a]	R.I. npol[b]	Odor Description	Identity	Observations	Reference
1807	1360	Cork, damp	2,4,6-Trichloroanisole (TCA)	Responsible for cork taint	Buser et al. 1982; Ferreira et al. 2009
1811	1318	Aldehyde, rancid	(*E*,*E*)-2,4-Decadienal		Ferreira et al. 2009
1818	1392	Bark, sweet, apple	β-Damascenone		
1837		Caramel	Cyclotene	Found in toasted oak wood, proposed as important in Chardonnay from Burgundy	Cutzach et al. 1997; Le Fur and Etievant 1998
1838	1258	Pleasant, flowery	2-Phenylethyl acetate		Campo et al. 2005
1854	1134	Vegetable, dry, grapefruit	3-Mercaptohexanol		Bouchilloux et al. 1998a, 1998b; Kotseridis and Baumes 2000
1868		Toasty, caramel	Dihydromaltol	Found in toasted oak wood	Cutzach et al. 1997
1869	1267	Citric, flowery	Geraniol		Lopez et al. 1999
1870	1020	Cheese, green	Hexanoic acid		Ferreira et al. 2002
1875		Strawberry, sulfur	4-Mercapto-2,5-dimethylthiophen-3-one	Key odor of some non-*Vitis vinifera* wines	Guedes De Pinho et al. 1997
1879	1434	Sweet, fruit	α-Ionone		Lopez et al. 1999
1882	1103	Phenolic, chemical	Guaiacol		
1889		Nice, flowery			Lopez et al. 1999
1900	1357	Strawberry, plum, flowery	Ethyl dihydrocinnamate		
1916		Citric, pleasant, jam	2-Phenylethyl isobutyrate		Lopez et al. 1999
1931	1146	Roses	β-Phenylethylalcohol		
1952	1489	Pleasant, berry	β-Ionone		Lopez et al. 1999
1957	1134	Coconut	(*Z*)-Whiskylactone		Cullere et al. 2004
1961		Caramel	Maltol	Found in toasted oak wood, proposed as important in Chardonnay from Burgundy	Cutzach et al. 1997; Le Fur and Etievant 1998

1965		Pleasant, lactone-like	Unknown		Lopez et al. 1999
1973	1332	Peach, coconut	(*E*)-Whiskylactone		Le Fur and Etievant 1998; Aznar et al. 2001
1985		Lactone-like, green, apple, jam	δ-Octalactone		Lopez et al. 1999
2025	1288	Phenolic	4-Ethylguaiacol	Maximum in Rioja red wines	Aznar et al. 2001
2031	1081	Cotton candy	Furaneol		
2038	1372	Coconut	γ-Nonalactone		
2047		Fruit, candy, vanilla	Unknown		Lopez et al. 1999
2060	1200	Rancid, harsh	Octanoic acid		Lopez et al. 1999
2074	1175	Cotton candy, peach	Homofuraneol	Maximum in Cabernet and Merlot	Kotseridis and Baumes 2000
2077		Roses, sweet	Unknown		Ferreira et al. 2009
2094	1103	Animal, woody, phenolic	*p*-Cresol		Ferreira et al. 2009
2088	1103	Woody, smoky, phenolic	*m*-Cresol		Dzhakhua et al. 1978
2113		Chlorine, humidity	Tetrachloroanisole	Responsible for cork taint	
2116	1404	Clove, phenolic	4-Propylguaiacol		Ferreira et al. 2001
2122	1474	Strawberry, cream	Ethyl cinnamate		
2130	1475	Peach	γ-Decalatone		Lopez et al. 1999
2140	1369	Clove	Eugenol	Highest in oxidized wines and in Rioja red wines	Escudero et al. 2000a
2155	1540	Barbecue, roasty	*bis*(2-Metil-3-furil)disulfuro		Ferreira et al. 2009
2180	1505	Peach	δ-Decalactone		Lopez et al. 1999
2184	1154	Phenolic, animal	4-Ethylphenol	Maximum in Rioja red wines	Aznar et al. 2001
2192	1455	Coconut	Wine lactone	Highest in Schreube and Gewürtraminer	Guth 1997
2200	1154	Phenolic, animal	3-Ethylphenol		Guth 1997; Ferreira et al. 2009
2210	1124	Curry, black pepper (spice)	Sotolon		

(*continued*)

TABLE 14.1 (Continued)
Odor Regions Detected in Wine Extracts by Different Researchers in Effluent of Gas Chromatography–Olfactometer

R.I. pol[a]	R.I. npol[b]	Odor Description	Identity	Observations	Reference
2211		Toasty, caramel	2,3-Dihydro-3,5-dihydroxy-6-methyl-4(*H*)-piran-4-one		Cutzach et al. 1997
2218		Farmed feed, phenolic	Unknown	Highest in oxidized wines	Escudero et al. 2000a
2220	1328	Phenolic, pleasant	4-Vinylguaiacol	Highest in white wines	Escudero et al. 2004
2221		Leather, coffee	Unknown	Detected in aged champagne	Escudero and Etievant 1999
2223	1309	Sweet	*o*-Aminoacetophenone	Found in muscadine grape juice	Baek et al. 1997
2223		Honey	Unknown	Proposed as important in Chardonnay from Burgundy	Le Fur and Etievant 1998
2229	1393	Fatty, unpleasant	Decanoic acid		Lopez et al. 1999
2233		Toasty, caramel	Hydroxymaltol	Found in toasted oak wood	Cutzach et al. 1997
2238	1586	Coconut, flowery	γ-Undecalactone	Tentatively found in aged champagne	Escudero and Etievant 1999
2239	1343	Coconut, flowery	Methyl anthranilate	Highest in Rioja red wines	Chisholm et al. 1995; Moio and Etievant 1995; Aznar et al. 2001
2240	1360	Floral	4-Phenyl-3-hydroxybutan-2-one		Kotseridis and Baumes 2000
2249	1345	Incense, phenolic, chemical	2,6-Dimethoxyphenol		Kotseridis and Baumes 2000
2251		Carnation	Unknown		Lopez et al. 1999

2255		Incense	Unknown		Lopez et al. 1999
2261		Flowery, rose	Unknown		Lopez et al. 1999
2275	1672	Cork, chlorine, humidity	2,4,6-Tribromoanisole	Responsible for cork taint	Chatonnet et al. 2004
2280	1425	Sweet, fruity	Ethyl anthranilate	Found in Pinot Noir wines	Kotseridis and Baumes 2000
2308		Medicinal, phenolic, gouache	4-Vinylphenol	Highest in white wines	Chatonnet et al. 1993
2352	1438	Woody, sweet	Isoeugenol		Kotseridis and Baumes 2000
2396	1657	Sweet	(*Z*)-6-Dodeceno-γ-lactone		Guth 1997
2415		Incense, almond, cypress, oak	Unknown		Aznar et al. 2001
2463		Lee, flowery	Indole		Buettner 2004
2566		Oak	Unknown		Aznar et al. 2001
2580	1249	Pollen, flowery, roses	Phenylacetic acid		Kotseridis and Baumes 2000
2581	1389	Vanilla, candy	Vanillin		Aznar et al. 2001
2676	1579	Pollen, flowery	Ethyl vanillate		Aznar et al. 2001
2683		Caramel	Acetovanillone		Escudero et al. 2000a

[a] R.I. pol, retention index in a polar column (DB-wax).

[b] R.I. npol, retention index in a nonpolar column (5% phenyl polymethylsiloxane).

studies that so far have been carried out. The table intends to constitute a useful tool for the identification of compounds detected by olfactometry, and hence odorants have been ranked according to the retention index on a polar column (typically a Carbowax 20 M polymer). All in all, the table contains nearly 180 odor zones, of which 147 have been positively identified along the years and remarkably, 35 out of the 147 identifications have been carried out only in the past 10 years. It must also be pointed out that most unidentified compounds were detected only in extracts obtained by direct extraction (liquid–liquid or liquid–solid) of wine, and not in headspace extracts. This fact, together with the in general low intensities at which most unidentified compounds were found, make us think that most likely they are not very important to wine aroma and to suspect that the most important wine aroma compounds are already known.

One question that should be kept in mind during aroma identification is that some of the odorants in the table, particularly those identified in the past 15 years, are present at really negligible levels in wine, often well below the 50 ng/L level. This fact, together with the huge levels of major volatile compounds produced during fermentation, make it nearly impossible to obtain a clear mass spectrometric signal of the compound unless a dual gas chromatographic system or very powerful enrichment and preseparation techniques are used. Researchers should therefore distrust of the identifications based only on the mass spectra recorded at the same time of the odor. In this context, the general guidelines proposed by Molyneux and Schieberle (2007) should be carefully fulfilled. An additional precaution just to avoid working unnecessarily in the identification of irrelevant aroma molecules is to rely on a dynamic headspace-trapping system for the preparation of extracts (Campo et al. 2005; San Juan et al. 2010), instead of extracts obtained by direct liquid–liquid or solid–liquid extraction. These extracts can be more concentrated, but they present two major problems. On one hand, these extracts tend to contain high levels of polar odorants, in any case higher than the levels that these highly soluble in wine compounds can reach in the headspaces above wine. This causes the olfactometric profiles obtained from these extracts to strongly overemphasize the relevance of these compounds, which can mislead the research. On the second hand, these extracts are so enriched in wine major volatiles (fusel alcohols and ethyl esters of major acids) that both the chromatographic separation and the MS signal can be seriously hampered if the extract is not previously fractionated.

From our personal experience, a very useful hint is to consider the question from the opposite side, that is, first use a reliable dynamic headspace enrichment technique, and second instead of trying to identify the odorants in the experiment, try to simply locate those ones that it is almost sure that will be present. For such a task, Table 14.2 can be of help since it only contains the odorants most likely found in the GC-O analysis of headspaces techniques obtained with the techniques described elsewhere (Campo et al. 2005; San Juan et al. 2010). The table has been adapted from previous studies (Campo et al. 2005; Petka et al. 2006; Gomez-Miguez et al. 2007; Campo et al. 2008; Cullere et al. 2008; Ferreira et al. 2009) and contains the 65 odorants most often found in the GC-O of wines including an indication of the likelihood of finding each odorant.

TABLE 14.2
Odorants Typically Detected in the GC-O Study of Extracts Obtained by a Dynamic-Headspace Solid Phase Extraction Trapping Technique

LRI DB-WAX	LRI DB-5	Odor Descriptor	Identity	Frequency	Observations
	663	Bread crust, aldehyde	2-Methylbutanal	Sometimes	Most typical of aged/oxidized wines
935	713	Fruity, alcoholic	Isopropyl acetate	Frequent	
957	600	Lactic, strawberry	2,3-Butanedione (diacetyl)	Nearly always detected	
1012	771	Sweet	Isobutyl acetate + Ni	Nearly always detected	
1033	800	Strawberry, lactic	Ethyl butyrate	Nearly always detected	
1042	914	Sweaty, garlic	Dimethyl disulfide	Sometimes	Most typical of aged/oxidized wines
1050	846	Fruity, anise, strawberry	Ethyl 2-methylbutyrate	Nearly always detected	
1054	695	Lactic	2,3-Pentanedione	Rare	
1068	856	Fruity, anise	Ethyl 3-methylbutyrate	Nearly always detected	
1099	621	Bitter, green	Isobutanol	Nearly always detected	
1124	860	Banana	Isoamyl acetate	Nearly always detected	
1141	941	Sweet, floral	Ethyl 2-methylpentanoate	Sometimes	Most typical of aged/fortified wines
1185	960	Sweet	Ethyl 3-methylpentanoate	Sometimes	Most typical of aged/fortified wines
1193	969	Lactic, fruity	Ethyl 4-methylpentanoate	Frequent	Typical of aged/fortified wines
1217	719	Fusel	Isoamyl alcohol	Nearly always detected	
1242	999	Fruity, anise	Ethyl hexanoate	Nearly always detected	
1291	952 1004	Lemon, orange, solvent	Furfuryl ethyl ether/octanal	Frequent	Typical of aged/oxidized wines
1303	975	Mushroom	1-Octen-3-one	Frequent	
1315	890	Fried, barbecue, toasted	2-Methyl-3-furanthiol	Frequent	
1366	872	Green, grass	1-Hexanol	Frequent	

(continued)

TABLE 14.2 (Continued)
Odorants Typically Detected in the GC-O Study of Extracts Obtained by a Dynamic-Headspace Solid Phase Extraction Trapping Technique

LRI DB-WAX	LRI DB-5	Odor Descriptor	Identity	Frequency	Observations
1383	942	Box tree	4-Mercapto-4-methyl-2-pentanone	Sometimes	Typical of young fresh wines
1394	852	Grass	(*Z*)-3-hexenol	Frequent	
1424	1130	Fruity	Ethyl cyclohexanoate	Frequent	Typical of aged/fortified wines
1433	1040	Wet cardboard	3,5-Dimethyl-2-methoxypyrazine	Very rare	Defective wine
1436	907	Toasted, coffee	2-Furfurylthiol	Sometimes	Most likely in wood-aged wines
1445	1093	Pepper, earthy	3-Isopropyl-2-methoxypyrazine	Sometimes	Most likely in young wines from unripe grapes
1452	905	Green beans, cooked potatoes	Methional	Sometimes	Typical of aged/oxidized wines
1452	600	Vinegar	Acetic acid	Nearly always present	
1467	829	Sweet wood	2-Furaldehyde (furfural)	Rare	Most likely in wood-aged wines
1506	1147	Green, metallic	(*Z*)-2-Nonenal	Frequent	
1508	1173	Pepper, earthy	3-*sec*-Butyl-2-methoxypyrazine	Rare	Most likely in young wines from unripe grapes
1532	1181	Pepper, earthy	3-Isobutyl-2-methoxypyrazine	Sometimes	Most likely in young wines from unripe grapes
1561	1099	Floral	Linalool	Frequent	
1592	1158	Bleach, unpleasant	2-Methylisoborneol	Very rare	Defective wine
1621	1022	Toasty, burnt	2-Acetylpyrazine	Nearly always detected	
1641	821	Cheese	Butyric acid	Frequent	
1655	1050	Honey	Phenylacetaldehyde	Sometimes	Typical of aged/oxidized wines
1668	1170	Fried, barbecue, toasted	2-Methyl-3-(methyldithio)furan	Sometimes	Most typical in wood aged wines
1675	898	Cheese	2-/3-Methylbutyric acid	Nearly always detected	

1707	977	Plastic, green, thiol, meat	Methionol	Frequent	Most typical in red wines
1719		Sweet, tea	Ni	Sometimes	Most typical of aged/fortified wines
1726	1252	Basil, box tree	3-Mercaptohexyl acetate	Sometimes	Most typical of white fresh wines
1734	1181	Floral, honey	3-Methyl-2,4-nonanedione	Sometimes	
1789		Earthy	Ni	Sometimes	
1806	1087	Cork	2, 4, 6-Trichloroanisole (TCA)	Very rare	Defective wine
1811	1318	Rancid chip	(*E*,*E*)-2,4-Decadienal	Rare	
1818	1392	Sweet, apple	β-Damascenone	Nearly always detected	
1856	1524	Sulfury, citrus	3-Mercaptohexanol	Sometimes	Most typical of white fresh wines
1864	1086	Phenolic, chemical	2-Methoxyphenol (guaiacol)	Frequent	Most typical in reds and wood aged wines
1886	1353	Sweet, pleasant	Ethyl dihydrocinnamate	Frequent	Most typical of aged/fortified wines
1916	1108	Roses	β-Phenethyl alcohol	Nearly always detected	
1957	1134	Sweet wood	(*Z*)-Whiskylactone	Frequent	Exclusively in wood aged wines
2010		Musty, sweat	*o*-Cresol	Sometimes	Most typical in wood aged wines
2034	1285	Clove	4-Ethylguaiacol	Sometimes	Exclusively in red wines
2045	1096	Candy cotton	2,5-Dimethyl-4-hydroxy-3(2*H*)-furanone (furaneol)	Sometimes	Most typical in wood aged wines
2091	1103	Animal, leather, phenolic	*p*-Cresol (*m*-cresol)	Frequent	Most typical in reds and wood aged wines
2115	1404	Tea, clove	4-Propylguaiacol	Sometimes	Most typical in reds and wood aged wines
2131	1460	Floral, sweet	Ethyl cinnamate	Frequent	Most typical in red-wood aged wines
2155	1540	Meal, popcorn, toasted, fried	*bis*(2-Methyl-3-furyl)disulfide	Rare	
2176	1365	Clove	4 Allyl-2-methoxyphenol (eugenol)	Frequent	Most typical in wood aged wines
2185	1168	Leather, animal	4-Ethylphenol	Sometimes	Exclusively in red wines
2194	1170	Leather, animal	3-Ethylphenol	Rare	Exclusively in red wines
2198	1328	Bitumen	4-Vinylguaiacol	Frequent	Mostly in white wines
2204	1114	Burnt, curry	4,5-Dimethyl-3-hydroxy-2-(5*H*)-furanone (sotolon)	Frequent	Most typical of aged/fortified wines

14.3 WINE AROMA QUANTITATIVE COMPOSITION

Data in Table 14.3 summarize quantitative data about the most important compounds in wine aroma. Data in the table have been organized according to the role that the compounds can play in the wines when they reach the maximum levels described in the table. As can be seen, there is a first set of aroma compounds that in some "healthy" wines can reach concentrations high enough to communicate to the wine their specific odor nuance, and hence to act as genuine impact compounds. These are, together with some compounds that can be grouped into "aroma families," the most important wine aroma chemicals, and are responsible for many important wine aroma nuances. These compounds can be further classified according to their genesis.

14.3.1 Grape-Derived Impact Aroma Compounds

This first group includes three terpenoids, one norisoprenoid, and three polyfunctional mercaptans.

Linalool. This was the first identified aroma component able to exert an impact on Muscat wines (Cordonnier and Bayonove 1974; Ribéreau-Gayon et al. 1975). It also contributes to the flowery or even citrus notes of many other white cultivars (Arrhenius et al. 1996; Lee and Noble 2003; Campo et al. 2005; Palomo et al. 2006), always in combination with the other terpenols, particularly with geraniol.

cis-Rose oxide. This terpene of pleasant flowery character was first identified as a characteristic impact aroma compound of wines made with Gewürztraminer (Guth 1997a, 1997b). Later, it was also found to be a key odorant in wines made with the varietal Devin (Petka et al. 2006), and was also detected in the hydrolyzed fractions from precursors obtained from different neutral grape varieties (Ibarz et al. 2006). As with linalool, it requires the presence of the other terpenols to be clearly perceived.

β-Damascenone. This compound is found in nearly all wines at concentrations of about 1–4 μg/L. At these concentrations, the compound acts mainly as aroma enhancer, promoting the fruity aroma of wine esters (Escudero et al. 2007; Pineau et al. 2007). However, this compound can be present at much higher concentrations in wines made from sun-dried grapes (Campo et al. 2008) or overripe grapes (Pons et al. 2008). It is a key odor compound in Pedro Ximénez wines (Campo et al. 2008).

Rotundone. This compound is a sesquiterpene responsible for the spicy notes of Shiraz wines and also of black and white pepper, and was recently reported in Australian wines (Wood et al. 2008). This compound elutes out of the chromatographic phases very late, which precluded its identification when the aroma composition of pepper was first addressed. Its odor threshold in water was found to be about 25 ng/L, although 25% of tasters were not sensitive to this compound. Concentrations in Shiraz grapes are highly variable, ranging from less than 10 to more than 600 ng/L (Wood et al. 2008).

4-Mercapto-4-methylpentan-2-one. This compound has a characteristic scent of the box tree (*Buxus* spp.), which can be perceived in some wines made with Sauvignon blanc (Darriet et al. 1991, 1993, 1995) or Scheurebe (Guth 1997a, 1997b). At lower

TABLE 14.3
Wine Odorants Found at Concentrations Higher or Very Close to Their Olfactory Threshold

Odorant	Concentration Range (ppb)	Olfactory Threshold[a] (ppb)	OUV Range	References
Genuine impact odorants				
Formed in grape (varietals)				
Linalool	1.7–1500	15 (G)	0.1–100	Ribéreau-Gayon et al. 1975; Ferreira et al. 2000
		25.2 (F1)	0.07–60	
(*Z*)-Rose oxide	0–21	0.2 (G)	0–105	Guth 1997
β-Damascenone	0.3–25	0.05 (G)	6–500	Guth 1997; Kotseridis et al. 1999; Ferreira et al. 2000; Campo et al. 2008
Rotundone	0–0.14	0.016 (W)	0–8.8	Wood et al. 2008
4-Mercapto-4-methyl-2-pentanone	0–0.4	0.0006 (G)	0–800	Guth 1997; Tominaga et al. 1998; Tominaga et al. 2000
		0.0008 (Tom1)	0–500	
3-Mercaptohexanol	0–12.8	0.060 (Tom1)	0–213	Tominaga et al. 1998; Tominaga et al. 2000
3-Mercaptohexyl acetate	0–0.80	0.004 (Tom1)	0–200	Tominaga et al. 1998; Tominaga et al. 2000
Formed by yeast/bacteria				
2,3-Butanodione (diacetyl)	200–2722	100 (G)	2–27	Hayasaka and Bartowsky 1999; Ferreira et al. 2000
Isoamyl acetate	118–7354	30 (G)	4–245	Ferreira et al. 1995, 2000
Wood/aging related				
(*Z*)-Whiskylactone	46–520	67 (Et)	0.7–7.8	Lopez et al. 2002
Dimethyl sulfide	7–53	10 (G)	0.7–5.3	Guth 1997; Rauhut et al. 1998
2-Furfurylthiol	0–0.14	0.0004 (Tom4)	0–350	Tominaga and Dubourdieu 2006
Benzenemethanethiol	0–0.04	0.0003 (Tom2)	0–133	Tominaga et al. 2003a, 2003b
Methylthiopropanaldehyde (methional)	0–140	0.50 (Et)	0–280	Escudero et al. 2000a
Phenylacetaldehyde	2.4–130	1.0 (Cu)	2.4–130	Cullere et al. 2007
Acetaldehyde	1000–160,000	500 (G)	2–140	Criddle et al. 1983; Guth 1997

(continued)

TABLE 14.3 (Continued)
Wine Odorants Found at Concentrations Higher or Very Close to Their Olfactory Threshold

Odorant	Concentration Range (ppb)	Olfactory Threshold[a] (ppb)	OUV Range	References
3-Hydroxy-4,5-dimethyl-2(5*H*)-furanone (sotolon)	5–207	5 (G)	1–54	Guichard et al. 1993; Cutzach et al. 1999
Relevant families of odorants				
Branched ethyl esters				
Ethyl isobutyrate	30–480	15 (G)	2–32	Baumes et al. 1986; Guth 1997; Ferreira et al. 2000
Ethyl 2-methylbutyrate	1–30	1 (G)	0–30	Guth 1997; Ferreira et al. 2000
		18 (F1)	0–1.7	
Ethyl isovalerate	2–36	3 (G, F1)	0.7–12	Guth 1997; Ferreira et al. 2000
Ethyl 4-methylpentanoate	0–1.43	0.75 (Np)	0–1.9	Campo et al. 2007
Ethyl cyclohexanoate	0–0.05	0.03 (Np)	0–1.7	Campo et al. 2007
Linear ethyl esters				
Ethyl acetate	2000–150,000	7500 (G)	0.27–20	Maarse and Visscher 1989
Ethyl butyrate	69–2194	125 (Np)	0.5–17.5	Ferreira et al. 1995; Guth 1997; Ferreira et al. 2000
Ethyl hexanoate	153–2731	62 (Np)	2.5–44	Baumes et al. 1986; Ferreira et al. 1995; Guth 1997; Ferreira et al. 2000
Ethyl octanoate	138–2636	2 (G)	28–1318	Ferreira et al. 1995; Guth 1997; Ferreira et al. 2000
		5 (F1)	69–527	
Ethyl decanoate	14–821	200 (F1)	0.1–4	Ferreira et al. 1995; Ferreira et al. 2000
Acetates				
Phenylethyl acetate	0.5–744	250 (G)	0–3	Ferreira et al. 1995, 2000
Ethyl phenylacetate		73 (T)		Tat et al. 2007b
Ethyl cinnamates				
Ethyl dihydrocinnamate	0.21–3	1.6 (F1)	0.1–2	Aubry et al. 1997; Ferreira et al. 2000

Ethyl cinnamate	0.11–8.9	5.1 (F1)	0–2	Aubry et al. 1997; Ferreira et al. 2000
γ-Lactones				
γ-nonalactone	3–41	30 (N)	0.1–1.4	Nakamura et al. 1988
Volatile phenols				
Guaiacol	1.1–15	10 (G)	0.1–0.5	Chatonnet and Boidron 1988; Ferreira et al. 2000
Eugenol	0.5–30	6 (F1)	0.1–5	Chatonnet and Boidron 1988
4-Vinylguaiacol	0.2–710	1100 (F1)	0–0.7	Versini and Tomasi 1983; Ferreira et al. 2000
4-Vinylphenol	3–1241	180 (Ch)	0–6.9	Chatonnet and Boidron 1988
4-Ethylguaiacol	0–400	33 (F1)	0–13	Schreier et al. 1980; Ferreira et al. 2000
4-Ethylphenol	0–6480	440 (Ch)	0–15	Chatonnet and Boidron 1988
Terpenols				
α-Terpineol	0.6–145	250 (F1)	0–0.6	Ribéreau-Gayon et al. 1975
Geraniol	0.9–1059	30 (G)	0–35	Ribéreau-Gayon et al. 1975
Ionones				
α-Ionone	0.02–0.54	2.6 (Et)	0–0.2	Kotseridis et al. 1998, 1999; Ferreira et al. 2000
β-Ionone	0.03–0.24	0.09 (F1)	0.3–2.7	Kotseridis et al. 1998, 1999; Ferreira et al. 2000
Burnt sugar compounds				
4-Hydroxy-2,5-dimethyl-3(2*H*)-furanone (furaneol)	0–623	5 (F2)	0–63	Guedes De Pinho et al. 1997; Cutzach et al. 1999
Isoaldehydes				
Methylpropanal	0.9–132	6 (Cu)	0.1–22	Cullere et al. 2007
2-Methylbutanal	3.3–105	16 (Cu)	0.2–6.6	Cullere et al. 2007
3-Methylbutanal	1–49	4.6 (Cu)	0.2–11	Cullere et al. 2007
Pyrazines				
2-Methoxy-3-isopropylpirazine	0–0.002	0.002 (S)	0–2	Lacey et al. 1991; Belancic and Agosin 2007
2-Methoxy-3-(2-methylpropyl)pyrazine	0.04	0.002 (S)	0–20	Lacey et al. 1991; Belancic and Agosin 2007
3-*sec*-butyl-2-methoxypyrazine	0–0.01	0.001 (S)	0–10	Sala et al. 2002

(continued)

TABLE 14.3 (Continued)
Wine Odorants Found at Concentrations Higher or Very Close to Their Olfactory Threshold

Odorant	Concentration Range (ppb)	Olfactory Threshold[a] (ppb)	OUV Range	References
Isoacids				
Isobutyric acid	430–4160	2300 (F1)	0.2–2	Guth 1997; Ferreira et al. 2000
Isovaleric acid	300–1150	33.4 (F1)	9–34	Ferreira et al. 2000
Fusel alcohols				
Isobutanol	25,700–108,900	40,000 (G)	0.6–2.7	Guth 1997; Ferreira et al. 2000
Isoamyl alcohol	72,000–318,000	30,000 (G)	2.4–10.6	Baumes et al. 1986; Ferreira et al. 2000
β-Phenylethanol	18,000–162,000	14,000 (F1)	1.3–11	Baumes et al. 1986; Guth 1997; Ferreira et al. 2000
Methionol	166–2400	500 (G)	0.3–5	Baumes et al. 1986; Ferreira et al. 2000
		1000 (F1)	0.1–2.4	
1-Hexanol	1000–13,200	8000 (G)	0.1–1.7	Baumes et al. 1986; Ferreira et al. 2000
Fatty acids				
Acetic acid	69,000–400,000	200,000 (G)	0.3–2	Ferreira et al. 2000
Butyric acid	434–4720	173 (F1)	2.5–27	Ferreira et al. 2000
Hexanoic acid	200–6200	420 (F1)	0.5–15	Marais and Pool 1980; Shinohara 1985; Ferreira et al. 2000
Octanoic acid	40–7900	500 (F1)	0.1–16	Marais and Pool 1980; Shinohara 1985
Decanoic acid	62–3400	1000 (F1)	0–3.4	Shinohara 1985; Ferreira et al. 2000

Miscellaneous				
Acetoine	600–159,000	150,000 (Et)	0–1	Ferreira et al. 2000)
Furfural	0.002–8.8	14.1 (F1)	0–0.5	Ho et al. 1999; Ferreira et al. 2000
Ethyl lactate	200–382,000	150,000	0–0.25	Simpson and Miller 1984
1,1-Diethoxyethane	0–500	50 (G)	0–10	Guth 1997
Dimethyl trisulfide	0.09–0.25	0.2 (G)	0.5–1.2	Guth 1997
4-Mercapto-4-methyl-2-pentanol	0–0.11	0.055 (Tom1)	0–2	Tominaga et al. 1998
3-Mercapto-2-methyl-propanol	0–10	3 (Bou)	0–3	Bouchilloux et al. 1998a, 1998b
2-Methyl-3-furanthiol	0–0.15	0.004 (Tom3)	0–37.5	Tominaga and Dubourdieu 2006
(*Z*)-3-Hexenol	7.2–850	400 (G)	0–2.1	Baumes et al. 1986; Ferreira et al. 2000
(*E*)-2-Nonenal	0.1–3.7	0.6 (Cu)	0.2–6.1	Cullere et al. 2007
Benzaldehyde	0–313			Escudero et al. 2002
c-6-Dodeceno-γ-lactone	0–3	0.1 (G)	0–3	Guth 1997
3*a*,4,5,7*a*-Tetrahydro-3,6-dimethylbenzofuran-2(3*H*)-one (wine lactone)	0–0.1	0.01 (G)	0–10	Guth 1997

[a] Bou (Bouchilloux et al. 1998a, 1998b); Ch (Chatonnet et al. 1992); Cu (Cullere et al. 2007); Et (Etievant 1991); G (Guth 1997); F1 (Ferreira et al. 2000); F2 (Ferreira et al. 2002); N (Nakamura et al. 1988); Np (not published); Pi (Pickering et al. 2007); Ro (Roujou de Boubee et al. 2000); S (Seibert et al. 1970); T (Tat et al. 2007b); Tom1 (Tominaga et al. 1998); Tom2 (Tominaga et al. 2003a); Tom3 (Tominaga and Dubourdieu 2006); Tom4 (Tominaga et al. 2000); W (Wood et al. 2008).

concentrations, the compound is not strictly speaking an impact compound, but a major contributor to the fresh fruity notes (Escudero et al. 2004; Mateo-Vivaracho et al. 2010).

3-Mercaptohexan-1-ol. This compound has a smell reminiscent of green mango and box tree with some rubbery notes. Its odor is very complex and changes with concentration and with the aroma environment in which it is found. It was first identified in wines from Sauvignon blanc, Cabernet-Sauvignon, and Merlot (Bouchilloux et al. 1998a, 1998b); however, it was later found in many others (Tominaga et al. 2000). It is an impact compound of some rosé wines (Murat et al. 2001; Ferreira et al. 2002), white wines made with Petit Arvine (Fretz et al. 2005), and Sauternes wines (Bailly et al. 2006; Sarrazin et al. 2007b; Campo et al. 2008; Mateo-Vivaracho et al. 2010).

3-Mercaptohexyl acetate. This compound was first found in wines from Sauvignon blanc (Tominaga et al. 1996), but it can also be found in many other wine types (Tominaga et al. 2000; Lopez et al. 2003; Cullere et al. 2004; Gomez-Miguez et al. 2007; Mateo-Vivaracho et al. 2010). It has been recently shown that it is the impact aroma compound of the wines made with the Spanish variety Verdejo, imparting the characteristic tropical fruit aroma nuance to the wine (Campo et al. 2005).

14.3.2 Fermentative Impact Compounds

Diacetyl. This compound is another odorant playing a complex role on wine aroma. It was one of the first identified wine aroma molecules (Fornacho and Lloyd 1965), and it has often been blamed as the cause of a defect when it is present at high concentrations (Clarke and Bakker 2004). Its sensory effect is extremely dependent on the type of wine (Martineau et al. 1995a; Bartowsky et al. 2002), and its concentration is also time dependent and related to the concentration of sulfur dioxide in the wine (Nielsen and Richelieu 1999). Diacetyl is responsible for the buttery note appreciated in some Chardonnay wines (Martineau et al. 1995b; Bartowsky et al. 2002), and its role in the sweet notes of some Port wines has also been suggested (Rogerson et al. 2001). Several authors agree on its ambiguous character (Lonvaud-Funel 1999; Bartowsky and Henschke 2004).

Isoamyl acetate. This is the only ester capable of imparting its characteristic aroma nuance to wines, sometimes too overtly. In wines made with Pinotage or Tempranillo varieties, it is a characteristic aroma compound (Van Wyk et al. 1979; Ferreira et al. 2000).

14.3.3 Age-Related Impact Aroma Compounds

(E)-Whiskylactone. This is an impact compound in wines aged in oak wood (Boidron et al. 1988). Above a given concentration, it can produce an excessive and unpleasant woody characteristic (Pollnitz et al. 2000).

Dimethyl sulfide (DMS). This compound was identified some time ago in aged wines (Marais 1979) and apparently plays an ambiguous role in wine aroma. Quite often it is related to a defect (sulfury odor) (Park et al. 1994; Ferreira et al. 2003c), but other groups have demonstrated that it exerts a powerful enhancing effect on the fruity note of some highly appreciated red wines (Segurel et al. 2004; Escudero et al. 2007).

Furfurylthiol (FFT, or 2-furanmethanethiol). This strong coffee-smelling compound is formed by reaction between furfural from the oak cask and sulfydric acid formed during the fermentation (Blanchard et al. 2001), and is able to transmit its aroma to some types of wine. There are no substantial analytical data on the occurrence of FFT because of difficulties in its determination, but it has been found at relatively high concentrations in aged wines from Champagne (Tominaga et al. 2003b) and in other wines (Tominaga and Dubourdieu 2006; Mateo-Vivaracho 2009).

Benzyl mercaptan (or benzenemethanethiol). This is a compound with a powerful toasty aroma, and together with FFT can impart smoky and empyreumatic nuances to some aged wines, such as Champagne or Chardonnay sur lie (Tominaga et al. 2003a, 2003b) but also to normal aged dry wines (Mateo-Vivaracho et al. 2010).

Methional (3-(methylthio)propanal). This compound also plays an ambiguous role. In young white wines, it causes unpleasant odors (Escudero et al. 2000c), but in complex wines, such as some Chardonnays or some great red wines, is a net contributor to some appreciated odor nuances (Ferreira et al. 2005).

Phenylacetaldehyde. This is also a compound with an ambiguous role. Its smell of honey is very pleasant but gives to the wine oxidation notes that are considered defective and that depress fruitiness (Aznar et al. 2003; Ferreira et al. 2003b). However, this compound can act as impact compound in Sauternes or Pedro Ximénez wines, in which it is found at very high concentrations (Sarrazin et al. 2007a; Campo et al. 2008).

Acetaldehyde is a well-known compound of Sherry-like wines (Criddle et al. 1983).

Sotolon (3-hydroxy-4,5-dimethyl-2(5H)-furanone). This is also an impact compound in wines made with botrytized grapes (Masuda et al. 1984), or wines from biological aging (Martin et al. 1990, 1992; Moreno et al. 2005), natural sweet wines (Cutzach et al. 1998, 1999), Pedro Ximénez (Campo et al. 2008), Oporto (Ferreira et al. 2003a), or Madeira (Camara et al. 2004). Its concentration, in general, increases with oxidation (Escudero et al. 2000a).

All the aforementioned aromas, at lesser concentrations, do not play a role of impact compound, but that of major, net, or even subtle contributor to some aroma nuance related to one of its more or less general or specific aroma descriptors.

In addition to these key compounds, wine also contains some groups of compounds sharing aromatic characteristics and share common formation pathways. These groups of compounds can act more or less additively, which means that even when they do not reach high Odor Activity Values, they can be important contributors to some wine aroma nuances. Because some of the compounds in the families are present even at subthreshold concentrations, they are not included in the table. Only some families will be briefly commented.

Ethyl esters of branched or cyclic fatty acids (ethyl isobutyrate, 2-, 3-methylbutyrate, 2-, 3- and 4-methylpentanoates, and ethyl cyclohexanoate) (Campo et al. 2006), some of which have been recently identified. The aroma of these compounds could act additively and contribute to the fruity notes of red wines, as has been recently suggested (Ferreira et al. 2009). Apparently, the linear/branched ratio has a strong influence on aroma quality (Pineau et al. 2009).

Ethyl esters of fatty acids, responsible for fruity notes (apple like, ester like) of some white wines (Ferreira et al. 1995).

Fusel alcohol acetates, which can contribute to the flowery and/or fruity notes of white wines (Campo et al. 2005).

Ethyl cinnamate and ethyl dihydrocinnamate, which can contribute to the sweet and floral notes of some wines, particularly Chardonnays (Loscos et al. 2007, 2009).

Aliphatic γ-lactones, which contribute to the peachy aroma of some reds (Ferreira et al. 2004; Jarauta 2004), but can also be contributors to the sweet nuances of many other wines (Loscos et al. 2007).

Volatile phenols such as guaiacol, eugenol, 2,6-dimethoxyphenol, *iso*-eugenol, and allyl-2,6-dimethoxyphenol, which are responsible for phenolic and toasted notes of wines (Escudero et al. 2007). Vinyl phenols and ethyl phenols constitute two specific families of compounds with deep and generally negative effects on the aroma of white and red wines, respectively (Chatonnet et al. 1992).

Some families able to add or contribute substantially to some wine odor nuances are the burnt-sugar compounds (furaneol, homofuraneol, maltol, the aliphatic and branched aldehydes, terpenols, ionones, or methoxypyrazines. Other relevant groups of compounds are acids (branched and linear) and fusel alcohols. The role of these compounds is more subtle, since together with ethanol and some ethyl esters, they constitute the base of wine aroma. They are not clearly perceived, but their aroma is integrated into the general vinous perception. The absolute and relative composition of this base has a deep, but not well understood, effect on the perception of the other compounds.

14.4 CONCLUSION

In summary, wine flavor is determined by the aroma compounds described in Table 14.3, which roughly can play three major roles: compounds in the base, families of aroma compounds, and potentially impact compounds. The base, together with the nonvolatile matrix, has a deep effect on the easiness with which a given aroma nuance will be perceived, whereas families and impact compounds constitute the different aroma nuances that can be perceived in wine aroma.

REFERENCES

Arrhenius, S.P., L.P. McCloskey, et al. 1996. Chemical markers for aroma of *Vitis vinifera* var Chardonnay regional wines. *J. Agric. Food Chem.* 44(4): 1085–1090.

Aubry, V., P.X. Etievant, et al. 1997. Quantitative determination of potent flavor compounds in Burgundy Pinot noir wines using a stable isotope dilution assay. *J. Agric. Food Chem.* 45(6): 2120–2123.

Aznar, M., R. Lopez, et al. 2003. Prediction of aged red wine aroma properties from aroma chemical composition. Partial least squares regression models. *J. Agric. Food Chem.* 51(9): 2700–2707.

Aznar, M., R. Lopez, et al. 2001. Identification and quantification of impact odorants of aged red wines from Rioja. GC-olfactometry, quantitative GC-MS, and odor evaluation of HPLC fractions. *J. Agric. Food Chem.* 49(6): 2924–2929.

Baek, H.H., K.R. Cadwallader, et al. 1997. Identification of predominant aroma compounds in muscadine grape juice. *J. Food Sci.* 62: 249–252.

Bailly, S., V. Jerkovic, et al. 2006. Aroma extraction dilution analysis of Sauternes wines. Key role of polyfunctional thiols. *J. Agric. Food Chem.* 54(19): 7227–7234.

Bartowsky, E.J., I.L. Francis, et al. 2002. Is buttery aroma perception in wines predictable from the diacetyl concentration? *Aust. J. Grape Wine Res.* 8(3): 180–185.

Bartowsky, E.J., and P.A. Henschke. 2004. The 'buttery' attribute of wine-diacetyl-desirability, spoilage and beyond. *Int. J. Food Microbiol.* 96(3): 235–252.

Baumes, R., R. Cordonnier, et al. 1986. Identification and determination of volatile constituents in wines from different vine cultivars. *J. Sci. Food Agric.* 37(9): 927–943.

Belancic, A., and E. Agosin. 2007. Methoxypyrazines in grapes and wines of *Vitis vinifera* cv. Carmenere. *Am. J. Enol. Vitic.* 58(4): 462–469.

Blanchard, L., T. Tominaga, et al. 2001. Formation of furfurylthiol exhibiting a strong coffee aroma during oak barrel fermentation from furfural released by toasted staves. *J. Agric. Food Chem.* 49(10): 4833–4835.

Boidron, J.N., P. Chatonnet, et al. 1988. Influence du bois sur certaines substances odorantes des vins. *Connaiss. Vigne Vin* 22(4): 275–294.

Bouchilloux, P., P. Darriet, et al. 1998a. Identification of a very odoriferous thiol, 2 methyl-3-furanthiol, in wines. *Vitis* 37(4): 177–180.

Bouchilloux, P., P. Darriet, et al. 1998b. Identification of volatile and powerful odorous thiols in Bordeaux red wine varieties. *J. Agric. Food Chem.* 46(8): 3095–3099.

Buettner, A. 2004. Investigation of potent odorants and afterodor development in two Chardonnay wines using the buccal odor screening system (BOSS). *J. Agric. Food Chem.* 52(8): 2339–2346.

Buser, H.R., C. Zanier, et al. 1982. Identification of 2,4,6-trichloroanisole as a potent compound causing cork taint in wine. *J. Agric. Food Chem.* 30(2): 359–362.

Camara, J.S., J.C. Marques, et al. 2004. 3-Hydroxy-4,5-dimethyl-2(5*H*)-furanone levels in fortified Madeira wines: Relationship to sugar content. *J. Agric. Food Chem.* 52(22): 6765–6769.

Campo, E., J. Cacho, et al. 2006. Multidimensional chromatographic approach applied to the identification of novel aroma compounds in wine—Identification of ethyl cyclohexanoate, ethyl 2-hydroxy-3-methylbutyrate and ethyl 2-hydroxy-4-methylpentanoate. *J. Chromatogr. A* 1137(2): 223–230.

Campo, E., J. Cacho, et al. 2007. Solid phase extraction, multidimensional gas chromatography mass spectrometry determination of four novel aroma powerful ethyl esters—Assessment of their occurrence and importance in wine and other alcoholic beverages. *J. Chromatogr. A* 1140(1–2): 180–188.

Campo, E., J. Cacho, et al. 2008. The chemical characterization of the aroma of dessert and sparkling white wines (Pedro Ximenez, Fino, Sauternes, and Cava) by gas chromatography–olfactometry and chemical quantitative analysis. *J. Agric. Food Chem.* 56(7): 2477–2484.

Campo, E., V. Ferreira, et al. 2005. Prediction of the wine sensory properties related to grape variety from dynamic-headspace gas chromatography–olfactometry data. *J. Agric. Food Chem.* 53(14): 5682–5690.

Campo, E., V. Ferreira, et al. 2006. Identification of three novel compounds in wine by means of a laboratory-constructed multidimensional gas chromatographic system. *J. Chromatogr. A* 1122(1–2): 202–208.

Chatonnet, P., C. Barbe, et al. 1993. Origines et incidences organoleptiques de phenols volatils dans les vins. Application à la maîtrise de la vinification et de l'élevage. *Revue Française Oenologie, Montpellier*: 279–287.

Chatonnet, P., and J.N. Boidron. 1988. Dosage de phenols volatils dans les vins par chromatographic en phaxe gazeuse. *Sci. Aliments* 8: 479–488.

Chatonnet, P., S. Bonnet, et al. 2004. Identification and responsibility of 2,4,6-tribromoanisole in musty, corked odors in wine. *J. Agric. Food Chem.* 52(5): 1255–1262.

Chatonnet, P., D. Dubourdieu, et al. 1992. The origin of ethylphenols in wines. *J. Sci. Food. Agric.* 60(2): 165–178.

Chisholm, M.G., L.A. Guiher, et al. 1995. Aroma characteristics of aged Vidal Blanc Wine. *Am. J. Enol. Vitic.* 46(1): 56–62.

Clarke, R.J., and J. Bakker. 2004. *Wine Flavour Chemistry*. Oxford, UK: Blackwell Publishing.

Cordonnier, R., and C.L. Bayonove. 1974. Mise en evidence dans la baie de raisin, var. Muscat d'Alexandrie, de monoterpenes lies revelables par une ou plusieurs enzymes du fruit. *C.R. Acad. Sci. Paris (Serie D)* 278: 3387–3390.

Criddle, W.J., R.W. Goswell, et al. 1983. The chemistry of sherry maturation. 2. An investigation of the volatile components present in standard sherry base wine. *Am. J. Enol. Vitic.* 34(2): 61–71.

Cullere, L., J. Cacho, et al. 2007. An assessment of the role played by some oxidation-related aldehydes in wine aroma. *J. Agric. Food Chem.* 55(3): 876–881.

Cullere, L., A. Escudero, et al. 2004. Gas chromatography–olfactometry and chemical quantitative study of the aroma of six premium quality Spanish aged red wines. *J. Agric. Food Chem.* 52(6): 1653–1660.

Cullere, L., A. Escudero, et al. 2008. 2-Methyl-3-(methyldithio)furan: A new odorant identified in different monovarietal red wines from the Canary Islands and aromatic profile of these wines. *J. Food Compos. Anal.* 21(8): 708–715.

Cutzach, I., P. Chatonnet, et al. 1999. Study of the formation mechanisms of some volatile compounds during the aging of sweet fortified wines. *J. Agric. Food Chem.* 47(7): 2837–2846.

Cutzach, I., P. Chatonnet, et al. 1997. Identification of volatile compounds with a "toasty" aroma in heated oak used in barrelmaking. *J. Agric. Food Chem.* 45(6): 2217–2224.

Cutzach, I., P. Chatonnet, et al. 1998. [Study in aroma of sweet natural non muscat wines. 2nd part: Quantitative analysis of volatile compounds taking part in aroma of sweet natural wines during ageing] Etude sur l'arôme des vins doux naturels non muscatés. 2e partie: Dosages de certains composés volatils intervenant dans l'arôme des vins doux naturels au cours de leur élevage et de leur vieillissement. *J. Int. Sci. Vigne Vin* 32(4): 211–221.

Darriet, P., V. Lavigne, et al. 1991. [Characterization of Sauvignon blanc wine varietal aroma by gas-chromatography–sniffing techniques] Caracterisation de l'arome varietal des vins de Sauvignon par couplage chromatographique en phase gazeuse–odometrie. *J. Int. Sci. Vigne Vin* 25(3): 167–174.

Darriet, P., T. Tominaga, et al. 1993. Mise en évidence dans le raisin de Vitis vinifera var. Sauvignon d'un précurseur de la 4-mercapto-4-méthylpentan-2-one. *C.R. Acad. Sci. Paris (Ser. D)* 316: 1332–1335.

Darriet, P., T. Tominaga, et al. 1995. Identification of a powerful aromatic component of *Vitis vinifera* L. var. Sauvignon wines: 4-Mercapto-4-methylpentan-2-one. *Flavour Fragrance J.* 10(6): 385–392.

Dufour, C., and C.L. Bayonove. 1999a. Influence of wine structurally different polysaccharides on the volatility of aroma substances in a model system. *J. Agric. Food Chem.* 47(2): 671–677.

Dufour, C., and C.L. Bayonove. 1999b. Interactions between wine polyphenols and aroma substances. An insight at the molecular level. *J. Agric. Food Chem.* 47(2): 678–684.

Dufour, C., and I. Sauvaitre. 2000. Interactions between anthocyanins and aroma substances in a model system. Effect on the flavor of grape-derived beverages. *J. Agric. Food Chem.* 48(5): 1784–1788.

Dzhakhua, M.Y., E.S. Drboglav, et al. 1978. Study of volatile phenol compounds in white wines. *Prikl. Biokhim. Mikrobiol.* 14(1): 156–158.

Escudero, A., E. Asensio, et al. 2002. Sensory and chemical changes of young white wines stored under oxygen. An assessment of the role played by aldehydes and some other important odorants. *Food Chem.* 77(3): 325–331.

Escudero, A., J. Cacho, et al. 2000a. Isolation and identification of odorants generated in wine during its oxidation: A gas chromatography–olfactometric study. *Eur. Food Res. Technol.* 211(2): 105–110.

Escudero, A., E. Campo, et al. 2007. Analytical characterization of the aroma of five premium red wines. Insights into the role of odor families and the concept of fruitiness of wines. *J. Agric. Food Chem.* 55(11): 4501–4510.

Escudero, A., M. Charpentier, et al. 2000b. Characterization of aged champagne wine aroma by GC-O and descriptive profile analyses. *Sci. Aliments* 20(3): 331–346.

Escudero, A., and P. Etievant. 1999. Effect of antioxidants on the flavor characteristics and the gas chromatography/olfactometry profiles of champagne extracts. *J. Agric. Food Chem.* 47(8): 3303–3308.

Escudero, A., B. Gogorza, et al. 2004. Characterization of the aroma of a wine from Maccabeo. Key role played by compounds with low odor activity values. *J. Agric. Food Chem.* 52(11): 3516–3524.

Escudero, A., P. Hernandez-Orte, et al. 2000c. Clues about the role of methional as character impact odorant of some oxidized wines. *J. Agric. Food Chem.* 48(9): 4268–4272.

Etievant, P.X. 1991. Wine. In *Volatile Compounds of Food and Beverages*. ed. H. Maarse. New York: Marcel Dekker.

Ferreira, V., M. Aznar, et al. 2001. Quantitative gas chromatography–olfactometry carried out at different dilutions of an extract. Key differences in the odor profiles of four high-quality Spanish aged red wines. *J. Agric. Food Chem.* 49(10): 4818–4824.

Ferreira, A.C.S., J.C. Barbe, et al. 2003a. 3-Hydroxy-4,5-dimethyl-2(5*H*)-furanone: A key odorant of the typical aroma of oxidative aged Port wine. *J. Agric. Food Chem.* 51(15): 4356–4363.

Ferreira, V., and J. Cacho. 2009. Identification of impact odorant of wines. *Wine Chemistry and Biochemistry*, ed. M.V. Moreno-Arribas and M.C. Polo, 393–415. Berlin: Springer.

Ferreira, V., P. Fernandez, et al. 1995. Investigation on the role played by fermentation esters in the aroma of young Spanish wines by multivariate-analysis. *J. Sci. Food Agric.* 67(3): 381–392.

Ferreira, A.C.S., T. Hogg, et al. 2003b. Identification of key odorants related to the typical aroma of oxidation-spoiled white wines. *J. Agric. Food Chem.* 51(5): 1377–1381.

Ferreira, V., I. Jarauta, et al. 2004. A simple strategy for the optimization of solid-phase-extraction procedures through the use of solid–liquid distribution coefficients. Application to the determination of aliphatic lactones in wine. *J. Chromatogr. A* 1025: 147–156.

Ferreira, V., R. López, et al. 2001. Olfactometry and aroma extract dilution analysis of wines. *Analysis of Taste and Aroma*, volume 21, ed. J. Jackson, 89–122. Berlin: Springer-Verlag.

Ferreira, V., R. Lopez, et al. 2000. Quantitative determination of the odorants of young red wines from different grape varieties. *J. Sci. Food Agric.* 80(11): 1659–1667.

Ferreira, V., N. Ortin, et al. 2002. Chemical characterization of the aroma of Grenache rose wines: Aroma extract dilution analysis, quantitative determination, and sensory reconstitution studies. *J. Agric. Food Chem.* 50(14): 4048–4054.

Ferreira, A.C.S., P. Rodrigues, et al. 2003c. Influence of some technological parameters on the formation of dimethyl sulfide, 2-mercaptoethanol, methionol, and dimethyl sulfone in port wines. *J. Agric. Food Chem.* 51(3): 727–732.

Ferreira, V., F. San Juan, et al. 2009. Modeling quality of premium Spanish red wines from gas chromatography–olfactometry data. *J. Agric. Food Chem.* 57(16): 7490–7498.

Ferreira, V., M. Torres, et al. 2005. Aroma composition and aromatic structure of red wines made with Merlot. *State of the art in Flavour Chemistry and Biology, Proceedings from the 7th Wartburg Symposium*. P.S.T. Hofman, Deutsche Forsch. Lebensm. Garching, pp. 292–299.

Fornacho, J.C., and B. Lloyd. 1965. Bacterial production of diacetyl and acetoin in wine. *J. Sci. Food Agric.* 16(12): 710–716.

Fretz, C.B., J.L. Luisier, et al. 2005. 3-Mercaptohexanol: An aroma impact compound of Petite Arvine wine. *Am. J. Enol. Vitic.* 56(4): 407–410.

Gomez-Miguez, M.J., J.F. Cacho, et al. 2007. Volatile components of Zalema white wines. *Food Chem.* 100(4): 1464–1473.

Guedes De Pinho, P., A.A. Beloqui, et al. 1997. Detection of a sulfur compound responsible for the typical aroma of some non *Vitis vinifera* wines. *Sci. Aliments* 17(4): 341–348.

Guichard, E., T.T. Pham, et al. 1993. Quantitative-determination of sotolon in wines by high-performance liquid-chromatography. *Chromatographia* 37(9–10): 539–542.

Guth, H. 1997a. Identification of character impact odorants of different white wine varieties. *J. Agric. Food Chem.* 45(8): 3022–3026.

Guth, H. 1997b. Quantitation and sensory studies of character impact odorants of different white wine varieties. *J. Agric. Food Chem.* 45(8): 3027–3032.

Hayasaka, Y., and E.J. Bartowsky. 1999. Analysis of diacetyl in wine using solid-phase microextraction combined with gas chromatography mass spectrometry. *J. Agric. Food Chem.* 47(2): 612–617.

Ho, P., T.A. Hogg, et al. 1999. Application of a liquid chromatographic method for the determination of phenolic compounds and furans in fortified wines. *Food Chem.* 64(1): 115–122.

Hufnagel, J.C., and T. Hofmann. 2008a. Orosensory-directed identification of astringent mouthfeel and bitter-tasting compounds in red wine. *J Agric. Food Chem.* 56(4): 1376–1386.

Hufnagel, J.C., and T. Hofmann. 2008b. Quantitative reconstruction of the nonvolatile sensometabolome of a red wine. *J Agric. Food Chem.* 56(19): 9190–9199.

Ibarz, M.J., V. Ferreira, et al. 2006. Optimization and evaluation of a procedure for the gas chromatographic–mass spectrometric analysis of the aromas generated by fast acid hydrolysis of flavor precursors extracted from grapes. *J. Chromatogr. A* 1116(1–2): 217–229.

Jarauta, I. 2004. Estudio analítico de fenómenos concurrentes en la generación del aroma durante la crianza del vino en barricas de roble con diferentes grados de uso. Nuevos métodos de análisis de importantes aromas y caracterización de su papel sensorial. *Analytical Chemistry*. Zaragoza: University of Zaragoza.

Komes, D., D. Ulrich, et al. 2006. Characterization of odor-active compounds in Croatian Rhine Riesling wine, subregion Zagorje. *Eur. Food Res. Technol.* 222(1–2): 1–7.

Kotseridis, Y., and R. Baumes. 2000. Identification of impact odorants in Bordeaux red grape juice, in the commercial yeast used for its fermentation, and in the produced wine. *J. Agric. Food Chem.* 48(2): 400–406.

Kotseridis, Y., R.L. Baumes, et al. 1999. Quantitative determination of free and hydrolytically liberated beta-damascenone in red grapes and wines using a stable isotope dilution assay. *J. Chromatogr. A* 849(1): 245–254.

Kotseridis, Y., A.A. Beloqui, et al. 1998. An analytical method for studying the volatile compounds of Merlot noir clone wines. *Am. J. Enol. Vitic.* 49(1): 44–48.

Lacey, M.J., M.S. Allen, et al. 1991. Methoxypyrazines in Sauvignon Blanc grapes and wines. *Am. J. Enol. Vitic.* 42(2): 103–108.

Le Fur, Y., and P. Etievant. 1998. Donneés relatives à l'étude de quatre composés volatils supectés d'intervenir sur la qualité d'odeur et d'arôme des vins de Chardonnay de Bourgogne: l'example d'une démarche analytique. *Rev. Oenol.* 88(13–16).

Lee, S.J., and A.C. Noble. 2003. Characterization of odor-active compounds in Californian Chardonnay wines using GC-olfactometry and GC-mass spectrometry. *J. Agric. Food Chem.* 51(27): 8036–8044.

Lonvaud-Funel, A. 1999. Lactic acid bacteria in the quality improvement and depreciation of wine. *Antonie Van Leeuwenhoek Int. J. Gen. Mol. Microbiol.* 76(1): 317–331.

Lopez, R., M. Aznar, et al. 2002. Determination of minor and trace volatile compounds in wine by solid-phase extraction and gas chromatography with mass spectrometric detection. *J. Chromatogr. A* 966(1–2): 167–177.

Lopez, R., V. Ferreira, et al. 1999. Identification of impact odorants of young red wines made with Merlot, Cabernet Sauvignon and Grenache grape varieties: A comparative study. *J. Sci. Food Agric.* 79(11): 1461–1467.

Lopez, R., N. Ortin, et al. 2003. Impact odorants of different young white wines from the Canary Islands. *J. Agric. Food Chem.* 51(11): 3419–3425.

Loscos, N., P. Hernandez-Orte, et al. 2007. Release and formation of varietal aroma compounds during alcoholic fermentation from nonfloral grape odorless flavor precursors fractions. *J. Agric. Food Chem.* 55: 6674–6684.

Loscos, N., P. Hernandez-Orte, et al. 2009. Comparison of the suitability of different hydrolytic strategies to predict aroma potential of different grape varieties. *J. Agric. Food Chem.* 57(6): 2468–2480.

Maarse, H., and C.A. Visscher. 1989. Volatile compounds in food, alcoholic beverages. Aj Zeist: TNO-CIVO Food Analysis Institute.

Marais, J. 1979. Effect of storage time and temperature on the formation of dimethyl sulfide and on white wine quality. *Vitis* 18(3): 254–260.

Marais, J., and H.J. Pool. 1980. Effect of storage time and temperature on the volatile composition and quality of dry white table wines. *Vitis* 19(2): 151–164.

Martin, B., P. Etievant, et al. 1990. The chemistry of sotolon: A key parameter for the study of a key component of Flor Sherry wines. *Flavour Science and Technology*, ed. Y. Bessière and A.F. Thomas, 53–56. Chichester, UK: Wiley.

Martin, B., P.X. Etievant, et al. 1992. More clues about sensory impact of sotolon in some flor sherry wines. *J. Agric. Food Chem.* 40(3): 475–478.

Martineau, B., T.E. Acree, et al. 1995a. Effect of wine type on the detection threshold for diacetyl. *Food Res. Int.* 28(2): 139–143.

Martineau, B., T. Henickkling, et al. 1995b. Reassessment of the influence of malolactic fermentation on the concentration of diacetyl in wines. *Am. J. Enol. Vitic.* 46(3): 385–388.

Masuda, M., E. Okawa, et al. 1984. Identification of 4,5-dimethyl-3-hydroxy-2(5*H*)-furanone (Sotolon) and ethyl 9-hydroxynonanoate in botrytised wine and evaluation of the roles of compounds characteristic of it. *Agric. Biol. Chem.* 48(11): 2707–2710.

Mateo-Vivaracho, L. 2009. Desarrollo de nuevos métodos de análisis cuantitativo de mercaptanos de alto impacto aromático en vino. Aplicación de novedosas estrategias analíticas de extracción, aislamiento y derivatización. *Analytical Chemistry*, PhD. Zaragoza: University of Zaragoza.

Mateo-Vivaracho, L., J. Zapata, et al. 2010. Analysis, occurrence and potential sensory significance of five polyfunctional mercaptans in white wines. *J. Agric. Food Chem.* 58: 10184–10194.

Moio, L., and P.X. Etievant. 1995. Ethyl anthranilate, ethyl cinnamate, 2,3-dihydrocinnamate, and methyl anthranilate—Important odorants identified in Pinot-Noir Wines of Burgundy. *Am. J. Enol. Vitic.* 46(3): 392–398.

Molyneux, R.J., and P. Schieberle. 2007. Compound identification: A journal of agricultural and food chemistry perspective. *J. Agric. Food Chem.* 55(12): 4625–4629.

Moreno, J.A., L. Zea, et al. 2005. Aroma compounds as markers of the changes in sherry wines subjected to biological ageing. *Food Contr.* 16(4): 333–338.

Murat, M., T. Tominaga, et al. 2001. Mise en évidence de composés clefs dans l'arôme des vins rosés et clairets de Bordeaux. *J. Int. Sci. Vigne Vin* 35(2): 99–105.

Nakamura, S., E.A. Crowell, et al. 1988. Quantitative analysis of g-nonalactone in wines and its threshold determination. *J. Food Sci.* 53: 1243–1244.

Nielsen, J.C., and M. Richelieu. 1999. Control of flavor development in wine during and after malolactic fermentation by *Oenococcus oeni*. *Appl. Environ. Microbiol.* 65(2): 740–745.

Palomo, E.S., M.S. Perez-Coello, et al. 2006. Contribution of free and glycosidically-bound volatile compounds to the aroma of muscat. A petit grains wines and effect of skin contact. *Food Chem.* 95(2): 279–289.

Park, S.K., R.B. Boulton, et al. 1994. Incidence of volatile sulfur-compounds in California wines—A preliminary survey. *Am. J. Enol. Vitic.* 45(3): 341–344.

Petka, J., V. Ferreira, et al. 2006. Sensory and chemical characterization of the aroma of a white wine made with Devin grapes. *J. Agric. Food Chem.* 54(3): 909–915.

Pickering, G.J., A. Karthik, et al. 2007. Determination of ortho- and retronasal detection thresholds for 2-isopropyl-3-methoxypyrazine in wine. *J. Food Sci.* 72: S468–S472.

Pineau, B., J.C. Barbe, et al. 2007. Which impact for beta-damascenone on red wines aroma? *J. Agric. Food Chem.* 55(10): 4103–4108.

Pineau, B., J.C. Barbe, et al. 2009. Examples of perceptive interactions involved in specific "Red-" and "Black-berry" aromas in red wines. *J. Agric. Food Chem.* 57(9): 3702–3708.

Pollnitz, A.P., K.H. Pardon, et al. 2000. 4-Ethylphenol, 4-ethylguaiacol and oak lactones in Australian red wines. *Aust. Grapegrow. Winemak.* 438(45): 47–50.

Pons, A., V. Lavigne, et al. 2008. Identification of volatile compounds responsible for prune aroma in prematurely aged red wines. *J. Agric. Food Chem.* 56(13): 5285–5290.

Rauhut, D., H. Kurbel, et al. 1998. Headspace GC-SCD monitoring of low volatile sulfur compounds during fermentation and in wine. *Analysis* 26(3): 142–145.

Ribéreau-Gayon, P., J.N. Boidron, et al. 1975. Aroma of muscat grape varieties. *J. Agric. Food Chem.* 23(6): 1042–1047.

Robinson, A.L., S.E. Ebeler, et al. 2009. Interactions between wine volatile compounds and grape and wine matrix components influence aroma compound headspace partitioning. *J. Agric. Food Chem.* 57(21): 10313–10322.

Rogerson, F.S.S., Z. Azevedo, et al. 2002. 1,3-Dimethoxybenzene, a newly identified component of port wine. *J. Sci. Food Agric.* 82(11): 1287–1292.

Rogerson, F.S.S., H. Castro, et al. 2001. Chemicals with sweet aroma descriptors found in Portuguese wines from the Douro region: 2,6,6-Trimethylcyclohex-2-ene-1,4-dione and diacetyl. *J. Agric. Food Chem.* 49(1): 263–269.

Roujou de Boubee, D., C. Van Leeuwen, et al. (2000). Organoleptic impact of 2-methoxy-3-isobutylpyrazine on red bordeaux and loire wines. Effect of environmental conditions on concentrations in grapes during ripening. *J. Agric. Food Chem.* 48(10): 4830–4834.

Saenz-Navajas, M.P., E. Campo, et al. 2010a. Effects of the nonvolatile matrix on the aroma perception of wine. *J. Agric. Food Chem.* 58(9): 5574–5585.

Saenz-Navajas, M.P., E. Campo, et al. 2010b. An assessment of the effects of wine volatiles on the perception of taste and astringency in wine. *Food Chem.* 121(4): 1139–1149.

Saenz-Navajas, M.P., V. Ferreira, et al. 2010c. Characterization of taste-active fractions in red wine combining HPLC fractionation, sensory analysis and ultra performance liquid chromatography coupled with mass spectrometry detection. *Anal. Chim. Acta* 673(2): 151–159.

Sala, C., M. Mestres, et al. 2002. Headspace solid-phase microextraction analysis of 3-alkyl-2-methoxypyrazines in wines. *J. Chromatogr. A* 953(1–2): 1–6.

San Juan, F., J. Pet'ka, et al. 2010. Producing headspace extracts for the gas chromatography–olfactometric evaluation of wine aroma. *Food Chem.* 123(1): 188–195.

Sarrazin, E., D. Dubourdieu, et al. 2007a. Characterization of key-aroma compounds of botrytized wines, influence of grape botrytization. *Food Chem.* 103(2): 536–545.

Sarrazin, E., S. Shinkaruk, et al. 2007b. Odorous impact of volatile thiols on the aroma of young botrytized sweet wines: Identification and quantification of new sulfanyl alcohols. *J. Agric. Food Chem.* 55(4): 1437–1444.

Schreier, J.P., P. Drawert, et al. 1980. Identification and determination of volatile constituents in Burgundy Pinot Noir. *Lebensm.-Wiss. Technol.* 13: 318–321.

Segurel, M.A., A.J. Razungles, et al. 2004. Contribution of dimethyl sulfide to the aroma of Syrah and Grenache Noir wines and estimation of its potential in grapes of these varieties. *J. Agric. Food Chem.* 52(23): 7084–7093.

Seibert, R.M., R.G. Buttery, et al. 1970. Syntesis of some 2-methoxy-3-alkylpyrazines with strong bell pepper-like odors. *J. Agric. Food Chem.* 18: 246–249.

Shinohara, T. 1985. Gas chromatography analysis of volatile fatty acids in wines. *Agric. Biol. Chem.* 49: 2211–2212.

Silva Ferreira, A.C., T. Hogg, et al. 2003. Identification of key odorants related to the typical aroma of oxidation-spoiled white wines. *J. Agric. Food Chem.* 51(5): 1377–1381.

Simpson, R.F., D.L. Capone, et al. 2004. Isolation and identification of 2-methoxy-3,5-dimethylpyrazine, a potent musty compound from wine corks. *J. Agric. Food Chem.* 52(17): 5425–5430.

Simpson, R.F., and G.C. Miller. 1984. Aroma composition of chardonnay wine. *Vitis* 23(2): 143–158.

Tat, L., F. Battistutta, et al. 2007a. Ethyl phenylacetate as the probable responsible of honey-like character in Aglianico del vulture wine. *Proceedings of the International Workshop on Advances in Grapevine and Wine Research*, pp. 557–562. V.G.P. G.C. Nuzzo.

Tat, L., P. Comuzzo, et al. 2007b. Sweet-like off-flavor in aglianico del vulture wine: Ethyl phenylacetate as the mainly involved compound. *J. Agric. Food Chem.* 55(13): 5205–5212.

Tominaga, T., R. Baltenweck-Guyot, et al. 2000. Contribution of volatile thiols to the aromas of white wines made from several *Vitis vinifera* grape varieties. *Am. J. Enol. Vitic.* 51(2): 178–181.

Tominaga, T., L. Blanchard, et al. 2000. A powerful aromatic volatile thiol, 2-furanmethanethiol, exhibiting roast coffee aroma in wines made from several *Vitis vinifera* grape varieties. *J. Agric. Food Chem.* 48(5): 1799–1802.

Tominaga, T., P. Darriet, et al. 1996. Identification of 3-mercaptoethanol acetate in Sauvignon wine, a powerful aromatic compound exhibiting box-tree odor. [Identification del'acétate de 3-mercaptohexanol, composé a forte odeur de buis, intervenant dans l'arôme des vins de Sauvignon.] *Vitis* 35(4): 207–210.

Tominaga, T., and D. Dubourdieu. 2006. A novel method for quantification of 2-methyl-3-furanthiol and 2-furanmethanethiol in wines made from *Vitis vinifera* grape varieties. *J. Agric. Food Chem.* 54(1): 29–33.

Tominaga, T., G. Guimbertau, et al. 2003a. Contribution of benzenemethanethiol to smoky aroma of certain *Vitis vinifera* L. wines. *J. Agric. Food Chem.* 51(5): 1373–1376.

Tominaga, T., G. Guimbertau, et al. 2003b. Role of certain volatile thiols in the bouquet of aged Champagne wines. *J. Agric. Food Chem.* 51(4): 1016–1020.

Tominaga, T., M.L. Murat, et al. 1998. Development of a method for analyzing the volatile thiols involved in the characteristic aroma of wines made from *Vitis vinifera* L. cv. Sauvignon Blanc. *J. Agric. Food Chem.* 46(3): 1044–1048.

Tsachaki, M., R.S.T. Linforth, et al. 2009. Aroma release from wines under dynamic conditions. *J. Agric. Food Chem.* 57(15): 6976–6981.

Van Wyk, C.J., O.P.H. Augustyn, et al. 1979. Isoamyl acetate, a key fermentation volatile of wines of *Vitis vinifera* cv. Pinotage. *Am. J. Enol. Vitic.* 30: 167–173.

Versini, G., and T. Tomasi. 1983. Confronto tra i componenti volatili dei vini rossi ottenuti con macerazione tradizionale e macerazione carbonica. *Enotecnico* 19: 595–600.

Wood, C., T.E. Siebert, et al. 2008. From wine to pepper: Rotundone, an obscure sesquiterpene, is a potent spicy aroma compound. *J. Agric. Food Chem.* 56(10): 3738–3744.

15 Flavor of Bread and Bakery Products

Salim-ur-Rehman and Javaid Aziz Awan

CONTENTS

15.1 INTRODUCTION

Flavors are the fundamentals of bakery products, which can play a vital role in promoting consumer sensory perceptions and the growth of baking industry. There are numerous factors that contribute to flavor during mixing of ingredients: their interactions and baking processes. Among cereals, wheat is extensively used for the production of bakery products. The unique bread-making properties of wheat flour are mainly due to its gluten proteins that form a viscoelastic network when mixed with water. The flour not only provides nutrition but also contributes towards flavoring compounds produced during fermentation and subsequent baking. For bread production, straight dough, sponge dough, liquid ferment, sourdough, and continuous bread-making processes are used. However, each process has its merits and demerits. In some relatively smaller bakery plants, straight doughs are used to make pan breads. In the traditional sponge system, part of the flour is mixed initially with the yeast and sufficient water to make a sponge that is permitted to ferment overnight. The sponge is then mixed with the remaining part of flour, water, salt, and fat to a desired consistency and is then allowed to ferment for a short period. This bread is richer in flavor due to bacterial souring of the sponge (Brummer and Lorenz 1991). The sourdoughs are dominated by a complex microflora composed of yeasts and lactic acid bacteria (LAB) that encourage the improvement of the bread flavor. The role of yeast in a dough starter is to leaven the bread, whereas LAB, which have trophic and nontrophic relationships with yeast types, play a significant role in the production of flavoring compounds. In the present era, sourdough has a paramount importance in the preparation of specialty breads. In this chapter, emphasis is laid on various dynamics such as fate of carbohydrates metabolism, role of ingredients, role of external and internal factors, and interactions of microorganisms for the generation of variety of volatile and nonvolatile compounds in bakery products. The sourdough flavoring potential could be achieved by focusing on process conditions and influence of exogenous agents on the retention of flavoring compounds during bread production.

15.2 BREAD FLAVOR AND SENSORY PERCEPTIONS

Apart from appearance, flavor in bread is a key factor affecting consumer perceptions of quality and consistency. Understanding the effect of ingredients and processing on bread flavor can facilitate bakers in maintaining uniformity, duplicate the appeal of traditional products, and develop more typical varieties. Flavor perception is a complex phenomenon concerning aroma, taste, and texture. Aroma relates to the volatile flavoring compounds that are sensed by smell. On the other hand, taste is mainly related to sweet, salt, sour, and bitter components that are perceived by taste buds on the tongue. In addition to these traditional components, there is a fifth component, "umami," which is also considered as a basic taste. Texture is the tangible sensation in the mouth that is perceived as chewing quality. The desirable flavor of fresh leavened bread is generally described as a yeasty/wheaty aroma with a sweet, salty, and somewhat sour taste. The crumb texture is soft and moist, whereas the crust is dry and crispy. Specialty breads have unique or more distinctive flavors that depend on the ingredients as well as the processing.

Sensory evaluation of food and food products is an important aspect of quality control. Color and texture are considered as two primary sensory characteristics of flat bread. The acceptability of any food product such as bread is subject to a number of factors such as effect of climate, geographical location, consumer's age, and level of income along with other parameters. Flat breads are mostly made from flour of high extraction rate. So, the sensory qualities of flat breads including color, texture, and chewability may be closely examined (Qarooni 1996) because as time passes, breads lose their commercial value through sensory changes that are due to physical and chemical deterioration (Park et al. 2006; Mueen-ud-Din 2009).

The acceptability of flat breads and chapatti can be related to a number of factors, such as behavior during preparation, rheological characteristics, and sensory qualities. The flat bread's characteristics such as color and appearance of dough, percentage of water absorption, and puffing, appearance, texture and taste are affected due to differences in wheat varieties (Rehman et al. 2006a, 2007a). Although, grain protein increases with fertilizer increase, yet no changes are observed in chapatti quality characters, such as water absorption, puffing, texture, and taste (Austin and Jhamb 1964).

The overall sensory characteristics of flat breads are affected by flour extraction rate and fermentation. In Iranian flat breads, by increasing the extraction rate, the quality of breads improves up to 90% extraction rate (Azizi et al. 2006). The texture profile is lower in sourdough bread as compared to controlled bread. There has been no significant difference in the sensory characteristics of bread except for the shape of the bread. The results suggest that the quality of bread is improved by the use of sourdough (Park et al. 2006).

A lexicon has been developed for describing the flavor of wheat sourdough bread (Lotong et al. 2000). A highly trained descriptive sensory panel identified, defined, and referenced 32 flavor attributes for wheat sourdough bread. Crumb and crust of wheat sourdough were evaluated differently. All 37 wheat sourdough breads showed different flavor attributes and intensities in both crumb and crust samples. The principal component analysis for crumb and crust indicate that the number of attributes could not be reduced into a smaller set of components that completely describe the wide range of breads used in that study. The crumb of wheat bread made from sourdough fermented with the heterofermentative *Lactobacillus sanfranciscensis* shows a pleasant, mild, and sour odor and taste, whereas sourdough bread fermented with the homofermentative *L. plantarum* has an unpleasant metallic sour taste. However, sourdough wheat bread supplemented with sourdough yeast, *Saccharomyces cerevisiae*, illustrates a more aromatic flavor (Katina et al. 2006). Sensory evaluation of sourdough bread crumb suggests that a most intense and bread-like flavor relates to propanone, 3-methylbutanal, benzyl alcohol and 2-phenylethanol (Hansen et al. 1989a).

15.3 FLAVOR PERCEPTION IN BISCUITS AND SENSORY ATTRIBUTES

The primary relationships between aroma released and the perception it evokes have mainly been studied using liquid and semisolid food systems because these are homogenous and the flavor components (aroma, taste, and viscosity) are relatively easy to manipulate. The liquid and semisolid systems also undergo minimal oral processing so that their physical properties do not change distinctly since the main oral process is dilution with saliva. From these systems, there are two primary findings. One is that the food system can influence aroma release by physicochemical mechanisms, for example, oil–water and water–air partition (Wright et al. 2003; Boland et al. 2004); the other is that the flavor perceived during eating is often the result of numerous different stimuli that demonstrate interactions at the cognitive level (Taylor and Hort 2004).

On the other hand, in bakery solid foods, there are additional factors to consider both in relation to aroma release and to potential crossmodal interactions. With a relatively dry matrix, the rate and extent of hydration during eating will be related to matrix composition and may affect aroma release (Brauss et al. 1999). Starch can bind some aroma compounds (Seuvre et al. 2006). Moreover, many cereal foods also contain significant amounts of fat, which affect partition. Therefore, the pattern of aroma release in vitro and in vivo may be changed (Miettinen et al. 2004).

The structure of the solid food is also related to sensory attributes such as crispness, hardness, and crumbliness that may affect perceived flavor through cross-modal interactions. Foods with complex structures reveal both physicochemical and cross-modal effects as a result of compositional variation. In a milk system with different fat levels, aroma release decreases with increasing fat content (Roberts et al. 2003) and the in vivo aroma release correlates well with perceived flavor at low and intermediate levels. However, perception decreases as fat level increases and is not well related to in-nose release at higher levels of fat in biscuits. This may be known as "perceptual masking" as a cause for this behavior, but it may also be attributed to crossmodality between the viscosity of higher fat milk samples and the aroma. Sensory sweetness is a function of biscuit composition and is also due to fat–sweetness interactions, as well as an effect of type of sugar on sweetness perception. Similarly, biscuit flavor is the result of Maillard reaction, which may vary as a function of composition and so a non-Maillard, but biscuit-congruent flavor (anethole), can be added to the dough before baking (Burseg et al. 2005).

15.4 EFFECT OF MAILLARD BROWNING REACTIONS ON BAKERY PRODUCTS

Aroma is mainly produced during the production process as a result of enzymes, fermentation, or thermal reactions during baking. The aroma compounds are from different chemical classes. Quantitatively, the most important chemical groups are aldehydes, alcohols, ketones, esters, acids, pyrazines, and pyrrolines, as well as other compounds such as hydrocarbons, furans, and lactones. There are several steps in which these compounds can be produced: through dough fermentation by yeast and lactic bacteria, enzymatic activity during the elaboration process (kneading), lipid oxidation reactions, and thermal reactions taking place during baking mainly through Maillard and caramelization reactions. Moreover, the recipe (ingredients and elaborating techniques) could greatly contribute to the final aroma. In the baking process, the moisture from the surface of the products evaporates, raises the temperature of the product to above boiling point of water, and facilitates caramelization and Maillard reactions. These nonenzymatic browning reactions assist to develop crusts in bakery products. Crusty bread has more flavor than noncrusty bread because of the migration of many of the Maillard reaction products from crust to crumb, where they attach to gelatinized starch. At lower temperatures for longer times and adding steam at the start of the process help to enhance crust formation. Steam plays a significant role in changing the crust color from yellow-brown to dark brown-reddish by dextrinizing of starch during baking and making it available for Maillard reaction. A total of 89 volatile compounds have been identified in steamed breads with ethanol and 3-methyl-1-butanol being the most abundant (Yangsoo et al. 2009).

Crust color development during baking is the effect of caramelization of sugars and Maillard reactions between reducing sugars and amino acids. Both require heat and are nonenzymatic, but Maillard reactions require less heat and are particularly motivating because of the large number of end products that result from a small number of reactants. Glucose, fructose, maltose, and lactose are common reducing sugars in bakery products. Sucrose is not a reducing sugar, but yeast or acid inverts sucrose to produce glucose and fructose. Flours contain a good number of the common amino acids, and yeast fermentation increases the levels of free amino acids, including lysine, alanine, proline, and cysteine. Moreover, the crust flavor is modified by enzymes and proline that generate several desirable compounds including 2-acetyl-1-pyrroline and 6-acetyltetrahydropyridine in the breads. However, sensory less desirable bitter-tasting and burnt-smelling compounds such as pyrrolizines and azepinones are also identified. These compounds are predominantly generated in breads containing relatively high levels of proline (Bredie et al. 2006). It is demonstrated that inulin is also responsible for the formation and release of volatiles in white bread during baking. It accelerates the formation of the bread crust and the Maillard reaction. It leads to breads with an overall quality similar to that of non-enriched breads, but is baked for a shorter time. Correlations between some crust properties and the amount of Maillard volatiles showed that crust water activity, moisture, and clearness are the first-class indicators of the Maillard reaction during the baking of bread (Pauline et al. 2010).

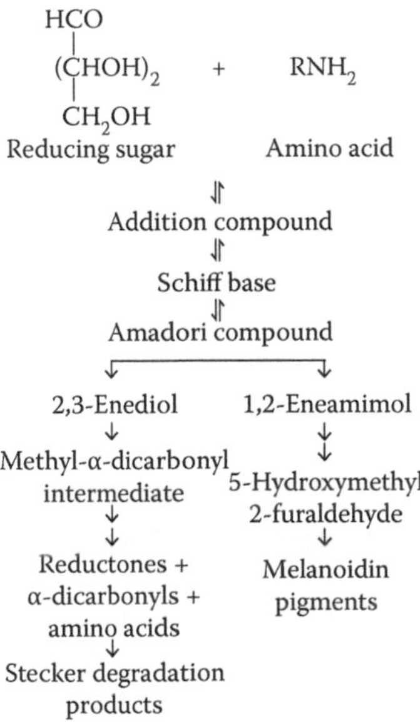

FIGURE 15.1 Pathways for production of flavoring compounds during Maillard reactions. (Adapted from Pozo-Bayón, M.A., Guichard E., Cayot, N., *Food Rev. Int.*, 22, 4, 335–379, 2006.)

Figure 15.1 illustrates that the Maillard reactions are taking place in different phases, beginning with the combination of an aldehyde group (–CHO) from a reducing sugar with the amino group ($-NH_2$) of an amino acid to form a Schiff base intermediate. Schiff bases are unstable and they are rearranged within a short time into what are known as Amadori compounds. The Schiff bases and Amadori compounds are colorless, and the reactions that form them are reversible. In the next phases, more than 100 different components are produced from Amadori compounds through various condensation, isomerization, polymerization, cyclization, and degradation reactions. The major pathway leads over the 1,2-eneamimol of the Amadori compound to 5-hydroxymethyl-2-furaldehyde into the formation of melanoidin pigments. A minor pathway leads over the 1,3-enediol and the methyl alpha-dicarbonyls intermediate to various *C*-methyl reductones and alpha-dicarbonyls. A third important branch of the Maillard reaction involves the Strecker degradation of alpha-amino acids into aldehydes. The products of these various pathways range from colorless to extremely colored and many are responsible for volatile aromatic compounds. They include heterocyclic compounds (mostly pyrazines), nonheterocyclic compounds (such as thiophenes, oxazoles, or oxazolines), and pyranones, furanones, and related compounds (Pozo-Bayón et al. 2006). Details of Maillard reaction mechanisms leading to formation of flavor compounds are discussed in Chapter 5.

15.5 EFFECT OF INGREDIENTS ON FLAVOR OF BAKERY PRODUCTS

Flour also plays a significant role in the generation of distinctive flavor either in leavened or unleavened breads. Different cereals and legume flours including whole wheat, rye, corn, soy, gram, malt, and other specialty flours add their unique flavors to variety breads. Addition of as little as 5% rye flour in wheat flour has identifiable impact on the flavor of bread. Also, water absorption of flour affects dough rheological character, which in turn affects flavor development in the breads. Slack doughs with high levels of unbound water, as in Italian ciabatta-type bread production, enhance crust color and flavor development during baking.

The wheat flour types influence the production of flavoring compounds in wheat sourdough. It has a significant effect on the production of ethyl acetate and ethanol in sourdoughs fermented with heterofermentative cultures, with the highest amount detected in sourdoughs made from whole meal flour and low grade flour. A high ash content of flour has been reported to increase the amount of volatile compounds in mixed fermentations (Czerny and Schieberle 2002; Hansen and Hansen 1994a).

Addition of other optional ingredients contribute to the flavor of a variety breads, including milk powder, shortening, butter, egg products, lecithin, sesame, poppy, caraway, fibers, spices, fruits, and nuts. However, dairy products are traditional ingredients for bread making. Skimmed milk and sweet whey help to improve bread quality. Besides nutrition aspects, the main benefits from the use of dairy ingredients in bread formulations include higher water absorption, better fermentation time tolerance, and nicer crumb grain, as well as crust color formation and pleasing taste and flavor of breads. However, milk replacers are becoming increasingly popular and less costly than skimmed milk (Doerry 1989).

Fermentation is an effective process in developing new dairy ingredients. There are four types of fermented dairy ingredients: traditional fermented dairy products (cheese or yogurt); form-modified fermented dairy ingredients (spray-dried yogurt); flavor-modified fermented dairy ingredients (enzyme-modified cheese); and functionally engineered fermented ingredients (cultured whey) (Main 1991). Yogurt may be used in bread formulations (Hill 1974) and has a positive effect on the flavor of bread, but it decreases loaf volume (Lehmann and Dreese 1981). Incorporation of acid whey (main by-product from cottage cheese manufacturing) in a bread recipe has markedly changed flavor, which may be used to produce sourdough-type bread (Shenkenberg et al. 1972). Other fermented dairy ingredients may be tailored to enhance bread flavor to speed up sourdough processing. The behavior of low moisture, part skim type and low browning Mozzarella cheese made by using strains of *Streptococcus* spp. and *L. helveticus* releases inappreciable amounts of galactose into the cheese curd and has been compared with high browning cheese made by using cultures that release galactose into the curd. Cheese composition has differed only for galactose content. The browning potentials of the cheeses are found to be significantly different. When used in pizzas, panelists could not distinguish between brown and low browned pizza regarding sensory properties (Beatriz et al. 1994).

In addition, these flavorants may be important in shortened bread-making processes, such as the no-time dough process, where fermentation times are kept at a minimum and cannot drastically contribute to flavor development. The choice of microbial cultures is of major importance in fermented milk technology. According to their optimal growth temperature, two classes of LAB are normally used for the preparation of fermented milks: mesophilics and thermophilics. Most mesophilics are used in cheese manufacturing and for the preparation of some fermented milks (cultured buttermilk). Adding citrate to some species produces diacetyl, a highly odoriferous component important in butter flavor. They are good aroma producers, but not very acidifying. Thermophilics, the other group of LAB offering some potential, are used for the production of fermented milks such as yogurt. Depending on their ability to produce mainly lactic acid, these cultures are considered either homofermentative or heterofermentative. They are good acid producers and develop flavorful compounds such as acetaldehyde and, possibly diacetyl (Dellagio 1988).

Sugars are added in bakery products as sucrose, glucose, lactose, and high fructose corn syrup, and are also contributed by enzymes acting on the starch in flour. The type of sugar used affects sweetening passion of the products and its ability to contribute in browning reactions. Table salt makes a significant contribution to the desired flavor of bread, and without it, breads have a flat and unpleasant taste. Moreover, the perception of sweetness is enhanced by the level of salt in the formula of the product.

15.6 EFFECT OF PROCESSING ON FLAVOR OF BREAD

Mixing develops the gluten network in the wheat flour dough, affecting gas retention and crumb texture. Bread produced by the continuous mixing process has a finer, more uniform texture than bread produced by conventional mixing. Some consumers identify the difference in texture as a lack of flavor, and this perception may remain even when a flour brew or preferment step is used to increase the amount of fermentation aroma.

In bakery products, fermentation is initiated because of the action of yeast and the bacteria normally allied with commercial yeast, malt, and flour. During the process, organic acids are produced, which lower the pH and increase the total titratable acidity. It also produces alcohols, esters, ketones, and aldehydes that make up the slightly pungent flavor of freshly baked bread crumb. Longer fermentation time given to water brews or flour brews increases the concentration of these flavor components to the point where they can become objectionable. Lowering the fermentation temperature permits longer fermentation times and produces a more appealing flavor.

The bread dough is stored under frozen conditions in order to increase its storage life. The difference in bread quality is explained by their variable aromatic compounds composition. By analyzing breads based on the same recipe but from different processes, it is demonstrated that adding a freezing stage before dough proofing or at the end of the conventional process, as well as after partial baking, do not influence bread aroma. Likewise, partial baking has no effect on bread odor and aromatic profile. Thus, the aromatic differences between commercial conventional, fully baked frozen, and frozen dough breads, on the one hand, and commercial partially baked breads, on the other hand, may be due to their different formulations. Concerning bread physical properties, the recipe influences bread crust/crumb ratio and density. Moreover, adding a partial baking stage to the process leads to breads with a more compact crumb (Pauline et al. 2008).

During storage, some flavoring components are lost faster than others. Sweet and salty tastes decrease with the passage of time and the remaining sourness starts to become unpleasant. The desirable alcohol smell of yeast is lost, the wheaty odor is reduced, and the leftover doughy or starchy aromas become unpleasant. The texture of the crumb becomes firmer and drier, whereas the texture of the crust becomes soft and leathery. Heating stale bread temporarily reverses some of the changes and releases aroma compounds that were entrapped by starch (Samuel et al. 2009). In bakery products, staling indicates decreasing consumer acceptance, which is caused by changes in crumb other than those resulting from the action of spoilage organisms (Bechtel et al. 1953). The application of LAB in the form of sourdough has a positive effect on bread staling. One such effect is an improvement in loaf specific volume, which is associated with the reduction in the rate of staling (Maleki et al. 1980; Bolourian et al. 2010). The breads containing sourdough can decrease the staling rate as measured by differential scanning colorimetric and enhance the shelf life of bread (Barber et al. 1991; Corsetti et al. 2000; Rehman et al. 2007b).

15.6.1 Sourdough Technology of Bread

Fermented foods are of great importance because they provide and preserve large quantity of nutritious foods with improved aroma and texture. These foods include alcoholic beverages, vinegar, pickles, sausages, cheese, yogurts, and sourdough breads. In recent years, sourdough bread has enjoyed renowned success because of increasing consumer demand in Europe for its natural taste and good health benefits

(Brummer and Lorenz 1991). A mixture of cereals in water resulting in the formation of dough, which is characterized by sour aroma when left alone for a while, is the first example of fermented food used by mankind. The sourdough is a mixture of flour and water that is fermented with LAB, mainly heterofermentative strains, elaborating lactic acid and acetic acid in the mixture, hence, resulting in a pleasant sour taste of the end product (Hammes and Ganzle 1995; De-Vuyst and Neysens 2004).

Sourdough fermentation is a traditional process for improving bread quality and producing different wheat and rye breads. At present, sourdough is used in the manufacturing of breads, cakes, and crackers. The typical characteristic of sourdough is mainly attributable to its microflora, basically represented by LAB and yeasts. Because of microbial community, such dough is metabolically active and can be reactivated. These microorganisms ensure acid production and leavening upon addition of flour and water (Ottogalli et al. 1996; Thiele et al. 2002).

The mechanisms of sourdough are complex. Various flour characteristics and process parameters contribute to exercise very particular effects on the metabolic activity of the sourdough microflora. During fermentation, biochemical changes occur in the carbohydrate and protein components of the flour due to the action of microbial and indigenous enzymes. Moreover, fermentation temperature also influences the growth and metabolism of LAB and yeast (Spicher et al. 1981).

15.6.1.1 Use of Sourdough in Cereal Products

Sourdough process has been used as a form of leavening and is the oldest biotechnological process in food production (Röcken and Voysey 1993). The use of sourdough in wheat breads has gained popularity as a means to improve the quality and flavor of wheat breads. To facilitate continuous production, one could save a portion of ripe sourdough to seed subsequent dough, a process that continued into the nineteenth century (Williams and Pullen 1998). A vast array of traditional products relies on the use of sourdough fermentation to yield baked goods with particular quality characteristics. Some examples include the well-known Italian products associated with Christmas, *Panettone*, which originated in Milan (Sugihara 1977). San Francisco sourdough French breads (Kline et al. 1970) and soda crackers (Sugihara 1985) are other examples of wheat products that rely on the process of souring. The same process is also used in the production of a number of flat breads, a typical example of which are the Egyptian baladi bread (Qarooni 1996) and naan bread (Mueen-ud-Din 2009).

15.6.1.2 Beneficial Effect of *Lactobacillus* Bacteria

Lactobacillus bacteria (LAB) have a long history of use in food and are generally regarded as safe organisms. Cereal grains such as wheat from which most breads are produced, are low in some of the essential amino acids, that is, lysine, threonine, methionine, tryptophan, and isoleucine. Such cereal grains can be considered low in quality protein. Thus, cereal grain-based diets, prevalent in many areas of the world, may be deficient in some essential amino acids. Lyophilized cultures of the microorganisms may be added to cereal grains such as wheat in bulk to increase the basic nutritive protein quality of the wheat (El-Megeed et al. 1989).

15.6.1.3 Contribution of Organic Acids in Bread

In order to control microbial spoilage of breads, organic acids including propionic acid, acetic acid (vinegar), and lactic acid are added. These are also contributed by yeast and bacterial fermentations. Traditional sour dough methods use prolonged fermentation by LAB to produce the unique flavor of rye bread, San Francisco sour dough, and French pain au levain. However, since these acids are a component of the sourdough flavor, adding them alone does not produce the same finished products.

Mold growth being the most important cause of bread spoilage, could be prevented by the use of homo- and heterofermentative LAB. This fungistatic effect is due to the production of acetic acid by LAB (Röcken and Voysey 1993). The antifungal activity of sourdough LAB is due to the production of a mixture of acetic acid, propionic acid, caproic acid, and butyric acid. The caproic acid, along with acetic acid, plays an important role in inhibiting fungal growth. The sourdough bread made from low pH and high concentration of acetic acid has large volume and lower rate of staling (Spicher 1983; Corsetti et al. 1998).

Hansen et al. (1989a) studied the effect of three heterofermentative and two homofermentative LAB strains on the production of organic acids in sourdough. The results show that acetic acid content is much higher in the dough acidified with heterofermentative strains as compared to homofermentative strains. The homofermentative strain produces only L-lactic acid, whereas other cultures produce both L-lactic acid and D-lactic acid. In another study (Hansen and Hansen 1994b), the effect of wheat flour types on the production of organic acid was also observed. The wheat flours have significant effect on the production of lactic and acetic acids. The highest lactic acid contents are observed in low grade and whole meal flours. The acetic acid is found to be 12% of the total acid contents in heterofermentative cultures, but it has not been detected in homofermentative cultures. However, sourdough made from straight grade flour fermented with homofermentative strain *L. plantarum* shows a small amount of acetic acid. The sourdough fermentation with *Candida milleri* results in high amounts of acetic acid and mannitol that improve the qualitative characteristics of leavened dough and the baked products (Vernocchi et al. 2004).

15.6.1.4 Sourdough vs. Straight Dough Systems

In modern sourdough technology, the choice of microbial starters is also of great importance. Homofermentative (*L. plantarum*, *L. acidophilus*) and heterofermentative (*L. brevis*, *L. fermenti*) LAB, as well as yeasts, are representative groups of microorganisms naturally found in sourdoughs or used as starters (Mueen-ud-Din 2009). Sourdough is an important intermediate product of fermentation of cereal flours and water (Vogel et al. 1999). This mixture is inoculated with microbial starter known as "mother culture," which is constantly renewed in a cyclical way, using the approved conditions of recipe and ripening (Hammes and Ganzle 1995; Onno and Rouseel 1994; Ottogalli et al. 1996). Two main systems are in practice for the production of bread. The sourdough system differs from the straight dough system in that only part of the flour is mixed at first with all the yeast and sufficient water to make a sponge that is allowed to ferment for some hours. The sponge is then mixed with the remainder of flour, water, and all the salt and fat to a required consistency, which is given a short fermentation time before proving and baking. This system is said to produce bread of richer flavor due to souring of sponge by bacteria, as compared to making bread with yeast alone using the straight dough system. Conventionally, 2%–5% of old dough is used as an inoculum for the new batch. Sourdough breads may vary in flavoring compounds from wheat breads to rye breads (Schieberle and Grosch 1994) and flat breads to San Francisco sourdough bread fermented with culture of *Lactobacillus* (Linko et al. 1997; Seitz et al. 1998). It contains metabolically active LAB (10^8–10^9 CFU/g) and yeasts (10^6–10^7 CFU/g) mainly responsible for acidification and leavening action of dough, respectively. However, the LAB that develop in the dough may either originate from selected natural contaminants in the flour or from a starter culture containing one or more known species of LAB (Vuystn and Neysens 2004). The metabolic activities of these microorganisms improve the sensory characteristics of bread and generate more flavoring components in the bread (Oura et al. 1982). These qualities of sourdough bread are also imparted from the interaction of endogenous and exogenous agents and types of flour sources because the flavor of native flour is mild. It needs some processing conditions such as heat treatment and carbon sources including sugars to enhance the flavor components (Hansen 1995; Martinez-Anaya 1996; Gobbetti 1998).

15.6.2 Sourdough Fermenting Microorganisms

The microbial ecology of the sourdough fermentation is based on ecological factors. Microbiological studies have revealed that more than 50 species of LAB, mostly of the genus *Lactobacillus*, and more than 20 species of yeasts, especially of the genera *Saccharomyces* and *Candida*, occur in this ecological niche. The sourdough microflora is composed of stable associations of lactobacilli and yeasts, in particular due to the metabolic interactions that contribute toward production of aromatic compounds (Vuystn and Neysens 2004; Hammes et al. 2005). Numerous species of LAB occur naturally in wheat flour, including members of the genera *Lactobacillus*, *Pediococcus*, *Enterococcus*, *Lactococcus*, and *Leuconostoc* (Plessas et al. 2005). Most of the species of LAB of the genus *Lactobacillus* are isolated from sourdoughs. *L. sanfranciscensis*, *L. brevis*, and *L. plantarum* are the most frequent lactobacilli isolated from sourdoughs (Hammes and Vogel 1997; Gobbetti 1998; Corsetti et al. 2001).

Homofermentative species do not produce any carbon dioxide; their function is acidification and flavor development. However, homofermentative species of LAB may be used in the majority of fermented food applications. Heterofermentative species play a major role in sourdough fermentation, especially when sourdoughs are prepared in a traditional manner. Heterofermentative LAB result in better taste and flavor of the sourdough breads, because only heterofermentative LAB can produce the considerable amount of acetic acid under anaerobic conditions, which is desired in sourdough. On the other hand, fermentation with homofermentative LAB results in high concentration of lactic acid, relative to acetic acid, resulting in a mild and flat sour taste. In rye bread prepared with pure culture of heterofermentative bacteria *L. brevis*, this provided the bread with desirable aroma but not an elastic crumb. They observed an opposite effect when they used homofermentative bacteria (*L. plantarum*). It was concluded that in order to obtain satisfactory aroma and crumb characteristics, both bacterial species must be incorporated (Spicher and Nierle 1984; Corsetti et al. 2001; Mueen-ud-Din 2009).

Cossignani et al. (1996) used *L. sanfranciscensis*, *L. plantarum*, and *S. cerevisiae* for leavening wheat sourdoughs. They found that the doughs fermented with starters had more balanced microbiological and biochemical characteristics than doughs started with *S. cerevisiae*, in which alcoholic fermentation end products largely predominated. By using starters, the greatest LAB cell number and acetic acid production was achieved. The starters resulted in more complete profiles of volatile compounds and greater structural stability.

Starting from glucose, homofermentative LAB mainly produce lactic acid through glycolysis (homolactic fermentation), whereas heterofermentative LAB produce—besides lactic acid—CO_2, acetic acid, and/or ethanol (depending on the presence of additional substrates acting as electron acceptors) (Axelsson 1998). LAB, both homofermentative and heterofermentative species, contribute most to the process of dough acidification, whereas yeasts are primarily responsible for the leavening; however, the heterofermentative LAB also contribute partly to the leavening process (Gobbetti et al. 1995a; Spicher 1983).

Gobbetti (1998) reported *L. sanfranciscensis* and *L. plantarum* association in Italian wheat sourdough. *L. plantarum* may be superseded by another facultative heterofermentative species, *L. alimentarius*, in its association with *L. sanfranciscensis* in sourdough made from durum wheat (Corsetti et al. 2001). *L. alimentarius* is capable of fermenting all four soluble carbohydrates (maltose, sucrose, glucose, and fructose), and it is possible that this reduces direct metabolic competition with *L. sanfranciscensis*. Most of the *L. alimentarius* strains, due to a phenotypical misidentification, probably belong to *L. paralimentarius*, a facultatively heterofermentative species first isolated from Japanese sourdough (Cai et al. 1999). *L. brevis* and *L. plantarum* have generally been found associated with *L. fermentum* in Russian sourdoughs (Kazanskaya et al. 1983). Gobbetti et al. (1994a) reported that *L. acidophilus* is common in Umbrian (Italian region) sourdoughs, even though it is rarely isolated from sourdoughs of different origins. Corsetti et al. (2005) described a new sourdough-associated species, *L. rossiae*, which seems to be widely diffused in sourdoughs of southern and central Italy (Settanni et al. 2005a). *L. rossiae* is often associated with the key sourdough *L. sanfranciscensis*. *L. rossiae* has been found in environments other than sourdough (De Angelis et al. 2006), whereas no other habitat is known for *L. sanfranciscensis* (Hammes et al. 2005). However, occurrence of LAB and yeasts in sourdoughs and the association between acidification and bacterial metabolism was first demonstrated in 1894 (Hammes and Ganzle 1995).

Association of yeasts and LAB are often used in the production of beverages and fermented foods (Gobbetti 1998). The vast majority of yeasts found in sourdoughs have been allotted to the species *C. milleri*, *C. holmii*, *S. exiguus*, and *S. cerevisiae* (Hammes and Ganzle 1995). Most of yeast preparations often contain LAB, especially lactobacilli rather than *Pediococcus*, *Lactococcus*, and *Leuconostoc* spp.

(Jenson 1998), which contributes a little to the aroma development acidification of dough because of the limbed processing time (Rothe and Ruttloff 1983).

15.6.2.1 Contribution of Yeasts in Sourdough Flavoring Compounds

Microorganisms are mostly found on the glumes and glumules of caryopsis, which is usually removed during milling or during the other treatments of the grain before milling, hence, yeasts are not very abundant in refined flour. Sourdough fermentation is a complex process caused by the combined effects of the metabolism of yeasts and LAB. The former is mainly responsible for the leavening, whereas the latter acidifies it. Under special conditions, some yeast and LAB act synergistically in the dough. Along with fermentation of sugars to carbon dioxide and ethanol, the yeast also produces some by-products that impart taste and flavor to the bread (Boraam et al. 1993). The yeasts found in sourdoughs belong to more than 20 species, and typical yeasts associated with LAB in sourdoughs are *S. exiguus*, *C. humilis* (formerly described as *C. milleri*), and *Issatchenkia orientalis* (*C. krusei*) (Spicher et al. 1982; Gullo et al. 2002; Succi et al. 2003). Other yeast species detected in sourdough ecosystem are: *Pichia anomala* as *Hansenula anomala*, *Saturnispora saitoi* as *Pichia saitoi*, *Torulaspora delbrueckii*, *Debaryomyces hansenii*, and *Pichia membranifaciens* (Succi et al. 2003; Foschino and Galli 1997; Gobbetti et al. 1994a). The presence of *C. humilis* in sourdough was reported by Barnett et al. (2000); however, the dominance of *C. humilis* in sourdough is a recent observation.

The available carbohydrates in wheat flour are maltose followed by sucrose, glucose, and fructose, along with some trisaccharides such as maltotriose and raffinose. Glucose increases during fermentation, whereas sucrose decreases in the presence of yeast because of the reaction of invertase (Gobbetti et al. 1994a). The yeasts present in sourdoughs are not able to ferment maltose, a sugar common in flour. However, it can develop because of glucose released into the medium by some LAB species, for example, *L. sanfranciscensis* (Foschino and Galli 1997; Boraam et al. 1993).

The optimal use of sourdough can improve the taste and flavor of the bread. The flavor of sourdough wheat bread is richer and more aromatic than wheat bread, a factor that can be attributed to the long fermentation time of sourdough (Brummer and Lorenz 1991). The concentration of 2-phyenylethanol, one of the most potent odorants of wheat bread crumb, increases in sourdough bread crumb (Gassenmeier and Schieberle 1995). The production of volatile flavor components in sourdough is strongly dependent on the starter culture, but the role played by the flour used has also been recognized (Hansen and Hansen 1994a). The main influence of microorganisms on sourdough flavor has been identified as their ability to enhance or reduce the amount of specific volatiles already present in the flour (Czerny and Schieberle 2002).

Yeast, when used as nonleavening yeast product or a yeast extract, has a distinct flavor profile and is used as a flavor enhancer in many savory food products including bread. However, when using very high levels of yeast (over 10% compressed yeast), the taste may become objectionable due to bitterness. When yeast is used as a leavening agent, fermentation affects bread flavor by converting sugars into carbon dioxide gas, ethanol, and lesser amounts of various other chemical compounds such as aldehydes, ketones, acids, and other alcohols. Some of these chemical compounds are further reacted to produce a variety of new flavor components during baking (Rehman et al. 2006b; Mueen-ud-Din 2009).

Several species of yeasts are found in sourdoughs, but *S. cerevisiae* is frequently present or is added for the production of bread (Corsetti et al. 2001) (Table 15.1). The amount of *S. cerevisiae* may be overestimated because of the lack of reliable systems for identifying and classifying yeasts from this habitat (Vogel 1997). In particular, *S. exiguus* (physiologically similar to *C. milleri*), *C. krusei*, *Pichia norvegensis*, and

TABLE 15.1
Volatile and Nonvolatile Compounds Present in Wheat Flour Sourdough Fermented with Various Yeast Strains

Compounds	A	B	C	D	E
Lactic acid	–	–	–	–	–
Acetic acid	–	–	–	–	–
Ethanol	+	–	–	+	+
1-Propanol	+	+	–	+	+
2-Methyl-1-propanol	+	–	–	+	+
Ethyl acetate	+	+	+	+	+
3-Methyl-1-butanol	+	–	–	+	+
2-Methyl-1-butanol	+	–	+	+	+
1-Pentanol	–	–	–	–	–
2-Methyl-1-pentanol	–	–	–	–	–
1-Hexanol	+	–	+	+	+
3-Hexen-1-ol	–	–	–	–	–
1-Heptanol	–	–	–	–	–
1-Octanol	–	–	–	–	–
Acetaldehyde	+	+	+	+	+
3-Methyl-1-butanal	+	+	+	+	+
2-Methyl-1-butanal	+	–	+	+	+
Hexanal	+	+	+	+	+
3-Methyl-hexanal	–	–	–	–	–
Heptanal	–	–	–	–	–
trans-2-Heptenal	–	–	–	–	–
Octanal	+	+	+	+	+
Nonanal	+	+	+	+	+
Benzaldehyde	–	–	–	–	–
Diacetyl	+	–	–	–	–
Hexane	+	+	+	+	+
Heptane	–	+	+	–	–
Octane	–	+	+	–	–

Source: Damiani, P. et al., *Lebensm.-Wiss. Technol.* 29, 63–70, 1996. With permission.
Note: +, present; – not present. A, *Saccharomyces cerevisiae;* B, *Candida krusei;* C, *Candida norvegensis;* D, *Saccharomyces exiguus;* E, *Hansenula anomala.*

H. anomala are yeasts associated with LAB in sourdoughs and the ratio of LAB to yeast should be 100:1 in sourdoughs for obtaining maximum results (Gobbetti et al. 1994a; Ottogalli et al. 1996). However, *C. milleri* is found as the dominant species in the sourdough, although it does not ferment maltose and grows fermenting the glucose released by heterofermentative LAB. It also supplies an electron source (fructose) to bacteria, which helps to proliferate their cell yield and acetic acid production. The production of acetic acid improves the qualitative characteristics of the leavened dough and the baked products (Vernocchi et al. 2004).

The interactive effects of LAB and yeasts can be identified through isolation of sourdough sponges. Seven kinds of bread have been made with the inoculation (1.5%) of *S. cerevisiae* and (1.5%) lactobacilli (*L. amylophilus*, *L. brevis*, *L. plantarum*, *L. sake*, and *L. acetotolerans*). In previous studies, *S. cerevisiae*, *S. delbrueckii*, *Torulopsis holmii*, and *Torulopsis unisporus* were isolated from sourdough sponges as yeasts (Gül et al. 2005). The conditions and substrates of the sourdough are also fundamental with respect to microbiological stability of the dough. Mixed commercial starters containing *L. brevis* and baker's yeast (*S. cerevisiae*) when fermented at 30°C for 20 h reduced yeast growth in dough and ethanol production, but more glycerol (80%) and acetic acid (55%) were formed without affecting the production of lactic acid (Meignen et al. 2001). However, yeasts isolated from sourdoughs generally used for the production of durum wheat bran flour breads enable us to affirm that more than 95% strains belong to the species *C. humilis*, whereas *C. krusei* and *S. cerevisiae* are the dominant yeasts in rice sourdough (Merotha et al. 2004). On the other hand, in *Triticum aestivum* wheat flour sourdough, 58 strains have been identified as *S. cerevisiae*, five as *C. colliculosa*, four as *C. lambica*, three as *C. krusei*, three as *C. valida*, and two as *C. glabrata* (Succi et al. 2003).

15.6.2.2 Contribution of Bacteria in Sourdough Flavor Compounds

LAB are present in different proportions in different sourdoughs used for the preparation of specialty breads. The sourdough's LAB chiefly produce L-lactic acid and acetic acid with lesser amounts of other acids such as citric and malic acids, but the ratio of lactic acid to acetic acid is important for the flavor of the final product (Linko et al. 1997). There are many kinds of LAB, but lactobacilli, obligately homofermentative and facultatively or obligately heterofermentative, are the typical sourdough bacteria. *L. sanfranciscensis* (Trüper and de Clari 1997) (synonym *L. brevis* subsp. *lindneri*), *L. plantarum*, and *L. brevis* are the most frequently isolated lactobacilli. On the basis of volatile compounds (Table 15.2), the heterofermentative LAB species can be discriminated. The presence of some aldehydes may not only be derived from enzymatic oxidation or autooxidation of the lipid fraction of the wheat (Hann and Morrison 1975; Frankel 1982), but are also contributed through metabolism. Some strains, initially classified as *L. brevis*, are recently allotted to the new species *L. pontis* (Vogel et al. 1994). The homofermentative LAB, with the exception of *L. delbrueckii* subsp. *delbruecki* strains, are primary contributors of ethyl acetate and diacetyl (Hansen and Hansen 1993; Hansen et al. 1989a). Other LAB strains, including *Carnobacterium divergens* (*L. divergens*), *L. brevis*, *L. amylophilus*, *L. sake*, *L. acetotolerans*, *L. plantarum*, *Pediococcus pentosaceus*, *P. acidilactici*, and *Tetragenococcus halophilus* (*P. halophilus*), have been isolated

TABLE 15.2
Volatile and Nonvolatile Compounds Present in Wheat Flour Sourdough Fermented with Hetero (A to E) and Homofermentative (F to J) LAB Strains

Compounds	A	B	C	D	E	F	G	H	I	J
Lactic acid	+	+	+	+	+	+	+	+	+	+
Acetic acid	+	+	+	+	+	+	+	–	–	–
Ehanol	+	+	+	+	+	–	–	–	–	–
1-Propanol	+	–	–	–	–	–	–	–	–	–
2-Methyl-1-propanol	–	–	–	–	–	–	–	–	–	–
Ethyl acetate	+	+	+	+	+	+	+	+	+	+
3-Methyl-1-butanol	–	–	–	–	–	–	–	–	–	–
2-Methyl-1-butanol	–	–	–	–	–	–	–	–	–	–
1-Pentanol	–	–	–	–	–	–	–	–	–	–
2-Methyl-1-pentanol	+	+	–	–	–	–	–	–	–	–
1-Hexanol	+	+	+	+	+	+	+	+	+	+
3-Hexen-1-ol	+	–	–	–	–	–	–	–	–	–
1-Heptanol	+	–	–	–	–	+	+	–	–	–
1-Octanol	+	+	–	–	–	–	+	–	–	–
Acetaldehyde	+	+	+	+	+	+	+	+	+	+
3-Methyl-1-butanal	+	–	–	–	–	+	+	–	–	–
2-Methyl-1-butanal	–	–	–	–	–	–	–	–	–	–
Hexanal	+	+	+	+	+	+	+	+	+	+
3-Methyl-hexanal	–	–	–	–	–	+	–	–	–	–
Heptanal	+	+	–	+	–	+	+	–	–	–
Trans-2-heptenal	+	+	–	–	–	+	+	–	–	–
Octanal	+	+	+	+	+	+	+	+	+	+
Nonanal	+	+	+	+	+	+	+	+	+	+
Benzaldehyde	+	–	–	–	–	+	+	–	–	–
Diacetyl	–	–	–	–	–	+	+	+	+	+
Hexane	+	+	+	+	+	+	+	–	+	+
Heptane	+	+	+	+	–	+	+	+	+	+
Octane	+	+	+	+	–	+	+	+	+	+

Source: Damiani, P. et al., *Lebensm.-Wiss. Technol.* 29, 63–70, 1996. With permission.

Note: + present; – not present. A, *Lactobacillus brevis lindneri;* B, *Lactobacillus brevis;* C, *Lactobacillus fructivorans;* D, *Lactobacillus fermentum;* E, *Lactobacillus cellobiosus;* F, *Lactobacillus plantarum;* G, *Lactobacillus farciminis;* H, *Lactobacillus alimentarius;* I, *Lactobacillus acidophilus;* J, *Lactobacillus delbrueckii.*

from sourdoughs (Gül et al. 2005). However, in case of wheat flour sourdough, more strains of LAB are identified in different proportions as *L. sanfranciscensis*, 20% as *L. alimentarius*, 14% as *L. brevis*, 12% as *Leuconostoc citreum*, 7% as *L. plantarum*, 6% as *Lactococcus lactis* subsp. *lactis*, 4% as *L. fermentum* and *L. acidophilus*, 2% as *Weissella confusa*, and 1% as *L. delbrueckii* subsp. *delbrueckii* (Corsetti et al. 2001). However, *L. sanfranciscensis* is considered a key sourdough lactic acid bacterium (Gobbetti and Corsetti 1997), whereas in the case of rice sourdough, *L. fermentum*, *L. gallinarum*, *L. kimchii*, *L. plantarum*, *L. pontis*, *L. paracasei*, and *L. paralimentarius* are proposed to be dominant (Merotha et al. 2004).

15.6.3 Factors Affecting Generation of Aromatic Compounds in Sourdough Breads

The composition of aromatic compounds of sourdough is not only influenced by microbial composition, but is also affected by the interactive effects among types of bread-making processes and ingredients (Collar 1996). Knowledge regarding ingredients exploitation and improvement of the stability of associated sourdough LAB and yeasts are necessary in order to prevent the loss of variety of regional specialties and to meet consumer and industry demands (Gobbetti 1998). Sourdoughs are mainly produced from *T. aestivum* and *Triticum durum* wheat flours (Succi et al. 2003). The number of LAB and yeasts range from ca. log 7.5 to log 9.3 colony forming units (CFU)/g and from log 5.5 to log 8.4 CFU/g, respectively (Corsetti et al. 2001). The dough yield variable is important in increasing yeast and lactic acid bacterial growth because slack doughs could produce more aromatic compounds. The stimulating effect of NaCl on yeasts can indirectly be attributed to a reduction in competition with LAB for the available sugar (Neysens et al. 2003) because of the low content of soluble carbohydrates in cereal flours. In wheat flour, the total concentration of maltose, sucrose, glucose, and fructose varies from 1.55% to 1.85% depending on the balance between starch hydrolysis, by the flour, as well as microbial enzymes and microbial consumption (Martinez-Anaya et al. 1993). Addition of sucrose in wheat dough has a stimulatory effect on both yeast and LAB growth, which increases the production of lactic and acetic acids by LAB associated with *S. cerevisiae* (Corsetti et al. 1994). However, titratable acidity of the dough increases with addition of sucrose in the range of 0%–6% and is largely due to the increase in acetic acid accumulation (Simonson et al. 2003). The lack of competition between *L. sanfranciscensis* and *S. exiguus* for maltose is fundamental for the stability of this association in San Francisco French bread (Sugihara et al. 1970), but LAB multiply and produce lactic and acetic acids more slowly in the mixtures with yeasts than in pure cultures (Merseburger et al. 1995).

Furthermore, bacterial growth and production of lactic and acetic acids might be decreased because of the faster consumption of maltose and, in particular, of glucose by *S. cerevisiae* when associated with *L. sanfranciscensis* in a synthetic medium containing these sugars (Collar 1996). However, the imbalance between yeast consumption and starch hydrolysis by flour enzymes leads to the rapid depletion of soluble carbohydrates during wheat sourdough fermentation which, in turn, decreases LAB acidification due to microbial competition (Rouzaud and Martinez-Anaya 1993). This situation is less pronounced in rye dough fermentation because of the greater flour enzyme activity, which increases the availability of soluble carbohydrates (Röcken and Voysey 1993). In case of wheat doughs fermented by yeasts and LAB, the concentration of maltose may remain between 2 g/kg and 5 g/kg

(Martinez-Anaya et al. 1993) because it is not metabolized by some yeasts until the available glucose and fructose supplies are depleted (Barber et al. 1991).

Cofermentations are another metabolic route that enables sourdough LAB to use nonfermentable substrates, thus increasing their adaptability. A cofermentation of fructose and maltose or glucose has been observed in a fructose-negative strain of *L. sanfranciscensis* (Gobbetti et al. 1995c, 1995d; Stolz et al. 1995). A cometabolism of citrate and maltose or glucose was also observed in the same strain of *L. sanfranciscensis* (Gobbetti and Corsetti 1996). Citrate and fructose modification can be used to optimize the relative ratios of acetic and lactic acids. These ingredients are responsible for changing the microstructural rheological features of the dough and bread organoleptic properties (Gianotti et al. 1997) and acetic acid production (Calderon et al. 2003). *L. sanfranciscensis* hydrolyzes maltose and accumulates glucose in the medium in a molar ratio of about 1 maltose:1 glucose (Gobbetti et al. 1994c; Stolz et al. 1993). The maltose uptake and glucose excretion in *L. sanfranciscensis* were analyzed previously (Neubauer et al. 1994). Once maltose is depleted, the consumption of the excreted glucose begins. Glucose excreted during sourdough fermentation may be used by maltose-negative yeasts such as *S. exiguus* or may prevent competitors from using maltose by glucose repression, thereby giving an ecological advantage to *L. sanfranciscensis* (Nout and Creemers-Molenaar 1987).

In addition, some strains of *S. cerevisiae* are sensitive to the acetic acid produced by LAB, especially at the normal sourdough pH (4.0–4.5), which favors the undissociated lipophilic and membrane-diffusible form of the organic acid. At this stage, it is necessary to use large amounts of baker's yeast to compensate for the poor survival of wild-type yeasts in consecutive sourdough fermentations, but in *L. sanfranciscensis* strains, the use of maltose is very effective and is not subject to glucose repression. However, some yeasts including *L. sanfranciscensis*, *L. pontis*, *L. reuteri*, and *L. fermentum* are unique among the Lactobacillaceae in that phosphorylate maltose and maltose phosphorylase may be considered key enzymes for lactobacilli growth during sourdough fermentation (Suihko and Makinen 1984; Vogel et al. 1994).

When *L. plantarum* is associated with *S. cerevisiae* or *S. exiguus* in the presence of sucrose (Gobbetti et al. 1994a), cell yield and lactic acid production increase. This might be due to the hydrolysis of sucrose by yeasts into glucose and fructose, which are then more rapidly depleted than the sucrose by LAB (Aksu and Kutsal 1986). Yeasts can hydrolyze sucrose about 200 times faster than the released hexoses are fermented (Martinez-Anaya 1996), causing the rapid disappearance of sucrose during sourdough fermentation (Seppi 1984). *S. exiguus* preferentially uses glucose or sucrose and has a high tolerance for the acetic acid produced by the heterolactic metabolism (Suihko and Makinen 1984).

In sourdough bread, generation of volatile compounds varies with the raw materials used for its production. Likewise, volatile compounds of germinated, sourdough fermented, and native rye are found to be substantially different, and they remained variable even after the second treatment (Heiniö et al. 2003). Addition of enzyme active soya flour can influence the volatile composition of bread (Luning et al. 1991). The release of flavoring compounds may also be affected by the environmental conditions such as temperature. Yeast and LAB growth can be increased with increase in temperature ranges between 15°C and 27°C. Optimum growth temperature of two

C. milleri strains is found to be between 26°C and 28°C in pure culture. However, decrease in temperature may affect the growth of yeast to largely the same extent as the growth of LAB (Neysens et al. 2003; Simonson et al. 2003).

15.6.4 Flavoring Compounds and Sourdough Fermentation

The bread flavor is composed of hundreds of volatile and nonvolatile compounds, that is, many alcohols, ketones, aldehydes, acids, esters, furan derivates, ether derivates, hydrocarbons, ketones, lactones, pyrazines, pyrrole derivates, and sulfur compounds that serve as flavor stimuli (Schieberle 1996). Chemical analyses of flavor compounds can be combined with the sensory analysis of bread. The compounds that have been positively correlated with flavor of wheat crumb are acetaldehyde, 2-methylpropanoic acid, 2/3-methyl-1-butanol, 3-methylbutanoic acid, isopentanal, 2-nonenal, benzylethanol, 2-phenylethanol, 2,3-butandione and 3-hydoxy-2-butanone, dimethyl sulfide, and 2-furfural (Hansen and Hansen 1996; Rothe 1974).

Abde-el-Malek et al. (1974) concluded that the microflora involved in fermentation of bladi bread are the LAB *L. brevis* and *L. fermenti*, and yeast. The lactobacilli are responsible for the typical flavor of bladi bread. In sourdough fermentation, organic acetic acid and volatile flavoring compounds produced are dependent on the microorganisms in the dough. The volatile compounds are produced both in lactic acid fermentation and in alcoholic fermentation, but the levels of these compounds are much higher in yeast fermentation (Hansen and Hansen 1994b; Meignen et al. 2001). However, Schieberle (1996) concluded that volatiles formed do not affect the final flavor of the bread. The compounds having a high flavor dilution factor would have a significant impact on the final odor. In a French yeast sourdough, more than 40 flavoring components have been identified: 20 alcohols, seven esters, six lactones, six aldehydes, three alkanes, and a single sulfur compound (Frasse et al. 1993).

There are two kinds of aromatic compounds that are produced during fermentation of sourdough. Nonvolatile compounds include organic acids produced by homo- (Gobbetti et al. 1995a) and heterofermentative bacteria (Gobbetti et al. 1995b) that acidify, decrease the pH, and produce aroma in the bread dough (Galal et al. 1978; Barber et al. 1985). Coculture sourdough fermentation is imperative in obtaining an acceptable flavor, since chemically acidified bread and breads prepared with pure commercial starter cultures have failed in sensory preference (Rothe and Ruttloff 1983; Lund et al. 1989). Volatile compounds of sourdough bread are composed of the alcohols, aldehydes, ketones, esters, and sulfur-containing compounds produced by biological and biochemical actions during fermentation and contribute flavor (Spicher 1983).

15.6.5 Nonvolatile Compounds

Interactions between LAB and exogenous enzymes affect the microbial kinetics of acidification, acetic acid production, and textural properties during sourdough fermentation. LAB can be used alone or in association with microbial glucose-oxidase, lipase, endo-xylanase, -amylase, or protease to obtain desirable results. Some of the *Leu. citreum* 23B, *Lac. lactis* subsp. *lactis* 11M, and *L. hilgardii* 51B may positively be influenced by such enzymes added and increased lactic acidification (Cagno et al. 2003). However, in a continuous sourdough fermentation, the association between *L. sanfranciscensis* and *S. cerevisiae* is found optimal for producing acetic acid, whereas the same effect is not found with yeast (Vollmar and Meuser 1992). *Torulopsis holmii* is found to improve dough acidification by *L. sanfranciscensis*, whereas *S. cerevisiae* is found to enhance acid production by *L. sanfranciscensis* and *L. plantarum* (Spicher et al. 1981, 1982). The temperature optima for the growth of *C. humilis* and *L. sanfranciscensis* are 28°C and 32°C. However, less production of acetate by *L. sanfranciscensis* has been observed at 35°C, although lactate and ethanol formation is not affected at this temperature (Brandt et al. 2004). Lactic acid production varies from 3.11 g/kg to 5.14 g/kg, and traces of acetic acid may be produced by some of the *L. plantarum* and *L. farciminis* strains during fermentation (Table 15.1).

15.6.6 Volatile Compounds

Microbial metabolisms affect the production of volatile compounds either generated by hetero- and homolactic or alcoholic sourdough fermentations. Table 15.2 demonstrates the effect of both types of bacteria on the production of flavoring compounds during sourdough fermentation of the breads. Segregation is primarily related to 2-methyl-1-propanol and 2,3-methyl-1-butanol, which are the main products of yeast fermentation. Heterofermentative LAB mainly produce ethyl acetate with some alcohols and aldehydes (Spicher et al. 1982) and homofermentative LAB synthetize diacetyl (Spicher et al. 1981) and other carbonyls, whereas isoalcohols may be produced by yeast fermentation (Hansen and Schieberle 2005). Sourdoughs started with bacterial associations may be characterized by their flavoring compound profiles. *L. brevis* subsp. *lindneri* and *L. plantarum* might have the most complete profiles of flavoring compounds. However, in association with yeasts (with the exception of *L. plantarum* DC400–*S. exiguus* M14 association), both hetero- and homofermentative LAB enhance the formation of the yeast volatile compounds (Damiani et al. 1996; Gobbetti et al. 1995b). Moreover, an interaction between the starter culture and the flour type has also been observed for some of the flavor compounds. Rye flour contains flavor precursors, such as amino acids, fatty acids, and phenolic compounds, which generate various flavoring compounds during processing of products (Schieberle and Grosch 1994; Hansen 1995). However, germinated and extruded rye may impart cereal and fresh flavor, and hard in texture. Dimethyl sulfide and 2-methylbutanal could highly be related to these sensory attributes. Rye sourdough fermented and extruded may contain a sour, intense flavor and porous texture that may be contributed due to the presence of furfural, ethyl acetate, 3-methylbutanol, and 2-methylbutanol. The extrudates of rye flour are rich in 2-ethylfuran, 2-methylfuran, hexanal, and pentanal (Heiniö et al. 2003), but the contributors to the overall crumb flavor of rye bread include 3-methylbutanal, 2-nonenal, 2,4-decadienal, hexanal, phenylacetaldehyde, methional, vanillin, 2,3-butanedione, 3-hydroxy-4,5-dimethyl-2(5*H*)-furanone, and 2- and 3-methylbutanoic acid (Kirchhoff and Schieberle 2001).

The concentration of volatile compounds in wheat bread may be influenced by types of wheat flour in addition to strains of LAB. Moreover, acidification of the sourdoughs and the production of lactic acid may also be primarily influenced by the type of flours, with most lactic acid produced in sourdoughs made from low-grade flour and whole meal wheat flours. The type of flour exerts a significant effect on the production of ethyl acetate and ethanol in sourdoughs fermented with heterofermentative cultures, with the highest amounts detected in sourdoughs made from whole meal flour than low-grade flour (Hansen and Hansen 1994a). Flour ash content from 0.55% to 1% has a positive influence on the total amount of volatiles. Wheat breads made with the addition of *L. plantarum* or *L. sanfranciscensis* had a higher content of 2,3-methyl-1-butanol; with the association of LAB and yeasts, the bread attained a higher flavor quality, which might be caused by the production of higher content of 2,3-methyl-1-butanol, 2-methyl-propanoic acid, 3-methyl-butanoic acid, and 2-phenylethanol (Hansen and Hansen 1996; Hansen et al. 1989b). However, the same effect is probably due to the combination of bacterial acidification and proteolysis (Gobbetti et al. 1994c; Levesque 1991) and might be attributed to *L. sanfranciscensis*. The association of *L. sanfranciscensis*, *L. plantarum*, and *S. cerevisiae* has been used to guarantee an equilibrated aroma in wheat sourdough breads (Hansen and Hansen 1994b).

The species of microorganisms differ and, in general, the strains differ within the species in their ecosystem. However, dough fermented with mixed starters produces more aroma as determined by bread sensory analyses, higher sugar-, acid-, and bitter-type flavors, but dough fermented in parallel with single starters for 15 h at 30°C, mixed and fermented for another 10 h, can enhance production of typical sourdough-type aroma (Merotha et al. 2004). Addition of exoenzymes in the sourdough can also enhance the amount of flavoring components of bread (Martinez-Anaya 1996). Likewise, addition of enzyme active soya flour has increased the concentrations of hexanal, 1-hexanol, 1-penten-3-ol, 1-pentanol, and 2-heptanone, whereas 2-heptenal and 1-octen-3-ol have only been detected in bread containing soya (Luning et al. 1991). More than 40 components have been identified, among which are 20 alcohols, seven esters, six lactones, six aldehydes, three alcanes, and one sulfur compound, in French bread dough prepared with yeast. Except for the aldehydes and the alcanes, all these classes of compounds are found to increase with the activity of *S. cerevisiae*, notably the alcohols due to the formation of fusel alcohols. However, the 2,3-butanedione, 3-methyl 1-butanol, 2-methyl 1-butanol, methional, 2-phenyl ethanol, and two unidentified components with a pungent and a mushroom-like odor are also generated during the fermentation of the dough (Frasse et al. 1993).

Production of volatile flavor compounds during fermentation with pure cultures of *S. cerevisiae*, *C. guilliermondii*, and *L. plantarum* has been investigated, using wheat doughs and several preferments as substrates including maltose and glucose (Stolz et al. 1993). Seven volatile compounds (acetaldehyde, acetone, ethyl acetate, ethanol, hexanal + isobutyl alcohol, and propanol) have been detected when using yeasts alone. Generally, *S. cerevisiae* can produce higher amounts of the different components than *C. guilliermondii*. However, these yeasts can produce larger amounts of volatile flavor compounds during fermentation in glucose and sucrose solutions than in maltose or wheat dough. However, these yeasts may produce more flavoring components than the lactobacilli, but the lactobacilli produce the highest number of volatile compounds in substrates containing flour (Vollmar and Meuser 1992; Torner et al. 1992; Hansen 1995).

The addition of fructose and citrate to the dough may enhance volatile synthesis by LAB. After baking, the ethanol disappeared; 2-methyl-1-propanal was synthesized without affecting lactic and acetic acid concentrations, but the total amount of volatiles is reduced to a level <12.5% of the initial amount, and an increase in the relative percentage of isoalcohols and aldehydes has been detected (Gobbetti et al. 1996).

Processing conditions such as proofing time, temperature, and slackness of sourdough may also affect the aroma volatiles. Low temperature (25°C) and sourdough firmness are considered appropriate for LAB souring activities but limited yeast metabolism. Raising the temperature to 30°C and semifluid sourdoughs can generate more complete volatile profiles. At 3 h, the sourdough may mainly be characterized by isoalcohols, but an increase of leavening time up to 9 h can produce volatiles about three times higher than that at 5 h as a result of LAB contribution (Lund et al. 1989).

There are also many other exo- and endogenous factors that influence the production of volatile compounds during processing and baking of sourdough bread. Production of free amino acids during fermentation as a result of hydrolysis of protein is responsible for changing the volatile compounds profile (Collar and Martinez 1993; Collar et al. 1991; Spicher and Nierle 1984; Gobbetti et al. 1994b, 1994d, 1996). Thiazolines, thiophenes, thiophenones, thiaziles, polythiacycloakanes, pyrroles, and pyrazines may be increased with increase in pH of wheat extrusion cooking (Hansen and Hansen 1994a), sourdough fermentation (Hansen et al. 1989a), proofing, baking, and the type of starters (Hansen et al. 1989b). More desirable results are obtained at pH 4.0–5.5 and at a temperature of 140°C (Przybylski and Kaminski 1983; Kaminski et al. 1981). The level of pyrroles, thiophenes, thiophenones, thiapyrans, and thiazolines may be increased at higher temperatures, whereas furans and aldehydes may be decreased (Bredie et al. 2002).

The addition of lipase, endo-xylanase, and -amylase may be responsible for enhancing the production of acetic acid by *L. hilgardii* 51B. Textural analyses pointed out that sourdoughs started with *L. hilgardii* 51B and individual enzymes might be characterized by higher stability and softening compared to doughs with enzymes added alone (Cagno et al. 2003). Even though the greatest amount of aroma substances is formed during baking due to browning reaction (Rothe and Ruttloff 1983), some compounds from lipid degradation and Maillard interactions are identified in cereal products (Bredie et al. 2002). Maillard reaction products such as pyrazines, pyrroles, furans and sulfur-containing compounds, and lipid degradation products such as alkanals, 2-alkenals and 2,4-alkadienals have been found in high temperature-processed cereal products (Parker et al. 2000).

15.7 SOLID-PHASE MICROEXTRACTION AND FLAVORING COMPOUNDS

Generally, bread quality is judged on the basis of its volume, color, and flavor, but bread aroma is the most important factor influencing the consumer acceptance. Previously, different extraction techniques have been introduced in flavor analysis. These techniques are organic solvent extraction, distillation, and purge-and-trap extraction method. All these techniques are complex, time consuming, and expensive (Ruiz et al. 2003). Solid-phase microextraction (SPME) is a modern, solvent-free sample preparation technique (Arthur and Pawliszyn 1990) that has been developed to combine sampling and sample preparation in one step. SPME technique, along with other applications, is used to measure the volatile flavor profiles of food stuff (Arthur et al. 1992a, 1992b; Young and Peppard 1994). The SPME is a fast, sensitive, solventless, and economical method of sample preparation before analysis through gas chromatography and in some cases, high-performance liquid chromatography (Hook et al. 2002).

The effectiveness of analyte preconcentration using the SPME technique depends on many factors such as type of fiber, sample volume, temperature and extraction time, salting, mode of extraction, desorption of analytes from the fiber, and derivatization (Muller et al. 1999). The SPME uses a short length of fused silica coated with adsorbent called as fiber. This fused silica-coated fiber is immersed directly into an aqueous sample or into the headspace above a liquid or solid sample with a general rule that "similar is dissolved in similar" applies to the SPME, too, that is, polar compounds are adsorbed on polar fibers and nonpolar on nonpolar ones (Pawliszyn 1997). However, the efficiency of preconcentration depends not only on the type of fiber but also on its thickness (coating volume). The type of fiber affects the amount and character of adsorbed species. A polydimethylsiloxanc fiber of 100 μm thickness is most frequently used in the analysis of fragrances and impurities in food products (Gorecki et al. 1999). An SPME method to analyze volatile compounds in bread crumb was developed. Three different fibers were used to determine the volatile compounds; carboxen/polydimethylsiloxane showed the best extraction efficiency (Ruiz et al. 2003).

Table 15.3 demonstrates the extraction and identification of volatile compounds in naan sourdough. The sourdough volatile compounds in naan sourdough have been collected by both SPME and dynamic headspace (DH) techniques and analyzed by gas chromatography with identification based on GC retention times for reference compounds and GC-mass spectrometry. The volatile compounds extracted using SPME technique are alcohols, esters, acids, carbonyls, and aromatic compounds. The alcohols produced are mainly ethanol, 1-propanol, 2-methyl, 1-butanol, 2-methyl-, 1-butanol, 3-methyl, 1-pentanol, 1-hexanol, and heptanol. Ethanol and ethyl acetate are present in naan sourdoughs fermented with cultures containing homofermentative strain *P. acidilacti* and yeast *S. cerevisiae* (LA-1) and mixed culture containing heterofermentative strain *L. brevis* and homofermentative strain *L. casei* with yeast *S. cerevisiae* (LA-5), whereas ethyl lactate is not observed in LA-5 fermented dough. The production of ethyl acetate increases with an increase in flour extraction rate. The carbonyls identified are acetaldehydes, benzaldehydes, and *N*-hexanal. The production of hexanal may be more affected by starter cultures rather than flour extraction rate. Both acetaldehydes and hexanal are produced in higher amounts in LA-1 fermented dough as compared to the one fermented with LA-5.

The volatiles detected using headspace technique are alcohols, acids, esters carbonyls, and many other aromatic compounds. Alcohols produced are mainly ethanol, isoamyl alcohol, 2-butanol, 1-butanol, 3-methyl, 1-pentanol, 1-hexanol, heptanol,

TABLE 15.3
Volatile Compounds Produced by LA-1 and LA-5 Cultures in Naan Sourdough Prepared with Different Extraction Rate Flours

		LA-1[a]		LA-5[a]			LA-1[b]		LA-5[b]	
Compounds	**Control[a]**	**76% ER**	**100% ER**	**76% ER**	**100% ER**	**Control[b]**	**76% ER**	**100% ER**	**76% ER**	**100% ER**
Ethanol	1200	1562	1683	1876	2200	57	80	62	53	186
2-Butanol	–	–	–	–	–	–	1344	714	230	109
Isoamyl alcohol	–	–	–	–	–	4565	–	–	–	–
1-Propanol, 2-methyl	31	100	80	–	–	–	–	–	–	–
1-Butanol, 2-methyl	–	378	299	40	22	–	–	–	–	–
1-Butanol, 3-methyl	77	94	129	376	175	6	67	32	1767	424
1-Pentanol	31	51	35	52	41	15	92	37	171	116
1-Hexanol	124	254	159	203	137	123	319	102	326	297
Heptanol	–	25	26	23	17	–	31	11	46	49
Propanol	–	–	–	–	–	56	77	40	39	44
1-Octanol	–	–	–	–	–	–	16	7	27	–
N-Octanol	–	–	–	–	–	–	16	7	–	–
Nananol	–	–	–	–	–	–	–	–	10	7
Ethyl acetate	231	244	274	444	480	920	1034	708	425	850
Ethyl lactate	–	117	121	–	–	–	9	–	–	–
Acetic acid	–	849	421	339	427	–	19	4	10	16
Butanoic acid	–	–	–	–	–	–	12	18	53	238

Propanoic acid	–		–	24	20	–	151	124	124	171
Hexanoic acid	–		25	–	–	–	38	–	–	–
Acetaldeyhde	41	30	16	19	47	4	234	340	34	50
Nonanal	–	–	–	–	–	–	8	4	10	–
Benzaldehyde	–	–	–	–	3	–	–	–	–	–
Butanal, 3-methyl	–	–	–	–	–	–	–	15	–	–
Decanal	–	–	–	–	–	7	–	–	–	–
2-Buten-1-ol, 2-methyl	–	–	–	–	–	–	–	–	16	–
N-Hexanal	–	15	23	–	–	–	–	–	–	–
Benzene ethanol	–	212	81	–	43	44	47	20	–	–
Furan, 2-pentyl	–	–	–	–	3					
1-Octen-3-ol	–	29	–	–	–	–	10	4	8	7
3-Buten-1-ol, 3-methyl	–	–	–	–	4	–	–	–	32	29

Source: Mueen-ud-Din, G., Effect of wheat flour extraction rates on physico-chemical characteristics of sourdough flat bread, PhD dissertation, National Institute of Food Science and Technology, University of Agriculture, Faisalabad, Pakistan, 2009. With permission.

Note: The amounts of volatile compounds are expressed as Relative Peak Area = (peak area of compound/peak area of internal standard) × 10.

Control Naan dough fermented with *Saccharomyces cerevisiae alone.* ER, extraction rate. *LA-1:* Starter culture containing *Pediococcus acidilacti* + *Saccharomyces cerevisiae*. LA-5: Starter culture containing *Lactobacillus brevis* + *Lactobacillus casei* + *Saccharomyces cerevisiae*.

[a] SPME technique.

[b] Dynamic headspace technique.

propanol, 1-octanol, *N*-octanol, and nonanol. Ethyl acetate is found as a dominating ester, followed by ethyl lactate depending on the type of cultures. It is, however, present in LA-1 sourdoughs, but the flour extraction rate does not affect it too much. The acids detected are acetic acid, butanoic acid, propanoic acid, and hexanoic acid. The acetic acid is found in LA-5 sourdough only, whereas butanoic acid and propanoic acid are detected in both sourdoughs. However, carbonyls are present in the highest amount in LA-1 sourdoughs (Mueen-ud-Din 2009).

REFERENCES

Abde-el-Malek, Y., M.A. El-Leithy, and Y.N. Awad. 1974. Microbiological studies on Egyptian bladi bread making—II. Microbiological and chemical changes during sourdough fermentation. *Chem., Mikrobiol., Technol. Lebensm.* 3: 148–153.

Aksu, Z., and T. Kutsal. 1986. Lactic acid production from molasses utilizing *Lactobacillus delbrueckii* and invertase together. *Biotechnol. Lett.* 8: 157–160.

Arthur, C.L., L.M. Killam, K.D. Buchholz, and J. Pawliszyn. 1992a. Automation and optimization of solid-phase microextraction. *Anal. Chem.* 64: 1960–1964.

Arthur, C.L., L.M. Killam, S. Motlagh, M. Lim, D.W. Potter, and J. Pawliszyn. 1992b. Analysis of substituted benzene compounds in groundwater using solid-phase microextraction. *Environ. Sci. Technol.* 26: 979–983.

Arthur, C.L., and J. Pawliszyn. 1990. Solid-phase microextraction with thermal desorption using fused silica optical fibers. *Anal. Chem.* 62: 2145–2148.

Austin, A., and V. Jhamb. 1964. Protein content and quality of chapatti of improved Indian wheats as affected by nitrogen application. *Indian J. Agron.* 9: 122–127.

Axelsson, L. 1998. Lactic acid bacteria: Classification and physiology. In *Lactic Acid Bacteria Microbiology and Functional Aspects*, ed. S. Salminen and A. von Wright, 1–72. New York: Marcel Dekker.

Azizi, M.H., S.M. Sayeddin, and S.H. Payghambardoost. 2006. Effect of flour extraction rate on flour composition, dough rheology characteristics and quality of flat breads. *J. Agric. Sci. Technol.* 8: 323–330.

Barber, S., R. Baguena, C. Benedito de Barber, and M.A. Martinez-Anaya. 1991. Evolution of biochemical and rheological characteristics and breadmaking quality during a multistage wheat sour dough process. *Z. Lebensm.-Unters. Forsch.* 192: 46–52.

Barber, S., C. Benedito De Barber, M.A. Martinez-Anaya, J. Martinez, and J. Alberola. 1985. Cambios en los acidos organicos volatiles C_2–C_5 durante la fermentacation de mases panarias preparados con masas madres comerciales y con cultivos puros de microorganismos. *Rev. Agroquim. Technol. Alimentos* 25: 223–232.

Barnett, J.A., R.W. Payne, and D. Yarrow. 2000. *Yeasts: Characteristics and Identification*, 3rd edn. Cambridge: Cambridge Univ. Press.

Beatriz, M., S.L. Cuppett, L. Keeler, and R.W. Hutkins. 1994. Browning of mozzarella cheese during high temperature pizza baking. *J. Dairy Sci.* 77: 2850–2853.

Bechtel, W.G., D.F. Meisner, and W.B. Bradley. 1953. The effects of crust on the staling of bread. *Cereal Chem.* 30: 160–168.

Boland, A.B., K. Buhr, P. Giannouli, and S.M. van Ruth. 2004. Influence of gelatin, starch, pectin and artificial saliva on the release of 11 flavour compounds from model gel systems. *Food Chem.* 86: 401–411.

Bolourian, S., M.H.H. Khodaparast, G.G. Movahhed, and M. Afshary. 2010. Effect of lactic acid fermentation (*Lactobacillus plantarum*) on physicochemical flavour, staling and crust properties of semi volume bread (Baguette). *World Appl. Sci. J.* 8: 101–106.

Boraam, F., M. Faid, J.P. Larpent, and A. Breton. 1993. Lactic acid bacteria and yeast associated with sour-dough traditional Moroccan bread. *Sci. Aliment.* 13: 501–509.

Brandt, M.J., W.P. Hammes, and M.G. Ganzle. 2004. Effects of process parameters on growth and metabolism of *Lactobacillus sanfranciscensis* and *Candia humillis* during rye. *Eur. Food Res. Technol.* 218: 333–338.

Brauss, M.S., B. Balders, R.S.T. Linforth, S. Avison, and A.J. Taylor. 1999. Fat content, baking time, hydration and temperature affect flavor release from biscuits in model-mouth and real systems. *Flavor Fragrance J.* 14: 351–357.

Bredie, W.L.P., M. Boesveld, M. Martens, and L. Dybdal. 2006. Modification of bread crust flavor with enzymes and flavor precursors. *Dev. Food Sci.* 43: 225–228.

Bredie, W.L.P., D.S. Mottram, and R.C.E. Guy. 2002. Effect of temperature and pH on the generation of flavor volatiles in extrusion cooking of wheat flour. *J. Agric. Food Chem.* 50: 1118–1125.

Brummer, J.M., and K. Lorenz. 1991. European developments in wheat sourdoughs. *Cereal Foods World* 36: 310–314.

Burseg K.M.M., J. Hort, J.R. Mitchell, and A.J. Taylor. 2005. Flavor perception of biscuits—biscuit composition, texture and sweetness perception. Paper presented at the 230th ACS National Meeting Washington, D.C.

Cagno, R.D., M.D. Angelis, A. Corsetti, et al. 2003. Interactions between sourdough lactic acid bacteria and exogenous enzymes: Effects on the microbial kinetics of acidification and dough textural properties. *Food Microbiol.* 20: 67–75.

Cai, Y., H. Okada, H. Mori, Y. Benno, and T. Nakase. 1999. *Lactobacillus paralimentarius* sp. *nov.*, isolated from sourdough. *Int. J. Syst. Evol. Microbiol.* 49:1451–1455.

Calderon, M., G. Loiseau, and J.P. Guyot. 2003 Fermentation by *Lactobacillus fermentum* Ogi E1 of different combinations of carbohydrates occurring naturally in cereals: Consequences on growth energetic and -amylase production. *Int. J. Food Microbiol.* 80: 161–169.

Collar, C. 1996. Biochemical and technological assessment of the metabolism of pure and mixed cultures of yeast and lactic acid bacteria in breadmaking applications. *Food Sci. Technol. Int.* 2: 349–367.

Collar, C., and C.S. Martinez. 1993 Amino acid profiles of fermenting wheat sour doughs. *J. Food Sci.* 58: 1324–1328.

Collar, C., A.F. Mascaros, J.A. Prieto, and C. Benedito de Barber. 1991. Changes in free amino acids during fermentation of wheat doughs started with pure culture of lactic acid bacteria. *Cereal Chem.* 68: 66–72.

Corsetti, A., M. Gobbetti, and J. Rossi. 1994. Sourdough fermentation in presence of added soluble carbohydrates. *Microbiol., Aliments, Nutr.* 12: 377–385.

Corsetti, A., L. Settanni, D. Van Sinderen, G.E. Felis, F. Dellaglio, and M. Gobbetti. 2005. *Lactobacillus rossii* sp. *nov.* isolated from wheat sourdough. *Int. J. Syst. Evol. Microbiol.* 55: 35–40.

Corsetti, A., M. Gobbetti, B. De Marco, et al. 2000. Combined effect of sourdough lactic acid bacteria and additives on bread firmness and staling. *J. Agric. Food Chem.* 48: 3044–3051.

Corsetti, A., M. Gobbetti, J. Rossi, and P. Damiani. 1998. Antimould activity of sourdough lactic acid bacteria: Identification of a mixture of organic acids produced by *Lactobacillus sanfrancisco* CB1. *Appl. Microbiol. Biotechnol.* 50: 253–256.

Corsetti, A., P. Lavermicocca, M. Morea, F. Baruzzi, N. Tosti, and M. Gobbetti. 2001. Phenotypic and molecular identification and clustering of lactic acid bacteria and yeasts from wheat (species *Triticum durum* and *Triticum aestivum*) sourdoughs of Southern Italy. *Int. J. Food Microbiol.* 64: 95–104.

Cossignani, L., M. Gobbetti, P. Damiani, A. Corseti, M.S. Simonetti, and G. Manfredi. 1996. The sourdough microflora. Microbiological, biochemical and bread making characteristics of doughs fermented with freeze dried mixed starters, freeze dried wheat sourdough and mixed fresh cell-starter. *Lebensm.-Unters. Forsch.* 203: 88–94.

Czerny, M., and P. Schieberle. 2002. Important aroma compounds in freshly grounded whole meal and white wheat flour—Identification and quantitative changes during fermentation. *J. Agric. Food Chem.* 50: 6835–6840.

Damiani, P., M. Gobbetti, L. Cossignani, A. Corsetti, M.S. Simonetti, and J. Rossi. 1996. The sourdough microflora. Characterization of hetero- and homofermentative lactic acid bacteria, yeasts and their interactions on the basis of the volatile compounds produced. *Lebens.-Wiss. Technol.* 29: 63–70.

De Angelis, M., S. Siragusa, M.G. Berloco, L. Caputo, L. Settanni, and G. Alfonsi. 2006. Selection of potential probiotic lactobacilli from pig faces to be used as additives in pelletted feedings. *Res. Microbiol.* 157: 792–801.

Dellagio, F. 1988. Starters for fermented milks. Thermophilic starters. *Int. Dairy Fed. Bull.* 227: 27–34.

De-Vuyst, L., and P. Neysens. 2004. The sourdough microflora: Biodiversity and metabolic interactions. *Trends Food Sci. Technol.* 16: 1–14.

Doerry, W. 1989. Nonfat dry milk in no-time bread doughs. *Am. Inst. Baking Tech. Bull.* 11: 1–8.

El-Megeed, E.A., Mohamed, and D.C. Sands. 1989. Method and compositions for improving the nutritive value of foods via *Lactobacillus feremenrum*. United States Patent Number 4889810.

Foschino, R., and A. Galli. 1997. Italian style of life: Pane, amore, lievito natural. *Tech. Alim.* 1: 42–59.

Frankel, E.N. 1982. Volatile lipid oxidation products. *Prog. Lipids Res.* 22: 4–33.

Frasse, P., S. Lambert, D. Richard-Molard, and H. Chiron. 1993. The Influence of fermentation on volatile compounds in french bread dough. *Lebensm.-Wiss. Technol.* 26: 126–132.

Galal, A.M., J.A. Johnson, and E. Varriano-Marston. 1978. Lactic acid and volatile (C_2-C_5) organic acids of San Francisco sourdough French bread. *Cereal Chem.* 55: 461–468.

Gassenmeier, K., and P. Schieberle. 1995. Potent aromatic compounds in the crumb of wheat bread (French-type). Influence of preferment and studies on the key odorants during dough processing. *Z. Lebensm.-Unters. -Forsch.* 201: 241–248.

Gianotti, A., L. Vannini, M. Gobbetti, A. Corsetti, F. Gardini, and M.E. Guerzoni. 1997. Modelling of the activity of selected starters during sourdough fermentation*1. *Food Microbiol.* 14: 327–337.

Gobbetti, M. 1998. The sourdough microflora: Interactions of lactic acid bacteria and yeasts. *Trends Food Sci. Technol.* 9: 267–274.

Gobbetti, M., and A. Corsetti. 1996. Co-metabolism of citrate and maltose by *Lactobacillus brevis* subsp. *lindneri* cb1 citrate-negative strain: Effect on growth, end-products and sourdough fermentation. *Z. Lebensm.-Unters. -Forsch.* 203: 82–87.

Gobbetti, M., and A. Corsetti. 1997. *Lactobacillus sanfranciscoa* key sourdough lactic acid bacterium: A review. *Food Microbiol.* 14: 175–188.

Gobbetti, M., A. Corsetti, and S. De Vincenzi. 1995a. The sourdough microflora. Characterization of homofermentative lactic acid bacteria based on acidification kinetics and impedance tests. *Ital. J. Food Sci.* 2: 91–102.

Gobbetti, M., A. Corsetti, and S. De Vincenzi. 1995b. The sourdough microflora. Characterization of heterofermentative lactic acid bacteria based on acidification kinetics and impedance tests. *Ital. J. Food Sci.* 2: 103–112.

Gobbetti, M., A. Corsetti, and J. Rossi. 1994a. The sourdough microflora. Interactions between lactic acid bacteria and yeasts: Metabolism of carbohydrates. *Appl. Microbiol. Biotechnol.* 41: 456–460.

Gobbetti, M., A. Corsetti, and J. Rossi. 1994b. The sourdough microflora. Interactions between lactic acid bacteria and yeasts: Metabolism of amino acids. *World J. Microbiol. Biotechnol.* 10: 275–279.

Gobbetti, M., A. Corsetti, and J. Rossi. 1995c. Maltose–fructose co-fermentation by *Lactobacillus brevis* subsp. *lindneri* CB1 fructose-negative strain. *Appl. Microbiol. Biotechnol.* 42: 939–944.

Gobbetti, M., A. Corsetti, J. Rossi, F. La Rosa, and S. De Vincenzi. 1994c. Identification and clustering of lactic acid bacteria and yeasts from wheat sourdoughs of central Italy. *Ital. J. Food Sci.* 1: 85–94.

Gobbetti, M., M.S. Simonetti, J. Rossi, L. Cossignani, A. Corsetti, and P. Damiani. 1994d. Free D- and L-amino acid evolution during sourdough fermentation and baking. *J. Food Sci.* 59: 881–884.

Gobbetti, M., E. Smacchi, P.F. Fox, L. Stepaniak, and A. Corsetti. 1996. The sourdough microflora. Cellular localization and characterization of proteolytic enzymes in lactic acid bacteria. *Lebensm.-Wiss. Technol.* 29: 561–569.

Gobbetti, M., M.S. Simonetti, A. Corsetti, F. Santinelli, J. Rossi, and P. Damiani. 1995d. Volatile compound and organic acid productions by mixed wheat sour dough starters: Influence of fermentation parameters and dynamics during baking. *Food Microbiol.* 12: 497–507.

Gorecki, T., X. Yu, and J. Pawliszyn. 1999. Theory of analyte extraction by selected polymer SPME fibers. *Analyst* 124: 643–652.

Gül, H., S. Özçelik, O. Sadıç, and M. Certel. 2005. Sourdough bread production with lactobacilli and *S. cerevisiae* isolated from sourdoughs. *Process Biochem.* 40: 691–697.

Gullo, M., A.D. Romano, A. Pulvirenti, and P. Giudici. 2002. *Candida humilis*-dominant species in sourdoughs for the production of durum wheat bran flour bread. *Int. J. Food Microbiol.* 80: 55–59.

Hammes W.P., and M. Ganzle. 1995. The genus *Lactobacillus*. In *The Genera of Lactic acid Bacteria*, Vol. II, ed. B.J.B. Wood and W.H. Holzapfel, 19–54. London: Blackie Academic and Professional.

Hammes, W.P., and R.F. Vogel. 1997. Sauerteig. In *Mikrobiolgie der Lebensmittel plfanzlicher Herkunf*, 1st edn., ed. G. Muller, W. Holzapfel, and H. Weber, 263–285. Hamburg: Beher Verlag.

Hammes, W.P., M.J. Brandt, K.L. Francis, M. Rosenheim, F.H. Seitter, and S. Vogelmann. 2005. Microbial ecology of cereal fermentations. *Trends Food Sci. Technol.* 16: 4–11.

Hann, D., and W.R. Morrison. 1975. Effects of ingredients on the oxidation of linoleic acid by lipoxygenase in bread doughs. *J. Sci. Food Agric.* 26: 493.

Hansen, A. 1995. Flavor compounds in rye sourdough and rye bread. In *International Rye Symposium: Technology and Products*, ed. K. Pountanen and K. Autio, 161–169. VTT Symposium, Helsinki, Finland.

Hansen, A., and B. Hansen. 1993. Flavor compounds in wheat sourdoughs. In *Proceedings of the European Conference on Food Chemistry*, ed. C. Benedito De Barber, C. Collar, M.A. Martinez-Anaya, and J. Morel. Valencia: IATA CSIC.

Hansen, Å., and B. Hansen. 1994a. Influence of wheat flour type on the production of flavor compounds in wheat sourdoughs. *J. Cereal Sci.* 19: 185–190.

Hansen, Å., and B. Hansen. 1994b. Volatile compounds in wheat sourdoughs produced by lactic acid bacteria and sourdough yeasts. *Z. Lebensm.-Unters.-Forsch.* 198: 202–209.

Hansen, Å., and B. Hansen. 1996. Flavor of sourdough wheat bread crumb. *Z. Lebensm.-Unters.-Forsch.* 202: 244–249.

Hansen, A., B. Lund, and M.J. Lewis. 1989a. Flavor of sourdough rye bread crumb. *Lebensm.-Wiss. Technol.* 22: 141–144.

Hansen, A., B. Lund, and M.J. Lewis. 1989b. Flavor production and acidification of sourdoughs in relation to starter culture and fermentation temperature. *Lebensm.-Wiss. Technol.* 22: 145–149.

Hansen, Å., and P. Schieberle. 2005. Generation of aroma compounds during sourdough fermentation: Applied and fundamental aspects. *Trends Food Sci. Technol.* 16: 1–10.

Heiniö, R., K. Katina, A. Wilhelmson, et al. 2003. Relationship between sensory perception and flavor-active volatile compounds of germinated, sourdough fermented and native rye following the extrusion process. *Lebensm.-Wiss. Technol.* 36: 533–545.

Hill, L.G. 1974. Yogurt-containing dough composition and baked product made therefrom. U.S. patent 3,846,561.

Hook, G.L., G.L. Kimm, T. Hall, and P.A. Smith. 2002. Solid-phase microextraction (SPME) for rapid field sampling and analysis by gas chromatography–mass spectrometry (GC–MS). *Trends Anal. Chem.* 21: 534–539.

Jenson, I. 1998. Bread and baker's yeast. In *Microbiology of Fermented Foods*, ed. B.J.B. Wood, 172–198. London: Blackie Academic and Professional.

Kaminski, E., R. Przybilski, and L. Gruchala. 1981. Thermal degradation of precursors and formation of flavor compounds during heating of cereal products: Part I. Changes of amino acids and sugars. *Die Nahrung* 25: 507–518.

Katina, K., R.L. Heinio, K. Autio, and K. Poutanen. 2006. Optimization of sourdough process for improved sensory profiles and texture of wheat bread. *LWT—Food Sci. Technol.* 39: 1189–1202.

Kazanskaya, L.N., O.V. Afanasyeva, and V.A. Patt. 1983. Microflora of rye sours and some specific features of its accumulation in bread baking plants of the USSR. In *Developments in Food Science. Progress in Cereal Chemistry and Technology*, ed. J. Holas and F. Kratochvil, 759–763. London: Elsevier.

Kirchhoff, E., and P. Schieberle. 2001. Determination of key aroma compounds in the crumb of a three-stage dilution assays and sensory studies. *J. Agric. Food Chem.* 49: 4304–4311.

Kline, L., T.F. Sugihara, and L.B. McCready. 1970. Nature of the San Francisco sourdough French bread process: 1. Mechanics of the process. *Baker's Dig.* 44: 48–50.

Lehmann, T.A., and P. Dreese. 1981. Functions of non fat dry milk and other milk products in yeast raised bakery foods. *Am. Inst. Baking Technol. Bull.* 3: 1–9.

Levesque, C. 1991. 'Etude des Potentialités de la Souche S47 de Saccharomyces cerevisiae à produire des alcools supérieurs à partir de composés organiques identifiés et des précurseurs de la farinè in thè se d'université, Nantes.

Linko, Y., P. Javanainen, and S. Linko. 1997. Biotechnology of bread baking. *Trends Food Sci. Technol.* 8: 339–344.

Lotong, V., I.V. Chambers, and D.H. Chambers. 2000. Determination of the sensory attributes of wheat sourdough bread. *J. Sens. Stud.* 15: 309–326.

Lund, B., Å. Hansen, and M.J. Lewis. 1989. The influence of dough yield on acidification and production of volatiles in sourdoughs. *Lebensm.-Wiss. Technol.* 22: 150–153.

Luning, P.A., J.P. Roozen, R.A.F.J. Moëst, and M.A. Posthumus. 1991. Volatile composition of white bread using enzyme active soya flour as improver. *Food Chem.* 41: 81–91.

Main, A. 1991. Fermented dairy products as food ingredients. *Food Res. Qual.* 51: 120–125.

Maleki, M., R.C. Hoseney, and P.J. Mattern. 1980. Effect of loaf volume, moisture content, and protein quality on the softness and staling rate of bread. *Cereal Chem.* 57: 138–140.

Martinez-Anaya, M.A. 1996. Enzymes and bread flavor. *J. Agric. Food Chem.* 44: 2469–2480.

Martinez-Anaya, M.A., B. Pitarch, and C. Benedito de Barber. 1993. Biochemical characteristics and breadmaking performance of freeze-dried wheat sour dough starters. *Z. Lebensm.-Unters.-Forsch.* 196: 360–365.

Meignen, B., B. Onno, P. Gélinas, M. Infantes, S. Guilois, and B. Cahagnier. 2001. Optimization of sourdough fermentation with *Lactobacillus brevis* and baker's yeast. *Food Microbiol.* 18: 239–245.

Merotha, C.B., W.P. Hammesa, and C. Hertela. 2004. Characterisation of the microbiota of rice sourdoughs and description of *Lactobacillus spicheri* sp. *nov. Syst. Appl. Microbiol.* 27: 151–159.

Merseburger, T., A. Ehret, O. Geiges, B. Baumann, and W. Schmidt-Lorenz. 1995. Microbiology of dough preparation: VII. Production and use of preferments to produce wheat-bread. *Mitt. Geb. Lebensm. Hyg.* 86: 304–324.

Miettinen, S.M., L. Hyvonen, R.S.T. Linforth, A.J. Taylor, and H. Tuorila. 2004. Temporal aroma delivery from milk systems containing 0–5% added fat, observed by free choice profiling, time intensity, and atmospheric pressure chemical ionization–mass spectrometry techniques. *J. Agric. Food Chem.* 52: 8111–8118.

Mueen-ud-Din, G. 2009. Effect of wheat flour extraction rates on physico-chemical characteristics of sourdough flat bread. PhD dissertation, National Institute of Food Science and Technology, University of Agriculture, Faisalabad, Pakistan.

Muller, J., T. Górecki, and J. Pawliszyn. 1999. Optimization of the SPME device design for field applications. *Fresenius' J. Anal. Chem.* 364: 610–616.

Neubauer, H., E. Glaasker, W.P. Hammes, B. Poolman, and W.N. Konings. 1994. Mechanism of maltose uptake and glucose excretion in *Lactobacillus sanfrancisco*. *J. Bacteriol.* 176: 3007–3012.

Neysens, P., W. Messens, and L.D. Vuyst. 2003. Effect of sodium chloride on growth and bacteriocin production by *Lactobacillus amylovorus* DCE 471. *Int. J. Food Microbiol.* 88: 29–39.

Nout, M.J.R., and T. Creemers-Molenaar. 1987. Microbiological properties of some wheatmeal sourdough starters. *Chem. Mikrobiol. Technol. Lebensm.* 10: 162–167.

Onno, B., and P. Rouseel. 1994. Technologie et microbiologie de le panification au levain. In *Bacteries lactiques*, ed. H. De Roissart and F.M. Luquet, Vol. 2, 293–321. France: Lorica Uriage.

Ottogalli, G., A. Galli, and R. Foschino. 1996. Italian bakery products obtained with sour dough: Characterization of the typical microflora. *Adv. Food Sci.* 18: 131–144.

Oura, E., H. Soumalainen, and R. Wiskari. 1982. Sourdough. In *Fermented Foods*, ed. A.H. Rose, 123–146. London: Academic Press.

Park, Y.H., L.H. Jung, and E.R. Jeon. 2006. Quality characteristics of bread using sourdough. *J. Food Sci. Nutr.* 33: 323–327.

Parker, J.K., G.M. Hassell, D.S. Mottram, and R.C. Guy. 2000. Sensory and instrumental analyses of volatile generated during the extrusion cooking of oat flour. *J. Agric. Food Chem.* 48: 3497–3506.

Pauline, P., G. Arvisenet, J. Grua-Priol, C. Fillonneau, A. Le-Bail, and C. Prost. 2010. Influence of inulin on bread: Kinetics and physico-chemical indicators of the formation of volatile compounds during baking. *Food Chem.* 119: 1474–1484.

Pauline, P., G. Arvisenet, J. Grua-Priol, D. Colas, C. Fillonneau, A. Le Bail, et al. 2008. Influence of formulation and process on the aromatic profile and physical characteristics of bread. *J. Cereal Sci.* 48: 686–697.

Pawliszyn, J. 1997. *Solid-Phase Microextraction. Theory and Practice*. New York: Wiley-VCH.

Plessas, S., I. Pherson, A. Bekatorou, P. Nigam, and A.A. Koutinas. 2005. Bread making using kefir grains as baker's yeast. *Food Chem.* 93: 585–589.

Pozo-Bayón, M.A., E. Guichard, and N. Cayot. 2006. Flavor control in baked cereal products. *Food Rev. Int.* 22(4): 335–379.

Przybilski, R., and E. Kaminski. 1983. Thermal degradation of precursors and formation of flavor compounds during heating of cereal products: Part II. The formation and changes of volatile flavor compounds in thermally treated malt extracts at different temperature and pH. *Die Nahrung* 27: 487–496.

Qarooni, J. 1996. *Flat Bread Technology*. New York: Chapman and Hall.

Rehman, S., A. Paterson, and J.R. Piggott. 2006a. Flavor in sourdough bread: A review. *Trends Food Sci. Technol.* 17: 557–566.

Rehman, S., A. Paterson, and J.R. Piggott. 2006b. Optimization of chapatti textural quality using British wheat cultivar flours. *Int. J. Food Sci. Technol.* 41(Suppl. 2): 30–36.

Rehman, S., A. Paterson, and J.R. Piggott. 2007a. Chapatti quality from British wheat cultivar flours. *LWT—Food Sci. Technol.* 40: 775–784.

Rehman, S., H. Nawaz, S. Hussain, M.M. Ahmad, M.A. Murtaza, and M.S. Ahmad. 2007b. Effect of sourdough bacteria on the quality and shelf life of bread. *Pak. J. Nutr.* 6: 562–565.

Roberts, D.D., P. Pollien, N. Antille, C. Lindinger, and C. Yeretzian. 2003. Comparison of nosespace, headspace and sensory intensity ratings for the evaluation of flavor absorption by fat. *J. Agric. Food Chem.* 51: 3636–3642.

Röcken, W., and P.A. Voysey. 1993. Sourdough fermentation in bread making. *J. Appl. Bacteriol.* 79(Suppl.): 38S–39S.

Rothe, M. 1974. *Aroma von Bort*. Belgium: Ackademie-Verlag.

Rothe, M., and H. Ruttloff. 1983. Aroma retention in modern bread production. *Die Nahrung* 27: 505–512.

Rouzaud, O., and M.A. Martinez-Anaya. 1993. Effect of processing conditions on oligosaccharide profile of wheat sourdoughs. *Z. Lebensm.-Unters. -Forsch.* 197: 434–439.

Ruiz, J.A., J. Quilez, M. Mestres, and J. Guash. 2003. Solid-phase microextraction method for haedspace analysis of volatile compounds in bread crumb. *Cereal Chem.* 80: 255–259.

Samuel P.H., J. Dufour, N. Hamid, W. Harvey, and C.M. Delahunty. 2009. Characterisation of fresh bread flavor: Relationships between sensory characteristics and volatile composition. *Food Chem.* 116: 249–257.

Schieberle, P. 1996. Intense aroma compounds—Useful tool to monitor the influence of processing and storage on bread aroma. *Adv. Food Sci.* 18: 237–244.

Schieberle, P., and W. Grosch. 1994. Potent odorants of rye bread crust—Differences from the crumb and from wheat bread crust. *Z. Lebensm.-Unters. -Forsch.* 198: 292–296.

Seitz, L.M., O.K. Chung, and R. Rengarajan. 1998. Volatiles in selected commercial breads. *Cereal Chem.* 75: 847–853.

Seppi, A. 1984. Studio sulla biochimica degli zuccheri nella fermentazione del pane. *Rev. Soc. Ital. Sci. Alimentos* 25: 223–232.

Settanni, L., D. Van Sinderen, J. Rossi, and A. Corsetti. 2005a. Rapid differentiation and *in situ* detection of 16 sourdough *Lactobacillus* species by multiplex PCR. *Appl. Environ. Microbiol.* 71: 3049–3059.

Seuvre, A.M., E. Philippe, S. Rochard, and A. Voilley. 2006. Retention of aroma compounds in food matrices of similar rheological behaviour and different compositions. *Food Chem.* 96: 104–114.

Shenkenberg, D.R., F.G. Barnes, and E.J. Guy. 1972. New process for sourdough bread improves uniformity and reduces process time. *Food Prod. Dev.* 6: 29–30, 32.

Simonson, L., H. Salovaar, and M. Korhola. 2003. Response of wheat sourdough parameters to temperature, NaCl and sucrose variations. *Food Microbiol.* 20: 193–199.

Spicher, G. 1983. Baked goods. In *Biotechnology*, ed. J.H. Rehm and G. Reed, 1–80. Weinheim: Verlag Chemie.

Spicher, G., and W. Nierle. 1984. The microflora of sourdough: XVIII. Communication: The protein degrading capabilities of the lactic acid bacteria of sourdough. *Z. Lebensm.-Unters.-Forsch.* 178: 389–392.

Spicher, G., E. Rabe, R. Sommer, and H. Stephan. 1981. The microflora of sourdough. xiv. Communication: About the behaviour of homofermentative sourdough bacteria and yeasts in mixed culture. *Z. Lebensm.-Unters. -Forsch.* 173: 291–296.

Spicher, G., E. Rabe, R. Sommer, and H. Stephan. 1982. Communication: On the behaviour of heterofermentative sourdough bacteria and yeasts in mixed culture. *Z. Lebensm.-Unters.-Forsch.* 174: 222–227.

Stolz, P., G. Böcker, W.P. Hammes, and R.F. Vogel. 1995. Utilization of electron acceptors by *Lactobacilli* isolated from sourdough. *Z. Lebensm.-Unters.-Forsch.* 201: 91–96.

Stolz, P., G. Böcker, R.F. Vogel, and W.P. Hammes. 1993. Utilization of maltose and glucose by *Lactobacilli* isolated from sourdough. *FEMS Microbiol. Lett.* 109: 237–242.

Succi, M., A. Reale, C. Andrighetto, A. Lombardi, E. Sorrentino, and R. Coppol. 2003. Presence of yeasts in southern Italian sourdoughs from *Triticum aestivum* flour. *FEMS Microbiol. Lett.* 225: 143–148.

Sugihara, T.F., L. Kline, and L.B. McCready. 1970. Nature of San Francisco sour dough French bread process: II. Microbiological aspects. *Baker's Dig.* 44: 51–56.

Sugihara, T.F. 1977. Nontraditional fermentation in the production of backed goods. *Baker's Dig.* 51, 76, 78, 80, 142.

Sugihara, T.F. 1985. Microbiology of bread making. In *Microbiology of Fermented Foods*, ed. B.J.B. Wood, 249–261. London: Elsevier Applied Science.

Suihko, M.L., and V. Makinen. 1984. Tolerance of acetate, propionate and sorbate by *Saccharomyces cerevisiae* and *Torulopsis holmii*. *Food Microbiol.* 1: 105–110.

Torner, M.J., M.A. Martínez-Anaya, B. Antuña, and C. Benedito de Barber. 1992. Headspace flavor compounds produced by yeasts and *Lactobacilli* during fermentation of preferments and bread doughs. *Int. J. Food Microbiol.* 15: 145–152.

Trüper, H.G., and L. de Clari. 1997. Taxonomic note: Necessary correction of specific epithets formed as substantives (nouns) 'in apposition'. *Int. J. Syst. Bacteriol.* 47: 908–909.

Taylor, A.J., and J. Hort. 2004. Measuring proximal stimuli involved in flavour perception. In *Flavour Perception*, ed. A.J. Taylor and D.D. Roberts, 1–38. Oxford: Blackwell.

Thiele, C., M.G. Ganzle, and R.F. Vogel. 2002. Contribution of sourdough lactobacilli, yeast and cereal enzymes to the generation of amino acids in dough relevant for bread flavor. *Cereal Chem.* 79: 45–51.

Vernocchi, P., S. Valmorri, V. Gatto, et al. 2004. A survey on yeast microbiota associated with an Italian traditional sweet-leavened baked good fermentation. *Food Res. Int.* 37: 469–476.

Vogel, R.F. 1997. Microbial ecology of cereal fermentations. *Food Technol. Biotechnol.* 35: 51–54.

Vogel R.F., G. Bocker, P. Stolz, et al. 1994. Identification of *Lactobacilli* from sourdough and description of *Lactobacillus pontis* sp. *nov*. *Int. J. Syst. Bacteriol.* 44: 223–229.

Vogel, R.F., R. Knorr, M.R.A. Müller, U. Steudel, M.G. Gänzle, and M.A. Ehrmann. 1999. Non-dairy lactic acid fermentations: The cereal world. *Antonie van Leeuwenhoek* 76: 403–411.

Vollmar, A., and F. Meuser. 1992. Influence of starter cultures consisting of lactic acid bacteria and yeast on the performance of a continuous sourdough fermenter. *Cereal Chem.* 69: 20–27.

Vuystn, L.D., and P. Neysens. 2004. The sourdough microflora: Biodiversity and metabolic interactions. *Trends Food Sci. Technol.* 15: 1–14.

Williams, T., and G. Pullen. 1998. Functional ingredients. In *Technology of Breadmaking*, ed. S.P. Cauvain and L.S. Young, 45–80. London: Blakie Acadamic and Professional.

Wright, K.M., B.P. Hills, T.A. Hollowood, R.S.T. Linforth, and A.J. Taylor. 2003. Persistence effects in flavour release from liquids in the mouth. *Int. J. Food Sci. Technol.* 38: 343–350.

Young, X., and T. Peppard. 1994. Solid-phase microextraction for flavor analysis. *J. Agric. Food Chem.* 42: 1925–1930.

Yangsoo, K., W. Huang, H. Zhu, and P. Rayas-Duarte. 2009. Spontaneous sourdough processing of Chinese Northern-style steamed breads and their volatile compounds. *Food Chem.* 114: 685–692.

16 Food Taints and Off-Flavors

Kathy Ridgway and S. P. D. Lalljie

CONTENTS

16.1 INTRODUCTION

The presence of a compound causing a taint or off-flavor may cause a food to be unfit for human consumption or may result in consumer non acceptance. Although the food with the taint or off-flavor is often not a safety risk to the consumer, the perception of low safety or quality, brand damage, and adverse publicity can be extremely costly to the food industry. The compounds responsible for taints are frequently only present at trace and ultra-trace (sub ppb, $\mu g\ kg^{-1}$) levels, and due to the complexity of food as a matrix, this presents a significant challenge to the analytical chemist. This chapter discusses the origins of food taints and outlines the approaches taken to both identify the origin of a taint and quantitative methods of analysis. This chapter aims to provide only a summary, but more detailed discussions on the mechanisms involved in taint and off-flavor formation in foods can be found in books dedicated to the subject (Baigrie 2003; Saxby 1993).

16.2 POTENTIAL SOURCES OF FOOD TAINTS AND OFF-FLAVORS

Taints and off-flavors in foods can originate from many sources (Whitfield 1998; Mottram 1998; Ridgway et al. 2010a), at all points in the supply chain, as illustrated in Table 16.1. The technical difference between a taint and an off-flavor is that a taint in food results from contamination by a foreign chemical from an external source, whereas an off-flavor is an atypical odor or taste resulting from a compound formed by internal deterioration of the food, from microbiological spoilage, or chemical reaction. In each part of the food supply chain as practices and processes are changed or developed, additional sources of compounds with the potential to cause taints or off-flavors may emerge. Some of the more common taints and off-flavors associated with raw materials, microorganisms, processing, packaging and storage are discussed below.

16.2.1 Raw Materials

Water can be responsible for food taints as any product produced using contaminated water is also likely to be tainted. Water containing a source of phenol (e.g., from soil) can then be chlorinated to produce chlorophenols and similarly the presence of bromine can lead to bromophenols. Most taints detected in fish originate from the aquatic environment (Tucker 2000; Whitfield 1999). An "earthy–musty" taint in water has been reported to originate from the presence of geosmin, 2-methylisoborneol, and haloanisoles (Zhang et al. 2005) and is generally associated with microorganisms, particularly bacteria (Watson et al. 2003). Various treatment processes are available to remove off-odors from water. In dairy products, "transmitted" off-flavors have been reported due to transfer of substances from the cow's feed or environment (Jeon 1993). Compounds can originate from poorly ventilated barns, feeding too close to milking time, or a change in feed. For example, grazing on weeds, such as wild garlic and onion, can produce a "weedy" off-flavor in milk. Taints can also be produced in the digestive system of animals, as is the case for a taint described as "fecal" or "urinous," attributed to the presence of indole and skatole, which accumulate in the animal tissue tainting the meat. This taint has been particularly observed and studied in uncastrated male pigs (boar taint) and has been linked to the presence of 5-α-androst-16-en-3-one (a pheromonal steroid) and also to levels of the amino acid tryptophan in the diet (Verheyden et al. 2007; Zabolotsky et al. 1995; Hansson et al. 1980; Patterson 1968).

16.2.2 Microorganisms

As microorganisms grow and metabolize, they produce chemicals, and the odors produced are indications of microbial spoilage of the food. The materials, as well as being responsible for taints and off-flavors in foods, could be harmful

TABLE 16.1
Examples of Taints and Their Possible Origins

Odor Descriptor	Compounds	Possible Origins
Acrid	Acrolein	Formed microbiologically in distillery mashes
Acrid/plastic	Ethyl and methyl acrylate	Industrial chemicals
	Methyl methacrylate	Industrial chemicals
Almond	Heptane-2-one	Oxidation of oils (rancid coconut), light-induced oxidation of fats
	1,4-Dichlorobenzene	Drain cleaners and moth-proofing agents
	Benzaldehyde	Packaging—reaction by product
Apple	Damascenone	Microbiological—produced by *Actinomycetes*
	Oct-1-en-3-one	Autooxidation of fats and sometimes found in plastics containing diisooctyl phthalate
	Acetaldehyde	Overproduction in milk cultures or yogurt (also described as green). Also can be a degradation product of PET packaging
Brine/seaside	Bromocresol (2-bromo-4-methylphenol)	Associated with corresponding bromophenol/anisole
	Dibromocresol (2,6-dibromo-4-methylphenol	Associated with corresponding bromophenol/anisole
Cabbage	Dimethyl sulfide	Reactions with methionine and the cause of off-flavor in beer
	Diphenyl sulfide	Photoinitiator for cationic inks
Cardboard	2,4-Nonadienal	Autooxidation of oils and fats
	Oct-1-en-3-one	Autooxidation of fats and sometimes found in plastics containing diisooctyl phthalate
	Hexanal	Lipid degradation associated with paper (decarboxylation and oxidation of lignin)
Catty/cats urine	4-Mercapto-4-methylpentan-2-one	Reaction of hydrogen sulfide (in foods) with mesityl oxide (solvent impurity found in some paints/varnishes)
Chemical	Chlorobenzene	Used as an antifungal agent in some glues
	2,4- or 2,6-dichlorophenol	Fungicides, biocides, and herbicide intermediates. Found in packaging—wood pulp that has been treated and cardboard
Cucumber	*trans*-2-*cis*-6-Nonadienal	Algae in water
Disinfectant	6-Chloro-*o*-cresol (2-methyl-6-chlorophenol)	Disinfectants and drain cleaners or impurity in some herbicides
	2-Chlorophenol	Chlorination of phenol (associated with 2-methyl-6-chlorophenol), e.g., from water containing phenol (e.g., from peat soil) that is chlorinated
	2,3-Dichlorophenol	Fungicides, biocides, and herbicide intermediates. Or from water containing phenol (e.g., from peat soil) that is chlorinated

(continued)

TABLE 16.1 (Continued)
Examples of Taints and Their Possible Origins

Odor Descriptor	Compounds	Possible Origins
	2,4,6-Trichlorophenol	Found in packaging—wood pulp that has been bleached and cardboard and polyvinyl acetate glues
	2-Bromophenol	Present in algae (major portion of the diet of prawns). Also can be formed by reactions—e.g., has been found as a taint in fish that has been bleached with hydrogen peroxide, treated with brine (containing a bromide impurity) in the presence of trace levels of phenol (in oak storage barrels)
Drains	2,6-Dimethyl-3-methoxypyrazine	Produced by certain bacteria
Earthy	Geosmin (*trans*-1,10-dimethyl-*trans*-9-decalol)	Microorganisms—particularly bacteria. Produced by actinomycetes, blue-green algae and cyanobacteria (can contaminate water supplies or soil)
	Pentachloroanisole	Microbial methylation of the corresponding chlorophenols—particularly in wood/pallets treated with a chlorophenol preservative
	2,3,4,6-Tetrachloroanisole	Microbial methylation of the corresponding chlorophenol—particularly in wood/pallets treated with a chlorophenol preservative or in corks treated with chlorophenol. Can be formed by degradation of pentachloroanisole
	2,3,6- and 2,4,6-Trichloroanisole	Microbial methylation of the corresponding chlorophenols—particularly in wood/pallets treated with a chlorophenol preservative or in corks treated with chlorophenol
	2-Methylisoborneol	Water contaminated with actinomycetes or cyanobacteria
Fecal	Indole (2,3-benzopyrrole)	Rotting potatoes and also associated with boar taint in male pigs
	Skatole (3-methylindole)	Bacterial metabolite of amino acids, found in mammalian feces and has been associated with taint in meat from male pigs
Fruity	Acetaldehyde	Overproduction in milk cultures or yogurt (also described as green). Also can be a degradation product of PET packaging
	2,4-Dichloroanisole	Microbial methylation of 2,4-dichlorophenol
	Ethyl butanoate, ethyl hexanoate, ethyl octanoate	Microorganisms in foods including dairy, fish, and meat
Geranium	*cis*-Octa-1,5-dien-3-one	Autooxidation of butterfat
	Benzophenone	Packaging—photoinitiator in UV inks and varnishes

(*continued*)

TABLE 16.1 (Continued)
Examples of Taints and Their Possible Origins

Odor Descriptor	Compounds	Possible Origins
Green	Decanal	Autooxidation of fats
Iodine	2-Bromophenol	Present in algae (major portion of the diet of prawns). Also can be formed by reactions—e.g., has been found as a taint in fish that has bleached with hydrogen peroxide, treated with brine (containing a bromide impurity) in the presence of trace levels of phenol (in oak storage barrels)
Iodoform	2,6-Dibromophenol	Aquatic environment—seafood, also can be present in some fungicides, biocides, and herbicide intermediates (wood treatment)
	2,4,6-Tribromophenol	Seafood, or reaction of biocide/bromination of phenol
Kerosene	1,3-Pentadiene	Degradation of sorbate by the *Penicillium* species (products treated with sorbic acid as a mold inhibitor)
Medicinal	2-Chlorophenol	Chlorination of phenol (associated with 2-methyl-6-chlorophenol), e.g., from water containing phenol (e.g., from peat soil) that is chlorinated
	6-Chloro-*o*-cresol	Disinfectants and drain cleaners or impurity in some herbicides
	2,6-Dichloroanisole	Microbial methylation of corresponding chlorophenol
	Guaiacol	Microbiological degradation of vanillin/ degradation product of lignin
	2-Iodo-4-cresol	Reaction of *p*-cresol (used in some flavors) with iodized salt
	Dichlorobenzene	Disinfectants, drain cleaner, fumigants
Metallic	1-Octen-3-ol	Fungal growth, autooxidation of fats, natural component of clover and fresh mushrooms
	Oct-1-en-3-one	Autooxidation of fats and sometimes found in plastics containing diisooctyl phthalate
	cis-Octa-1,5-dien-3-one	Autooxidation of butterfat
Moldy	1-Octen-3-ol	Fungal growth, autooxidation of fats, natural component of clover, and fresh mushrooms
	Geosmin (*trans*-1,10-dimethyl-*trans*-9-decalol)	Produced by actinomycetes and blue–green algae (can contaminate water supplies or soil)
Musty	Pentachlorophenol	Used as a biocide in wood treatment and adhesive glues
	Pentachloroanisole	Microbial methylation of the corresponding chlorophenols—particularly in wood/pallets treated with a chlorophenol preservative

(*continued*)

TABLE 16.1 (Continued) Examples of Taints and Their Possible Origins

Odor Descriptor	Compounds	Possible Origins
	2,3,4,6-Tetrachloroanisole	Microbial methylation of the corresponding chlorophenol—particularly in wood/pallets treated with a chlorophenol preservative or in corks treated with chlorophenol. Can be formed by degradation of pentachloroanisole
	2,3,6- and 2,4,6-Trichloroanisole	Microbial methylation of the corresponding chlorophenols—particularly in wood/pallets treated with a chlorophenol preservative or in corks treated with chlorophenol
	2,4- and 2,6-Dichloroanisole	Microbial methylation of corresponding chlorophenol
	Geosmin (*trans*-1,10-dimethyl-*trans*-9-decalol)	Produced by actinomycetes and blue–green algae (can contaminate water supplies or soil)
	2-Methylisoborneol	Water contaminated with actinomycetes or cyanobacteria
	2,4,6-Tribromoanisole	Reaction of some biocides with phenol, followed by microbial methylation to form the anisole
	1-Octen-3-ol	Fungal growth, autooxidation of fats, natural component of clover and fresh mushrooms
	Octa-1,3-diene	Metabolite of *Anabaena oscillarioides* and autooxidation of fats
	α-Terpineol	Disinfectants
	4,4,6-Trimethyl-1,3-dioxan	Reaction of 2-methyl-2,4-pentanediol in packaging film with formaldehyde during storage
	Trimethylanisole	Contaminant in rubber seals
Paint	Heptane-2-one	Oxidation of oils and fats
	trans, *trans*-Hepta-2,4-dienal	Autooxidation of fats
	trans-1,3-Pentadiene	Degradation of sorbate by the *Penicillium* species (products treated with sorbic acid as a mold inhibitor)
Paraffin	*trans*-1,3-Pentadiene	Degradation of sorbate by the *Penicillium* species (products treated with sorbic acid as a mold inhibitor)
Pear-like	Acetaldehyde	Degradation product sometimes formed during processing of PET packaging
	Butyl actetate	Printing inks
Petroleum	Dimethylsulfide	Formed from sulfur containing precursors in the aquatic environment such as plankton
	Xylenes	Residual solvents from varnishes/lacquers—can migrate through packaging

(*continued*)

TABLE 16.1 (Continued)
Examples of Taints and Their Possible Origins

Odor Descriptor	Compounds	Possible Origins
Phenolic	2-Bromophenol	Present in algae (major portion of the diet of prawns). Also can be formed by reactions—e.g., has been found as a taint in fish that has been bleached with hydrogen peroxide, treated with brine (containing a bromide impurity) in the presence of trace levels of phenol (in oak storage barrels)
	p-Cresol (4-methylphenol)	Microbiological degradation
	2,4- or 2,6-Dichlorophenol	Impurities in herbicides and in packaging from bleaching of wood pulp. Or from water containing phenol (e.g., from peat soil) that is chlorinated
	2,6-Dichloroanisole	Microbial methylation of corresponding chlorophenol
	Guaiacol	Microbiological degradation of vanillin/ degradation product of lignin
Piney	α-Terpineol	Disinfectants
Plastic	Styrene	Migration from polystyrene containers or formed from cinnamaldehyde (in cinnamon)
	Benzothiazole	Butyl rubbers
	trans-1,3-Pentadiene	Degradation of sorbate by the *Penicillium* species (products treated with sorbic acid as a mold inhibitor)
Rancid	*cis*-Oct-2-enal	Metabolite of *Anabaena oscillarioides* and autooxidation of fats
Smoky	Guaiacol	Microbiological degradation of vanillin/ degradation product of lignin
	4-Vinylguaiacol	Degradation product in orange juice
Soapy	Decanoic acid	Lipolysis of lipids (palm kernel oil, coconut oil)
	Lauric acid (dodecanoic acid)	Lipolysis of lauryl glycerides (palm kernel oil, coconut oil, butter)
Sulfury	Methanethiol (methyl mercaptan)	Degradation of sulfur-containing proteins
Sweet	2,4-Dichloroanisole	Microbial methylation of 2,4-dichlorophenol
	Cyclohexane	Screen-printing solvent
TCP	6-Chloro-*o*-cresol	Disinfectants and drain cleaners or impurity in some herbicides
Turpentine	*para*-Cymene (1-isopr+opyl-4-methylbenzene)	Degradation product of lemon oil and limonene and γ-terpinene in soft drinks
	Nonan-2-one	Rancid coconut
Urine	5α-Androst-16-en-3-one	Meat from uncastrated male pigs
Woody	1,4-Dichlorobenzene	Drain cleaners and also used in moth-proofing agents

Source: Ridgway, K. et al., *Food Addit. Contam. Part A*, 27(2), 146–168, 2010a. With permission.

if consumed. Microbial activity can produce taints by either production of primary metabolites, metabolism of food components, or residual enzyme activity. Several off-flavors in food can be related to microorganisms, such as bacteria, yeasts, and fungi, as reviewed by Whitfield (1998, 2003) and Springett (1993). Growth of microorganisms in food is dictated by the physical environment and chemical composition of the food. Key parameters include water activity (a_w), pH, and temperature, as well as storage conditions. The addition of preservatives can control the growth of some organisms, but may also act as precursors for some taints.

In meat and meat products, surface microbial contamination is the major cause of spoilage taints in raw meat. The compounds and odors produced depend on the level of spoilage and whether the meat has been stored in vacuum packs or aerobically. Sulfur compounds are principally responsible for the "putrid" odor associated with spoiled meats, but others may be detected before this stage, such as those described as buttery, sweet, and fruity, which indicate changes in microflora and chemicals as storage time is increased.

Microbial methylation of halophenols to haloanisoles can produce compounds that cause taints at extremely low concentrations in food products, and such compounds are reported to be responsible for the "cork taint" observed in some wines. Chloroanisoles have also been reported as being responsible for musty taints in eggs and poultry where wood shavings containing chlorophenols were used as bedding and methylation occurred due to endogenous fungi. Off-odors produced by bacteria in fish depend on species and origin—the spoilage of temperate climate marine fish is characterized by ammoniacal rotten and sulfurous odors and flavors, whereas tropical and freshwater species spoilage is characterized by fruity and sulfhydryl odors and flavors. The most common compound associated with spoiled fish is trimethylamine, which is formed by the bacterial reduction of trimethylamine oxide, a natural constituent of fish muscle. Taints described as earthy and muddy in fish and water supplies have been shown to be due to the presence of geosmin and 2-methylisoborneol, which can be produced by cyanobacteria. In crustaceans, bis-(methylthio)-methane was identified as being responsible for a "garlic-like" odor, although the organisms responsible were not identified. Microbial off-flavors in milk and dairy products are often due to enzymatic activity of microorganisms (Jeon 1993).

trans-1,3-Pentadiene has been reported to be responsible for a kerosene-like taint in dairy products (Loureiro and Querol 1999; Pinches and Apps 2007). Microorganisms, particularly the *Penicillium* species, are able to decarboxylate the sorbic acid, used as a preservative in foods such as cheese and convert it to 1,3-pentadiene. The production of styrene in foods has also been linked to the action of a specific yeast on cinnamaldehyde, although the presence of cinnamon or cinnamon flavors is not a prerequisite for styrene production (Pinches and Apps 2007). The production of guaiacol from vanillin and ethyl vanillin has led to smoky/phenolic off-flavors in yogurt and ice cream and although the organisms responsible were not identified, the proposed pathway involves oxidation to vanillic acid followed by decarboxylation. Fruits and vegetables have more microbial resistance than high protein animal products and therefore are less prone to produce taints from microbial contamination. Those reported included off-flavors in orange juice due to diacetyl and acetoin and fruit containing *Penicillium* species producing an earthy odor due to the presence of geosmin.

16.2.3 Chemical Changes/Lipid Oxidation

The lipid content of foods is the main source of off-flavors, primarily due to oxidation. Common compounds associated with the rancid off-flavors produced include aldehydes, ketones, lactones and furans, carboxylic acids, and hydrocarbons, with hexanal frequently being used as a marker.

The chemical mechanisms involved result in the formation of intermediates, which then break down to give odorous compounds. Oxidation can occur through autooxidation (which can occur in the dark and at room temperature through the production of free radicals), photooxidation (occurs in the light when certain sensitizers are present), and also can be induced by enzymes.

Dairy products are particularly susceptible to lipid-derived off-flavors and examples include the production of alkanones, such as 2-heptanone in UHT milk.

An example of an off-flavor in foods from enzymatic activity is a "rancid" off-flavor in milk due to lipoprotein lipase. Lipases in milk are endogenous, but can also originate from lipolytic bacteria although microbial lipases are heat stable. Although some aspects are well studied, the interrelationship between microbial enzymes, endogenous enzymes, and off-flavors can be complex (Jeon 1993).

The inclusion of more unsaturated (healthier) fats in foods means that they may be more readily oxidized and better protection (through the use of additives, processing, or storage conditions) is needed.

Some Maillard reaction products have a role to play in preventing lipid oxidation, because of their antioxidant properties, but the Maillard reaction can also be a source of off-flavors. A typical example is the browning and off-flavor produced in fruit juices during storage, due to the formation of compounds such as substituted furfurals and furans. Deterioration of milk flavor during storage can also be attributed to the Maillard reaction.

A good example of a chemical reacting with food components causing a taint is the compound 4-mercapto-4-methylpentan-2-one. This compound has an odor described as "catty" and is produced by the reaction of hydrogen sulfide (naturally present in foods) with mesityl oxide (present as a solvent impurity in paint or can lacquers). As manufacturers have become aware of such taints, the use of solvents containing mesityl oxide in lacquer formulations have been reduced and taints due to this reaction are now rare.

16.2.4 Processing/Cleaning Products

Processing of food can reduce off-flavors but may also increase opportunities for taint or off-flavors to be introduced. Thermal processing and the Maillard reaction are responsible for many food flavors and can be responsible for off-flavors in foods such as furfurals and pyrroles as seen in fruit juices (Handwerk and Coleman 1988). The location of food processing plants has also been the cause of taints due to airborne pollutants, for example, from nearby agrochemical plants. Foods with a high lipid content or large surface area are the most susceptible to volatile compounds from the environment. Cases reported included taints due to the compound 6-chloro-*o*-cresol, which is a by-product in the manufacture of a herbicide and has particularly high volatility. The airborne contaminant was then absorbed by the food leading to a "disinfectant" taint. This illustrates that all aspects must be considered to avoid food taints and that air purification systems should be installed where necessary. The particularly low sensory threshold of 6-chloro-*o*-cresol has also led to several taints through its use in certain disinfectants and premises handling food should avoid using cleaning products containing this compound.

A large number of reported taints each year originate from cleaning product or disinfectants (Olieman 2003), either accidentally from poor rinse procedures or from direct contact if no-rinse products are used. In particular disinfectants based on active chlorine or oxygen can react with food components (such as phenols) to form compounds that can produce a taint (e.g., chlorophenols).

16.2.5 Packaging

Taints originating from packaging can occur due to two mechanisms—transfer of volatiles, or migration due to direct contact. The latter is more prevalent for fatty food substances or those with a high surface area. Although direct food contact packaging is generally controlled and tested for taint and odor, secondary packaging can also be responsible for taints in food. A wide variety of materials are used in food packaging, including not only the principle components, but impurities and additives. Paper and carton board often form part of a multilayer packaging consisting of adhesives, varnishes, and plastics. Compounds responsible for taints can also be formed by reaction products formed during manufacturing or environmental contamination of primary or secondary packaging. Inks or compounds used in secondary packaging or on the outer surfaces of packing may migrate into products if no sufficient barrier is present. Examples of compounds originating from packaging that may cause a taint in food include residual monomers from plastic packaging, such as styrene, solvents from residual solvents from inks and varnishes, and photoinitiators such as benzophenone. Degradation or oxidation products from processing such as acetaldehyde have also been reported to be responsible for taints in food.

Odors originating from paper and board packaging may be due to bacteria molds, oxidation of residual resins, or degradation of processing chemicals. Decarboxylation and oxidation of lignin can produce vanillic acid, which subsequently can degrade to guaiacol, a compound responsible for a taint described as smoky. There are several volatile compounds present in pulp, which may also be responsible for taints in food, in particular after oxidation of lipids, catalyzed by the presence of metallic ions. Such compounds include aldehydes, alcohols, and esters, for example hexanal, which is often present in paper and board at low levels. Although mostly these compounds are not present at a high enough level to give rise to a taint, some paper and board can become more odorous on storage and complexing agents are frequently added to reduce the level of metallic ion catalysts.

16.2.5.1 Recycled Packaging

The use of recycled materials for food contact applications could lead to more potential contaminants, from inks or previous usage. Material sources and recycling processes must therefore be strictly controlled and monitored to minimize the risk of a compounds that may cause a taint in the packaging and therefore potentially the packaged food product.

16.2.6 Storage

As discussed, taints and off-flavors caused by reactions, in particular oxidation, can worsen because of inappropriate storage conditions. Taints in food may also originate from the storage environment itself, such as a building or container. One of the most common taints reported in foods is due to the presence of halophenols and haloanisoles. One of the major sources of these compounds is wooden storage pallets. Chlorophenols have been used industrially as fungicides, biocides, and herbicide intermediates, most commonly in the treatment of wooden storage pallets and can be microbially methylated by numerous organisms to the corresponding chloroanisoles (Parr et al. 1974). Pallets made from soft wood that has been treated with certain fungicides can therefore be responsible for taints due to the migration of chlorophenols or chloroanisoles into ingredients or products during storage. Since a reduction in the use of chlorinated phenols (in particular pentachlorophenol), bromophenols have been used, and these can also lead to the formation of bromoanisoles through microbial methylation, which can be responsible for taints when present at extremely low levels (ppt, ng kg^{-1}). Airborne contamination of products or ingredients either during storage or transportation can also cause taints. New floors, diffusion within warehouses, or transit containers have all been reported as sources of tainting compounds.

16.3 CONSUMER PERCEPTION AND THRESHOLD VALUES

The possibility of someone detecting a taint is dependent on the concentration of the compound(s) responsible, although some individuals are more sensitive to certain off-flavors, odors, or specific compounds. Threshold values are used for the sensory analysis of taints and are generally defined as the probability of detection being 0.5, that is, 50% of the general population will detect a taint (Figure 16.1). The values for most compounds are measured in air or water and therefore may not represent detection in a real food sample (in which the probability of detection may be higher or lower).

As well as differences in the levels of tainting compounds that can be perceived by consumers, descriptors of taints can also vary. Sensory analysis is the key to more quantitative methods of analysis and therefore an accurate description can provide key information to the analyst investigating the origin of a taint or off-flavor. It may

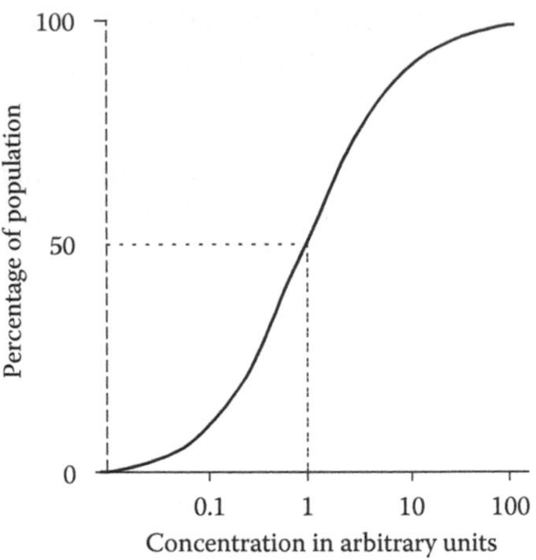

FIGURE 16.1 Variations in taste thresholds. (Reproduced from Springer Science and Business Media, *Food Taints and Off-Flavours*, ed. M.J. Saxby, 1993. With permission.)

be that more than one compound is responsible for a taint or off-flavor in food and this further complicates the sensory descriptors.

16.4 METHODS OF ANALYSIS FOR DETERMINATION OF FOOD TAINTS AND OFF-FLAVORS

16.4.1 Sampling

Taking a representative sample is important for any analytical procedure, and for taint analysis, this is especially the case as the compound responsible for the taint may not be homogenously distributed throughout the food ingredient or product. This is particularly the case where a taint is due to solvents or migration of compounds from packaging as "hot spots" can occur. Because of the complexity of most food matrices, a "good" sample is often used as a control to enable identification of differences in both the sensory and chromatographic profiles of the suspect sample compared to one with acceptable taste/flavor characteristics. Care should be taken to avoid crosscontamination of control and suspect samples and also contamination in transit to the laboratory or on storage.

16.4.2 Sensory Analysis

Following the report of a taint in food, sensory analysis is invariably the first step in the investigation. As discussed, consumer descriptions of taints are frequently unreliable, due to lack of knowledge of the chemicals involved and a trained sensory panel will give objective assessments and descriptions that can be matched to reference guides (Baigrie 2003; Saxby et al. 1992; Saxby 1993), or specialized websites (www.odour.org.uk and www.flavornet.org).

Artificial taste sensors have been developed in an attempt to replace or support the use of human panelists, but as discussed in a recent review by Citterio and Suzuki (2008), currently no absolute models can correlate the taste that a human perceives with the chemical composition of a sample.

Even once an accurate descriptor is obtained, there may still be a number of compounds that could be responsible for a taint or off-flavor and chemical analysis is required to confirm the identity and levels of the compound(s) present.

16.4.3 Chemical Analysis

Determination of taints in foods generally requires a more investigative approach than determination of known chemical contaminants as illustrated in Figure 16.2. It is often necessary to predict the compounds responsible for a taint to allow for more targeted analysis as qualitative approaches may not give the sensitivity required for some compounds present at trace and ultra-trace (sub ppb, $\mu g\ kg^{-1}$) concentrations. If sensory descriptors do not provide sufficient information or when the cause of a taint is unknown, a generic screening procedure may be used as a first step in analysis. As the majority of taints are detected through odor, most of the compounds that cause taints in food are volatile and can therefore be analyzed using gas chromatography. The initial step in a taint investigation typically involves comparison of volatile profiles [gas chromatography–mass spectrometry (GC-MS)] between a control and a suspect sample. Figure 16.3 shows a comparison of a "control" and "suspect" samples and illustrates the difficulty in observing additional chromatographic peaks using the total ion chromatography (TIC). The extracted ions (*m/z*) show the peak corresponding to dichlorophenol present in the suspect sample (at approximately 0.1 $\mu g\ kg^{-1}$), but not the control. In this example, because of the information provided, it was possible to predict the compounds that may be responsible for the taint and examine the chromatograms accordingly. However, this is not always the case. If additional chromatographic peaks are observed in the suspect sample, then initial identification of the compounds can be performed using library spectral searches [such as National Institute of Standards and Technology (NIST) mass spectral library]. Such libraries, however, should be used with care and the use of analytical standards is preferred to enable unequivocal identification of a chemical compound. Initial identification of the compounds responsible for the taint should always be compared with sensory analysis and any other information available regarding the origins of the taint to ensure that accurate conclusions are drawn.

Once the compound potentially responsible for the taint has been identified, a more fit for purpose approach can be used for targeted quantitative analysis. The method chosen will depend on the compound's sensory threshold, that is, the sensitivity required from the predicted levels in the samples. For some compounds, direct headspace GC-MS may be sufficient for both identification and quantification, but more frequently, methods providing an enrichment/concentration step will be utilized.

16.4.3.1 Direct Headspace

Direct headspace GC-MS can be used for both the initial screen (comparison of chromatographic profiles) and more targeted quantitative analysis, once the identity

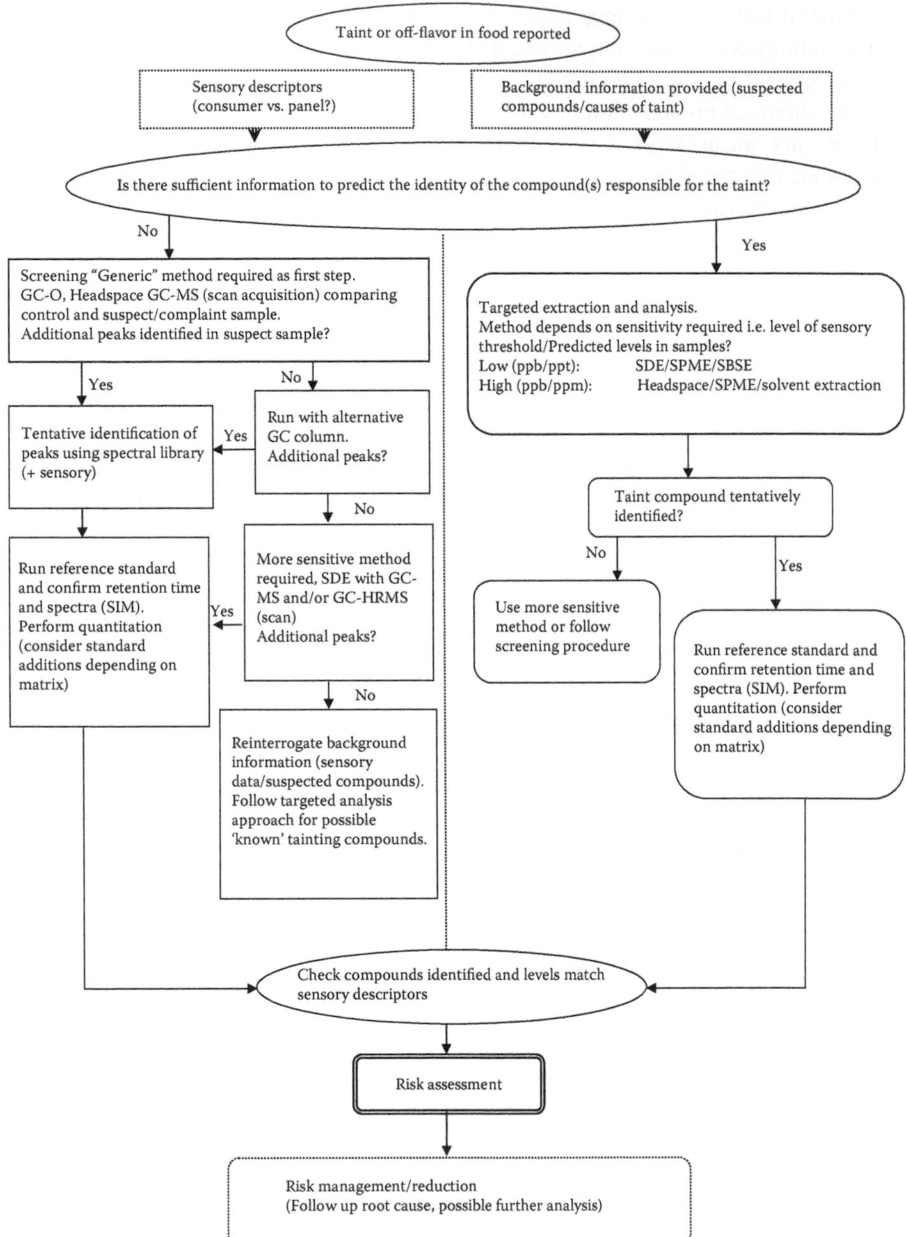

FIGURE 16.2 A sample analytical approach for investigation of food taints. (From Ridgway, K. et al., *Food Addit. Contam.* 27, 146–168, 2010. With permission.)

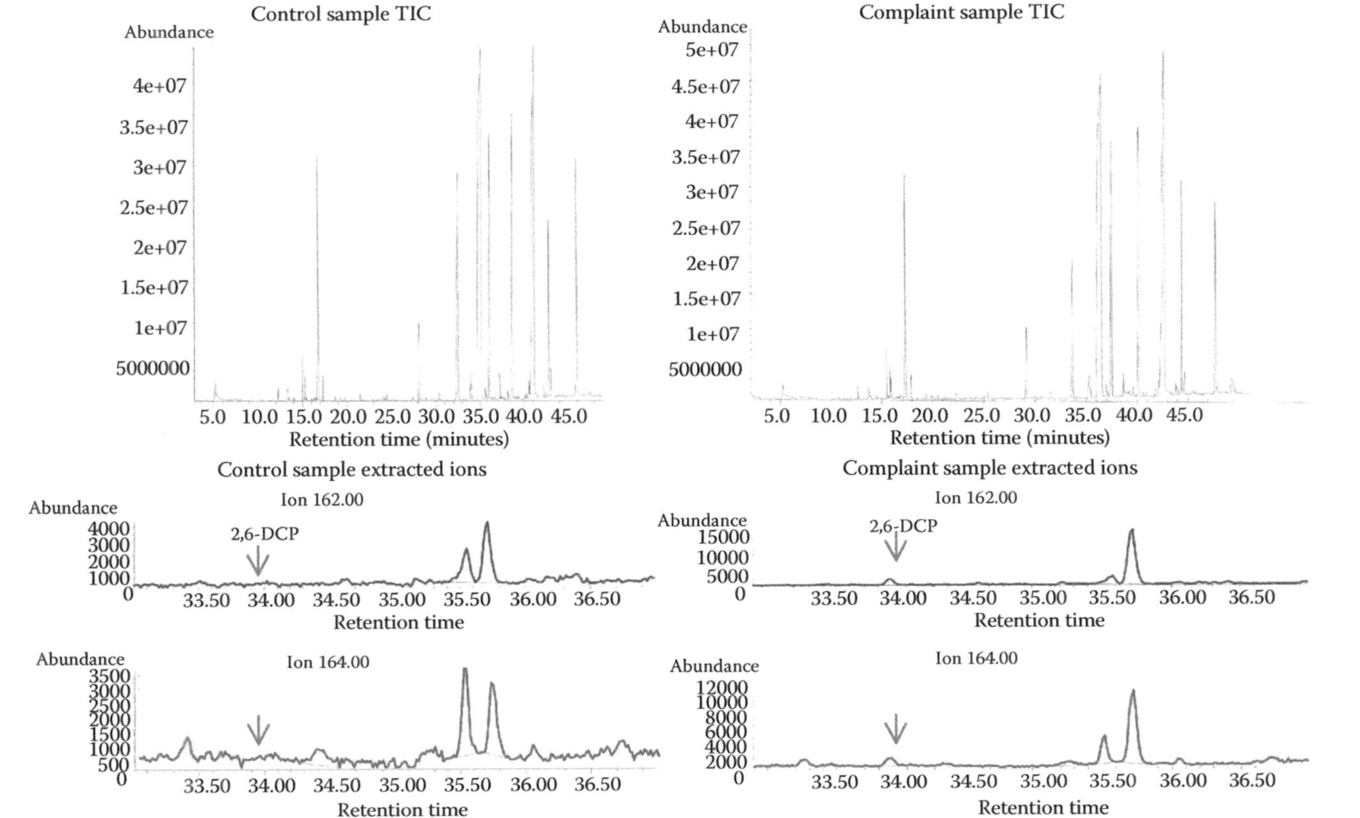

FIGURE 16.3 TIC chromatograms and extracted ions (*m/z* 162 and 164) for 2,6-dichlorophenol following SBSE of control and suspect samples. (From Ridgway, K. et al., *Anal. Chim. Acta*, in press, 2010b. With permission.)

of the tainting compound is known. For accurate quantitation, the method of standard additions, or the use of an internal standard is recommended. As only the headspace above the sample is analyzed, it can be used for most matrix types, including direct analysis of packaging, although it can lack the sensitivity required for some tainting compounds. Dynamic systems, such as purge and trap can help concentrate the extracts, but as no further selectivity is provided, they do little to increase sensitivity in most cases and issues can arise with carry over.

16.4.3.2 Solvent Extraction

Direct solvent extraction is sometimes used for analysis of food taints, for example, for the determination of indole and skatole in meat from male pigs, for which supercritical fluid extraction has also been used. Soxhlet extraction, pressurized liquid extraction, and microwave extraction have all been used for determination of trichloroanisole in corks. However, isolation from matrix components can be a challenge and further clean up and enrichment steps are generally required [such as solid phase extraction (SPE)]. Steam distillation extraction (SDE), using apparatus such as Likens–Nickerson (Likens and Nickerson 1964), is still widely used in the food industry for extraction of taints from food. This technique involves the codistillation of the sample between two solvents (Figure 16.4). Generally, water containing the sample is boiled in one flask, whereas the extracting solvent (such as diethyl ether) is boiled in another. As the vapors mix in the central chamber, the volatile compounds are extracted into the solvent, leaving the nonvolatile food matrix components in the sample flask. As the extraction is continuous, relatively small volumes of solvents can be used, but a concentration step is often still required for trace analysis. Various modifications of the apparatus have been made (Maarse 1993) and now include vacuum distillation to reduce artifact formation. The advantage of the technique is that it is suitable for most matrix types and produces a clean extract of volatile components. The disadvantage of this technique is the extraction time

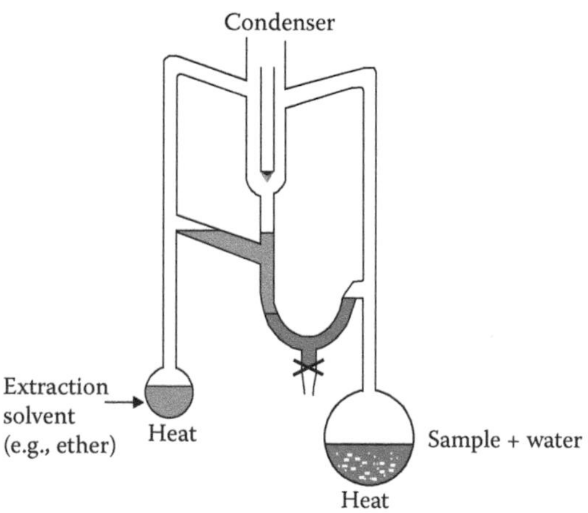

FIGURE 16.4 Likens–Nickerson apparatus.

(which is typically more than an hour) and the need for specialist glassware. The possibility of cross contamination and losses during the concentration step should also be considered.

16.4.3.3 Microextraction Techniques

More recently miniaturized extraction techniques have been used, as these use minimal solvent and can provide high concentration factors/enrichment of trace components. Solid-phase microextraction (SPME) in particular can be used to increase the sensitivity compared to direct headspace sampling and headspace–SPME extraction is being increasingly used for determination of food taints and off-flavors (Citterio and Suzuki 2008). As SPME is an equilibrium technique, experimental conditions need to be controlled and results can depend on the matrix. The use of labeled internal standards or the method of standard additions is often required for accurate quantitation. Fibers and extraction conditions, such as temperature, must be chosen to suit the target compounds to get the best sensitivity. For example, very volatile compounds will often give a better response using direct headspace and an increase in temperature will not always result in an increase in response using headspace-SPME, because of the partitions involved between both the sample and fiber and the fiber and headspace. The technique can provide low detection limits and has been used for targeted analysis of a range of tainting compounds (Chatonnet 2007), but needs to be optimized for each matrix/analyte combination.

Stir bar-sorptive extraction (SBSE) has also been used for determination of tainting compounds, in particular chlorophenols and chloroanisoles in cork (Callejon et al. 2007; Hayasaka et al. 2003; Chatonnet et al. 2004), but also for a range of tainting compounds (Ridgway et al. 2010b). The use of SBSE for the determination of compounds known to cause taints in food offers several advantages compared to either static headspace or steam distillation extraction. SBSE offers a larger volume of coating (extraction phase) compared to SPME and, therefore, larger enrichment factors can be achieved, resulting in low limits of detection for many compounds. As with other microextraction techniques, it offers a reduction in the use of organic solvents compared to SDE and also as specialized glassware, such as the Likens–Nickerson equipment is not required, a larger number of samples can be extracted simultaneously. An added benefit of the simplicity of SBSE, particularly relevant for determination of taints in foods, is that there is much less potential for contamination from external laboratory sources. For "dirty" matrices following sampling, the stir bar can be washed and dried as necessary before desorption and GC-MS analysis.

Similarly to SPME, as well as direct immersion sampling, SBSE can also be used to sample the headspace above the sample after heating (known as headspace sorptive extraction, HSSE).

Currently only one extracting phase is commercially available (PDMS), but samples can be modified to enhance extraction, for example, by modification of the pH to enable extraction of phenols into the nonpolar PDMS phase (Chatonnet et al. 2004; Zalacain et al. 2004). Derivatization of chlorophenols is also regularly used to improve chromatographic performance and reduce peak tailing and this can be done *in situ* with SBSE.

16.5 CONCLUSIONS

Once a taint in food is detected, there are few options to remove the tainting compounds and restore the quality of the food. The course of action will depend on several factors, such as whether the product is already on the market, the number of batches affected, and whether the contamination poses a potential risk to human health. Even where the tainting compound represents no risk to consumers, a silent recall may be undertaken as the perception of low quality, brand damage, and adverse publicity can be extremely costly to the food industry. Analytical determination of tainting compounds in foods is often a complex procedure and a flexible approach needs to be taken as the compounds responsible for taints are frequently only present at trace levels (low μg kg^{-1}). Each case must be viewed individually and the analytical methods used will depend on many factors, including instrument availability and analyst experience. Therefore, the prevention of taints and off-flavors in foods by controlling raw materials, processes, packaging, and storage conditions is paramount to ensure food quality and potentially avoid food safety issues.

REFERENCES

Baigrie, B. 2003. *Taints and Off-Flavours in Food*. Cambridge: Woodhead Publishing Limited.

Callejon, R.M., A.M. Troncoso, and M.L. Morales. 2007. Analysis for chloroanisoles and chlorophenols in cork by stir bar sorptive extraction and gas chromatography–mass spectrometry. *Talanta* 71: 2092–2097.

Chatonnet, P., S. Bonnet, S. Boutou, and M.D. Labadie. 2004. Identification and responsibility of 2,4,6-tribromoanisole in musty, corked odors in wine. *J. Agric. Food Chem.* 52: 1255.

Citterio, D., and K. Suzuki. 2008. Smart taste sensors. *Anal. Chem.* 80: 3965–3972.

Handwerk, R.L., and R.L. Coleman. 1988. Approaches to the citrus browning problem. A review. *J. Agric. Food Chem.* 36: 231–236.

Hansson, K.E., K. Lundstrom, S. Fjelkner-Modig, and J. Persson. 1980. The importance of androstenone and skatole for boar taint. *Swed. J. Agric. Res.* 10: 167.

Hayasaka, Y., K. MacNamara, G.A. Baldock, R.L. Taylor, and A.P. Pollnitz. 2003. Application of stir sorptive extraction for wine analysis. *Anal. Bioanal. Chem.* 375: 948–955.

Hocking, A.D., K.J. Shaw, N.J. Charley, and F.B. Whitfield. 1998. Identification and relations to other growth characteristics. *J. Food Mycol.* 1: 23–30.

Jeon, I.J. 1993. Undesirable flavors in dairy products. In *Food Taints and Off-Flavours*, ed. M.J. Saxby, 122. Blackie Academic & Professional (Chapman and Hall).

Likens, S.T., and G.B. Nickerson. 1964. Detection of certain hop oil constituents in brewing products. *Am. Soc. Brew. Chem. Proc.* 5: 5–13.

Loureiro, V., and A. Querol. 1999. The prevalence and control of spoilage yeasts in food and beverages. *Trends Food Sci. Technol.* 10: 356–365.

Maarse, H. 1993. Analysis of taints and off-flavours. In *Food Taints and Off-Flavours*, ed. M.J. Saxby, 63. Blackie Academic & Professional (Chapman and Hall).

Mottram, D.S. 1998. Chemical tainting of foods. *Int. J. Food Sci. Technol.* 33: 19–29.

Olieman, C. 2003. In *Taints and Off-Flavours in Food*, ed. Baigrie, 189. Cambridge: Woodhead Publishing Limited.

Parr, L.J., R.F. Curtis, D. Robinson, D.G. Land, and M.G. Gee. 1974. Chlorophenols from wood preservatives in broiler house litter. *J. Sci. Food Agric.* 25: 835–841.

Patterson, R.L.S. 1968. 5α-Androst-16-ene-3-one: Compound responsible for taint in boar fat. *J. Sci. Food Agric.* 19: 31–38.

Pinches, S.E., and P. Apps. 2007. Production in food of 1,3-pentadiene and styrene by *Trichoderma* species. *Int. J. Food Microbiol.* 116: 182–185.

Ridgway, K., S.P.D. Lalljie, and R.M. Smith. 2010a. Analysis of food taints and off-flavours: A review. *Food Addit. Contam.* 27: 146–168.

Ridgway K., S.P.D. Lalljie, and R.M. Smith. 2010b. An alternative method for analysis of food taints using stir bar sorptive extraction. *Anal. Chim. Acta*, in press.

Saxby, M J, W.J. Reid, and S. Wragg. 1992. *Index of Chemical Taints*. Leatherhead, UK: Leatherhead Food RA.

Saxby, M.J., ed. 1993. *Food Taints and Off-Flavours*, p. 37. Blackie Academic & Professional (Chapman and Hall).

Springett, M.B. 1993. In *Food Taints and Off-Flavours*, ed. M.J. Saxby, 244. Blackie Academic & Professional (Chapman and Hall).

Tucker, C.S. 2000. Off-flavor problems in aquaculture. *Rev. Fish. Sci.* 8: 45–88.

Verheyden, K., H. Noppe, M. Aluwe, S. Millet, J. Vanden Bussche, and H.F. De Brabander. 2007. Development and validation of a method for simultaneous analysis of the boar taint compounds indole, skatole and androstenone in pig fat using liquid chromatography–multiple mass spectrometry. *J. Chromatogr. A.* 1174: 132–137.

Watson, S.B., J. Ridal, B. Zaitlin, and A. Lo. 2003. Odours from pulp mill effluent treatment ponds: The origin of significant levels of geosmin and 2-methylisoborneol (MIB). *Chemosphere* 51: 765–773.

Whitfield, F.B. 1998. Microbiology of food taints. *Int. J. Food Sci. Technol.* 33: 31–51.

Whitfield, F.B. 1999. Biological origins of off-flavours in fish and crustaceans. *Water Sci. Technol.* 40: 265–272.

Whitfield, F.B. 2003. *Taints and Off-Flavours in Food*, ed. B. Baigrie, 112. Cambridge: Woodhead Publishing Limited.

Zabolotsky, D.A., L.F. Chen, J.A. Patterson, J.C. Forrest, H.M. Lin, and A.L. Grant. 1995. Supercritical carbon dioxide extraction of androstenone and skatole from pork fat. *J. Food Sci.* 60: 1006–1008.

Zalacain, A., G.L. Alonso, C. Lorenzo, M. Iniguez, and M.R. Salinas. 2004. Stir bar sorptive extraction for the analysis of wine cork taint. *J. Chromatogr. A* 1033: 173–178.

Zhang, L., R. Hu, and Z. Yang. 2005. Simultaneous pictogram determination of "earthy–musty" odours compounds in water using solid phase microextraction and gas chromatography–mass spectrometry coupled with initial cool programmable temperature vaporizer inlet. *J. Chromatogr. A* 1098: 7–13.

17 Volatile Compounds in Food Authenticity and Traceability Testing

Tomas Cajka and Jana Hajslova

CONTENTS

17.1 FOOD AUTHENTICITY, FRAUD, AND TRACEABILITY

Food authenticity refers to whether a food purchased by a consumer matches its description. Mislabeling can occur in many ways, from the undeclared addition of water or other cheaper components, or the wrong declaration of the amount of a particular ingredient in the product, to making a false statement about the source of

FIGURE 17.1 Logos of geographical indications to a food product.

ingredients, that is, geographical, botanical, or animal origin. On the other hand, *food fraud* is committed when food is deliberately placed on the market, for financial gain, with the intention of deceiving the consumer. There are two main types of food fraud: (1) the sale of food that is unfit and potentially harmful, and (2) the deliberate misdescription of food, which—although not necessarily unsafe—deceives the consumer as to the nature of the product (Food Standard Agency 2010; Mannina and Di Tullio 2009).

To protect consumers' health, the European Union (EU), introduced in 2002, a regulation (EU Regulation 178/2002) laying down the general principles and requirements of food law, establishing the European Food Safety Authority, and laying down procedures in matters of food safety. Among others, this regulation also introduces *traceability*, which means the ability to track any food, feed, food-producing animal, or substance that will be used for consumption, through all stages of production, processing, and distribution (European Commission 2002a).

In recent years, there has been a growing interest in the authenticity and traceability of food products among consumers. In particular, there is an increasing focus on the geographical and/or botanical origin of raw materials and finished products, for various reasons, including specific sensory properties, perceived health values, confidence in locally produced products, and, finally, media attention (Lees 2003; Oliveri et al. 2010).

Subsequently, the EU has supported the differentiation of quality products on a regional basis, introducing an integrated framework for the protection of geographical origin for agricultural products and foodstuffs by a specific regulation (EU Regulation 510/2006) (European Commission 2006). This regulation permits the application of the following geographical indications to a food product: protected designation of origin (PDO), protected geographical indication, and traditional specialty guaranteed (Figure 17.1).

17.2 TESTING OF FOOD AUTHENTICITY USING FINGERPRINTING AND PROFILING TECHNIQUES

A wide range of approaches has been developed for authenticity assessment of food. The older studies were based mainly on the composition of various physicochemical and biochemical measurements and, when needed, morphological features (e.g., pollen analysis for honey origin) of a respective matrix were also examined. Nowadays, there is a trend towards fingerprinting and profiling techniques. Fingerprinting of samples is performed for comparative analyses aimed at detection of differences, whereas profiling of samples is focused on individual sample components (including components originating during food processing) to be identified for further analyses (Wishart 2008). In both cases, the analytical effort is to collect the maximum information (features, markers) on the sample composition as compared to previous focus on a single or only a few markers.

Fingerprinting and profiling techniques are based on the measurement of the composition of foodstuffs either in a nonselective or selective way. The fingerprinting and profiling data can be chromatograms, spectroscopic, or spectrometric measurements (specific signals or complete spectra) obtained by infrared spectroscopy, Raman spectroscopy, nuclear magnetic resonance (NMR), mass spectrometry (MS), ultraviolet–visible (UV–Vis), or fluorescence spectrophotometry (Mannina and Di Tullio 2009). For MS, either previous chromatographic separation [gas (GC) or liquid chromatography (LC)] or ambient MS analysis without separation of sample components can be considered. The nature of fingerprints/profiles may largely differ depending on the ionization technique used (Figure 17.2) (Cajka et al. 2008).

The procedure of food analysis by fingerprinting/profiling techniques is summarized in Figure 17.3.

Generally, samples are analyzed directly or after a simple extraction procedure. The generated data are processed using suitable statistical methodologies to the specific problem. In this way, identity confirmation of a given food/commodity as well as of a product obtained using specific technological processes can be obtained.

From the point of view of complexity, foods represent typically complicated matrices containing a sum of tens or hundreds of (semi-)volatile and nonvolatile compounds. Besides their vapor pressures, odor thresholds of particular compounds can vary considerably: some strong odorants might be present at very low concentrations, thus, creating a challenge for the instrumental analysis.

Depending on the approach used, various sample components make a contribution to the recorded signals. In the paragraphs below, attention will be paid mainly

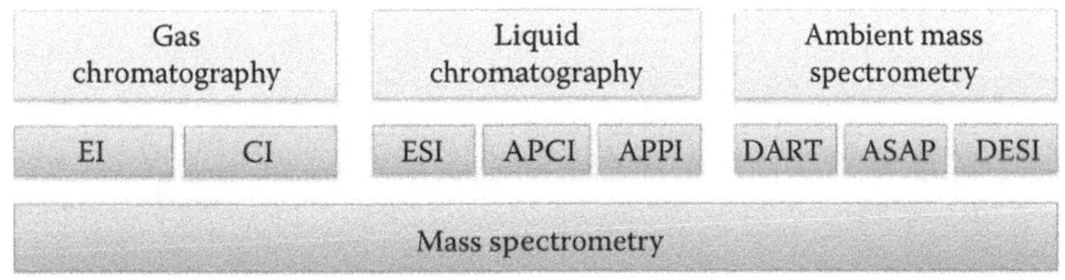

FIGURE 17.2 Overview of mass spectrometric-based techniques available for profiling. (EI, electron ionization; CI, chemical ionization; ESI, electrospray ionization; APCI, atmospheric pressure chemical ionization; APPI, atmospheric pressure photoionization; DART, direct analysis in real time; DESI, desorption electrospray ionization; ASAP, atmospheric pressure solids analysis probe.)

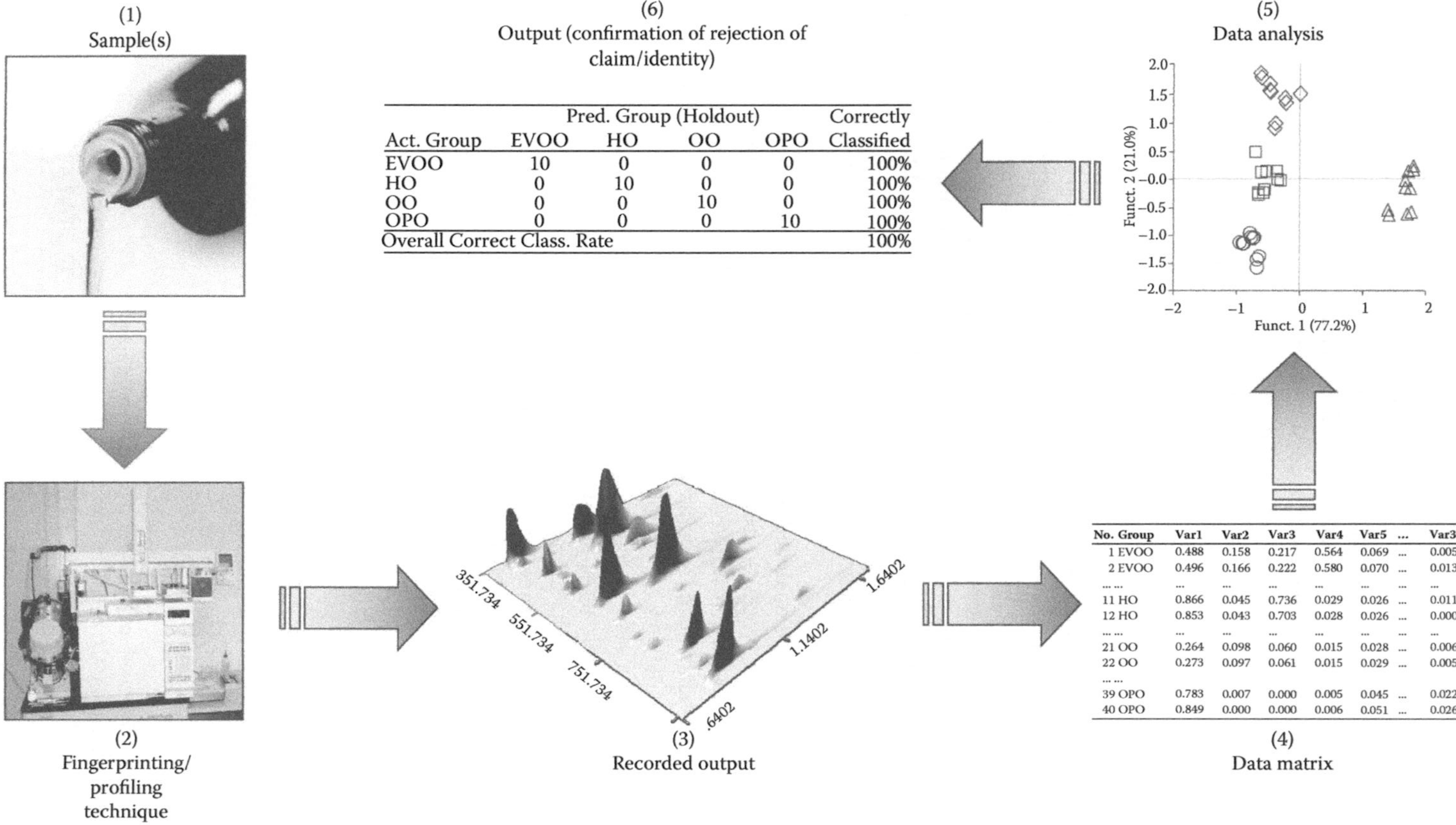

FIGURE 17.3 Food analysis by fingerprinting and profiling techniques (example of HS-SPME–GC × GC–TOFMS analysis of olive oil volatiles).

to (semi-)volatile substances with low molecular weight. They are either metabolome components (products of biosynthesis), or their formation is induced by various processing techniques.

To achieve the sensitivity required for the analysis of food volatiles, appropriate sample preparation and preconcentration steps are required. Headspace methods (for the measurement of components in the vapor phase above the sample) comprise *static headspace*, *dynamic headspace* (purge and trap), and *solid-phase microextraction* (SPME). Other sample preparation methods, such as *solid-phase extraction* (SPE), *liquid/liquid extraction*, and *simultaneous distillation–extraction*, are used for isolation of volatiles and their use is determined by the type of compounds to be isolated. Subsequently, isolated volatiles are detected after previous GC separation, or after ionization during which fingerprint of ions/fragments is formed (Jelen and Majcher 2009).

In more detail, the following analytical strategies can be considered as the most promising for the comprehensive and/or rapid analysis of volatiles in food matrices for the purpose of authenticity assessment:

- Gas chromatography–mass spectrometry or flame ionization detection (GC–MS, GC–FID) with isolation/preconcentration of volatiles using direct headspace or SPME
- Headspace–mass spectrometry (HS–MS) [electronic nose (e-nose)]
- Direct analysis in real time–mass spectrometry (DART–MS)

17.2.1 SPME–GC–MS

SPME in combination with GC–MS has been implemented as the key fingerprinting technique for the analysis of food volatiles. SPME represents an inexpensive, solvent-free sampling technique that enables an isolation of low molecular weight analytes by their extraction from the headspace (HS) or liquid phase (direct immersion, DI) into a fiber coating (Kataoka et al. 2000). The SPME method involves selection of the fiber as well as sampling time and temperature, and addition of salts, to analyze a particular food matrix. However, it should be emphasized that the relative concentration of components in the headspace does not reflect accurately the concentration in the sample because of the differences in volatility of compounds. Also, the profile of collected volatiles is dependent on the type, thickness, and length of the fiber used, as well as on the sampling time and temperature. Therefore, it is essential to analyze the samples under well defined and constant conditions.

In the next step, thermal desorption of absorbed/adsorbed components in a hot GC injector port follows. Among the GC methods for the separation of volatiles, conventional *one-dimensional GC* (1D-GC) is frequently used. However, due to the complexity of volatiles of some food, the chromatographic resolution of certain compounds is not sufficient (Figure 17.4) (Cajka et al. 2007). Under these conditions, *comprehensive two-dimensional gas chromatography* (GC × GC) permits an efficient separation of the entire sample. In GC × GC, two GC separations with different separation mechanisms are applied for sample characterization. Once separated, the components enter the mass spectrometer for ionization and detection.

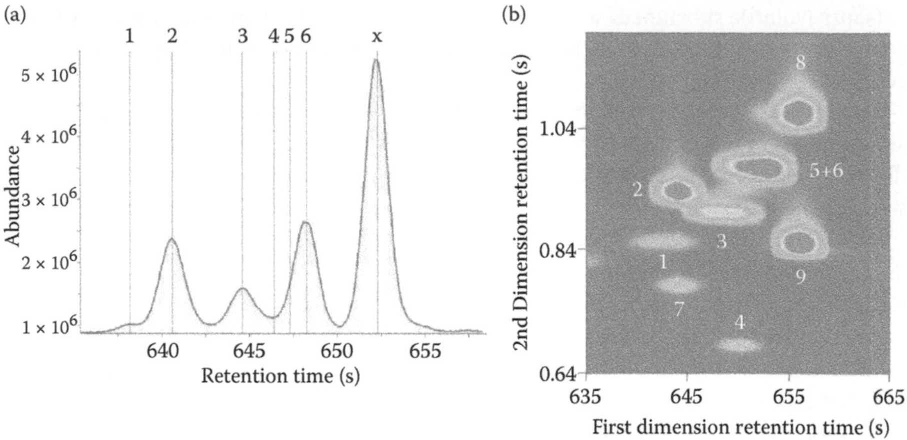

FIGURE 17.4 Comparison of separation of selected honey volatiles in two GC systems: (a) 1D-GC–TOFMS, and (b) GC × GC–TOFMS (DB-5ms × Supelcowax 10 columns). Marked compounds: (1) nonan-2-one; (2) linalool oxide; (3) dehydro-*p*-cymene; (4) undecane; (5) nonan-2-ol; (6) linalool; (7) terpinolen; (8) hotrienol; (9) nonanal; (X) complete coelution of hotrienol and nonanal in 1D-GC system. Compound (7) not identified in 1D-GC. Compounds (5) and (6) partially coeluted in GC × GC. (Reproduced from Cajka, T., et al., *J. Sep. Sci.*, 30, 534–546, 2007. With permission.)

Nowadays, MS is considered as a primary detection tool. There are various types of mass analyzers (e.g., quadrupole, ion trap, time-of-flight) available for the detection of ions. In the case of GC × GC, TOFMS is the only viable option for recording of very narrow chromatographic peaks produced by this separation technique (Hajslova and Cajka 2008). In addition, because the TOF is a nonscanning mass analyzer, all ions are recorded simultaneously, thus, there are no changes in the ratios of analyte ions across the peak during the acquisition of the mass spectrum and, consequently, no spectral skewing (observed commonly by scanning instruments) is encountered. This allows automated deconvolution of partially overlapped peaks on the basis of increasing/decreasing ion intensities in collected spectra and background subtraction followed by identification using library search (Figure 17.5).

Although the deconvolution function (using software correction for spectral skewing) is currently available also for scanning instruments, for example, in the AMDIS software (Automated Mass Spectral Deconvolution and Identification System; provided by the NIST free of charge), the low signal intensity during full spectra acquisition as well as relatively low acquisition rate of common scanning instruments can be a drawback that limits the potential of this function when coupled to fast GC separations. Very narrow chromatographic peaks (<2 s at the baseline) generated under these conditions require a detector with high acquisition rates and, at the same time, with high sensitivity during the acquisition of full mass spectra, such as TOF, to fully utilize the advantage of spectral deconvolution (Cajka et al. 2008).

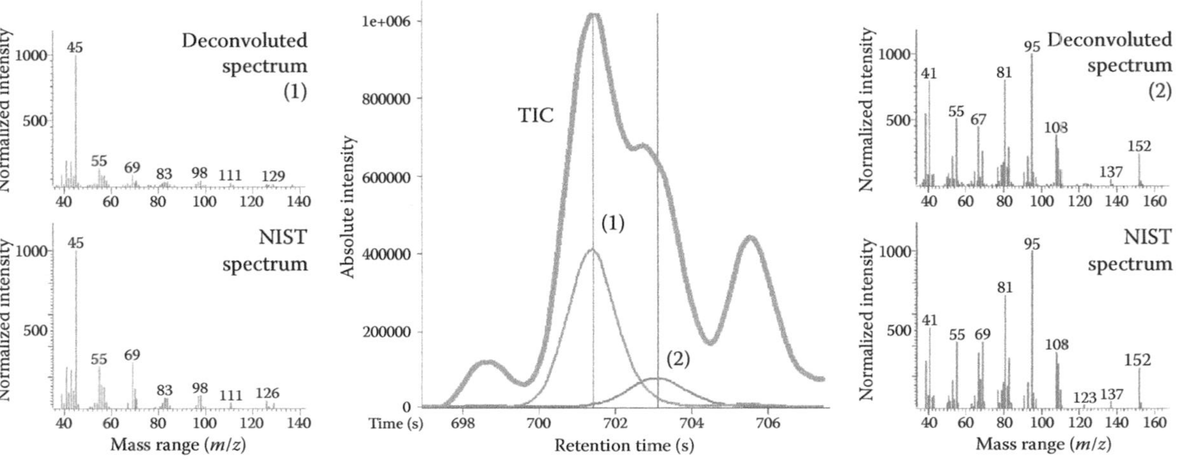

FIGURE 17.5 Example of spectral deconvolution of two closely eluted beer markers of chromatographic fingerprint (HS-SPME–GC–TOFMS analysis). (1) Nonan-2-ol, *m/z* 45 displayed; (2) (1*S*)-1,7,7-trimethylnorbornan-2-one (camphor), *m/z* 95 displayed. (Reproduced from Cajka, T., et al., *J. Chromatogr. A*, 1217, 4195–4203, 2010. With permission.)

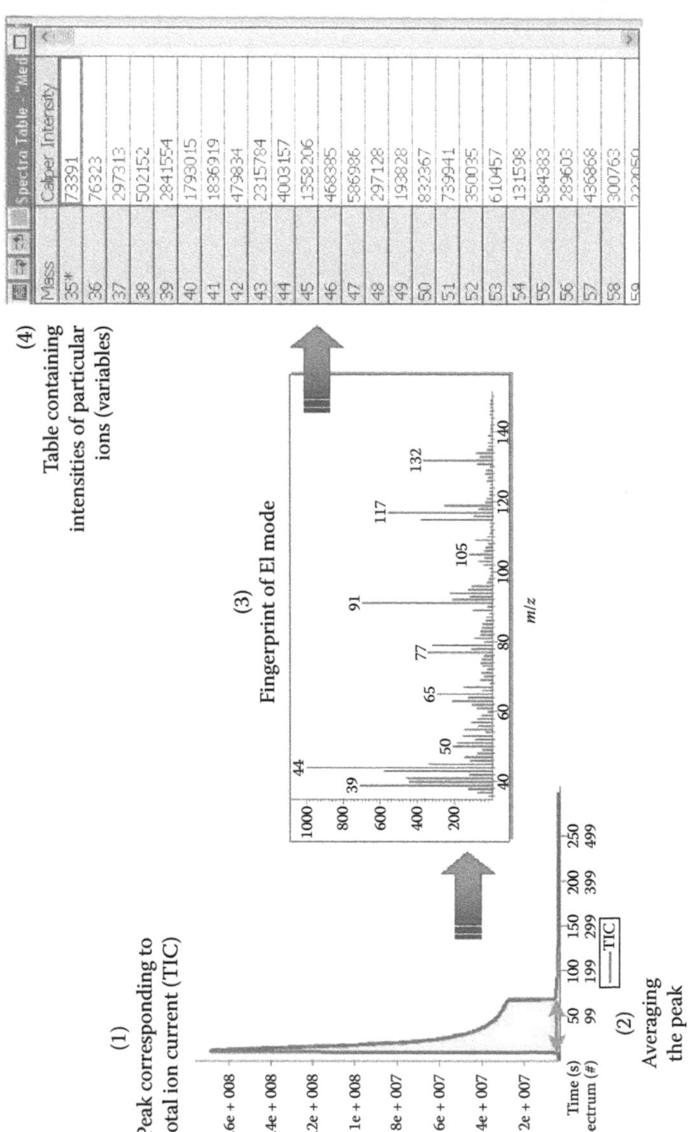

FIGURE 17.6 Fingerprinting using HS-SPME–MS e-nose.

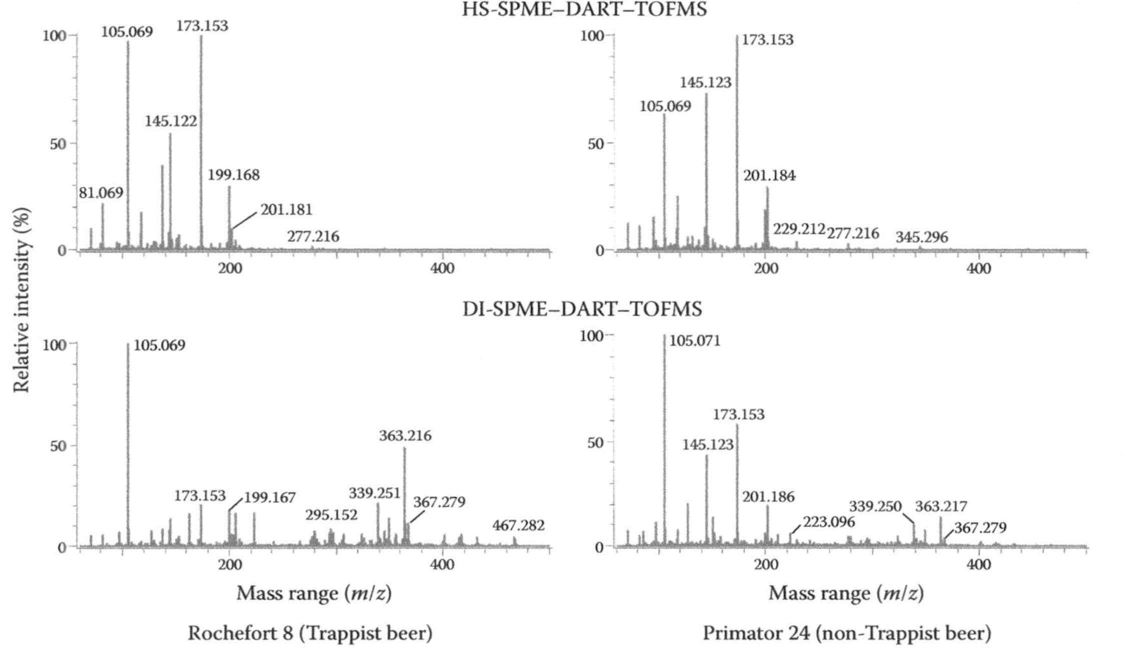

FIGURE 17.7 Comparison of MS spectra (fingerprints) of Rochefort 8 (Trappist) and Primator 24 (non-Trappist) beers under SPME–DART–TOFMS conditions. (Reproduced from Cajka, T., et al., *J. Chromatogr. A*, 1217, 4195–4203, 2010. With permission.)

17.2.2 HS–MS E-Nose

The *electronic nose* based on the mass detector (HS–MS e-nose, also called chemical sensor) is able to carry out analyses in very short times and with minimum sample preparation. In this instrument, the volatile compounds are extracted from the samples by static or dynamic headspace, or SPME. After that, they are introduced in the ion source of a mass spectrometer, in which they are fragmented [typically using electron ionization (EI) at 70 eV]. The fragments of all the volatile compounds are recorded as the abundance of each ion of different mass-to-charge ratios (*m/z*) (Figure 17.6) (Vera et al. 2011).

17.2.3 Ambient Mass Spectrometry

Over recent years, a large number of novel ambient desorption ionization techniques, such as desorption electrospray ionization (DESI), atmospheric pressure solids analysis probe (ASAP), direct analysis in real time (DART), and many others, have become available. Their main advantages compared to conventional techniques, involve the possibility of direct sample examination in the open atmosphere, minimal, or no sample preparation requirements, and, remarkably high sample throughput.

DART represents one of atmospheric pressure chemical ionization (APCI)-related techniques using a glow discharge for the ionization. Metastable helium atoms, which originated in the plasma, react with ambient water, oxygen, or other atmospheric components to produce the reactive ionizing species. DART ion source was shown to be efficient for soft ionization of a wide range of both polar and non-polar compounds. DART produces relatively simple mass spectra characterized in most cases by $[M + H]^+$ in positive-ion mode, and $[M - H]^-$ in negative-ion mode. Coupling the DART ion source with a mass spectrometer with a high mass resolving power [e.g., high-resolution time-of-flight MS (HRTOFMS), orbitrap] with a capability of accurate mass measurements allows confirmation of the target analyte identity and the calculation of elemental composition of "unknowns" (Hajslova et al. 2011). Figure 17.7 illustrates differences in MS profiles obtained by HS- and DI-SPME–DART analyses. While the headspace SPME mode provided profiles with ions in a lower mass region with most *m/z* values <200 Da, the direct immersion SPME mode allowed operators to obtain a fingerprint for higher molecular weight, less volatile, components up to approximately *m/z* 500 Da (Cajka et al. 2010).

17.3 CHEMOMETRICS IN DATA ANALYSIS

Nowadays, the fingerprinting/profiling techniques can generate great amounts of information (variables or features) for a large number of samples (objects) in a relatively short time. Therefore, smart chemometric tools are required in order to efficiently extract the maximum useful information from experimental data.

17.3.1 Data Pretreatment

Data pretreatment typically includes the following steps (Berrueta et al. 2007):

- *Basic and descriptive statistics* [mean, standard deviation, variance, skewness, kurtosis, correlation matrix, *t*-test, *F*-test, analysis of variance (ANOVA), box and whisker plots, and checking the normality]
- *Presence of outliers* (i.e., observations that appear to break the pattern or grouping shown by the vast majority of the samples)
- *Scaling* to modify the relative influences of the variables on a model

The data obtained by fingerprinting/profiling techniques are typically present in the form of a data matrix (samples × signals). In the initial phase, the pretreatment of the data is carried out using scaling methods such as *constant row sum* (each variable is divided by the sum of all variables for each sample); *normalization variable* (variables are normalized with respect to a single variable for each sample); and *range transformation* (the minimum value for a variable is set to "0", the maximum value to "1", and all intermediate values lie along a linear range between 0 and 1 for each sample). This data pretreatment is a very important step since the sensitivity of the detector may slightly fluctuate during longer time period; nevertheless, such data transformation allows us to obtain, in a reproducible way, characteristic relative abundances of markers obtained from peak or ion abundances (area or height). These scaling methods "normalize" the particular variables in each sample (i.e., "row normalization"). After this, scaling of each variable present in the whole sample set is conducted using centering, autoscaling, Pareto scaling, range scaling, vast scaling, level scaling, log transformation, or power transformation (i.e., "column normalization"). The most popular methods are autoscaling (all volatiles have a standard deviation of 1) and Pareto scaling (instead of the standard deviation used in autoscaling, the square root of the standard deviation is used as the scaling factor) (Berrueta et al. 2007; van den Berg et al. 2006).

17.3.2 Exploratory and Unsupervised Pattern Recognition Techniques

Unsupervised pattern recognition techniques, represented mainly by principal component analysis (PCA) and cluster analysis (CA), are often the first step of the data analysis in order to detect patterns in the measured data. By the reduction of the data dimensionality, the PCA allows visualization, retaining as much as possible the information present in the original data. Thus, PCA transforms the original measured variables into new uncorrelated principal components. Each principal component is a linear combination of the original measured variables. Typically, only those principal components that account for a large percentage of the total variance (typically with eigenvalue >1) are retained. The correlation coefficients between the original variables and the principal components are called component loadings. The values that represent the sample in the space defined by the principal components are the component scores, which can be used as input to other multivariate techniques, instead of the original measured variables (Berrueta et al. 2007; Kemsley 1996).

CA represents other unsupervised pattern recognition techniques for preliminary evaluation of the information contents in the data matrices. In CA, samples are grouped on the basis of similarities without taking into account the information about the class membership. CA groups samples according to a similarity metric, which can be distance, correlation, or combination of both (Berrueta et al. 2007).

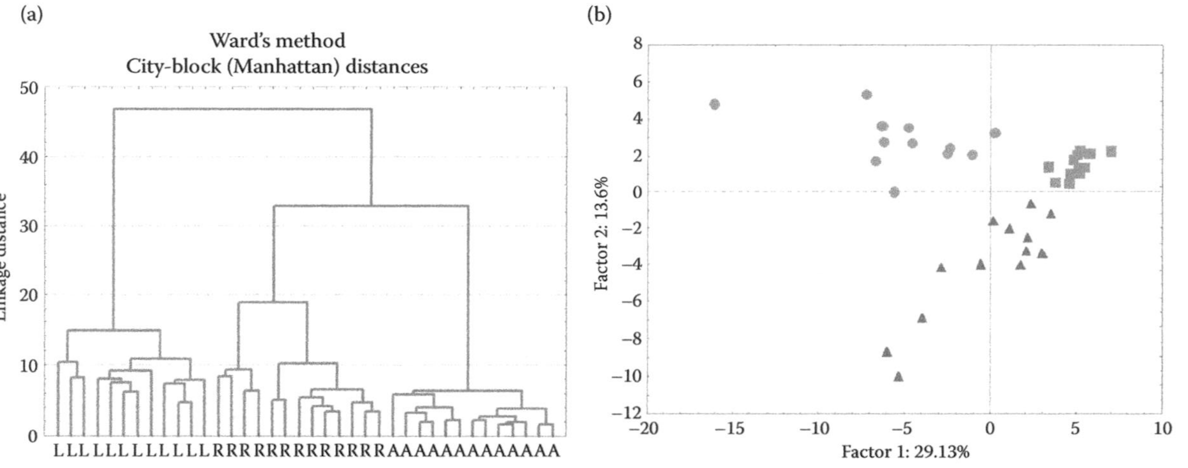

FIGURE 17.8 An example of graphs generated by (a) cluster analysis and (b) principal component analysis. Data matrix consisted of 38 honey samples (L, ●…lime; R, ▲…rape; A, ■…acacia) and 89 volatiles determined by HS-SPME–GC × GC–TOFMS. (Authors' unpublished data.) In the case of CA, dendrogram is read from top to bottom. Horizontal lines show joined clusters. Position of line on the scale indicates distance at which clusters are joined. In this example, there are two clusters of samples [no. 1: lime (L); no. 2: rape (R) and acacia (A)] and second cluster consists of two subclusters [rape (R) and acacia (A)]. In PCA, score plot (from first and second scores) indicates three clusters [lime (●), rape (▲), and acacia (■)], but clusters consisting of rape and acacia honeys are close together (similarly as in the case of CA). First principal compound accounts for 29.1% variance, whereas second PC contributes 13.7%.

Figure 17.8 shows application of both unsupervised pattern recognition techniques on the example of clustering of honey samples according to their botanical origin.

17.3.3 Supervised Pattern Recognition Techniques

Supervised pattern recognition techniques use the information about the class membership of the samples to a given group (class or category) in order to classify a new "unknown" sample in one of the known classes on the basis of its pattern of measurement. The most frequently used supervised pattern recognition techniques are linear discriminant analysis, partial least-squares discriminant analysis, and artificial neural networks (Berrueta et al. 2007; Massart et al. 1997).

Linear discriminant analysis (LDA) determines linear discriminant functions, which maximize the ratio of between-class variance and minimize the ratio of within-class variance. Contrary to PCA that selects a direction retaining a maximal structure among the data in a lower dimension, LDA selects a direction that achieves maximum separation among the given classes. The latent variable obtained in LDA is a linear combination of the original variables. This function is called canonical variate, and its values are roots. However, LDA is not applicable if the number of variables exceeds the number of objects.

Partial least-squares discriminant analysis (PLS-DA) models a relationship between dependent variables (***Y***, i.e., sample origin) and independent variables (***X***, i.e., intensities of markers). The main aim of PLS is to find the components, in the input matrix (***X***), which describe as much as possible the relevant variations in the input variables and at the same time have maximal correlation with the target value in ***Y***, thus, PLS models both ***X*** and ***Y*** simultaneously to find the latent variables in ***X*** that will predict the latent variables in ***Y***. In contrast to PCA that uses the information of matrix ***X*** only, PLS-DA also takes into account the information in matrix ***Y***. Also, PLS-DA gives less weight to the variations that are irrelevant or noisy. Figure 17.9 shows application of PLS-DA and LDA on the example of grouping of honey samples according to their botanical origin.

Artificial neural networks (ANN) represent structures comprising artificial neurons (simple processing elements) that are capable of performing parallel computations for data processing and knowledge representation. The neurons are sorted in: (1) an input layer containing one neuron for each independent variable; (2) one or more hidden layers for data processing; and (3) an output layer with one neuron for each dependent variable. The data are from the input layer propagated through the network via synapses associated with different weight. The learning process tunes the weights that produce the best fit of predicted outputs of the training data set. The hidden layers are important to cope with nonlinear classification problems. A main advantage of ANN is that causal knowledge of the relationship between the input and the output variables is not required. Instead, networks learn these relationships through successive trainings.

17.3.4 Variable Selection and Reduction

Variable selection is a preliminary step used in multivariate data analysis, particularly if (1) the number of objects is relatively small, (2) the number of variables is large, (3) and

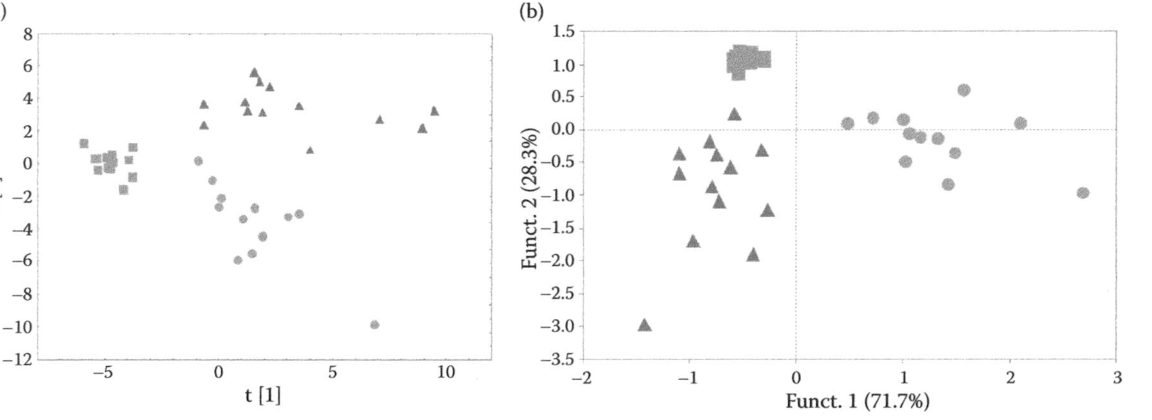

FIGURE 17.9 An example of graphs generated by (a) partial least-squares discriminant analysis (PLS-DA) and (b) linear discriminant analysis (LDA). Data matrix consisted of 38 honey samples (L, ●...lime; R, ▲...rape; A, ■...acacia) and 89 volatiles determined by HS-SPME–GC × GC–TOFMS. (Authors' unpublished data.) In the case of PLS-DA, score plot (from first and second scores) indicates three groups [lime (●), rape (▲), and acacia (■)]. In LDA, data were compressed using PCA first because LDA does not work if number of variables is higher than number of samples. Obtained scores were subsequently submitted to LDA for both grouping and classification. Score plot of two discriminant functions of LDA was calculated from six principal components [the optimal number of PCs for LDA was obtained using leave-one-out cross-validation (LOOCV)].

many of these variables contain redundant or noisy information. In these cases, a variable or feature selection procedure is required in order not to fall into the overfitting problem. Variable selection for discriminant analysis selects a subset of variables that are the most discriminating (Berrueta et al. 2007).

The preferred variable selection method is a stepwise selection, which is based on a search that sequentially adds or deletes variables from the pool of candidate variables. The addition or deletion of a single variable is performed with regard to the largest improvement in the classification, and the process goes on until the search gets trapped in the first local optimum. The addition or removal of a variable is considered simultaneously based on probability or Fisher criteria (p or F-values). Best subset selection is a variable selection procedure that performs a search of all possible subsets of variables that fulfill the criterion for choosing the best one (Wilk's lambda, rate of misclassification, etc.) (Berrueta et al. 2007).

17.3.5 Validation of the Model

Model validation process allows demonstrating that the models obtained by the supervised pattern recognition techniques are good enough to perform classification of unknown samples. This can be done by observing how successful the model is at classifying known objects, that is, by evaluating the recognition and prediction abilities of the model. Recognition ability represents a percentage of the samples in the training set successfully classified. Prediction ability is a percentage of the samples in the test set correctly classified by using the model developed during the training step. In addition to these abilities, sensitivity and specificity can be evaluated. Sensitivity of a class model is the percentage of objects belonging to a given class, which is correctly identified by the model. Specificity is the percentage of objects foreign to the modeled class that are

(a)

	Pred. Group (Std)			Correctly
Act. Group	Acacia	Lime	Rape	Classified
Acacia	13	0	0	100.0%
Lime	0	12	0	100.0%
Rape	1	0	12	92.3%
Overall Correct Class. Rate				97.4%

(b)

	Pred. Group (Holdout)			Correctly
Act. Group	Acacia	Lime	Rape	Classified
Acacia	13	0	0	100.0%
Lime	1	11	0	91.7%
Rape	1	0	12	92.3%
Overall Correct Class. Rate				94.7%

FIGURE 17.10 An example of classification tables containing (a) recognition ability and (b) prediction ability of the model. Each group of samples shows correct classification score (in %) and number of samples classified by the model to each group. At the end, an overall correct classification rate is calculated.

classified as foreign (Berrueta et al. 2007). Figure 17.10 shows an example of the classification tables, typically obtained during the validation of the model containing recognition and prediction abilities.

The validation procedure can be performed using (1) external validation, (2) k-fold cross-validation, and (3) leave-one-out cross-validation. In external validation, the test set is completely independent from the model building process. The k-fold cross-validation consists of assigning samples randomly to a training set and a test set, the latter containing $1/k$ of the samples (k = 3–5). The leave-one-out cross-validation removes only one sample at a time from the training set and considers it as a test set. In general, the recognition ability of a model is better than the prediction ability (see an example in Figure 17.11). However, if they are substantially different, this indicates that the decision rule depends too much on the actual objects in the training set, thus, the classification achieved is not stable and therefore not reliable.

17.3.6 Overoptimistic Classification

Overoptimistic classification results can be obtained under certain conditions. This may occur mainly when (1) the number of samples is too low, (2) the model is overfitted, (3) only the recognition ability is determined, and (4) clusters occur inside a category. Regarding these considerations, it should be noted that replicates are definitely not independent observations. If clusters are observed in a class, splitting the group into two or more classes should be considered. Furthermore, as a rule of thumb, the onset of overfitting should be strongly suspected when the number of variables exceeds $(n - g)/3$, where n is the number of objects and g is the number of categories (Berrueta et al. 2007; Defernez and Kemsley 1997).

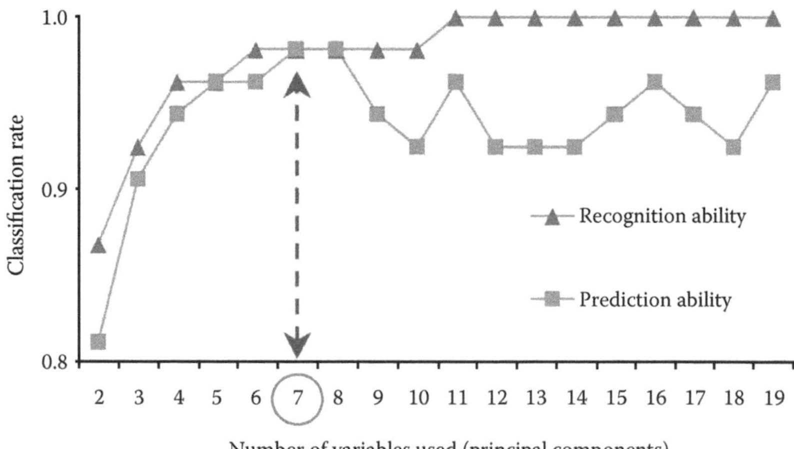

FIGURE 17.11 An example of classification rates obtained during model validation (LOOCV). LDA models were constructed using increasing number of variables (principal components) in particular case obtained from original variables (normalized intensities of volatiles). Optimal number of variables is indicated by a circle. (Authors' unpublished data.)

17.4 CASE STUDIES

The following case studies published in scientific literature summarize information about the possibility of using volatile compounds as markers for authenticity/traceability purposes (Tables 17.1 through 17.4, 17.6, 17.8, and 17.9). As noted earlier, the goal of these studies is to develop a model to classify food samples according to different criteria, such as the geographical or botanical origin, technological process, quality state, and detection of adulteration.

17.4.1 Beer

Beer represents a widely popular alcoholic beverage with a high world production rate (approximately 1.6×10^{11} L/year). This very complex mixture of constituents, varying widely in nature and concentration levels, is brewed from raw materials including malt, water, yeast, and hops and contains a broad range of different chemical components that may react and interact at all stages of the brewing process. The volatile compounds, which have been identified in beer and are associated to its flavor, belong to various chemical groups, including several aliphatic and aromatic alcohols, esters, acids, carbonyl compounds, terpenic substances, and others. A better understanding of the key volatile compounds is of paramount importance for modern brewing technology, helping the selection of raw materials and yeast strains, as well as for routine quality control (Vera et al. 2011; Cajka et al. in press; da Silva et al. 2008) (Table 17.1).

Cajka et al. (2010) used an automated HS-SPME-based sampling procedure coupled to GC–TOFMS for obtaining fingerprints of beer volatiles (45 markers). In total, 265 specialty beer samples were collected over a one-year period (covering possible seasonal variability of the products) with the aim of distinguishing, based on analytical (profiling) data: (1) the beers labeled as Rochefort 8; (2) a group consisting of Rochefort 6, 8, 10 beers; and (3) Trappist beers. For the chemometric evaluation of the data, PLS-DA, LDA, and ANN with multilayer perceptrons (MLP) were tested. The best prediction ability was obtained for the model that distinguished a group of Rochefort 6, 8, 10 beers from the rest of the beers. In this case, all chemometric tools used provided 100% prediction ability. In addition to this approach, the same set of beers was also examined, using ambient MS with a DART ion source (Cajka et al. in press). Using positive and negative DART ion mode, 37 ions/markers (including volatiles, but also polar compounds such as amino and organic acids) were selected for subsequent chemometric analysis using PLS-DA, LDA, and ANN-MLP. The best prediction ability (98%) was obtained for the model that differentiated the group of Rochefort 6, 8, 10 beers from the rest of the beers (ANN-MLP). However, considering the sample throughput rate, it should be emphasized that SPME–GC–TOFMS required some, although quite simple, sample preparation steps with subsequent sample incubation/extraction (10 min) and separation/detection of beer volatiles (20 min). On this basis, the time requirement for sample analysis was in fact one order of magnitude higher as compared to the direct measurement of beer samples using DART–TOFMS (<1 min is required for a single analysis). The study showed that DART–TOFMS metabolomic fingerprinting/profiling is a powerful analytical strategy

TABLE 17.1
Examples of Studies Dedicated to Authenticity of Beers

Origin of Samples	Characterization Studies	Instrumental Techniques	Input Data	Data Pretreatment	Pattern Recognition Techniques	Validation	Remarks	Ref.
Belgium, The Netherlands, Czech Republic	Classification of beers according to brand	HS-SPME–GC–TOFMS	265 samples 45 volatiles	Range transformation (each sample)	(1) PCA (2) LDA, PLS-DA, ANN-MLP	External	100% prediction ability for model distinguishing a group of Rochefort 6, 8, 10 beers from the rest of specialty beers	Cajka et al. 2010
Belgium, The Netherlands, Czech Republic	Classification of beers according to brand	DART–TOFMS	265 samples 37 ions	Range transformation (each sample)	(1) PCA (2) LDA, PLS-DA, ANN-MLP	External	>95% prediction ability for model distinguishing a group of Rochefort 6, 8, 10 beers from the rest of specialiy beers	Cajka et al. in press
Brazil	Classification of beers according to brand	HS-SPME–GC–FID	20 samples 32 volatiles	Autoscaling	Kohonen NN		Grouping of six sets of Brazilian Pilsner beers	Vera et al. 2011
Spain	Classification and characterization of a series of beers according to production site	HS–MS (e-nose)	67 samples 101 fragments (*m*/*z*)		(1) PCA (2) LDA, Bayesian LDA, Fisher LDA	*k*-Fold CV	90% prediction ability for model distinguishing beers according to four different factories	Silva et al. 2008

for monitoring of batch-to-batch fluctuation of characteristic component profiles to assess the constancy of the beer production process. Figure 17.12 shows separation of classes using PLS-DA according to the origin of beer samples and instrumental technique used.

Vera et al. (2011) conducted a study to classify and characterize a series of beers according to their production site and chemical composition using HS–MS (e-nose). The analyzed beer samples were of the same brand but obtained from four different factories. The collected variables (101 *m/z* fragments) were compressed using PCA. Using seven principal components (PCs) as an input for Bayesian LDA, a prediction ability of 90% was achieved, demonstrating the difference of aroma characteristics of beers from the four different factories. In addition, application of Fisher LDA with 18 of the most discriminant *m/z* fragments (selected with the stepwise LDA algorithm) revealed a relationship between volatile alcohols/sulfur compounds/esters and factories. The results obtained in this study enable consideration of the HS–MS (e-nose) as a potential aroma sensor because it is capable of discriminating and characterizing the samples according to their predominant aromas with the help of multivariate analysis.

da Silva et al. (2008) examined a set of Brazilian Pilsner beers using HS-SPME–GC–FID and Kohonen NN. The identified markers belonged to alcohols, esters, organic acids, phenolic compounds, ketone, and other volatiles. Analysis of Kohonen map showed that the 20 different brands of beer could be grouped into six sets. Using this approach, it was possible to infer which samples presented analogous volatile profiles, even if they were produced by different breweries, through the assessment of the groups formed in Kohonen map.

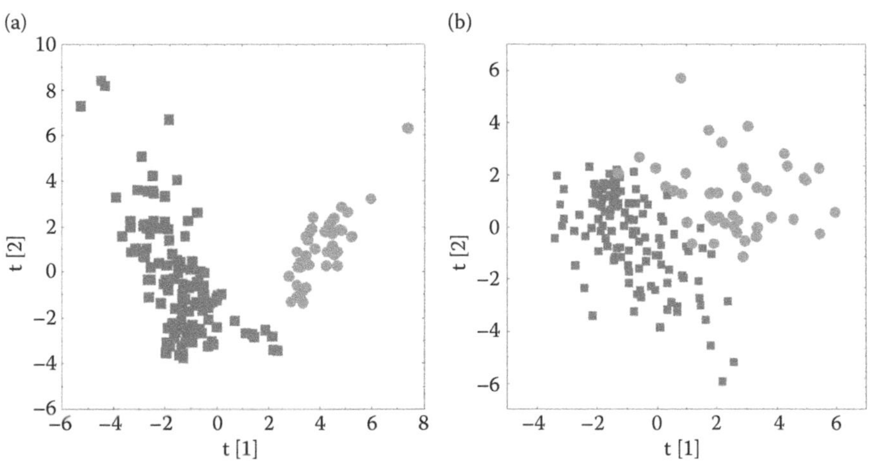

FIGURE 17.12 First and second PLS scores for Rochefort 6, 8, 10 (●) vs. rest of beers (■). Graphs constructed using calibration data set (n = 166); (a) HS-SPME–GC–TOFMS, (b) DART–TOFMS. (Reproduced from (a) Cajka, T., et al., *J. Chromatogr.* A, 1217, 4195–4203, 2010; (b) Cajka, T., et al., 2009, Paper presented at 4th International Symposium on Recent Advances in Food Analysis, Prague, Czech Republic. With permission.)

17.4.2 Wine

Although the world production rate (approximately 2.8×10^{10} L/year) of wine is lower than that of beer, wine authenticity is a subject of ongoing concern and has been extensively investigated. Seemingly similar wines from different origins might be of substantially different price. The use of geographical indications allows producers to obtain market recognition, often reflected in a premium price. False use of geographical indications by unauthorized parties may be detrimental to consumers or legitimate producers. From this point of view, the development of new and sophisticated techniques for determining the geographical origin of agricultural products is highly desirable for consumers, producers, retailers, and administrative authorities (Berna et al. 2009) (Table 17.2).

Aznar and Arroyo (2007) used purge and trap–gas chromatography–quadrupole mass spectrometry (P&T–GC–QMS) for differentiation of red and white wines from seven different Spanish regions. For data interpretation, PCA was used. Results show that red and white wines were grouped separately. Red wines had, in general, a higher concentration of 2-methylpropan-1-ol, 2-methylbutan-1-ol, 3-methylbutan-1-ol, ethyl 2-hydroxypropanoate (ethyl lactate), and diethyl butanedioate (diethyl succinate). White wines were linked to a higher concentration of ethyl hexanoate and ethyl octanoate. It was also observed that most red and white wines were also grouped on the basis of their origin. White wines from the region Castillala Mancha had higher concentrations of 3-methylbutyl acetate (isoamyl acetate), propan-1-ol, and butan-1-ol, and white wines from the Badajoz region were linked mainly to higher concentrations of ethyl 3-methylbutanoate (ethyl isovalerate). Using this approach, most of the wines were correctly grouped on the basis of their origin and kind of wine (red/white wine). These results indicated that the origin of wines had an influence on the volatiles analyzed and, therefore, on the wine aromas linked to them.

Gil et al. (2006) analyzed a large set (194) of wines from Spain ("Vinos de Madrid") according to type (white/rosé/red) using GC–FID (major volatiles) and GC–QMS (minor volatiles). LDA showed that the isoamylic alcohols (2-methylbutan-1-ol and 3-methylbutan-1-ol), higher major alcohols (propan-1-ol, 2-methylpropan-1-ol, and isoamylic alcohols), ethyl hexanoate, acetates, ethyl octanoate, 4-ethenyl-2-methoxyphenol (4-vinyl guaiacol), and decanoic acid were the most significant compounds in the differentiation among the white, rosé, and red young wines. The same analysis showed a prediction ability of 92% according to their type (white, rosé, or red). Diaz-Reganon et al. (1998) also focused on the differentiation of wines from Spain (Madrid region) according to type (white/rosé/red) using GC–FID (major volatiles). Using LDA, 90%, 95%, and 80% classifications were obtained for white, rosé, and red wines, respectively, according to geographical origin (Arganda, Navalcarnero, San Martin).

Diaz et al. (2003) used GC–FID for differentiation of wines from the Canary Islands according to type (white/rosé/red). Using PCA, it was observed that the red wines tend to separate from the white and rosé wines. This was due to differences in the process of fermentation that generate a higher amount of volatile compounds in red wines. Applying PCA to the samples of red wines and to the samples of white wines in an independent manner, the red wines were clearly separated according to vintage. The behavior of white wines was different, showing no separation according

TABLE 17.2
Examples of Studies Dedicated to Authenticity of Wines

Origin of Samples	Characterization Studies	Instrumental Techniques	Input Data	Data Pretreatment	Pattern Recognition Techniques	Validation	Remarks	Ref.
Spain	Differentiation of red and white wines from different regions of Spain	GC–QMS	40 samples (17 white wines, 23 red wines) 26 volatiles	Autoscaling	PCA		Most of the wines were correctly grouped on the basis of their origin and kind of wine (red/white)	Aznar and Arroyo 2007
Spain	Differentiation of wines from Spain ("Vinos de Madrid") according to type (white/rosé/red)	GC–FID (major volatiles), GC–QMS (minor volatiles)	194 samples (71 white, 53 rosé, and 70 red wines) 25 volatiles		LDA	LOOCV	92% prediction ability for model distinguishing wines labeled as "Vinos de Madrid" according to type (white/rosé/red)	Gil et al. 2006
Spain	Differentiation of white/rosé/red wines according to geographical origin	GC–QMS	88 samples (white/ rosé/red wines from regions Arganda, Navalcarnero, San Martín) 29 volatiles		LDA		90%, 95%, and 80% classification for white, rosé, and red wines, respectively, according to geographical origin (Arganda, Navalcarnero, San Martín)	Diaz-Reganon et al. 1998

(continued)

TABLE 17.2 (Continued)
Examples of Studies Dedicated to Authenticity of Wines

Origin of Samples	Characterization Studies	Instrumental Techniques	Input Data	Data Pretreatment	Pattern Recognition Techniques	Validation	Remarks	Ref.
Spain	Differentiation of wines from Canary Islands according to type (white/rosé/red)	GC–FID	153 samples (64 white, 29 rosé, and 60 red wines) 6 volatiles		PCA		Red wines tend to separate from the white and rosé wines	Diaz et al. 2003
South Africa	Differentiation of wines according to grape variety	SBSE–GC–QMS	334 samples (65 Sauvignon Blanc, 45 Chardonnay, 41 Pinotage, 64 Shiraz, 60 Cabernet Sauvignon, 59 Merlot) 37 volatiles		PCA, LDA	External	100% prediction ability for model distinguishing white and red wines. 100% prediction ability for model distinguishing Sauvignon Blanc and Chardonnay 97% prediction ability for model distinguishing Pinotage and the rest of red wines	Weldegergis et al. 2011
Portugal	Classification and differentiation of Madeira wines according the grape varieties	HS-SPME–GC–ITMS	36 samples 42 volatiles	Normalization (each sample)	PCA, LDA	LOOCV	96% prediction ability for model distinguishing wines from Madeira according to wine variety (Boal, Malvasia, Sercial, Verdelho)	Camara et al. 2006

Spain, Australia	Classification of Tempranillo wines according to geographical origin	HS–MS (e-nose)	60 samples (35 Australian and 25 Spanish Tempranillo) 6 fragments (*m/z*)	Autoscaling	PCA, PLS-DA, LDA	LOOCV	>80% prediction ability for model distinguishing Tempranillo wines from two different geographical origins	Cynkar et al. 2010
Canada, Czech Republic	Differentiation of ice wines according to geographical origin	HS-SPME–GC–TOFMS	137 samples (121 samples from Canada, 16 samples from the Czech Republic) 58 volatiles	Log transformations (each sample) Centered	KSOM		Clear discrimination of all samples according to their Canadian and Czech origins	Giraudel et al. 2007
Spain	Differentiation of white wines according to certified brands of origin (Condado de Huelva, Ribeiro, Rueda, Penedés)	GC–QMS, GC–FID	66 samples 17 volatiles		PCA, LDA, ANN-MLP	LOOCV, External	95%, 91%, 100%, and 81% prediction abilities for model (LDA) distinguishing wines according to brand (Condado de Huelva, Ribeiro, Rueda, Penedés) 100% prediction ability for model (ANN-MLP) distinguishing wines according to brand (Condado de Huelva, Ribeiro, Rueda, Penedés)	Jurado et al. 2008

(*continued*)

TABLE 17.2 (Continued)
Examples of Studies Dedicated to Authenticity of Wines

Origin of Samples	Characterization Studies	Instrumental Techniques	Input Data	Data Pretreatment	Pattern Recognition Techniques	Validation	Remarks	Ref.
Slovak Republic	Classification of white wines according to variety, producer, and vintage	GC–QMS	87 samples 20 volatiles		ANN-MLP	External	Prediction abilities of 100%, 100%, and 93% to classify white wines according to variety, producer, and vintage, respectively	Kruzlicova et al. 2009
Spain	Differentiation of sweet wines according to grape variety (Muscat, Pedro Ximénez)	HS-SPME–GC–FID	20 samples (10 samples Muscat, 10 samples Pedro Ximénez) 30 volatiles		CA, PCA, LDA	LOOCV	100% prediction ability for model distinguishing sweet wines according to grape variety (Muscat, Pedro Ximénez)	Marquez et al. 2008
China	Differentiation of red wines according to grape variety (Cabernet Sauvignon, Merlot, Cabernet Gernischt)	HS-SPME–GC–ITMS	39 samples (test set) + 36 samples (external set) 30 volatiles		PCA, LDA	LOOCV, External	97% prediction ability (LDA-LOOCV, 11 variables) 100% prediction ability (LDA-LOOCV, six principal components) 65% prediction ability (LDA, external set) for model distinguishing red wines according to grape variety	Zhang et al. 2010

to vintage. Thus, the authors concluded that the climatic conditions and the timing of the grape harvest had a greater influence on the red wines than on the white wines.

Weldegergis et al. (2011) analyzed a large set of wines (334) that originated from South Africa according to grape variety (Sauvingnon Blanc, Chardonnay, Pinotage, Shiraz, Cabernet Sauvignon, Metlot), using stir bar sorptive extraction (SBSE)–GC–QMS. Using LDA, a prediction ability of 100% was obtained for the model distinguishing white and red wines. The same prediction ability was reached for the model distinguishing Sauvingnon Blanc and Chardonnay. A prediction ability of 97% was obtained for the model distinguishing Pinotage and the rest of the red wines; 3-methylbutyl acetate and ethyl octanoate were seen as influential in this differentiation.

Camara et al. (2006) analyzed 36 monovarietal Madeira wine samples elaborated from Boal, Malvasia, Sercial, and Verdelho white grape varieties using HS-SPME–GC–ion trap MS (ITMS) and PCA. The most important contributions to the differentiation of Boal wines were phenylmethanol and (*E*)-hex-3-en-1-ol. Ethyl octadecanoate, (*Z*)-hex-3-en-1-ol, and benzoic acid were the major contributions in Malvasia wines and 2-methylpropan-1-ol was associated to Sercial wines. Verdelho wines are most correlated with 5-(ethoxymethyl)furan-2-carbaldehyde, nonan-2-one, and ethyl dec-9-enoate. A prediction ability of 96% was obtained by the application of stepwise LDA using the 19 variables for the model distinguishing wines from Madeira according to wine variety (Boal, Malvasia, Sercial, Verdelho).

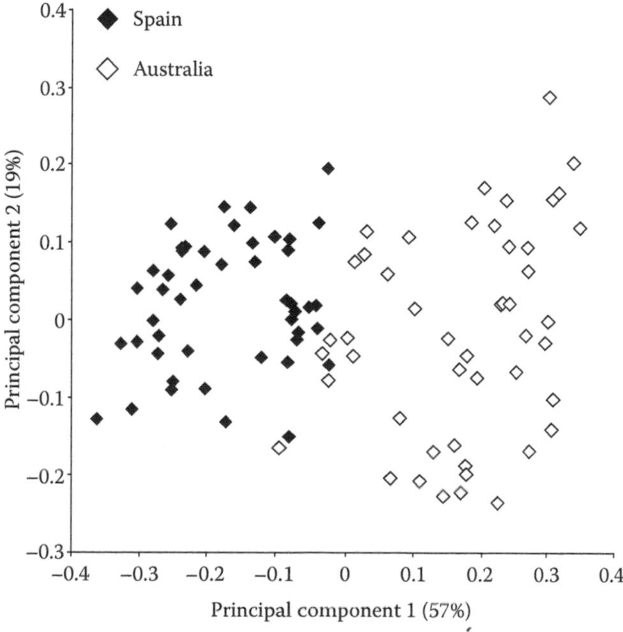

FIGURE 17.13 Score plot of first two principal components of Tempranillo wines measured by HS–MS (e-nose). (Reproduced from Cynkar, W., et al., *Anal. Chim. Acta*, 660, 227–231, 2010. With permission.)

Cynkar et al. (2010) focused on the classification of Tempranillo wines according to geographical origin using HS–MS (e-nose). The PCA showed a clear separation between wines according to their geographical origin (Figure 17.13). However, samples from two labels of Australian wines overlapped with the Spanish wine samples. The PLS-DA models produced an overall rate of correct classification of 85%. Seventy-four percent of Australian wines were correctly classified, whereas for Spanish wines, the correct classification reached 100%. The stepwise LDA classification based on the first three scores from PCA, which account for more than 70% of the variance, gave a similar overall rate of correct classification (86%). More Australian wines (88%) and less Spanish wines (85%) were correctly classified.

Giraudel et al. (2007) used HS-SPME–GC–TOFMS method for the analysis of volatile and semivolatile components of ice wine that originated from Canada and the Czech Republic. Using Kohonen self-organizing maps (KSOM), a clear discrimination of the 137 samples, according to their Canadian and Czech origins, was obtained from a 300-cell trained map, without any outlying sample or analysis constituent.

Jurado et al. (2008) used HS-SPME–GC–QMS for differentiation of white wines according to certified brands of origin (Condalo de Huelva, Ribeiro, Rueda, Penedés). For the interpretation of data set, LDA and ANN were used. Using stepwise LDA, the most discriminant variables were the amounts of ethyl acetate, ethyl hexanoate, hexyl acetate, ethyl octanoate, ethyl decanoate, ethyl dodecanoate, ethyl tetradecanoate, ethyl hexadecanoate, 3-methylbutyl acetate, 3-methylbutyl octanoate, 3-methylbutan-1-ol, and 2-phenylethanol. Prediction abilities of 95%, 91%, 100%, and 81% were reported for the model distinguishing wines according to brand (Condalo de Huelva, Ribeiro, Rueda, Penedés). Figure 17.14 shows the distribution of the samples in the plane of the two calculated discriminant functions. Although all classes

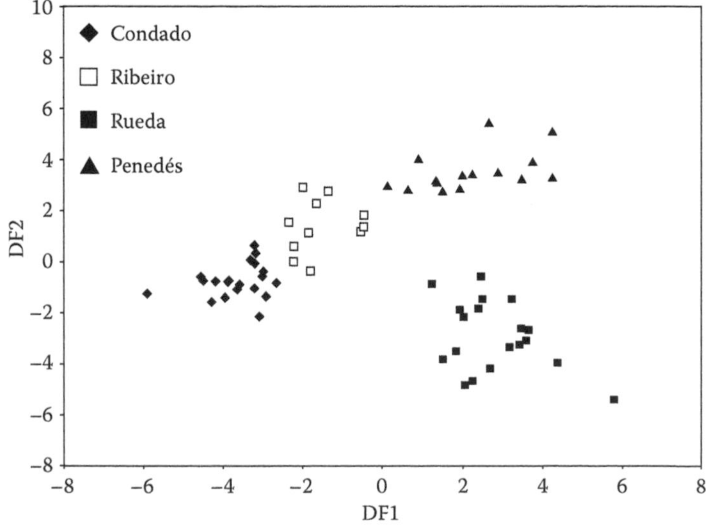

FIGURE 17.14 Scatter plot (LDA) of wines belonging to four certified brands of origin in the plane of the two first discriminant functions. (Reproduced from Jurado, J.M., et al., *Anal. Bioanal. Chem.*, 390, 961–970, 2008. With permission.)

appear separated from the others, prediction abilities have shown that LDA did not enable complete sorting of the samples from the four brands of origin. Consequently, a nonlinear model created by ANN was applied to differentiate the classes. Using ANN, a prediction ability of 100% for the model distinguishing wines according to brand (Condalo de Huelva, Ribeiro, Rueda, Penedés) was obtained, indicating that ANN modeled better than LDA.

Kruzlicova et al. (2009) demonstrated the possibility to use GC–QMS and ANN for the classification of white wine varietal wines that originated from the Slovak Republic. Prediction abilities of 100%, 100%, and 93% to classify white wines according to variety, producer, and vintage, respectively, were achieved using this approach.

Marquez et al. (2008) used HS-SPME–GC–FID method for the differentiation of sweet wines according to grape variety (Muscat, Pedro Ximénez). CA formed three main clusters: one cluster for Pedro Ximénez sweet wines, and two clusters for Muscat sweet wines (Figure 17.15). Clusters were formed exactly according to the grape variety. The groups obtained demonstrate that the variables possess sufficient explanatory power to detect the grape variety used.

Using PCA, the loadings of each compound on PC1 and PC2 showed that ethyl esters and fatty acids are mainly responsible for PC1, whereas some terpenic compounds (limonene, *p*-cymene, and linalool), isoamyl alcohols, and 5-methylfuran-2-carbaldehyde are mainly responsible for PC2. It can be observed that two types of wines are identified according to their aromatic composition (Figure 17.16): first there are the wines from Pedro Ximénez variety (negative values of PC1 and positive values of PC2), and then, a broad group composed of wines obtained from Muscat grapes (negative values of PC2).

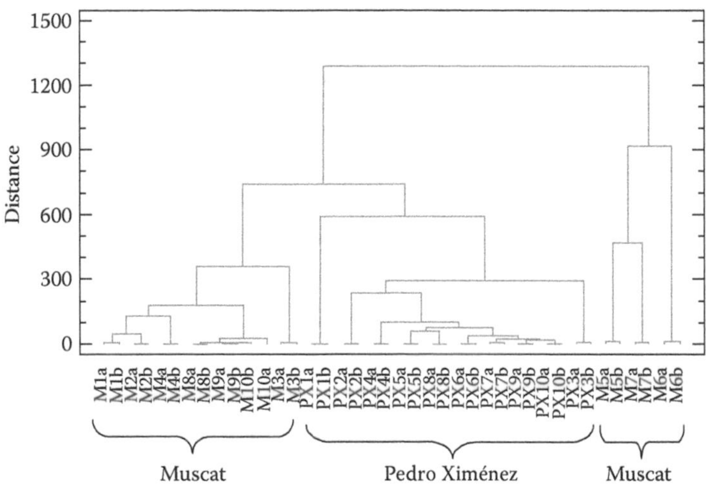

FIGURE 17.15 Dendrogram obtained after a hierarchical agglomerative CA performed on volatile compounds of all samples studied. M, Muscat wine; PX, Pedro Ximénez wine. (Reproduced from Marquez, R., et al., *Eur. Food Res. Technol.*, 226, 1479–1484, 2008. With permission.)

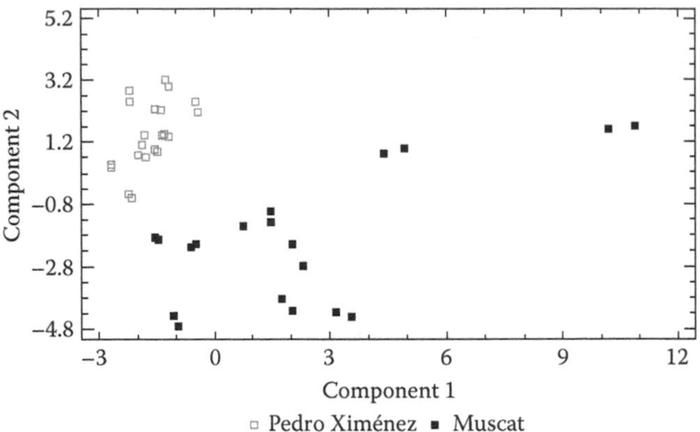

FIGURE 17.16 Score plot on first and second principal components for samples studied. (Reproduced from Marquez, R., et al., *Eur. Food Res. Technol.*, 226, 1479–1484, 2008. With permission.)

Finally, using LDA, a prediction ability of 100% for the model distinguishing sweet wines according to grape variety (Muscat, Pedro Ximénez) was obtained. The variables with the highest discrimination power were α-terpineol, 5-methylfuran-2-carbaldehyde, isoamyl alcohols, ethyl decanoate, hexanal, and *p*-cymene.

Zhang et al. (2010) used HS-SPME–GC–ITMS method for differentiation of red wines according to grape variety (Cabernet Sauvignon, Merlot, Cabernet Gernischt). Stepwise LDA allowed 100% prediction ability for Cabernet Sauvignon and Cabernet Gernischt wines, but only 92% for Merlot wines. A more valid and robust way was to use the PCA scores for the discriminant analysis. When stepwise LDA was performed in this way, 100% prediction abilities were obtained. The authors also demonstrated the models using commercial wines; the models showed 100% recognition ability for the wines collected directly from winery and without aging, but only 65% for the others. Therefore, the varietal factor was currently discredited as a differentiating parameter for commercial wines in China. Nevertheless, the authors concluded that this method could be applied as a screening tool and as a complement to other methods for grape base liquors that do not need aging and blending procedures.

17.4.3 Spirits

Whereas beer and wine are typically limited to a maximum alcohol content of approximately 15%, the spirits refer to a distilled beverage that contains no added sugar and at least 20% of ethanol. The most popular spirits include brandy, fruit brandy, gin, rum, tequila, vodka, and whisky. An important authenticity issue in this context represents the use of raw materials for particular spirits, which should be specified on the label (e.g., vodka), or is connected only with one type of raw material (e.g., use of sugar cane molasses for rum production). Also, possible frauds, such as adulteration of certain type of spirits by cheaper raw material, are of a high concern (Jelen et al. 2010; Lachenmeier et al. 2006) (Table 17.3).

TABLE 17.3
Examples of Studies Dedicated to Authenticity of Spirits

Origin of Samples	Characterization Studies	Instrumental Techniques	Input Data	Data Pretreatment	Pattern Recognition Techniques	Validation	Remarks	Ref.
Poland	Discrimination of raw spirits used for alcoholic beverage production	HS-SPME–MS	138 samples, 261 fragments (*m/z*)	Log transformation Autoscaling	PCA, LDA	External	96% prediction ability achieved for model distinguishing raw spirit samples produced from rye, corn, and potato 23 ions selected by stepwise LDA with the best discrimination power reported	Jelen et al. 2010
Mexico	Quantification of volatiles in Mexican Agave spirits (Tequila, Mezcal, Sotol, Bacanora)	GC–FID	95 samples, 18 volatiles		ANOVA		Composition of Mexican Agave spirits was found to very over a relatively large range	Lachenmeier et al. 2006
Cyprus	Discrimination of Cypriot spirit "Zivania"	GC–FID	42 samples, 10 volatiles (+16 other analytes/ parameters)		PCA, CA, DA, classification and regression trees (CART)	LOOCV	88% prediction ability achieved for model distinguishing five groups of alcoholic beverages (zivania, zivania red, various distillates, Greek distillates, and eau de vie)	Kokkinofta and Theocharis 2005

Jelen et al. (2010) focused on the determination of the botanical origin of raw spirits used for alcoholic beverage production. A large set of raw spirit samples produced from rye, corn, and potato were analyzed using HS-SPME–MS (e-nose). Using PCA, the first PC separated rye spirits from the others, whereas the third PC separated the corn from the potato spirits. Using stepwise LDA, ions with the best prediction power were obtained (*m/z* 50, 51, 53, 54, 58, 72, 85, 91, 93, 94, 95, 96, 104, 107, 108, 116, 119, 133, 145, 147, 159, 172, 191). Also, the LDA allowed classification of raw spirits based on the material they were produced from with a prediction ability of 96%. Figure 17.17 illustrates the separation among the classes using this approach.

Lachenmeier et al. (2006) analyzed a large collection of Mexican *Agave* spirits with protected appellations of origin (Tequila, Mezcal, Sotol, and Bacanora) using GC–FID. The two Tequila categories ("100% Agave" and "mixed") showed differences in the methanol, 2-/3-methylbutan-1-ol, and 2-phenylethanol concentrations with lower concentrations in the mixed category. Mezcal showed no significant differences in any of the evaluated parameters that would allow a classification. Sotol showed higher nitrate concentrations and lower 2-/3-methylbutan-1-ol concentrations. Bacanora was characterized by exceptionally high acetaldehyde concentrations and a relatively low ethyl lactate content. As a permanent problem in food control, the basic categories of Tequila (100% *Agave* and mixed Tequila) need to be distinguished. For the high-quality category 100% Agave, only pure Agave juice is allowed to be fermented and distilled. A mixed Tequila is manufactured by adding up to 49% (w/v) of sugar, mainly from sugar cane. This lower-end Tequila is usually shipped out in bulk containers for bottling in the importing countries, and there is,

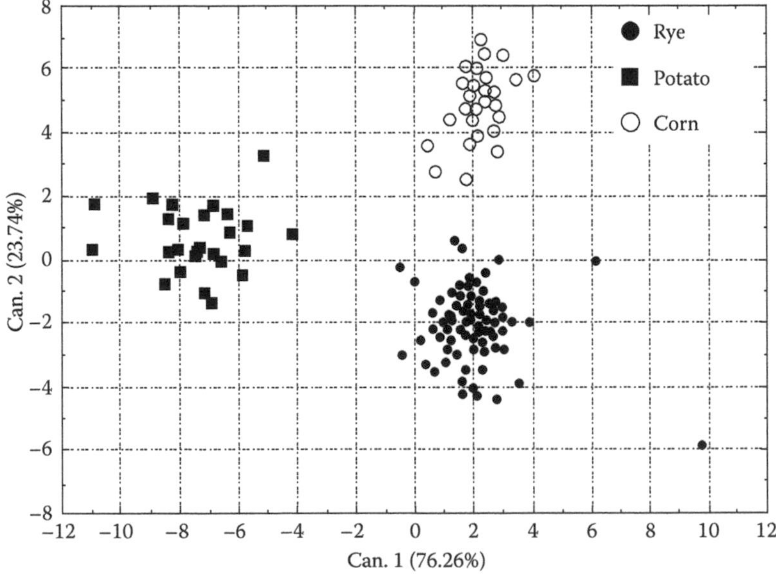

FIGURE 17.17 Score plot of two dicriminant functions of LDA calculated from SPME–MS data. (Reproduced from Jelen, H.H., et al., *J. Agric. Food Chem.*, 58, 12585–12591, 2010. With permission.)

of course, a high economic incentive for labeling fraud. The authors concluded that the concentrations of methanol and 2-/3-methylbutan-1-ol were the most suitable analytical approach to differentiate 100% Agave and mixed Tequila.

Kokkinofta and Theocharis (2005) used 26 chemical and physicochemical variables obtained by HPLC, GC, ^{1}H NMR, and ICP-MS for characterization of the Cypriot spirit "Zivania." Using regularized DA, a prediction ability of 88% was achieved for the model distinguishing five groups of alcoholic beverages (zivania, zivania red, various distillates, Greek distillates, and eau de vie). 3-Methylbutan-1-ol was the most important analytical variable for this classification, and it was the feature that contained the most discriminatory information for the classification of the Zivania samples.

17.4.4 Coffee

Coffee is the world's most popular beverage after water, with a high world production rate (approximately 7.7 million tons/year). The coffee aroma is characterized by the presence of a wide range of volatiles belonging to several classes of compounds such as furans, pyrazines, ketones, alcohols, aldehydes, esters, pyrroles, thiophenes, sulfur compounds, benzenic compounds, phenols, pyridines, thiazoles, oxazoles, lactones, alkanes, alkenes, and acids. These components are present in variable concentrations and each of them contributes uniquely to the final aroma quality.

One of the most commonly utilized strategies in coffee authenticity is the possibility to discriminate coffees based on the coffee bean variety. From the 66 species of the genus *Coffea* L. that have been identified so far, only two varieties with pronounced economical and commercial importance are extensively cultivated in coffee-producing regions: *Coffea arabica* L. (arabica coffee) and *Coffea canephora* Pierre (robusta coffee). Arabica coffee beans command higher commercial value prices and are therefore subject to fraudulent practices because they are characterized by more pronounced quality and consumers' acceptance.

Another important aspect of coffee authenticity is geographical origin declaration. Considering that increasing practices associated with selling coffees on the basis of their geographical origin and falsifying the product declaration have been detected in recent years, the need for analytical methodologies capable of verifying the geographical origin of coffee is therefore outstanding (Risticevic et al. 2008) (Table 17.4).

Risticevic et al. (2008) used 29 volatiles determined by HS-SPME–fast GC–TOFMS in differentiation of *Arabica* coffee samples of different origins. Using PCA, the corresponding geographical origin discriminations of coffees originating from South and Central America, Africa, and Asia were successfully established. In addition to successful geographical discrimination of: (1) authentic sample collections from Brazil and Colombia, and (2) nonauthentic sample collections from South America, Central America, Africa, and Asia, this classification study was also successful in detecting potential compositional changes that coffee undergoes due to the limited shelf-life stability over extensive storage conditions.

Costa Freitas and Mosca (1999) used HS-SPME–GC–FID with the support of PCA for differentiation of coffees on the bases of botanical and geographical origins. While arabica and robusta varieties were clearly separated, the results did not show any separation according to geographical origin (probably due to the reduced number

TABLE 17.4
Examples of Studies Dedicated to Authenticity of Coffee

Origin of Samples	Characterization Studies	Instrumental Techniques	Input Data	Data Pretreatment	Pattern Recognition Techniques	Validation	Remarks	Ref.
Asia, America, Africa	Discrimination of Arabica coffee samples originating from Asia, America, Africa	HS-SPME–GC–TOFMS	26 samples 29 volatiles		PCA		Differentiation among coffees originating from Asia, America, and Africa	Risticevic et al. 2008
America, Africa	Differentiation of arabica and robusta coffee	HS-SPME–GC–FID	9 samples 30 volatiles		PCA		Lack of geographical origin separation probably due to the reduced number of samples analyzed	Costa Freitas and Mosca 1999
Asia, America, Africa	Characterization of roasted coffee	GC–FID	16 samples 23 volatiles		PCA, Hierarchical clustering		Green and roasted coffee as well as arabica and robusta coffee differentiated	Costa Freitas et al. 2001
Different origin	Differentiation and characterization of arabica and robusta coffee	HS-SPME–GC–QMS	30 samples 15 volatiles	Centered, log-ratio transformation	CA, PCA		Six major volatiles chosen as the most relevant markers	Korhonova et al. 2009

Asia, South America, Africa	Differentiation of arabica and robusta coffee	GC–FID	31 samples 13 volatiles		CDA	Differentiation among classes: Arabica/freshly roasted, Arabica/10 days old, Robusta/freshly roasted, Robusta/10 days old	Holscher and Steinhart 1992
Brazil	Discrimination of coffee samples with different overall qualities	HS-SPME–GC–FID	58 samples 70 volatiles		PLS-DA	Prediction of the notes conferred by the cuppers for flavor, body, cleanliness, and overall quality of Brazilian arabica coffees	Ribeiro et al. 2009
Brazil	Detection of coffee adulteration with roasted barley	HS-SPME–GC–ITMS	Group of light/medium/dark roasted coffee/barley and admixtures Over 250 volatiles		PCA	Detection of adulteration 1% (w/w) roasted barley in roasted coffee samples (for the darkest degrees of roast)	Oliveira et al. 2009

of samples used in this work). Costa Freitas et al. (2001) also examined coffees from eight different origins produced by the same manufacturer using HS-SPME–GC–FID and subsequent chemometric analysis with PCA and HCA. The aroma fraction was successfully used to characterize roasted coffees of different origins. The effect of roasting was emphasized when the same method was applied to green coffees of the same origin and the same manufacturer.

Korhonova et al. (2009) applied HS-SPME–GC–QMS to the determination of volatile compounds in 30 commercially available coffee samples. In order to differentiate and characterize arabica and robusta coffee, six major volatile compounds (acetic acid, 2-methylpyrazine, furan-2-carbaldehyde, furan-2-ylmethanol, 2,6-dimethylpyrazine, 5-methylfuran-2-carbaldehyde) were chosen as the most relevant markers. Performing centered log ratio transformation (instead of standard statistical procedure) and CA, the considered three groups (arabica, robusta, and blends) were distinguished. Similarly, PCA with centered log ratio transformation illustrates very good resolution of groups of coffee and the reliability of the result is supported by high proportion of explained variability in the data set (96%) using the first two principal components.

Holscher and Steinhart (1992) attempted to discriminate roasted coffee beans with respect to botanical variety and freshness based on headspace profile analysis. Despite of the fact that only 13 peaks from the headspace profile were taken into account, the canonical DA allowed an exact differentiation between four distinctive classes: Arabica/freshly roasted and Arabica/10 days old, as well as Robusta/freshly roasted and Robusta/10 days old. Methanethiol had the biggest influence on the discriminant analysis both on coffee origin and on age.

Ribeiro et al. (2009) analyzed volatile compounds in 58 arabica roasted coffee samples from Brazil by SPME–GC–FID. The results were compared with those from sensory evaluation of roasted coffees. Calibration models for each sensory attribute based on chromatographic profiles were developed by using PLS regression. The PLS-DA carried out on the chromatographic profiles of sound beans indicated that the compounds 2-methylpropanal, 2-methylfuran, furan-2-carbaldehyde, furan-2-ylmethyl formate, 5-methylfuran-2-carbaldehyde, 4-ethyl-2-methoxyphenol (4-ethylguaiacol), 3-methylthiophene, furan-2-ylmethyl acetate, 3-ethyl-2,5-dimethylpyrazine, 1-(furan-2-yl)butan-2-one, and three other unidentified compounds could be considered as possible markers for the overall differentiation of coffee beverages. The regression models (PLS) using chromatographic profiles predicted very well the notes conferred by the cuppers for flavor, body, cleanliness, and overall quality of Brazilian arabica coffees. From the results obtained in this study, the methodology proposed is a promising alternative tool for monitoring coffee beverage evaluation.

Oliveira et al. (2009) focused on the feasibility of detection of coffee adulteration with roasted barley by a comparative analysis of the volatile profiles of both coffee and barley, pure and mixed, at several roasting degrees. The separation of the nonadulterated and adulterated samples was accomplished by application of PCA to the chromatographic data obtained by HS-SPME–GC–ITMS. It was observed that, the higher the degree of roast, the more easily discriminated were the adulterated samples, allowing for detection of adulterations with as low as 1% (w/w) roasted barley in dark roasted coffee samples.

17.4.5 Honey

EU legislation (2001/110/EC) defines honey as "the natural sweet substance produced by *Apis mellifera* bees from the nectar of plants or from secretions of living parts of plants or excretions of plant-sucking insects on the living parts of plants, which the bees collect, transform by combining with specific substances of their own, deposit, dehydrate, store and leave in honeycombs to ripen and mature" (European Commission 2002b). Besides water, honey consists mainly of monosaccharides (fructose and glucose) and many other substances such as organic acids, oligosaccharides, enzymes, vitamins, minerals, pigments, a wide range of aroma compounds, and solid particles derived from honey collection.

Honey, with its high world production rate (approximately 1.4 million tons/year), is popular not only as a source of energy but also for its potentially health-promoting properties provided by prebiotic, antioxidant, antibacterial, and/or antimutagenic functionalities of certain constituents. The price of honey is usually dictated by its botanical and/or geographical origin. In the case of botanical origin, the most expensive are unifloral honeys, whereas in the latter case, higher price arises when honey is produced in a specific geographical location. Up to now, the EU has specified 18 PDO regions for honey (one Greek, one Italian, one Luxemburgian, one Polish, two French, three Spanish, and nine Portuguese) (Cajka et al. 2009). In general, the adulteration techniques of honey are based on various principles: (1) water addition and extension with sugar and/or syrups; (2) bee feeding with sugars and/or syrups or artificial honey; and (3) mislabeling as a result of mixing of honeys originating from different floral or geographical origins (Cajka et al. 2009).

For the authenticity/traceability purpose, examination of the volatile profiles represents an important analytical strategy enabling honey authentication because composition of volatiles is known to vary widely with the floral origin and way of processing (Table 17.5).

Ampuero et al. (2004) used three different sampling modes: (1) static HS–MS, (2) HS-SPME–MS, and (3) inside-needle dynamic extraction (INDEX)–MS for differentiation of unifloral honeys. In all cases, fingerprints (*m/z* fragments) were generated as variables. Using discriminating factor analysis (DFA), the correct classification rates of 92%, 98%, and 98% for HSH, SPME, and INDEX, respectively, were achieved for the model that distinguished acacia, chestnut, fir, and rape honeys. The discriminant variables (*m/z* fragments) were *m/z* 44, 47, 54, 58, 94 for HSH, *m/z* 59, 76, 94, 104, 144, 145, 157 for SPME, and *m/z* 48, 57, 58, 64, 74, 94, 104 for INDEX.

Aliferis et al. (2010) used HS-SPME–GC–QMS with subsequent combining of MS spectra for the discrimination and classification of honey samples of different botanical origins. Obtained fragments were assigned to various groups of volatiles (Table 17.6).

The developed model based on combined MS spectra of headspace volatiles of 77 samples of honey from seven of the most common botanical origins provided a strong discrimination applying orthogonal partial least squares-discriminant analysis (OPLS-DA) and a percentage of correct classification of samples higher than 98% performing orthogonal partial least squares-hierarchical cluster analysis

TABLE 17.5
Examples of Studies Dedicated to Authenticity of Honeys

Origin of Samples	Characterization Studies	Instrumental Techniques	Input Data	Data Pretreatment	Pattern Recognition Techniques	Validation	Remarks	Ref.
Switzerland	Differentiation of unifloral honeys (acacia, chestnut, dandelion, lime, fir, rape)	(1) HS–MS (2) HS-SPME–MS	(1) 100 samples 5 fragments (*m/z*) (2) 96 samples, 7 fragments (*m/z*)	Normalization to *m/z* 40	PCA, DFA		(1) 92% and (2) 98% correct classification for acacia, chestnut, fir, and rape	Ampuero et al. 2004
Greece	Differentiation of unifloral honeys (chestnut, cotton, fir, heather, pine, thyme, citrus)	HS-SPME–GC–QMS	77 samples 154 fragments (*m/z*)	Pareto	PCA, OPLS-DA, SIMCA, OPLS-HCA		Correct classification >98% using OPLS-HCA	Aliferis et al. 2010
Spain	Differentiation and characterization of unifloral honeys (citrus, rosemary, eucalyptus, lavender, thyme, heather)	GC–QMS	49 samples 106 volatiles		HCA, PCA		Differentiation among unifloral honeys with high-level clusters; using PCA, compounds involved in differentiating were provided	Castro-Vazquez et al. 2009

Greece, Turkey	Differentiation of pine honeys from two different geographical areas	GC–QMS	44 samples 77 volatiles		KSOM		Correct classification 100% using KSOM	Tananaki et al. 2007
France, Italy, Austria, Ireland, Germany	Discrimination of honey samples labeled as "Corsica" (PDO) and other European countries	HS-SPME–GC × GC–TOFMS	374 samples 26 volatiles	Range and log transformations (each sample)	PCA, ANN-MLP	External	95% prediction ability for model distinguishing honeys labeled "Corsica" (PDO) and the rest of honeys	Cajka et al. 2009
France, Italy, Austria, Ireland, Germany	Discrimination of honey samples labeled as "Corsica" (PDO) and other European countries	HS-SPME–GC × GC–TOFMS	374 samples 26 volatiles	Row-closure	LDA, PLS-DA, SIMCA, SVM	External	85%, 87%, 64%, and 92% prediction ability for model distinguishing honeys labeled "Corsica" (PDO) and the rest of honeys using LDA, PLS-DA, SIMCA, and SVM, respectively	Stanimirova et al. 2010

TABLE 17.6
Fragment Ions Assigned to Volatiles Isolated from Honey Samples by HS-SPME

Volatiles	Fragments
Phenolics	*m/z* 43, 65, 91, 92
Terpenoides	*m/z* 41, 43, 55, 67, 71, 93
Organic acids	*m/z* 74, 87
Esters	*m/z* 41, 43, 55, 56, 57
Alcohols	*m/z* 41, 43, 55, 56, 57
Aldehydes	*m/z* 41, 43, 55, 56, 57, 69, 70, 81, 82, 84, 95, 96, 110

Source: Aliferis, K.A., et al., *Food Chem.*, 121, 856–862, 2010. With permission.

(OPLS-HCA). A grouping of honey samples achieved according to their botanical origin is shown in Figure 17.18.

Castro-Vazquez et al. (2009) focused on differentiation and characterization of Spanish unifloral honeys (citrus, rosemary, eucalyptus, lavender, thyme, heather) via GC–QMS and sensory analysis. Using HCA, honeys from the same floral origin were assigned to the same cluster, thus highlighting their similarity. Successive higher-level clusters indicated similarities between different types of honey: rosemary honeys were fairly similar to thyme honeys, but rather less similar to lavender and citrus honeys. Heather and eucalyptus were placed further away from this group, and there were even some subgroups, indicating small "intragroup" differences. To obtain more detailed information on the volatile compounds involved in differentiating the monofloral honeys studied, factorial PCA was applied to the whole data matrix. Volatiles involved in differentiating the honeys of different botanical origins were also reported.

Tananaki et al. (2007) analyzed honeys from Greece and Turkey using P&T–GC–QMS. From 77 compounds, two of them [car-3-ene and an unidentified compound (*m/z* 55, 79, 91, 107, 123, 165)] were found to be specific to Turkish honeys and may be considered as markers characterizing the Turkish origin of samples. Beyond this result, and by using the KSOM, a clear differentiation was obtained between the Turkish and Greek pine honey samples on the bases of volatile constituents.

Cajka et al. (2009) used an automated HS-SPME-based sampling procedure coupled to GC × GC–TOFMS for obtaining profiles of honey volatiles (26 markers). In total, 374 samples were collected, over two production seasons, in Corsica and other European countries with the aim to distinguish, based on analytical (profiling) data, the honeys labeled as "Corsica" (PDO) from all the other honeys. Using ANN-MLP, the prediction ability of the models for the first and second harvest was 93% and 92%, respectively. The model created for the first harvest (2006) was used to predict samples from the second harvest (2007). The prediction ability using this external (validation) set containing 2007 data was somewhat lower (81%), probably due to a large variation in the profiles of honey volatiles between these two years, possibly caused by different weather conditions in the two harvest years. A low prediction ability

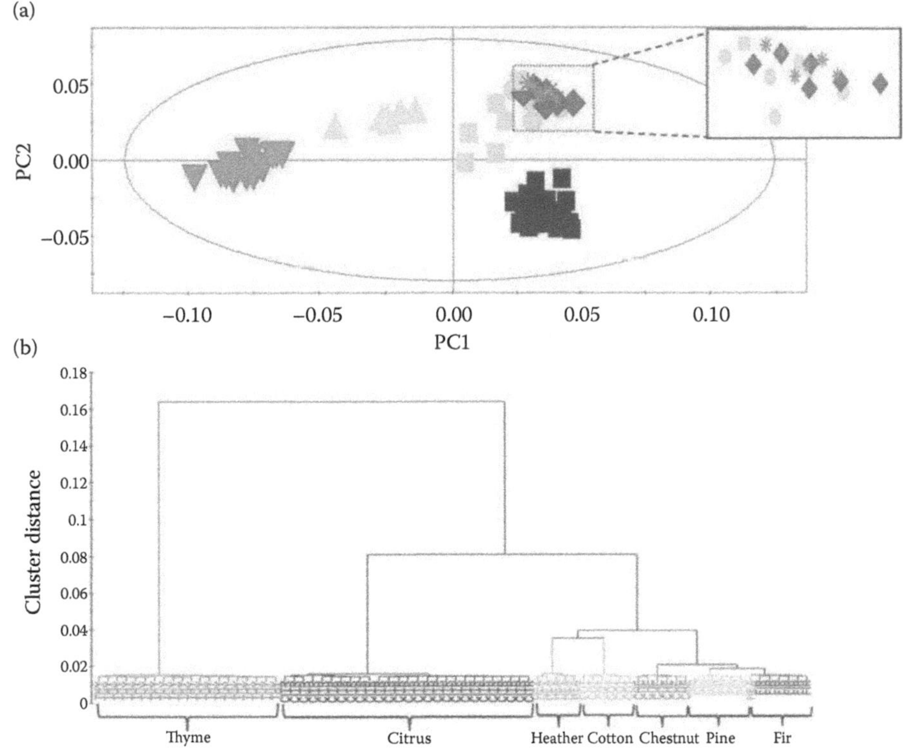

FIGURE 17.18 (a) OPLS-DA PC1/PC2 score plot for the whole data set. Ellipse represents Hotelling T2 with 95% confidence, and (b) OPLS-dendrogram illustrating OPLS-HCA using Ward's method. Honey samples are grouped according to their botanical origin [chestnut (*), citrus (■), cotton (▲), fir (◆), heather (■), pine (●), and thyme (▼)]. (Reproduced from Aliferis, K.A., et al., *Food Chem.*, 121, 856–862, 2010. With permission.)

(82%) was also obtained when the model from the second harvest (2007) was used to predict samples from the first harvest (2006). A more reliable approach was when the data of a two-year sampling were merged. Under these conditions, subsets used for training and testing contained representative samples from both harvests. Using this strategy, a prediction ability of 95% was obtained. The same data sets of honeys were also statistically evaluated by Stanimirova et al. (2010) employing LDA, PLS-DA, soft independent modeling of class analogy (SIMCA), and support vector machine (SVM). Using merged data sets from two harvests, a prediction ability of 85%, 87%, 64%, and 92% was obtained for LDA, PLS-DA, SIMCA, and SVM, respectively.

17.4.6 Olive Oil

Olive oil, obtained from the fruit of olive trees (*Olea europaea* L.), represents a very important food commodity with a world production rate of approximately 2.8 million tons/year. The popularity of olive oil is not only because of its delicious flavor, but also for its health-promoting potential provided by polyunsaturated fatty acids and other compounds (e.g., phenolics, squalene, and oleic acid) (Cajka et al. 2010).

According to the International Olive Council trade standard, the best brand, extra virgin olive oil, should meet specified physicochemical characteristics, including low free acids content (<0.8 g/100 g, expressed as oleic acid) (International Olive Council 2008). The higher price of extra virgin olive oil compared to virgin olive oil is due to a limited production rate and strict requirements for its high quality parameters [Commission Regulations (EC) No. 1019/2002] (European Commission 2002c).

The studies dealing with olive oil traceability are usually focused on investigation of the botanical or geographical origin. However, the concept of geographical traceability, in which the objective is the geographical location of the olive tree, is slightly different from the concept of botanical traceability, in which case the olive used for the olive oil production is the aim. In both cases, the selection of the markers (i.e., compounds with discrimination power) to be studied is complicated because the composition of extra virgin olive oils is the result of complex interaction among olive variety, environmental conditions, fruit ripening, and oil extraction technology (Montealegre et al. 2009).

One of the factors that are important for the consumers' choice is the geographical area of olive oil production. Up to now, the EU has specified 85 protected denomination of origin (PDO) regions for olive oil: seven French, 15 Greek, 37 Italian, six Portuguese, one Slovenian, and 19 Spanish (Cajka et al. 2010). Unfortunately, economic fraud, such as false claims of geographical origin on product labels, cannot be fully avoided.

Volatile compounds of olive oils (including carbonyl compounds, alcohols, esters, and hydrocarbons) can be used as an effective way to evaluate oil authenticity. In most cases, the analysis of volatiles involves enrichment techniques that can be applied to isolate low concentration compounds (Table 17.7).

Cajka et al. (2010) used an automated HS-SPME-based sampling procedure coupled to GC–ITMS for obtaining profiles of olive oil volatiles (44 markers). In total, 914 samples were collected, over three production seasons, in northwestern Italy—Liguria and other regions—in addition to the rest of Italy, Spain, France, Greece, Cyprus, and Turkey, with the aim to distinguish, based on analytical (profiling) data, the olive oils labeled as "Ligurian" (PDO) from all the others ("non-Ligurian"). PCA showed relatively large interannual variability in sample composition from particular regions; there was some overlap of the three harvests (3 years) data (Figure 17.19). The model developed for a single year data set was not fully applicable for other sampling years. A more reliable approach appeared to be a model that consisted of the three-year sampling data. Using this strategy, the prediction ability obtained for ANN-MLP was 81%, whereas somewhat lower prediction ability was achieved for LDA (62%).

Casale et al. (2007) utilized HS–MS for the analysis of 46 olive oil samples from three regions of Liguria (PDO). Using all variables (205) obtained in this fingerprinting technique, a prediction ability as low as 57% was obtained using LDA (after PCA dimension reduction). Subsequently, a strategy to eliminate the redundant and noisy variables was conducted (feature selection). Using 20 selected variables, the prediction ability as high as 91% was achieved for the model distinguishing olive oils from three regions of Liguria.

TABLE 17.7
Examples of Studies Dedicated to Authenticity of Olive Oils

Origin of Samples	Characterization Studies	Instrumental Techniques	Input Data	Data Pretreatment	Pattern Recognition Techniques	Validation	Remarks	Ref.
Italy, Spain, France, Greece, Cyprus, Turkey	Discrimination of olive oil samples labeled as "Liguria" (PDO) and other regions	HS-SPME–GC–ITMS	914 samples 44 volatiles	Range scaling	PCA, LDA, ANN-MLP	External	62% and 81% prediction ability for model distinguishing olive oils labeled "Liguria" (PDO) and the rest of olive oils using LDA and ANN-MLP, respectively	Cajka et al. 2010
Italy	Discrimination of olive oil samples that originated from three regions of Liguria (PDO)	HS–MS	46 samples 205 fragments (*m/z*)		LDA, SIMCA	*k*-Fold CV	57% (205 variables, LDA), 91% (20 selected variables, LDA), and 63% (16 selected variables, SIMCA) prediction abilities for model distinguishing olive oils from three regions of Liguria	Casale et al. 2007

(*continued*)

TABLE 17.7 (Continued)
Examples of Studies Dedicated to Authenticity of Olive Oils

Origin of Samples	Characterization Studies	Instrumental Techniques	Input Data	Data Pretreatment	Pattern Recognition Techniques	Validation	Remarks	Ref.
Italy	Discrimination of olive oil samples labeled as "Liguria" (PDO) and other regions	GC–QMS	105 samples 8 volatiles		PCA, SIMCA	External	95% and 97% prediction abilities for model distinguishing olive oils from Liguria and other regions	Zunin et al. 2005
Italy, Greece, Spain, Tunisia	Discrimination of olive oil samples labeled as "Liguria" (PDO) and other regions	HS–MS	105 samples 206 fragments (*m/z*)		PCA, LDA	External, LOOCV	91% (LOOCV) and 80% (external) prediction ability for model distinguishing olive oils from five regions (Liguria, Greece, Apulia, Spain, Tunisia)	Oliveros et al. 2005

Greece, Spain, Syria, Tunisia, Italy, Morocco	Discrimination of olive oil samples (geographical origin)	GC–FID	39 samples 64 volatiles	LDA	100% classification for olive oil samples that originated from Spain, Italy, and Greece on the bases of LDA with nine volatiles	Luna et al. 2006
Spain, Italy, Greece	Discrimination of olive oil samples (geographical origin)	GC–FID	32 samples 56 volatiles	ANOVA, LDA	91% classification for olive oil samples that originated from Spain, Italy, and Greece on the bases of LDA with three volatiles	Morales et al. 1994
Italy, Tunisia	Discrimination of olive oil samples (geographical origin)	HS-SPME–GC–QMS	40 samples 22 volatiles	PCA	Not possible to obtain a clear discrimination of samples of different geographical origin on the basis of volatiles	Kotti et al. 2011

(continued)

TABLE 17.7 (Continued)
Examples of Studies Dedicated to Authenticity of Olive Oils

Origin of Samples	Characterization Studies	Instrumental Techniques	Input Data	Data Pretreatment	Pattern Recognition Techniques	Validation	Remarks	Ref.
France	Discrimination of olive oil samples (PDO)	HS-SPME–GC–FID	53 samples 72 volatiles	.	PCA, SIMCA		Distinguishing among different qualities of olive oils and highlighting the specificity of PDO (Nice, France)	Berlioz et al. 2006
Italy	Oil quality assessment in order to predict panel test scores	GC–FID	204 samples 114 volatiles	Range scaling	ANN	External	Sensory evaluation (panel test methods) can be replaced by the dynamic head-space analysis and ANN	Angerosa et al. 1996
Different origin	Differentiation between virgin olive oils adulterated with hazelnut oil	(1) HS-SPME–GC–QMS (2) HS-SPME–QMS	9 samples (1) 50 volatiles (2) *m/z* fragments		PCA		Admixtures of EVOO and HO [5% (v/v)] were distinguished from "pure" EVOO samples	Mildner-Szkudlarz and Jelen 2008

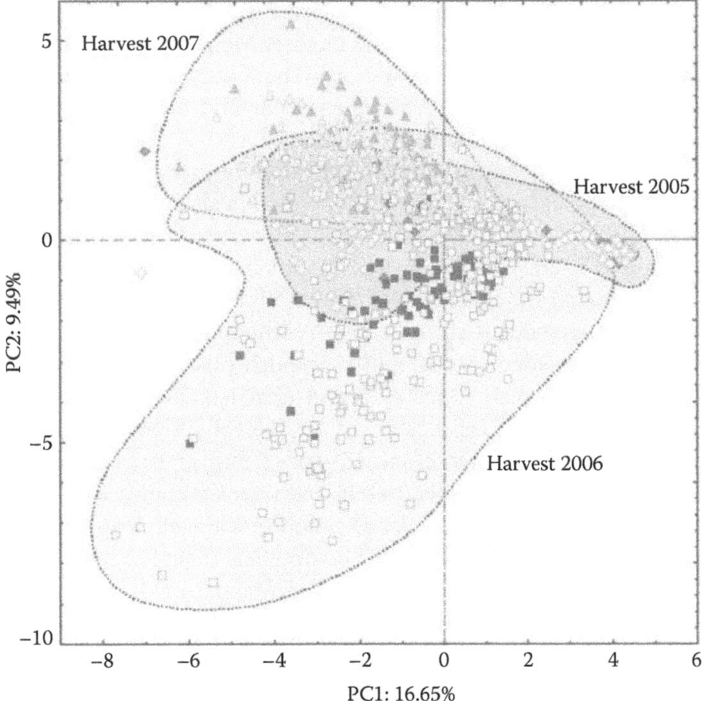

FIGURE 17.19 PCA clustering: year 2005 (◆ Liguria, ◇ non-Liguria); year 2006 (■ Liguria, □ non-Liguria); year 2007 (▲ Liguria, △ non-Liguria). (Reproduced from Cajka, T., et al., *Food Chem.*, 121, 282–289, 2010. With permission.)

Zunin et al. (2005) evaluated if the volatile terpenoid hydrocarbons of olive oils from West Liguria could trace their geographical origin. The obtained results showed that the analysis of volatile terpenoid hydrocarbons is a straight and powerful tool for tracing the geographical origin of extra virgin olive oils from West Liguria. As far as the five considered geographical areas were concerned, just three individual terpenoid hydrocarbons (i.e., α-copaene, α-muurolene, and α-farnesene) allowed the distinction of West Liguria oils by a simple decision tree. Moreover, the multivariate analysis of the eight detected terpenoids (α-pinene, limonene, *(Z)*-β-ocimene, (3*E*)-4,8-dimethylnona-1,3,7-triene, α-copaene, eremophyllene, α-muurolene, α-farnesene) allowed building class models with high predictive ability (>95%).

Oliveros et al. (2005) used HS–MS instrumentation for the analysis of 105 olive oils coming from five different Mediterranean areas. To visualize and rationalize the information of the analytical results, PCA was performed. The first PCs did not show a significant differentiation among geographical origin and the LDA results showed a poor predictive ability. This was probably due to the large number of fragments revealed by the spectral fingerprints. Actually, many of the 206 measured variables could be uninformative and noisy, so it was necessary to select the relevant *m*/*z* values. Thus, in order to improve the prediction results, the stepwise LDA method of feature selection was applied using two different criteria: (1) minimum Wilk's lambda and (2) selection of the variables producing the greatest Mahalanobis distance between a selected category (Liguria) and the closest one. The first criterion did not give advantage in any category, whereas the second one benefited the Ligurian samples (the selected features are reported in Table 17.8, respectively). After this feature selection procedure, it was possible to discriminate the different aromas of olive oils coming from several geographical areas with a mean prediction ability of 80%.

Luna et al. (2006) studied differentiation of olive oils from six different regions by using dynamic HS–GC–FID. Stepwise LDA analysis showed that the compounds selected to differentiate the Spanish and Greek olive oils were hexyl acetate, ethyl benzene, and a partially identified ester. These compounds allowed 100% of correct classifications. The equation to distinguish Spanish oils from the Italian oils was based on 2-methylbutyl acetate, 2-methylpent-4-enal, hexan-1-ol, and an ester. These compounds allowed correct classification of 100% of the Spanish samples but only 71% of the Italian samples. Pentan-3-one, (*Z*)-hex-3-enal, and a hydrocarbon allowed observers to distinguish between the Italian and Greek samples with 100% correct classifications. In the simultaneous classification of the samples from Spain, Italy, and Greece, the most discriminant volatile compounds (hexyl acetate, ethylbenzene, 2-methylpent-4-enal, an ester, 2-ethylfuran, (*E*)-pent-2-enal, 1,2,4-trimethylbenzene, 3-methylbutan-1-ol, and a hydrocarbon) allowed to obtain 100% correct classifications, although the level of strictness was not high enough.

Morales et al. (1994) applied dynamic HS–GC–FID profiling of olive oil samples from Spain, Italy, and Greece. Stepwise LDA was applied to the 17 selected volatiles in order to discriminate the samples on the basis of their origin, and 91% of correct classifications were obtained by only three volatiles: (3*E*)-hexa-1,3-dien-5-yne, 2-methylpropan-1-ol, and [(*Z*)-hex-3-enyl] acetate.

Kotti et al. (2011) analyzed olive oils that originated from Italy and Tunisia using HS-SPME–GC–MS followed by PCA. Although some differences between Italian and Tunisian samples were evidenced, it was not possible to obtain a clear discrimination of samples of different geographical origins on the bases of volatile compounds. Italian olive oil samples were generally characterized by higher amounts of C6 and C5 compounds, particularly (*E*)-hex-2-enal, in comparison with Tunisian oils.

Berlioz et al. (2006) studied the specificity of PDO virgin olive oils produced in a southern French region based on their volatiles analyzed by HS-SPME–GC–FID. The method developed was able to distinguish different qualities of olive oils and highlights the specificity of PDO of Nice.

Angerosa et al. (1996) conducted a comparative study with the aim of predicting panel test scores of the sensory quality evaluation of virgin olive oils differing in variety, ripeness, sanitary state, and geographical origin using dynamic headspace analysis coupled to GC–FID. Subsequently, the use of ANN allowed researchers to generalize and assign the sensory evaluations with a good degree of accuracy, thus, the panel test could be replaced by the dynamic HS–GC–FID with ANN.

Mildner-Szkudlarz and Jelen (2008) investigated the effectiveness of rapid methods (SPME–fast GC–FID, SPME–MS, HS–MS) for the analysis of volatiles with subsequent PCA treatment of data for differentiation between olive oil samples adulterated with

TABLE 17.8
Fragment Ions Assigned to Olive Oil Volatiles and Selected by Two Criteria of Stepwise LDA

Fragment[a]	Chemical Interpretation	Fragment[b]	Chemical Interpretation
67	(3*E*)-4,8-Dimethylnona-1,3,7-triene, [(*Z*)-hex-3-enyl] acetate, α-copaene, farnesene, hexanal, (*E*)-hex-2-enal, (*E*)-hex-2-en-1-ol, (*E*)-β-ocimene	67	(3*E*)-4,8-Dimethylnona-1,3,7-triene, [(*Z*)-hex-3-enyl] acetate, α-copaene, farnesene, hexanal, (*E*)-hex-2-enal, (*E*)-hex-2-en-1-ol, (*E*)-β-ocimene
92	α-Copaene, 2-phenylethanol, farnesene, methyl 2-hydroxybenzoate, (*E*)-β-ocimene, toluene	92	α-Copaene, 2-phenylethanol, farnesene, methyl 2-hydroxybenzoate, (*E*)-β-ocimene, toluene
79	(3*E*)-4,8-Dimethylnona-1,3,7-triene, α-copaene, (*E*)-hex-2-enal, (*E*)-β-ocimene	149	α-Copaene, eremophilene, farnesene, muurolene
72	Hexanal, nonanal, (*E*)-hex-2-en-1-ol	94	(3*E*)-4,8-Dimethylnona-1,3,7-triene, eremophilene, farnesene, limonene, (*E*)-β-ocimene
69	(3*E*)-4,8-Dimethylnona-1,3,7-triene, farnesene, nonanal, (*E*)-hex-2-enal	76	α-Copaene, farnesene, naphthalene, (*E*)-hex-2-enal, (*E*)-β-ocimene
76	α-Copaene, farnesene, naphthalene, (*E*)-hex-2-enal, (*E*)-β-ocimene	55	Hexanal, (*E*)-hex-2-enal, (*E*)-hex-2-en-1-ol
201	Not interpreted	61	[(*Z*)-hex-3-enyl] acetate, hexyl acetate, (*E*)-hex-2-enal
100	Pent-1-en-3-ol, (*Z*)-hex-3-en-1-ol, heptane, hexanal, nonanal, (*E*)-hex-2-enal, (*E*)-hex-2-en-1-ol	74	[(*Z*)-hex-3-enyl] acetate, (*E*)-hex-2-en-1-ol, hexyl acetate, (*E*)-hex-2-enal, (*E*)-hex-2-en-1-ol
70	(*Z*)-2-Hexen-1-ol, (*Z*)-hex-3-enal, (*Z*)-hex-3-en-1-ol, hexan-1-ol, nonanal, (*E*)-hex-2-enal		
74	[(*Z*)-Hex-3-enyl] acetate, (*E*)-hex-2-en-1-ol, hexyl acetate, (*E*)-hex-2-enal, (*E*)-hex-2-en-1-ol		

Source: Oliveros, C.C., et al., *J. Chromatogr. A*, 1076, 7–15, 2005. With permission.

[a] Stepwise LDA by minimum Wilk's lambda criterion.

[b] Stepwise LDA by selection of the variables producing the greatest Mahalanobis distance between the category Liguria and the closest one.

TABLE 17.9
Examples of Studies Dedicated to Authenticity of Cheese

Origin of Samples	Characterization Studies	Instrumental Techniques	Input Data	Data Pretreatment	Pattern Recognition Techniques	Validation	Remarks	Ref.
Spain	Influence of the season of manufacture on the levels of volatiles	P&T–GC–QMS	48 samples 90 volatiles		LDA		Correct classification of cheeses according to the dairy of manufacture Partial classification of cheeses according to the season of manufacturing	Fernandez-Garcia et al. 2004
Italy	Assessment of cheese aging	PTR–MS	20 samples (each 3 replicate) 240 fragments (m/z)		PLS-DA, UNEQ	*k*-Fold CV	85% and 100% prediction ability for model distinguishing young and ripened cheeses using PLS-DA and UNEQ, respectively	Aprea et al. 2007

Switzerland, Spain, Italy	Discrimination of PDO cheeses	SPME–GC–FID, P&T–GC–FID	18 samples 81 volatiles (SPME), 68 volatiles (P&T)		LDA		Sample correctly classified on the bases of their PDO (Gruyere, Switzerland; Manchego, Spain; and Ragusano, Italy)	Mallia et al. 2005
France	Discrimination of Camembert type cheeses	HS-SPME–MS	5 samples (each 8 portions), 5 fragments (*m*/*z*) after stepwise LDA	Normalization	LDA	LOOCV	100% prediction ability for model distinguishing Camembert-type cheeses	Peres et al. 2001
Switzerland, France, Austria	Differentiation of Gouda and Emmental cheeses	GC–QMS	6 samples 27 volatiles		PCA		Gouda cheeses from different producers characterized by similar aroma patterns, whereas more differences were observed between the Emmental cheeses	Dirinck and De Winne 1999

hazelnut oil. Using this approach, olive oil samples to which as little as 5% (v/v) of hazelnut oil was added, were distinguishable from remaining ones using all three methods.

17.4.7 Cheese

Cheeses also represent a commodity with a high production rate (approximately 18 million tons/year). Their styles, textures, and flavors depend on (1) the origin of the milk (including the animal's diet), (2) pasteurization, (3) the butterfat content, (4) the presence of bacteria and mold, (5) the processing, and (6) the aging. The PDO label is designed for cheeses manufactured and ripened in a defined geographical area possessing unique human and natural characteristics. To help the preservation of their history, tradition, and diversity, a total of 147 cheeses have been given the PDO label by the EU. In this context, chemical, microbial, and sensory properties are very important for characterizing and differentiating these foods from similar foods without this label. Flavor is one of the key characteristics that determine cheese quality (Fernandez-Garcia et al. 2004; Aprea et al. 2007; Mallia et al. 2005; Peres et al. 2001) (Table 17.9).

Fernandez-Garcia et al. (2004) investigated the influence of the season of manufacture and the dairy of origin for ewes raw milk Zamorano cheese (Spain) using P&T–GC–QMS. All the cheeses at 4 and 8 months of ripening were classified correctly for the dairy of manufacture (Figure 17.20a). Variables with more weight in this function were prop-2-en-1-ol and 2-methylpropan-2-ol with positive coefficient, and butyl acetate, pentan-2-ol and 2-butylhexanoate with negative coefficient. The cheeses were only partially classified by the season of manufacture (Figure 17.20b).

Aprea et al. (2007) used proton transfer reaction (PTR)–MS for the assessment of Trentingrana (Italy) cheese aging. PLS-DA indicated average behavior and possible outliers, but was not able to correctly classify all samples. The analysis of the first loading revealed that the separation along the first PLS component was mainly due to a group of masses related to esters (*m/z* 89, 103, 117, 131, 145, and 173). On the other

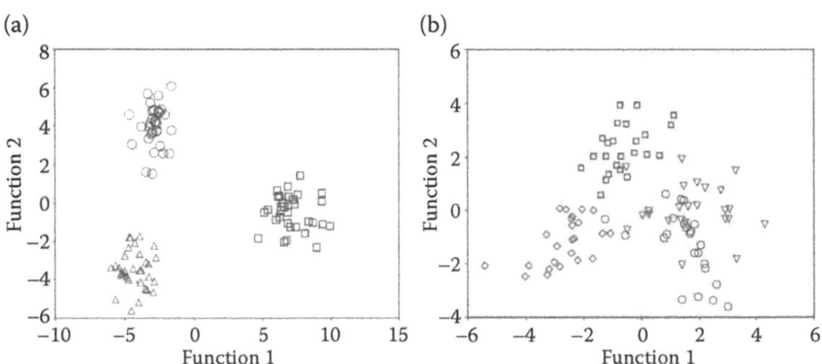

FIGURE 17.20 (a) Plot of sample distribution using two canonical discriminant functions. Cheeses made in artisan dairy A (△), artisan dairy B (○), and industrial dairy (□). (b) Plot of sample distribution using two canonical discriminant functions. Cheeses made in spring (▽), summer (○), autumn (◇), and winter (□). (Reproduced from Fernandez-Garcia, E., et al., *Int. Dairy J.*, 14, 701–711, 2004. With permission.)

hand, soft class modeling [unequal dispersed classes (UNEQ)] performed better and allowed a 100% correct classification. PLS calibration predicted the aging time of each loaf with reasonable accuracy with a maximum cross-validation error of 3.5 months.

Mallia et al. (2005) studied the volatile aroma of three European PDO hard cheeses (Gruyere Switzerland, Manchego Spain, and Ragusano Italy) using SPME–GC–FID and P&T–GC–FID. The LDA carried out separately for each extraction method made it possible to classify correctly all the samples according to their PDO. The most important volatile compounds extracted by SPME used for the classification of the cheese varieties were acetaldehyde, ethanol, butane-2,3-dione, prop-2-en-1-ol, butan-2-yl butanoate, and the fatty acids. The main volatile compounds extracted by P&T, which determine the classification of the cheese varieties, were 2-methylpropanal, butane-2,3-dione, butan-2-ol, toluene, pentane-2,3-dione, propyl butanoate, ethyl pentanoate, and 2,6-dimethylpyrazine.

Peres et al. (2001) developed rapid fingerprinting method for characterization of cheeses using HS-SPME–MS (e-nose). A prediction ability of 100% for the model distinguishing Camembert-type cheeses was reported. The stepwise LDA carried out on normalized data obtained at 20°C selected mass fragments 47, 58, 49, 55, and 48 in that order as best variables to calculate the canonical discriminant variables. Combining the data from HS-SPME–MS and HS-SPME–GC–MS, the ion *m/z* 47 was abundantly present in the spectra of sulfur-containing components (methanethiol, methylsulfanylmethane, methyldisulfanylmethane), which were strongly represented in the headspace of the products. Fragment *m/z* 58 characterizes compounds such as acetone, 3-methylbutanal, pentanal, and hexan-2-one. Ion *m/z* 49 is typical of chlorinated derivatives such as dichloromethane. Fragment *m/z* 55 is observed in the spectra of oct-1-en-3-one, pentan-2-one, and 3-methylbutan-1-ol. Lastly, fragment *m/z* 48 is characteristic of sulfur dioxide. Some of the above compounds, including the sulfur-containing compounds, 3-methylbutanal, and oct-1-en-3-one have in addition been recognized as responsible for the main odor characteristics of Camembert cheese.

Dirinck and De Winne (1999) analyzed Gouda and Emmental cheeses from Switzerland, France, and Austria using GC–QMS. Using PCA, it was observed that the Gouda cheeses from different producers had similar aroma patterns, whereas more differences were observed between the Emmental cheeses. Austrian Emmental was clearly differentiated from the French and Swiss products. Gouda cheeses were dominated by lactones [6-pentyloxan-2-one (δ-decalactone) and 6-heptyloxan-2-one (δ-dodecalactone)] and acetoine, which corresponded to the higher buttery note in these cheeses compared to the Emmental types. The Swiss and French Emmental were also closely related and were characterized by high levels of acids and of methyl-ketones. Some highly volatile acids were typical of Emmental: propionic, butanoic, and 2-methylbutanoic acid. The Austrian Emmental was clearly differentiated from the two other Emmental products and characterized by high levels of 5-octyloxolan-2-one (γ-dodecalactone) and 2-phenylacetaldehyde.

17. 5 CONCLUSIONS

In recent decades, issues related to authenticity as well as product tracing have become of enormous importance for diverse stakeholder groups, not only for companies and scientific professionals but also for consumers. Measurement of food volatile fingerprints in a nonselective way together with profiling approach, as introduced in this chapter, represent, undoubtedly, one of the challenging options to obtain the information on samples composition. In addition to these applications, control of volatiles fingerprints/profiles may enable industry monitoring or technological steps, thus keeping standardized quality of their products. With regard to generation of laboratory data and their follow-up processing, the key messages obtained within this state-of-the-art comprehensive review can be summarized as follows.

1. Solventless sampling techniques represented by static/dynamic headspace and SPME are well suited for isolation of food volatiles for fingerprinting/profiling purposes. It should be noted, that the relative concentrations of components occurring in headspace (measured signals) do not necessarily reflect their concentration in a respective sample. The partition of volatile components between solid/liquid and gaseous phase largely varies depending not only on their physicochemical properties but also on the nature (composition) of the matrix in which they are contained.
2. MS coupled to 1D-GC/GC × GC is a very suitable tool for obtaining a comprehensive fingerprints/profiles of volatiles occurring in the examined samples. If needed, characteristic markers can be identified.
3. Ambient MS when using a DART ion source coupled to a high-resolution mass spectrometer enables rapid fingerprinting/profiling of ionizable sample/sample extract components that can be volatilized (thermo desorbed) under experimental conditions. The information (fingerprint/profile components) obtained is fairly different from that measured by headspace techniques. Compounds with molecular weight as high as 1200 Da can be recorded.
4. Multivariate data analysis involving preliminary data inspection (using unsupervised pattern recognition techniques) and development of classification models (using supervised pattern recognition techniques) represents common practice in interpretation of measured data.
5. The data used for training should be sufficiently large to cover the possible variation of new "unknown" samples. In addition, a careful cross-validation and preferably an external validation should always be performed.

ABBREVIATIONS

1D-GC, one-dimensional gas chromatography; ANN, artificial neural networks; ANN-MLP, artificial neural networks with multilayer perceptrons; ANOVA, analysis of variance; APCI, atmospheric pressure chemical ionization; ASAP, atmospheric pressure solids analysis probe; CA, cluster analysis; CART, classification and regression trees; CI, chemical ionization; CV, cross-validation; DART, direct analysis in real time; DESI, desorption electrospray ionization; DI, direct immersion; e-nose, electronic nose; EI, electron ionization; ESI, electrospray ionization; EU, European Union; FID, flame ionization detection; FT, Fourier transform;

FTIR, Fourier transform infrared; GC, gas chromatography; GC × GC, comprehensive two-dimensional gas chromatography; HPLC, high-performance liquid chromatography; HRTOFMS; high-resolution time-of-flight mass spectrometry; HS, headspace; ICP, inductively coupled plasma; IR, infrared spectroscopy; ITMS, ion trap mass spectrometry; KSOM, Kohonen self-organizing maps; LLE, liquid/liquid extraction; LOOCV, leave-one-out cross-validation; MS, mass spectrometry; *m/z*, mass-to-charge ratio; NIR, near infrared spectroscopy; NMR, nuclear magnetic resonance; OPLS-DA, orthogonal partial least squares-discriminant analysis; OPLS-HCA, orthogonal partial least squares-hierarchical cluster analysis; P&T, purge & trap; PCA, principal component analysis; PC1, PC2, ..., principal component 1, 2, ...; PDO, protected denomination of origin; PGI, protected geographical indication; PLS, partial least squares; PLS-DA, partial least squares discriminant analysis; QMS, quadrupole mass spectrometry; RDA, regularized discriminant analysis; SBSE, stir bar sorptive extraction; SDE, simultaneous distillation–extraction; SIMCA, soft independent modeling of class analogy; SPE, solid phase extraction; SPME, solid-phase microextraction; SVM, support vector machine; TOFMS, time-of-flight mass spectrometry; TSG, traditional speciality guaranteed; UNEQ, unequal dispersed classes: a simple class-modeling technique based on the multivariate normal distribution; UV, ultraviolet; Vis, visible.

ACKNOWLEDGMENTS

This chapter was supported by the Ministry of Education, Youth and Sports of the Czech Republic [projects MSM 6046137305 and Specific University Research (MSMT No. 21/2011)] and the Ministry of Agriculture of the Czech Republic (NAZV QH72144).

REFERENCES

Aliferis, K.A., P.A. Tarantilis, P.C. Harizanis, and E. Alissandrakis. 2010. Botanical discrimination and classification of honey samples applying gas chromatography/mass spectrometry fingerprinting of headspace volatile compounds. *Food Chem.* 121: 856–862.

Ampuero, S., S. Bogdanov, and J.-O. Bosset. 2004. Classification of unifloral honeys with an MS-based electronic nose using different sampling modes: SHS, SPME and INDEX. *Eur. Food Res. Technol.* 218: 198–207.

Angerosa, F., L. Di Giacinto, R. Vito, and S. Cumitini. 1996. Sensory evaluation of virgin olive oils by artificial neural network processing of dynamic head-space gas chromatographic data. *J. Sci. Food Agric.* 72: 323–328.

Aprea, E., F. Biasioli, F. Gasperi, D. Mott, F. Marini, and T.D. Mark. 2007. Assessment of Trentingrana cheese ageing by proton transfer reaction–mass spectrometry and chemometrics. *Int. Dairy J.* 17: 226–234.

Aznar, M., and T. Arroyo. 2007. Analysis of wine volatile profile by purge-and-trap–gas chromatography–mass spectrometry: Application to the analysis of red and white wines from different Spanish regions. *J. Chromatogr. A* 1165: 151–157.

Berlioz, B., C. Cordella, J.-F. Cavalli, L. Lizzani-Cuvelier, A.-M. Loiseau, and X. Fernandez. 2006. Comparison of the amounts of volatile compounds in French protected designation of origin virgin olive oils. *J. Agric. Food Chem.* 54: 10092–10101.

Berna, A.Z., S. Trowell, D. Clifford, W. Cynkar, and D. Cozzolino. 2009. Geographical origin of Sauvignon Blanc wines predicted by mass spectrometry and metal oxide based electronic nose. *Anal. Chim. Acta* 648: 146–152.

Berrueta, L.A., R.M. Alonso-Salces, and K. Heberger. 2007. Supervised pattern recognition in food analysis. *J. Chromatogr. A* 1158: 196–214.

Cajka, T., J. Hajslova, J. Cochran, K. Holadova, and E. Klimankova. 2007. Solid phase microextraction–comprehensive two-dimensional gas chromatography–time-of-flight mass spectrometry for the analysis of honey volatiles. *J. Sep. Sci.* 30: 534–546.

Cajka, T., J. Hajslova, and K. Mastovska. 2008. Mass spectrometry and hyphenated instruments in food analysis. In *Handbook of Food Analysis Instruments*, ed. S. Otles. Boca Raton, FL: CRC Press, Taylor & Francis Group.

Cajka, T., K. Riddellova, J. Hajslova. 2009. Authentication of beer and wine using advanced mass-spectrometric techniques. Paper presented at 4th International Symposium on Recent Advances in Food Analysis, Prague, Czech Republic.

Cajka, T., J. Hajslova, F. Pudil, and K. Riddellova. 2009. Traceability of honey origin based on volatiles pattern processing by artificial neural networks. *J. Chromatogr. A* 1216: 1458–1462.

Cajka, T., K. Riddellova, E. Klimankova, M. Cerna, F. Pudil, and J. Hajslova. 2010. Traceability of olive oil based on volatiles pattern and multivariate analysis. *Food Chem.* 121: 282–289.

Cajka, T., K. Riddellova, M. Tomaniova, and J. Hajslova. 2010. Recognition of beer brand based on multivariate analysis of volatile fingerprint. *J. Chromatogr. A* 1217: 4195–4203.

Cajka, T., K. Riddellova, M. Tomaniova, and J. Hajslova. Ambient mass spectrometry employing a DART ion source for metabolomic fingerprinting/profiling: A powerful tool for beer origin recognition. *Metabolomics*, in press, doi: 10.1007/s11306-010-0266-z.

Camara, J.S., M.A. Alves, and J.C. Marques. 2006. Multivariate analysis for the classification and differentiation of Madeira wines according to main grape varieties. *Talanta* 68: 1512–1521.

Casale, M., C. Armanino, C. Casolino, and M. Forina. 2007. Combining information from headspace mass spectrometry and visible spectroscopy in the classification of the Ligurian olive oils. *Anal. Chim. Acta* 589: 89–95.

Castro-Vazquez, L., M.C. Diaz-Maroto, M.A. Gonzalez-Vinas, and M.S. Perez-Coello. 2009. Differentiation of monofloral citrus, rosemary, eucalyptus, lavender, thyme and heather honeys based on volatile composition and sensory descriptive analysis. *Food Chem.* 112: 1022–1030.

Costa Freitas, A.M., and A.I. Mosca. 1999. Coffee geographic origin—An aid to coffee differentiation. *Food Res. Int.* 32: 565–573.

Costa Freitas, A.M., C. Parreira, and L. Vilas-Boas. 2001. The use of an electronic aroma-sensing device to assess coffee differentiation—Comparison with SPME gas chromatography–mass spectrometry aroma patterns. *J. Food Compos. Anal.* 14: 513–522.

Cynkar, W., R. Dambergs, P. Smith, and D. Cozzolino. 2010. Classification of Tempranillo wines according to geographic origin: Combination of mass spectrometry based electronic nose and chemometrics. *Anal. Chim. Acta* 660: 227–231.

da Silva, G.A., F. Augusto, and R.J. Poppi. 2008. Exploratory analysis of the volatile profile of beers by HS–SPME–GC. *Food Chem.* 111: 1057–1063.

Defernez, M., and E.K. Kemsley. 1997. The use and misuse of chemometrics for treating classification problems. *TrAC, Trends Anal. Chem.* 16: 216–221.

Diaz, C., J.E. Conde, J.J. Mendez, and J.P.P. Trujillo. 2003. Volatile compounds of bottled wines with denomination of origin from the Canary Islands (Spain). *Food Chem.* 81: 447–452.

Diaz-Reganon, D.H., R. Salinas, R. Masoud, and G. Alonso. 1998. Adsorption-thermal desorption–gas chromatography applied to volatile compounds of Madrid region wines. *J. Food Compos. Anal.* 11: 54–69.

Dirinck, P., and A. De Winne. 1999. Flavour characterisation and classification of cheeses by gas chromatographic–mass spectrometric profiling. *J. Chromatogr. A* 847: 203–208.

European Commission. 2002a. Regulation (EC) No. 178/2002 of The European Parliament and of The Council of 28 January 2002. Laying down the general principles and requirements of food law, establishing the European Food Safety Authority and laying down procedures in matters of food safety. *Off. J. Eur. Communities* L31: 1–24.

European Commission. 2002b. Council Directive 2001/110/EC of 20 December 2001 relating to honey (2002). *Off. J. Eur. Communities* L10: 47–52.

European Commission. 2002c. Commission Regulations (EC) No. 1019/2002 of 13 June 2002 on the marketing standards for olive oil (2002). *Off. J. Eur. Communities* L155: 27–31.

European Commission. 2006. Council Regulation (EC) No. 510/2006 of 20 March 2006 on the protection of geographical indications and designations of origin for agricultural products and foodstuffs. *Off. J. Eur. Union* L93: 12–25.

Fernandez-Garcia, E., M. Carbonell, P. Gaya, and M. Nunez. 2004. Evolution of the volatile components of ewes raw milk Zamorano cheese. Seasonal variation. *Int. Dairy J.* 14: 701–711.

Food Standard Agency. 2010. Available online at http://www.food.gov.uk/. Accessed on Dec. 12, 2010.

Gil, M., J.M. Cabellos, T. Arroyo, and M. Prodanov. 2006. Characterization of the volatile fraction of young wines from the Denomination of Origin "Vinos de Madrid" (Spain). *Anal. Chim. Acta* 563: 145–153.

Giraudel, J.L., L. Setkova, J. Pawliszyn, and M. Montury. 2007. Rapid headspace solid-phase microextraction–gas chromatographic–time-of-flight mass spectrometric method for qualitative profiling of ice wine volatile fraction: III. Relative characterization of Canadian and Czech ice wines using self-organizing maps. *J. Chromatogr. A* 1147: 241–253.

Hajslova, J., and T. Cajka. 2008. Gas chromatography in food analysis. In *Handbook of Food Analysis Instruments*, ed. S. Otles. Boca Raton, FL: CRC Press, Taylor & Francis Group.

Hajslova, J., T. Cajka, and L. Vaclavik. 2011. Challenging applications offered by direct analysis in real time (DART) in food-quality and safety analysis. *Trends Anal. Chem.* 30: 204–218.

Holscher, W., and H. Steinhart. 1992. Investigation of roasted coffee freshness with an improved headspace technique. *Z. Lebensm.-Unters.-Forsch. A* 195: 33–38.

International Olive Council. 2008. Trade standard applying to olive oils and olive–pomace oils, COI/T.15/NC No. 3/Rev. 3, 24 November 2008.

Jelen, H., and M. Majcher. 2009. Food aroma compounds: Tools and techniques for detection. *Food Engineering & Ingredients*. Available online at http://www.fei-online.com/featured-articles/food-aroma-compounds-tools-and-techniques-for-detection/. Accessed on Dec. 12, 2010.

Jelen, H.H., A. Ziolkowska, and A. Kaczmarek. 2010. Identification of the botanical origin of raw spirits produced from rye, potato, and corn based on volatile compounds analysis using a SPME–MS method. *J. Agric. Food Chem.* 58: 12585–12591.

Jurado, J.M., O. Ballesteros, A. Alcazar, et al. 2008. Differentiation of certified brands of origins of Spanish white wines by HS-SPME–GC and chemometrics. *Anal. Bioanal. Chem.* 390: 961–970.

Kataoka, H., H.L. Lord, and J. Pawliszyn. 2000. Applications of solid-phase microextraction in food analysis. *J. Chromatogr. A* 880: 35–62.

Kemsley, E.K. 1996. Discriminant analysis of high-dimensional data: A comparison of principal components analysis and partial least squares data reduction methods. *Chemom. Intell. Lab. Syst.* 33: 47–61.

Kokkinofta, R.I., and Theocharis, C.R. 2005. Chemometric characterization of the Cypriot spirit "Zivania". *J. Agric. Food Chem.* 53: 5067–5073.

Korhonova, M., K. Hron, D. Klimcikova, L. Muller, P. Bednar, and P. Bartak. 2009. Coffee aroma—Statistical analysis of compositional data. *Talanta* 80: 710–715.

Kotti, F., L. Cerretani, M. Gargouri, E. Chiavaro, and A. Bendini. 2011. Evaluation of the volatile fraction of commercial virgin olive oils from Tunisia and Italy: Relation with olfactory attributes. *J. Food Biochem.* 35: 681–698.

Kruzlicova, D., J. Mocak, B. Balla, J. Petka, M. Farkova, and J. Havel. 2009. Classification of Slovak white wines using artificial neural networks and discriminant techniques. *Food Chem.* 112: 1046–1052.

Lachenmeier, D.W., E.-M. Sohnius, R. Attig, and M.G. Lopez. 2006. Quantification of selected volatile constituents and anions in Mexican *Agave* spirits (Tequila, Mezcal, Sotol, Bacanora). *J. Agric. Food Chem.* 54: 3911–3915.

Lees, M., ed. 2003. *Food Authenticity and Traceability*. Boca Raton, FL: CRC Press.

Luna, G., M.T. Morales, and R. Aparicio. 2006. Characterisation of 39 varietal virgin olive oils by their volatile compositions. *Food Chem.* 98: 243–252.

Mallia, S., E. Fernandez-Garcia, and J.O. Bosset. 2005. Comparison of purge and trap and solid phase microextraction techniques for studying the volatile aroma compounds of three European PDO hard cheeses. *Int. Dairy J.* 15: 741–758.

Mannina, L., and V. Di Tullio. Food analysis by fingerprinting techniques. Centro Stampa De Vittoria (2009) Italy. Available online at http://www.trace.eu.org/img/BOOKLET_Trace_150dpi+final.pdf (Accessed on Dec. 12, 2010).

Marquez, R., R. Castro, R. Natera, and C. Garcia-Barroso. 2008. Characterisation of the volatile fraction of Andalusian sweet wines. *Eur. Food Res. Technol.* 226: 1479–1484.

Massart, D.L., B.G.M. Vandeginste, L.M.C. Buydens, S. De Jong, P.J. Lewi, and J. Smeyers-Verbeke. 1997. *Handbook of Chemometrics and Qualimetrics*, Part A. Amsterdam: Elsevier.

Mildner-Szkudlarz, S., and H.H. Jelen. 2008. The potential of different techniques for volatile compounds analysis coupled with PCA for the detection of the adulteration of olive oil with hazelnut oil. *Food Chem.* 110: 751–761.

Montealegre, C., M.L.M. Alegre, and C. Garcia-Ruiz. 2009. Traceability markers to the botanical origin in olive oils. *J. Agric. Food Chem.* 58: 28–38.

Morales, M.T., R. Aparicio, and J.J. Rios. 1994. Dynamic headspace gas chromatographic method for determining volatiles in virgin olive oil. *J. Chromatogr. A* 668: 455–462.

Oliveira, R.C.S., L.S. Oliveira, A.S. Franca, and R. Augusti. 2009. Evaluation of the potential of SPME–GC–MS and chemometrics to detect adulteration of ground roasted coffee with roasted barley. *J. Food Compos. Anal.* 22: 257–261.

Oliveri, P., V. Di Egidio, T. Woodcock, and G. Downey. 2010. Application of class-modelling techniques to near infrared data for food authentication purposes. *Food Chem.* 125: 1450–1456.

Oliveros, C.C., R. Boggia, M. Casale, C. Armanino, and M. Forina. 2005. Optimisation of a new headspace mass spectrometry instrument—Discrimination of different geographical origin olive oils. *J. Chromatogr. A* 1076: 7–15.

Peres, C., C. Viallon, and J.-L. Berdague. 2001. Solid-phase microextraction–mass spectrometry: A new approach to the rapid characterization of cheeses. *Anal. Chem.* 73: 1030–1036.

Ribeiro, J.S., F. Augusto, T.J.G. Salva, R.A. Thomaziello, and M.M.C. Ferreira. 2009. Prediction of sensory properties of Brazilian Arabica roasted coffees by headspace solid phase microextraction–gas chromatography and partial least squares. *Anal. Chim. Acta* 634: 172–179.

Risticevic, S., E. Carasek, and J. Pawliszyn. 2008. Headspace solid-phase microextraction–gas chromatographic–time-of-flight mass spectrometric methodology for geographical origin verification of coffee. *Anal. Chim. Acta* 617: 72–84.

Stanimirova, I., B. Ustun, T. Cajka, et al. 2010. Tracing the geographical origin of honeys based on volatile compounds profiles assessment using pattern recognition techniques. *Food Chem.* 118: 171–176.

Tananaki, C., A. Thrasyvoulou, J.L. Giraudel, and M. Montury. 2007. Determination of volatile characteristics of Greek and Turkish pine honey samples and their classification by using Kohonen self organising maps. *Food Chem.* 101: 1687–1693.

van den Berg, R.A., H.C.J. Hoefsloot, J.A. Westerhuis, A.K. Smilde, and M.J. van der Werf. 2006. Centering, scaling, and transformations: Improving the biological information content of metabolomics data. *BMC Genomics* 7: 142–156.

Vera, L., L. Acena, J. Guasch, R. Boque, M. Mestres, and O. Busto. 2011. Characterization and classification of the aroma of beer samples by means of an MS e-nose and chemometric tools. *Anal. Bioanal. Chem.* 399: 2073–2081.

Weldegergis, B.T., A. de Villiers, and A.M. Crouch. 2011. Chemometric investigation of the volatile content of young South African wines. *Food Chem.* 128: 1100–1109.

Wishart, D.S. 2008. Metabolomics: Applications to food science and nutrition research. *Trends Food Sci. Technol.* 17: 482–493.

Zhang, J., L. Li, N. Gao, D. Wang, Q. Gao, and S. Jiang. 2010. Feature extraction and selection from volatile compounds for analytical classification of Chinese red wines from different varieties. *Anal. Chim. Acta* 662: 137–142.

Zunin, P., R. Boggia, P. Salvadeo, and F. Evangelisti. 2005. Geographical traceability of West Liguria extra virgin olive oils by the analysis of volatile terpenoid hydrocarbons. *J. Chromatogr. A* 1089: 243–249.

18 Mapping the Combinatorial Code of Food Flavors by Means of Molecular Sensory Science Approach

Peter Schieberle and Thomas Hofmann

CONTENTS

18.1 INTRODUCTION

The perception of aroma and taste, that is, the flavor of a given food, is induced by bioactive molecules interacting with chemoreceptor proteins located in the nose and on the tongue, respectively. On the basis of current knowledge in flavor research, the presence of certain structural elements (olfactophores, gustophores) as well as specific concentrations exceeding the sensory thresholds is an important prerequisite of the molecules to become "flavor-active." Moreover, it is well accepted today that not a single flavor impact molecule, but a combinatorial code of multiple odor- and taste-active key compounds, each in its specific concentration, is needed to reflect the typical aroma and taste profile of a given food. Thus, comparing flavor with sound perception, it is not a single instrument, but it is the instrumental ensemble of an orchestra and the well-balanced interplay between the single instruments that make music a unique experience. Thus, the number and the concentrations of such bioactive constituents determine whether, for example, a wine has an enjoyable overall flavor or whether it has developed an awful cork taint.

The attractive smell and typical taste of freshly brewed coffee, a tea infusion, freshly baked bread, the delicious taste of oak-matured red wine, or a barbecued beef steak are highly appreciated by consumers. However, what is the reason for the clearly different flavor quality of the same products, for example a freshly brewed coffee, or a coffee stored for an hour in a thermos flask? The reason is that, although the molecular decoding of aroma and taste is a necessary prerequisite for a knowledge-driven optimization of the sensory quality of food products by technological means, the combinatorial code of many food flavors has not yet been unraveled on the molecular level.

To sense the odor-active molecules, the regio olfactoria in the human nose is equipped with an array of olfactory neurons each expressing one representative of the repertoire of ~390 different odorant receptor proteins. Because ~5 million cells are available in the olfactory epithelium, it can be calculated that the same type of receptor protein is expressed in ~70,000 cells of the same type. To come into contact with the olfactory receptors, the odor-active ligands must be volatile, because they have to be transported orthonasally, that is, via the nostrils, to these receptor cells. However, during eating and, in particular, after swallowing, the odor-active molecules are also retronasally transported with the so-called after-swallow breath to the regio olfactoria as previously shown by magnetic resonance imaging techniques (Buettner and Schieberle 2000; Buettner et al. 2002).

Odor-active molecules can be characterized and classified by means of two compound-specific criteria: each molecule exhibits its own odor threshold concentration as well as a specific odor quality, respectively. As shown in Figure 18.1 for a couple of alkyl pyrazines, the odor threshold is significantly influenced by the chemical structure. For example, just substituting the methyl group in position 2 of 2,3,5-trimethylpyrazine (P5) by an ethyl group (P7) leads to a decrease in the odor threshold by a factor of ~50,000 (Wagner et al. 1999).

The odor note of a molecule is determined by its chemical structure, for example, simply changing the (*S*)-configured isopropyl group in the structure of carbon to the (*R*)-configuration swaps the odor note from caraway-like to peppermint-like (Figure 18.2). The same can be observed for a number of other stereoisomeric pairs, such as the (*E*)- and (*Z*)-8-methyl-α-ionone.

Compared to the volatile odorants, taste-active molecules commonly show low volatility or are not volatile at all. After dissolving in the saliva, the tastants interact with chemoreceptor proteins on the human tongue and, strongly depending on their chemical structure, induce one or the other of the five basic taste modalities: sour, salty, bitter, sweet, or umami. The last mentioned taste quality is a Japanese word for "delicious" and is induced by activation of the heterodimeric T1R1/T1R3 taste receptor by the amino acid L-glutamate (Li et al. 2002). It is known to be drastically potentiated by the taste-enhancing 5′-ribonucleotides, such as inosine-5′-monophosphate (IMP) and guanosine-5′-monophosphate (GMP). On the other hand,

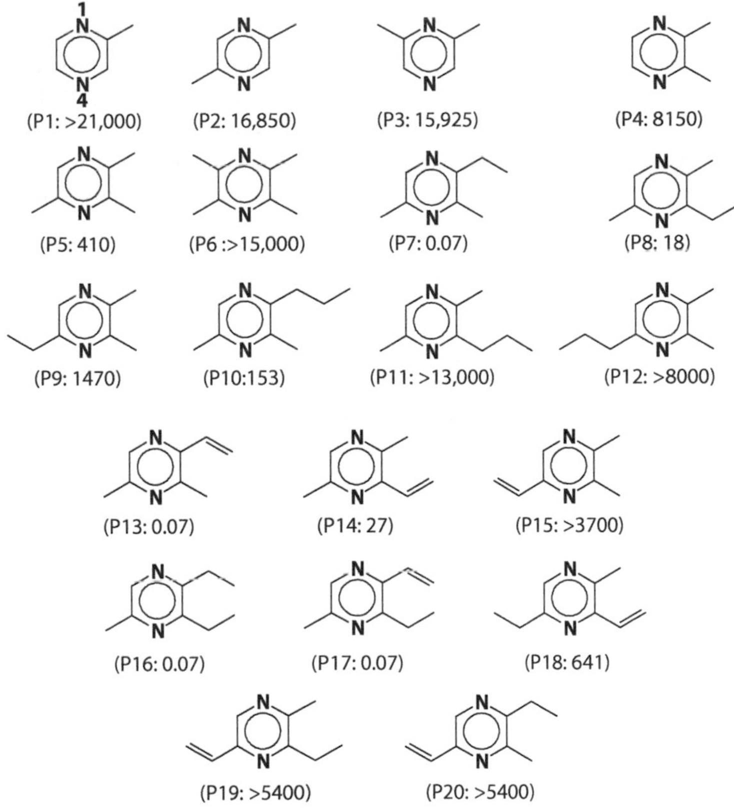

FIGURE 18.1 Odor thresholds (nanogram per liter in air) of synthesized alkylpyrazines.

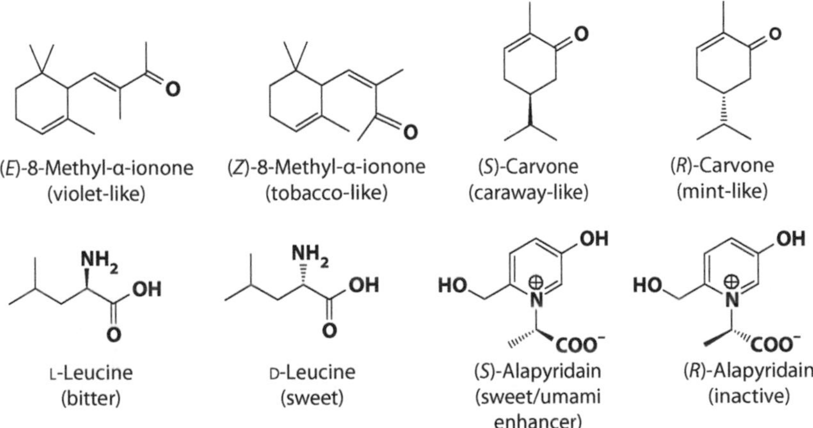

FIGURE 18.2 Influence of structure of selected odorants, tastants, and taste modulators on aroma or taste attributes, respectively.

the T1R3 protein unit combines with the protein T1R2 to form a functional general sweet taste receptor heterodimer sensing carbohydrates, high-potency sweeteners, as well as sweet proteins such as brazzein. On the contrary, the oral sensing of bitter molecules is mediated by a family of 25 recently discovered TAS2R bitter-taste receptor proteins found to be expressed in the taste buds of circumvallate and foliate papillae as well as in the palate, respectively (Chandrashekar et al. 2000). The sensory activity of many tastants as well as taste enhancers is strongly dependent on their stereochemistry: for example, the amino acid L-leucine exhibits a pronounced bitter taste, whereas D-leucine induces a sweet taste impression (Figure 18.2). The (*S*)-configured taste enhancer alapyridaine was shown to significantly decrease the human sensitivity of sweet and umami tastes (Soldo et al. 2003, 2004), whereas (*R*)-alapyridaine is entirely inactive (Figure 18.2).

Besides the five main taste qualities, additional sensory impressions have been recorded in human sensory studies and found to be induced by chemical stimuli. Such orosensations are astringency, pungency, cooling, and tingling, and were very recently extended by fatty taste and the so-called kokumi impression. Multiple polyphenols, but also nonphenolic compounds, such as the 3-indol acetic acid *N*-glucopyranoside from red currents, (no. 1 in Figure 18.3) were reported to impart an intense puckering astringent sensation (Schwarz and Hofmann 2007), but the biological mechanism underlying the sensing of these astringent stimuli is still orphan. In contrast, some other orosensations were identified to be mediated by specific chemoreceptor events: hot compounds such as capsaicin (no. 2) were found to activate the heat-sensitive vanilloid receptor TRPV1 (Liu and Simon 2000); (–)-menthol (no. 3) induces its physiological cooling activity by the activation of cold–sensitive receptor channel TRPM8

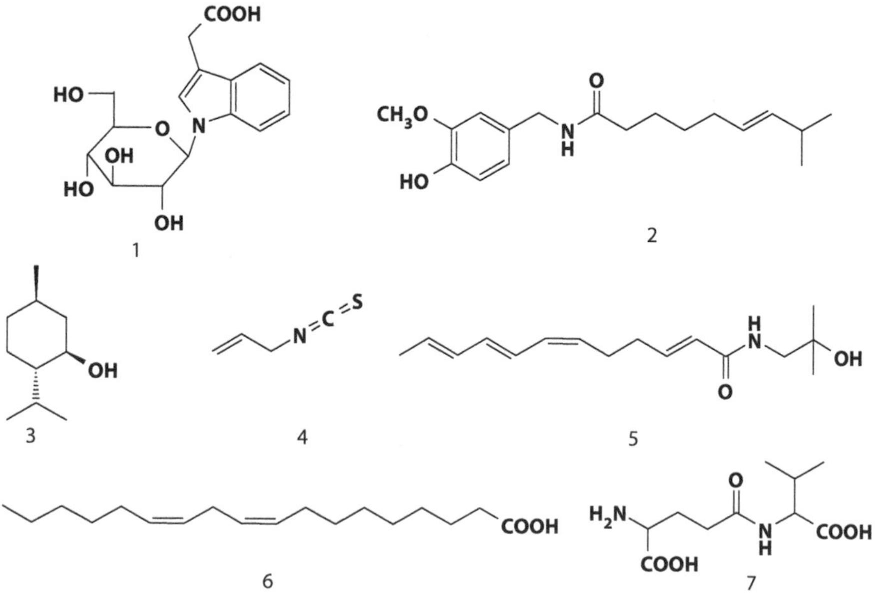

FIGURE 18.3 Selected taste-active compounds causing astringent (1), hot (2), cooling (3), pungent (4), tingling (5), fatty (6), and kokumi (7) sensations.

(Bautista et al. 2007) besides TRPA1 (Hinman et al. 2006), the latter of which is also mediating the pungent sensation of isothiocyanates such as allyl isothiocyanate, the pungent principle of the seeds of black mustard, horseradish, and wasabi (no. 4 in Figure 18.3). In addition, tingling compounds such as hydroxy α-sanshool (no. 5) from Szechuan pepper activate KCNK channels (Bautista et al. 2008), and free fatty acids such as linoleic acid have been identified as stimulus for fat taste in humans involving the G-protein coupled receptors GPR40 and GPR120 as potential gustatory fat taste receptors (Cartoni et al. 2010) (no. 6 in Figure 18.3). Moreover, some molecules enhancing mouthfulness, thickness, and complexity, or increasing the continuity of food taste perception were found in foods and coined about 10 years ago by the Japanese as "kokumi" compounds (Ueda et al. 1994). A series of kokumi-active γ-glutamyl-di/tripeptides were identified in garlic (Ueda et al. 1997), in common beans (Dunkel et al. 2007), as well as in matured Gouda cheese (Toelstede et al. 2009). Although being entirely tasteless on their own up to levels of about 10 mM/L, hypothreshold amounts of these γ-glutamyl peptides were found to increase mouthfulness, complexity, and palate length not only of savory solutions containing sodium chloride and L-glutamic acid, but also those of an artificial cheese base. Very recently, various kokumi-active γ-glutamyl peptides including γ-glutamyl-valine (no. 7 in Figure 18.3) and γ-glutamyl-ysteinyl-glycine (glutathione) were demonstrated to activate the extracellular calcium-sensing receptor (CaSR), a close relative of the class C G-protein-coupled sweet and umami receptors (Ohsu et al. 2010). However, a distinct function of the CaSR in human kokumi perception still needs to be confirmed.

Although the recognition of odorants and tastants occurs in different organs, both perceptions are well accepted to clearly influence each other. For example, a salty fruit juice will hardly be acceptable to a consumer, whereas a sweetened fruit juice will be preferred versus the unsweetened one. However, because the overall flavor quality of a food is often described rather generically as a "good taste" in sensory science, the clear differentiation between odor qualities and taste modalities has been complicated for a long time. Today, sophisticated high-vacuum distillation techniques, such as the solvent-assisted flavor evaporation (SAFE) method (Engel et al. 1999), allow the careful separation of the volatile fraction containing the odor-active molecules and the nonvolatile fraction containing the key tastants of foods. Once odorants and tastants have been separated, the combined application of analytical sensory evaluation tools involving educated human subjects with modern instrumental techniques enables the identification of the most important aroma and taste compounds and the scientific decoding of their combinatorial interplay.

18.2 WHY RESEARCH ON FOOD FLAVORS?

Today, the preparation of foods has moved more and more from the traditional style in the kitchen to an industrial production of finished foods, the so-called "convenience products." In addition, many raw materials can only be harvested once a year, whereas consumers demand foods and beverages such as juices, coffee, or tea in a premium quality everywhere on the planet at any time over the entire year. Thus, processing and storage are important steps in modern food production.

To meet consumers' demand on the continuous availability of certain foods, it is important to know the "flavor blueprint" of a golden standard, that is, a premium quality food product, in order to be able to modify the processing conditions of manufacturing with respect to obtaining an excellent flavor quality. "Flavor blueprint" means the combinatorial code of the entire set of odor- and taste-active food components in their natural concentrations in the food.

It must be considered that the characteristic flavor of foods prepared by means of enzymatic and/or thermal processing is generated as a result of the chemical formation of flavor-active compounds by a transformation of odor- or tasteless precursors. Industrial processes commonly differ from the food preparation in the kitchen and, thus, different odor and taste profiles may result. However, although fermentations and or thermal processes, such as cooking or baking, have been known to humans for thousands of years, the influence of the composition of the raw material as well as of the processing conditions on the formation of key aroma or taste compounds is often still unclear. Thus, industrial processes are usually modified on the basis of "trial and error" rather than on the basis of scientific knowledge. Furthermore, by law, the flavor of many foods is not permitted to be adjusted by adding synthetic flavor compounds. Furthermore, the trend in the food industry moves to a production with as much "natural" ingredients as possible. Thus, it remains a challenge for the food industry to develop innovative convenience products with high flavor quality. This, however, can only be achieved, for example, by promoting the formation of flavor-active molecules from precursors via selecting the appropriate processing conditions.

18.3 WHY IS THE COMBINATORIAL CODE OF AROMA AND TASTE OF FOODS OFTEN UNKNOWN?

With the introduction of gas chromatography (GC) into aroma research in the 1960s, the number of volatiles identified in foods has significantly increased over the last few years (Figure 18.4). Although several groups have quite early pointed to the fact that volatility is not the most important attribute of an odor-active food constituent (Rothe 1975; Teranishi et al. 1974; Acree et al. 1976), it can be estimated that today, approximately 12,000 volatiles have structurally been characterized in foods and beverages. However, as already indicated in Figure 18.1, the concentration of a flavor compound is not the key criterion for its flavor contribution, but its biological activity, which is clearly correlated with the flavor threshold. Another example is the [Z]-heptadeca-1,9-dien-4,6-diyn-3,8-diol, a key tastant contributing to the bitter off-taste of carrots and carrot products (Czepa and Hofmann 2003). Independent synthesis of the four possible stereoisomers, followed by cochromatography using chiral high-performance liquid chromatography (HPLC) analysis, revealed the naturally occurring 3(*R*),8(*S*)-stereoisomer to exhibit the lowest bitter threshold of 40 μM/kg, whereas the 3(*S*),8(*S*)-isomer showed a value above 450 μM/kg, and the 3(*S*),8(*R*)- and the 3(*R*),8(*R*)-isomer only induced a bitter taste at levels above 1000 μM/kg. These and many other examples known today indicate a coevolution of the odor- and taste-active natural products and their corresponding olfactory and taste receptors,

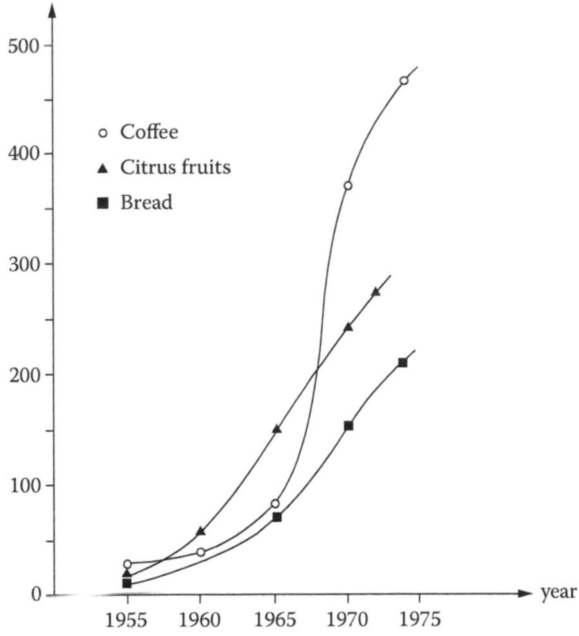

FIGURE 18.4 Increase in number of volatiles identified in foods: tremendous influence of GC–MS (Rothe 1975).

respectively, and also nicely highlight the enormous analytical challenge in the identification of such trace, but highly bioactive compounds.

Although hundreds of odor- and taste-active molecules have already been identified in many foods and beverages such as coffee or tea (Table 18.1), a bunch of low-abundant, but highly flavor-active constituents may still remain to be discovered. In consequence, it is of prime importance to start the analysis of aroma and taste compounds by means of screening methods to locate only those compounds in such complex extracts, which are able to interact with human odor or taste receptors, because it would make no sense to identify with a high analytical effort those food constituents that do not contribute to the flavor.

In the following chapters, the concept of molecular sensory science aimed at identifying key food aroma and taste compounds on the molecular level will be exemplified using black tea as an example. In the meantime, this concept has also been successfully applied to unravel the aromas of many foods, such as orange juice (Buettner and Schieberle 1999, 2001), fermented sausage (Söllner and Schieberle 2009), Bourbon whiskey (Poisson and Schieberle 2008a, 2008b), or apricots (Greger and Schieberle 2007). Because this strategy helps to map the combinatorial code of aroma- and taste-active key molecules, which are sensed by human chemosensory receptors and are then integrated by the brain, the concept is also assigned as "sensomics."

TABLE 18.1
Number of Volatile Constituents Identified in Black Tea Powders or Infusions

Substance Class	No.
Esters	89
Ketones	81
Aldehydes	75
Acids	74
Alcohols	68
Basic compounds	64
Hydrocarbons	52
Furans	21
Sulfur compounds	19
Lactones	18
Phenols	13
Others	17
Total	591

18.4 AROMA COMPOUNDS IN BLACK TEA INFUSIONS

After water, tea is the next most consumed beverage worldwide. Today, about 3.1 million metric tons of different types of tea are produced, with China being its main producer. The common process of black tea manufacturing consists of four stages: withering, rolling, fermentation, and drying/firing. Besides being differentiated by origin, three main types of tea are available in the trade: unfermented (green), semifermented (oolong), and fermented (black) tea. The typical astringent, bitter taste, as well as the characteristic aroma of the tea beverage, is an important criterion in the evaluation of the tea quality. However, as indicated above, the key aroma and taste compounds of black tea have only scarcely been investigated by means of sensory science concepts.

Thus, because numerous enzymatic reactions are taking place during fermentation and thermal processing, knowledge on the key flavor compounds would offer the opportunity to influence the tea flavor on the basis of scientific data. To establish a kind of golden standard, in the first step, 20 black tea samples were purchased on the local market, and the flavor quality of the beverages was judged by a group of 20 panelists using a hedonic approach. Based on this evaluation, a Darjeeling tea from India (Darjeeling Gold Selection; DGS) was selected for flavor analysis. In the next step, the volatiles were isolated from the extract by SAFE (Figure 18.5). An aliquot of the distillate was spread on a strip of filter paper, and the odor was checked by the panel against that of the beverage in order to prove that both odors were identical.

To locate the odor-active constituents, the distillate was separated on a GC stationary phase. At the end of the column, the effluent was split into two equal parts: one was led to the flame ionization detector (FID), the other to a sniffing port at which a panelist evaluated the eluting odors. By applying this "sniffing" procedure (GC-olfactometry) on 1:1 dilutions of the distillate, odor-active compounds could be

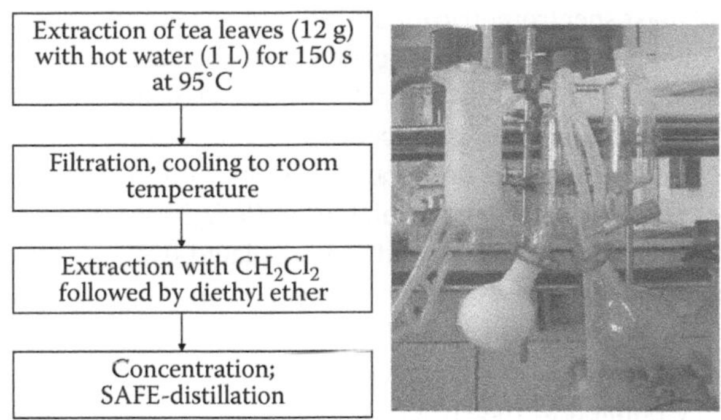

FIGURE 18.5 Procedure used for volatile isolation from tea infusion via SAFE method (Engel et al. 1999).

located, their odor qualities were assigned, and their relative odor potencies were determined (Figure 18.6). This screening procedure, assigned as aroma extract dilution analysis (AEDA) (Schieberle 1995), is a direct coupling of the human sensory detector (the nose) and a machine (the gas chromatograph), allowing the location of odor-active compounds in complex aroma distillates. The flavor dilution (FD) factor defines the dilution step in which the respective odorant was detected the last time, implying that the higher the FD factor, the higher the odor potency of the respective compound. The results of the application of the AEDA on a distillate from the black tea beverage showed a total of 47 odor-active areas (Figure 18.6) among which seven compounds with flowery (no. 19), oat flakes-like (no. 35), violet-like (no. 37), caramel-like (no. 41), seasoning-like (no. 45), honey-like (no. 46), and vanilla-like odors (no. 47) showed the highest FD factors.

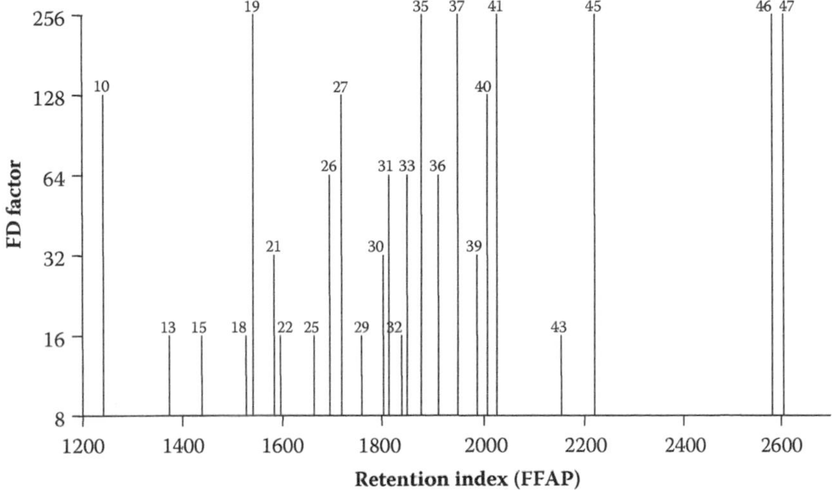

FIGURE 18.6 Flavor dilution chromatogram obtained by applying AEDA on an extract/distillate obtained from tea infusion.

On the basis of mass spectrometric measurements, retention indices, and comparison of the odor qualities and odor potencies, six of the seven compounds could be identified as linalool, β-ionone, 4-hydroxy-2,5-dimethyl-3(2*H*)-furanone, 3-hydroxy-4,5-dimethyl-2(5*H*)-furanone, 2-phenylacetic acid, and vanillin (Figure 18.7).

However, the identification of compound no. 35 with an intense oat flake-like odor needed more effort (Schuh and Schieberle 2005, 2006). First, due to its presence in trace amounts, the compound had to be isolated from an extract of ~25 L of the tea beverage, followed by separation using column chromatography. Isolation of the relevant area by preparative GC led to three clearly separated peaks showing an identical mass spectrum. Interpretation of the results from high-resolution mass spectrometry pointed to the structure of 2,4,6-nonatrienal. Because eight different geometrical isomers are possible with this structure, but no isomer was commercially available, a synthetic route was planned and performed for two isomers (Figure 18.8). However, although (*E*,*E*,*E*)-2,4,6- and (*E*,*Z*,*E*)-2,4,6-nonadienal were obtained following this route, both isomers did not coelute with compound 35, but, finally, substituting (*E*)-2-pentenal by (*Z*)-2-pentenal in the synthetic route (Figure 18.8), resulted in the generation of (*E*,*E*,*Z*)-2,4,6-nonadienal. This isomer showed a perfect match in all analytical data with those of compound 35, and was also by far the most potent aroma compound among the nonadienal isomers occurring in the tea beverage (Table 18.2). By means of this approach, finally, all key odorants of the black tea beverage could be identified (Schuh and Schieberle 2006).

Now the question arose on whether these compounds are able to generate the tea flavor: Did the FD-chromatogram display the importance of the single aroma compounds in decreasing order? The answer is no, because in the beverage, the

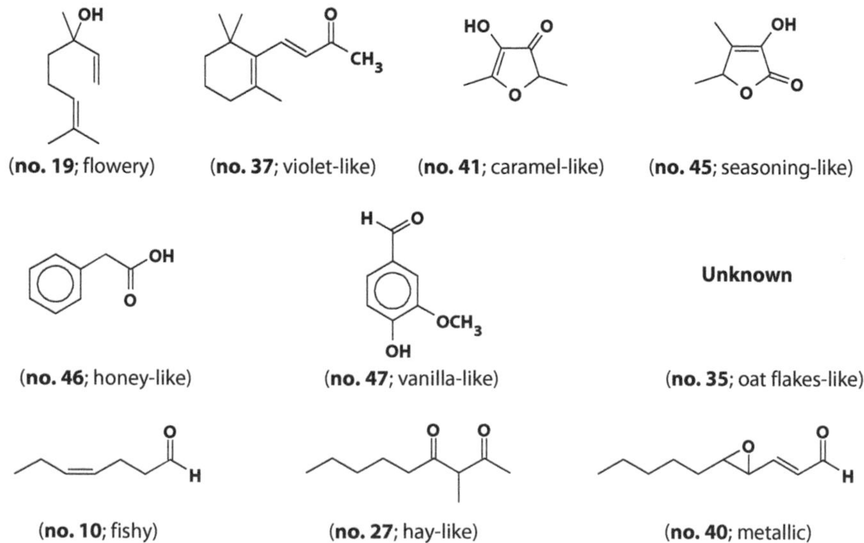

FIGURE 18.7 Structures of the most odor-active compounds in tea beverage (Darjeeling Gold Selection).

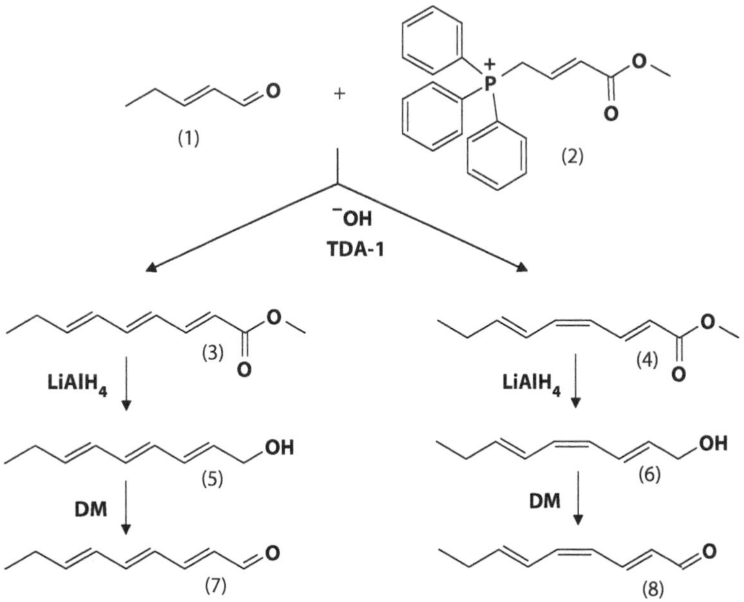

FIGURE 18.8 Synthetic route used in preparation of (*E*,*E*,*E*)- and (*E*,*Z*,*E*)-2,4,6-nonatrienal.

volatility of each odorant determines its available amounts in the headspace above the solution, whereas during GC–olfactometry, the entire amount of each compound is vaporized and, thus, the entire amount gets into contact with the odorant receptors. One approach to tackle this problem is the calculation of odor activity values, that is, the ratio of concentration to odor threshold (Schieberle 1995):

$$\text{OAV} = \frac{\text{Concentration}}{\text{Odor threshold}}.$$

Several methods are available to isolate volatiles for quantitation by GC–FID or GC–MS, for example, solid-phase microextraction and steam distillation or

TABLE 18.2
Retention Indices and Odor Thresholds of Three Nonatrienal Isomers Occurring in Black Tea Infusion

	RI on			Odor Threshold
Isomer	**FFAP**	**DB-5**	**DB-1701**	**(ng/L air)**
E,E,E	1900	1286	1452	4.4
E,E,Z	1877	1276	1440	0.0002
E,Z,E	1869	1271	1433	0.1

extraction/high-vacuum distillation. However, in any case, "losses" of volatiles during the isolation and/or concentration steps must be taken into account. The best approach to address these losses is the application of stable isotope dilution assays (SIDAs) (Schieberle 1995). The biggest challenge, however, is the synthesis of the isotopically labeled standards, which are often commercially not available. The synthesis of deuterium-labeled linalool is shown in Figure 18.9 as an example. After all internal standards have been synthesized, they are directly added to the tea before extraction and distillation. This way, any losses are ideally compensated. The differentiation between analyte and internal standard is finally performed by mass spectrometry, that is, by monitoring the respective molecular ions obtained by mass spectrometry–chemical ionization (MS–CI) of the analyte and the internal standard (Figure 18.9).

On the basis of the quantitative data measured and odor thresholds in water, determined by means of reference compounds using a sensory panel (Czerny et al. 2008), linalool, geraniol, and the newly identified (*E*,*E*,*Z*)-2,4,6-nonatrienal were identified as the key contributors to the overall aroma of the black tea beverage (Table 18.3).

To confirm the identification experiments, the last step of the molecular sensory science approach is the preparation of an aroma recombinate containing all key odorants identified in the concentrations occurring in the "natural" tea beverage. As shown in Figure 18.10, the recombinate mimicked the overall aroma of the tea beverage. Because none of the aroma compounds elicited an odor like the tea beverage itself, it must be concluded that the perception of the overall aroma is generated in the brain by putting together the single odorant/receptor events like a "puzzle." The fact that the human olfactory system is not able to resolve an entire flavor perception

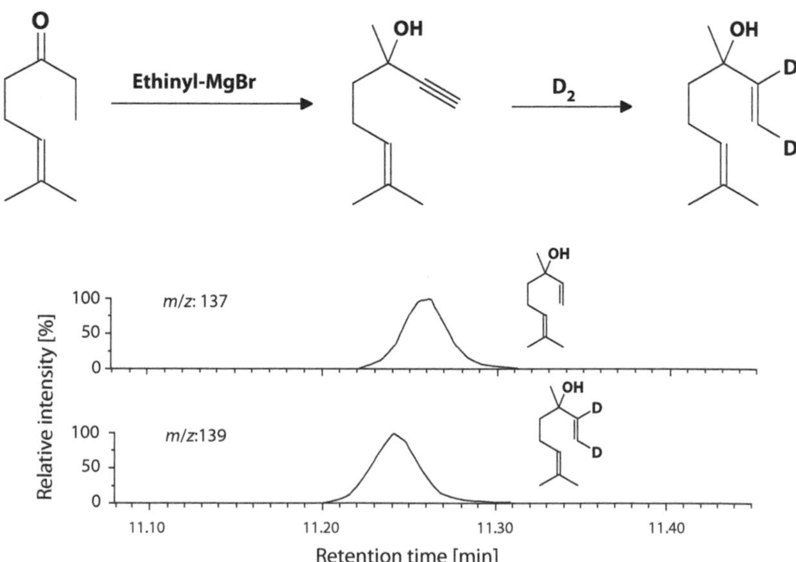

FIGURE 18.9 Synthesis of deuterium-labeled linalool for a stable isotope dilution assay (upper). Differentiation between analyte and internal standard by mass spectrometry (lower).

TABLE 18.3
Odor Thresholds and Odor Activity Values (OAV) of 16 Character Impact Odorants in Black Tea Infusion

Odorant	Odor Threshold (μg/L)	Conc. (μg/L)	OAV
R/*S*-Linalool	0.6	142	237
Geraniol	3.2	142	45
(*E*,*E*,*Z*)-2,4,6-Nonatrienal	0.025	1.1	41
(*E*)-β-Damascenone	0.004	0.15	38
Methylpropanal	1.9	69	37
3-Methylbutanal	1.2	42	37
2-Methylbutanal	4.4	82	37
3-Methyl-2,4-nonandione	0.01	0.48	37
(*E*,*Z*)-2,6-Nonadienal	0.03	0.56	22
(*E*,*E*)-2,4-Decadienal	0.2	2.9	18
(*Z*)-3-Hexenol	13	95	7
(*Z*)-4-Heptenal	0.06	0.66	11
Phenylacetaldehyde	6.3	57	9
β-Lonone	0.2	1.5	7
Hexanal	10	55	5
(*E*,*E*)-2,4-Nonadienal	0.2	0.45	3

into the odors of single components was previously found in many experiments (Laing and Francis 1989). For example, more than three compounds in a mixture can hardly be differentiated by our olfactory organ (Figure 18.11).

Although many questions remain open on how the overall aroma perception occurs in the human brain, on the basis of the analytical data, science-based food manufacturing is now possible, for example, the assessment of flavor differences between green and black tea, or simply the influence of addition of water on the release of volatiles from the fermented tea leaves.

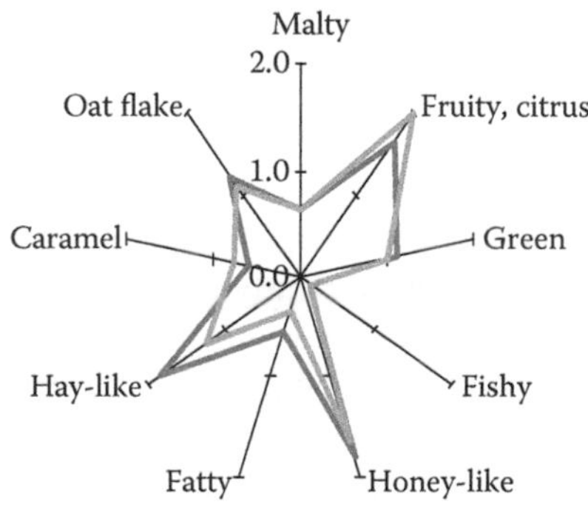

FIGURE 18.10 Aroma profiles of a black tea infusion (black line) and a tea aroma recombinate (gray line).

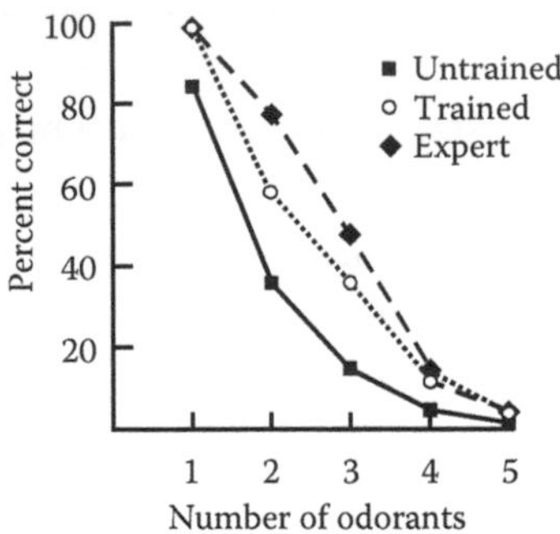

FIGURE 18.11 Percentage of correct responses of humans when identifying the components in mixtures containing up to five odorants (Laing and Francis 1989).

18.4.1 Differences in Odorant Composition between Black Tea Leaves and Tea Beverage

It is known that certain tea aroma compounds, such as linalool and geraniol, are released during processing of green tea by an enzymatic hydrolysis of the respective glycosides. Because the final firing process should inactivate all enzymes, the hot water extraction of the black tea leaves might only influence the extraction yields because of the different polarity of the odorants. However, no additional amounts of aroma compounds should be formed from precursors in the leaves. To investigate the influence of the hot water extraction on changes in odorant concentrations, first, the odor-active compounds present in the same black tea leaves used for the preparation of the infusion were characterized as described above. A comparison of the key aroma compounds with those identified in the beverage revealed that these were mostly identical from a qualitative point of view (Schuh and Schieberle 2006). However, the FD factors showed clear differences, suggesting differences in the concentrations (data not shown). Therefore, 22 of the key aroma compounds in the beverage were also quantified in the tea leaves. To monitor differences in the "yields" caused by the hot water extraction, the amounts of each odorant present in 12 g of leaves used per liter of water to prepare the tea beverage are contrasted to the amounts measured in 1 L of the tea beverage. As expected, some of the less polar compounds, such as (*E*,*E*)-2,4-nonadienal, 3-methyl-2,4-nonadione, or (*E*,*E*,*Z*)-2,4-6-nonatrienal, were obviously not fully extracted during the hot water treatment (Figure 18.12a), because their amounts in the beverage amounted to only 40% to 60% as compared to the amounts in the tea leaves.

On the contrary, nine odorants were clearly increased in the tea infusion (Figure 18.12b). The most significant increase was found for geraniol, the concentration of which was by a factor of 32 higher as compared to the amounts in the leaves. The same behavior was found for (*E*)-2-hexenol, 4-hydroxy-2,5-dimethyl-3(2*H*)-furanone, and 2-phenylethanol.

It is also interesting to note that aldehydes, such as 3-methylbutanal, known to be formed by a *Strecker*-type reaction, also showed a remarkable increase. Because a *Strecker* reaction involving carbohydrate degradation is very unlikely during a

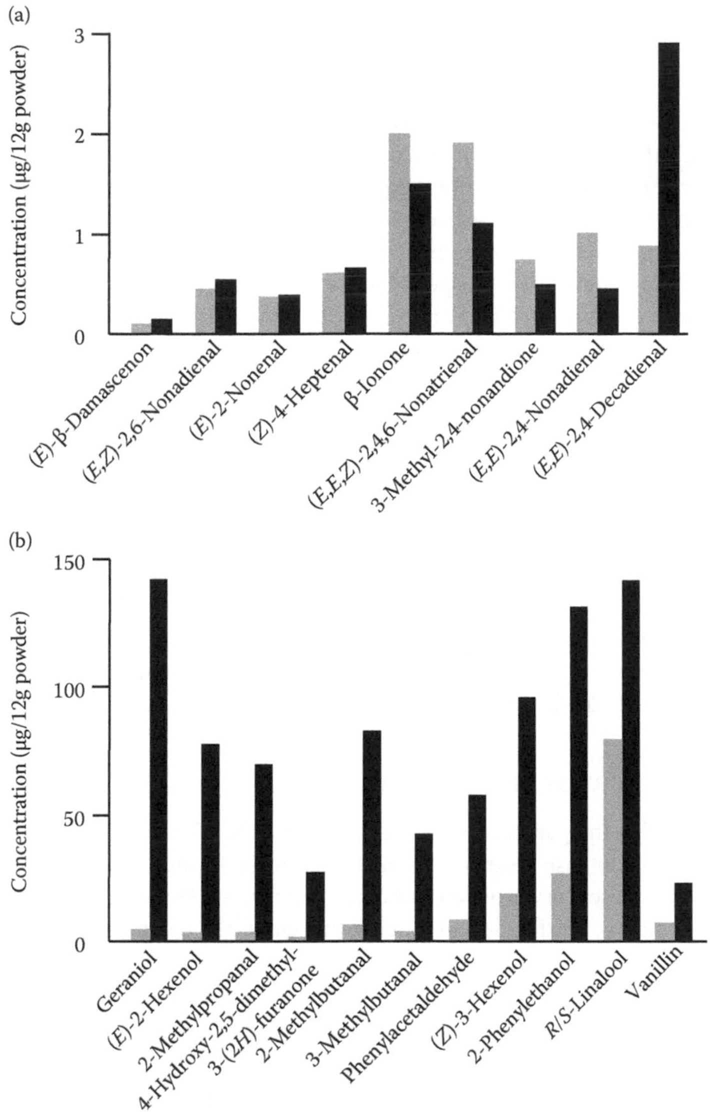

FIGURE 18.12 Comparison of concentrations of key aroma compounds in black tea powder (12 g; gray bars) and in infusion (1 L) prepared from 12 g powder (black bars). (a) Less polar odorants. (b) More polar odorants.

150-s hot water treatment, obviously yet unknown precursors are present in the black tea leaves, which are able to release the odorants from a precursor susceptible to hydrolysis. A similar release of flavor compounds was recently found by us in cocoa and chocolate, respectively (Buhr et al. 2010). Knowledge on the structure of such precursors would be an important task to be able to steer the formation of such aldehydes during tea manufacturing.

18.5 TASTE COMPOUNDS IN BLACK TEA INFUSIONS

Besides the aroma, the taste quality is also one of the key criteria used by tea tasters to describe the quality of tea liquors. Multiple attempts have been made to correlate the sensory results of the tea tasters and the molecules exhibiting the typical taste of tea infusions, but the data reported so far on the key tastants are rather contradictory. For example, the orange, low-molecular weight theaflavins as well as the red-brown polymeric thearubigins, both generated during tea fermentation upon flavan-3-ol oxidation (Roberts and Smith 1961; Millin et al. 1969; Scharbert et al. 2004), are believed to be responsible for the astringency of black tea infusions and have been recommended as an analytical measure of tea quality (Hilton and Ellis 1972). In contrast, other researchers could not find any statistical correlation between the overall astringent taste of tea infusions and the theaflavin concentration, but indicated a relationship between oral astringency and some flavan-3-ols such as, for example, epigallocatechin-3-gallate (Ding et al. 1992). Besides these phenols, 5-*N*-ethyl-L-glutamine (theanine) is reported to exhibit sweet-brothy and/or umami-like taste quality and is believed to contribute to the taste profile of tea infusions (Ekborg-Ott et al. 1997).

In order to evaluate the taste profile of the Darjeeling tea infusion (Darjeeling Gold Selection), the trained sensory panel was asked to rate the intensity of individual taste qualities on a scale from 0 (not detectable) to 3 (strongly detectable). By far, the highest scores of 2.2 and 1.6 were observed for the intensity of the astringent, mouth-drying taste quality, and the bitterness, respectively (Table 18.4).

18.5.1 Identification of Taste-Active Molecules

In order to bridge the gap between pure structural chemistry and human taste perception, the most intense taste compounds need to be located in the black tea infusion by means of a sensory-guided fractionation approach (Frank et al. 2001). To first gain insights into the astringent and bitter compounds, the tea infusion was separated

TABLE 18.4
Taste Profile Analysis of Tea Infusion and Artificial Taste Recombinate

	Intensities of Individual Taste Qualities[a]	
Taste Quality	**Tea Infusion**	**Taste Recombinate**
Astringent, mouth-drying	2.2	2.1
Bitter	1.6	1.5
Sour	0.5	0.5
Sweet	0.4	0.4
Salty	0	0
Umami	0	0

[a] Intensities were judged on a scale from 0 (not detectable) to 3 (strongly detectable).

by means of sequential ultrafiltration using filters with cut-offs of 10 kDa and 1 kDa, respectively. Three fractions were obtained, for example, the deeply brown colored, but tasteless fraction containing thearubigin-type polymers with molecular weights above 10 kDa; a red-brown, tasteless fraction containing the compounds with molecular weights between 1 kDa and 10 kDa; and a nearly colorless fraction containing the tea's low molecular weight compounds (LMW; <1 kDa) and reflecting the typical taste profile of the tea (Scharbert et al. 2004). These experiments clearly pointed out that the LMW compounds are the main contributors to the taste of the tea infusion. In consequence, this fraction was further analyzed by means of RP8-HPLC (Figure 18.13, left side). Aimed at rating the tea compounds in their relative taste impact, the effluent was separated into 43 fractions, which were freed from solvent and then used for the taste dilution analysis (TDA) (Frank et al. 2001) by using the recently developed half-tongue test (Figure 18.13, right side). Due to their high taste dilution (TD) factor of 8192, fractions no. 33 and 34 were evaluated with by far the highest taste impacts for astringency, closely followed by fraction nos. 30–32 and 23 judged with TD factors between 1024 and 4096 (Figure 18.13). In comparison, fraction nos. 6, 7, 10–12, and 16 were evaluated with lower taste impacts, whereas the low TD factors determined for all the other fractions excluded major contributions to the perception of tea astringency.

First, the identification was focused on those phenolic substances that were already discussed in literature as astringent compounds in black tea: the catechins and the theaflavins. Analysis of the tea infusion via high-performance liquid chromatography–diode array detection (HPLC–DAD) and HPLC–MS/MS as well as cochromatography with reference compounds led to the identification of gallocatechin (fraction 11), epigallocatechin (fraction 15), catechin (fraction 15), epigallocatechin-3-gallate (EGCG, fraction 20), epicatechin (fraction 22), gallocatechin-3-gallate (fraction 24), epicatechin-3-gallate (fraction 25), catechin-3-gallate (fraction 26), theaflavin (fraction 37), theaflavic acid (fraction 37), theaflavin-3-gallate (fraction 38), theaflavin-3′-gallate (fraction 38), and theaflavin-3,3′-digallate (fraction 39) as the key taste compounds in the individual HPLC fractions (Figure 18.13) (Scharbert et al. 2004). By comparing these data with the results of the TDA (Figure 18.13), it was obvious that those fractions containing the catechin- and theaflavin-type compounds were evaluated with TD factors below 128, whereas the unknown compounds in fraction nos. 30–34 were judged with TD factors of up to 8192 and, therefore, these were expected to be of major importance for the taste of black tea. Isolation of the tastants from fraction nos. 30–34 by means of polyamide chromatography and preparative reverse phase HPLC (RP-HPLC), followed by LC–MS/MS and 1D/2D-NMR spectroscopic structure determination, led to the unequivocal identification of 14 highly taste-active flavon-3-ol glycosides (Table 18.5), among which quercetin-3-*O*-[α-L-rhamnopyranosyl-(1→6)-β-D-glucopyranoside], known as rutin, was the most predominant representative (Scharbert et al. 2004).

In order to study the sensory impact of these tastants, the oral recognition thresholds were determined in water using the half-tongue test (Scharbert et al. 2004). The oral sensation imparted by the catechins was described as astringent with threshold concentrations between 190 μM/L and 930 μM/L, whereas the theaflavins induced a mouth-drying, rough-astringent sensation at threshold levels between 13 μM/L and

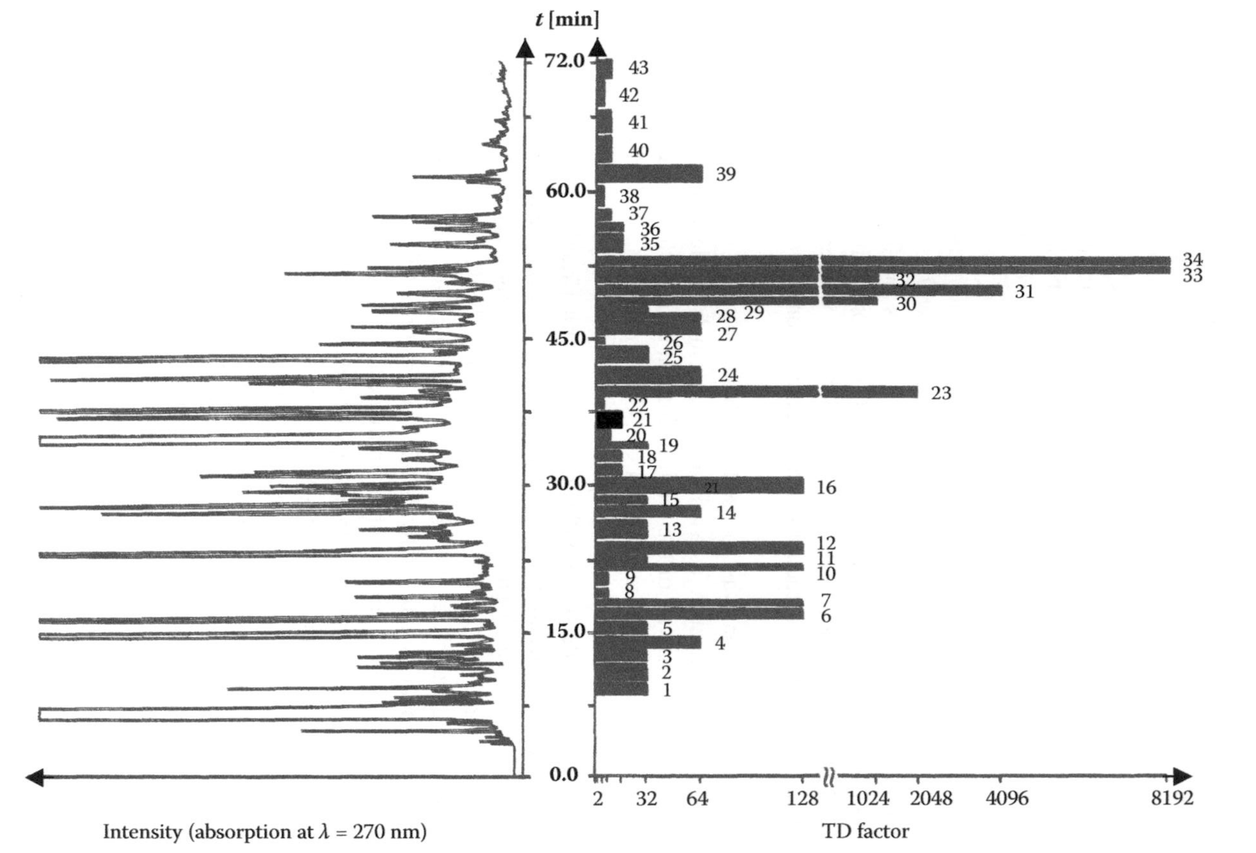

FIGURE 18.13 HPLC chromatogram (left side) and TD chromatogram (for astringency) of the LMW compounds isolated from tea infusion.

TABLE 18.5
Thresholds, Concentrations, and Dose-Over-Threshold (DoT) Factors of Tea Tastants

Tastant	Threshold [μM/L]	Conc. [μM/L]	DoT Factor[a]
Group I: Compounds Imparting Puckering Astringency and Rough Oral Sensation			
Epigallocatechin-3-gallate (EGCG)	190.0	328.0	1.7
Theaflavin	16.0	11.0	0.7
Catechin	410.0	221.0	0.5
Theaflavin-3,3′-digallate	13.0	6.7	0.5
Epicatechin-3-gallate	260.	113.0	0.4
Theaflavin-3-gallate	15.0	6.4	0.4
Theaflavin-3′-gallate	15.0	4.3	0.3
Epigallocatechin	520.0	131.0	0.3
Gallocatechin	540.0	131.0	0.3
Epicatechin	930.0	84.0	0.1
Catechin-3-gallate	250.0	11.0	<0.1
Gallocatechin-3-gallate	390.0	11.0	<0.1
Theaflavic acid	24.0	0.009	<0.1
Group II: Compounds Imparting Mouth-Drying and Velvety-Like Astringency			
Q-3-*O*-[α-L-rha-(1→6)-β-D-glc] (rutin)[b]	0.0015	11.1	9652.0
K-3-*O*-[α-L-rha-(1→6)-β-D-glc][b]	0.25	6.5	26.0
Q-3-*O*-β-D-gal[b]	0.43	5.4	12.6
Q-3-*O*-β-D-glc[b]	0.65	6.0	9.2
K-3-*O*-β-D-glc[b]	0.67	4.9	7.3
M-3-*O*-β-D-glc[b]	2.10	9.3	4.4
Q-3-*O*-[β-D-glc-(1→3)-*O*-α-L-rha-(1→6)-*O*-β-D-gal][b]	1.36	3.3	2.4
M-3-*O*-β-D-gal[b]	2.70	6.5	2.4
K-3-*O*-β-D-gal[b]	6.70	3.0	0.4
K-3-*O*-[β-D-glc-(1→3)-*O*-α-L-rha-(1→6)-*O*-β-D-glc][b]	19.80	8.6	0.4
Q-3-*O*-[β-D-glc-(1→3)-*O*-α-L-rha-(1→6)-*O*-β-D-glc][b]	18.40	7.1	0.4
A-8-*C*-[α-L-rha-(1→2)-β-D-glc][b]	2.80	0.9	0.3
M-3-*O*-[α-L-rha-(1→6)-β-D-glc][b]	10.50	2.2	0.2
K-3-*O*-[β-D-glc-(1→3)-*O*-α-L-rha-(1→6)-*O*-β-D-gal][b]	5.80	0.8	0.2
5-*N*-Ethyl-L-glutamine (theanine)	6000.00	281.0	<0.1
Group III: Bitter-Tasting Compounds			
Caffeine	500.0	990.0	2.0
Epigallocatechin-3-gallate (EGCG)[a]	380.0	328.0	0.9

[a] The dose-over-threshold (DoT) factor is calculated as the ratio of concentration and taste threshold.

[b] Q, quercetin; K, kaempferol; M, myricetin; A, apigenin; rha, rhamnose; glc, glucose; gal, galactose.

26 μM/L (Table 18.5). In contrast, the flavon-3-ol glycosides induced a mouth-drying and velvety astringent mouthfeel at extraordinarily low threshold concentrations spanning from 0.001 μM/L to 19.8 μM/L (Table 18.5).

18.5.2 Quantification of Taste Compounds and Calculation of Dose-over-Threshold Factors

Aimed at evaluating their taste contribution, all the polyphenols, caffeine as well as theanine, were quantified in the tea infusion, and rated in their sensory impact based on the ratio of the concentration and the taste threshold of a compound (Scharbert and Hofmann 2005). Calculation of dose-over-threshold (DoT) factors revealed that from the group of catechins and theaflavins, only the concentration of EGCG exceeded its taste threshold concentrations for astringency in the tea infusion by a factor of 1.7 (Table 18.5). In contrast, the concentration of the other catechins and theaflavins was found to be below their taste thresholds. Calculation of the DoT factors for the flavonol-3 glycosides revealed high values for most of these glycosides, among which rutin showed the highest DoT factor of 9652. Although theanine was present in quite huge concentrations, its high threshold concentration of 6000 μM/L (water) ruled out any taste contribution of this amino acid. In contrast, the bitter caffeine showed a DoT factor of 2.0, thus demonstrating the alkaloid as a contributor to tea's bitterness (Table 18.5).

18.5.3 Taste Recombination

To confirm the results obtained so far, an aqueous taste reconstitute was prepared using "natural" concentrations of caffeine as well as those flavonol-3 glycosides and catechins that have been evaluated with DoT factors >0.5 (Table 18.5). Theaflavins were not considered for these experiments, because preliminary investigations already excluded these compounds as taste contributors. The sensory analysis revealed that the taste profile of this taste recombinant did not differ significantly from that of the tea infusion (data not shown). In conclusion, the eight flavonol-3 glycosides evaluated with DoT factors >0.5, catechin, epigallocatechin-3-gallate, and caffeine have been successfully identified as the key taste compounds of the Darjeeling tea infusion (Scharbert and Hofmann 2005).

To validate the correct identification and quantitative analysis of all the key odorants and tastants in the Darjeeling tea infusion, a full flavor recombinant was made from the 22 most important odorants and tastants in their "natural" concentrations (Table 18.6). After adjusting the color of the recombinant solution to that of the Darjeeling tea infusion by adding small amounts of sugar couleur, a trained sensory panel was asked to identify the artificial flavor recombinant in a triangle of the recombinant and two samples of the authentic tea infusion. Because eight out of 10 subjects were not able to detect the correct sample, the full flavor recombinant was not significantly different from the authentic tea sample, thus demonstrating that all the key odor- and taste-active molecules in the Darjeeling tea infusion had been successfully identified.

TABLE 18.6
Full Flavor Recombinant of a Darjeeling Tea Infusion Containing 11 Odorants and 11 Tastants

Odorant[a]	Conc. (μg/L)
R/*S*-Linalool	142
Geraniol	142
2-Methylbutanal	82
2-Methylpropanal	69
Phenylacetaldehyde	57
3-Methylbutanal	42
(*E*,*E*)-2,4-Decadienal	2.9
β-Lonone	1.5
(*E*,*E*,*Z*)-2,4,6-Nonatrienal	1.1
(*E*,*Z*)-2,6-Nonadienal	0.6
3-Methyl-2,4-nonandione	0.5
Tastant[a]	**Conc. (mg/L)**
Caffeine	176.2
Epigallocatechin-3-gallate	150.2
Catechin	64.1
Q-3-*O*-[α-L-rha-(1→6)-β-D-glc] (rutin)[b]	6.8
M-3-*O*-β-D-glc[b]	4.7
K-3-*O*-[α-L-rha-(1→6)-β-D-glc][b]	3.9
M-3-*O*-β-D-gal[b]	3.2
Q-3-*O*-β-D-glc[b]	2.9
Q-3-*O*-β-D-gal[b]	2.6
Q-3-*O*-[β-D-glc-(1→3)-*O*-α-L-rha-(1→6)-*O*-β-D-gal][b]	2.6
K-3-*O*-β-D-glc[b]	2.3

[a] The 11 odorants and 11 tastants were dissolved in 1 L water (pH) and the color of the solution was adjusted to that of the Darjeeling tea by adding small amounts of sugar couleur before sensory analysis.

[b] Q, quercetin; K, kaempferol; M, myricetin; A, apigenin; rha, rhamnose; glc, glucose; gal, galactose.

18.5.4 Modulation of Caffeine Bitterness by Flavonol-3-*O*-Glycosides

Although the flavonol-3 glycosides did not exhibit any bitter taste on their own, recent studies demonstrated that the omission of these compounds from the taste recombinate led to a reduction of the bitterness intensity by about 50% (Scharbert and Hofmann 2005). Aimed at investigating the molecular drivers for the bitterness of black tea, the ability of flavanol-3 glycosides to modulate the bitter taste intensity of caffeine was studied. Human dose/response functions were recorded for EGCG and caffeine as well as for a solution of caffeine containing rutin in its natural ratio of 90:1 (Figure 18.14). Measurement of the bitter intensity of these solutions in 1:2

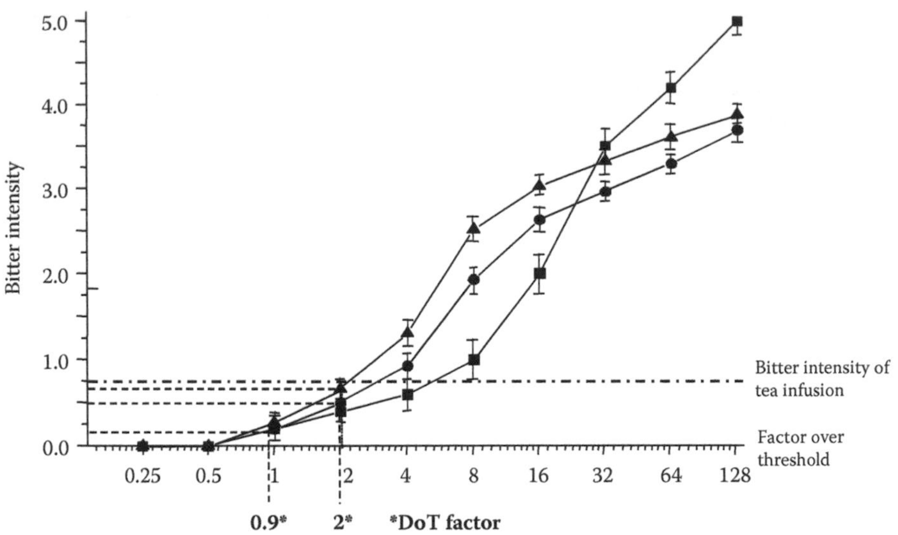

FIGURE 18.14 Human dose/response functions for bitterness of aqueous solutions of EGCG (■), caffeine (●), and caffeine + rutin (90/1; ▲).

dilutions revealed that at a level of 128-fold over the threshold, the EGCG exhibited the most intense bitter taste with a score of 5.0. However, at the DoT factor of 0.9 determined for EGCG in tea, the bitter perception was just at the recognition threshold judged with an intensity score of <0.2. In comparison, the caffeine solution at the DoT-level (2.0) induced a bitter perception with an intensity of 0.45. In the presence of "natural" amounts of rutin, the intensity was significantly increased from 0.45 to 0.70 (Figure 18.14). On that 5-point scale, the bitterness of the caffeine/rutin solution nearly matched that of the tea infusion (0.75), thus demonstrating that the key bitter principles had been successfully identified.

^{1}H NMR titration studies revealed a heteroassociation of caffeine and quercetin-3-*O*-glycosides such as rutin by means of a face-to-face π,π-stacking complex formation as observed by the anisotropic shielding of the aromatic protons H-C(6) and H-C(8) of the A-ring of the polyphenol and the anisotropic shielding of H-C(8) and the methyl protons of caffeine (Figure 18.15). Whether such 1:1 complexes do function as a high-affinity ligand for a TAS2R bitter taste receptor, or whether the caffeine-induced TAS2R activation is amplified by nonbitter-tasting quercetin-3-*O*-glycosides by means of an allosteric modulation mechanism, needs to be clarified in future investigations.

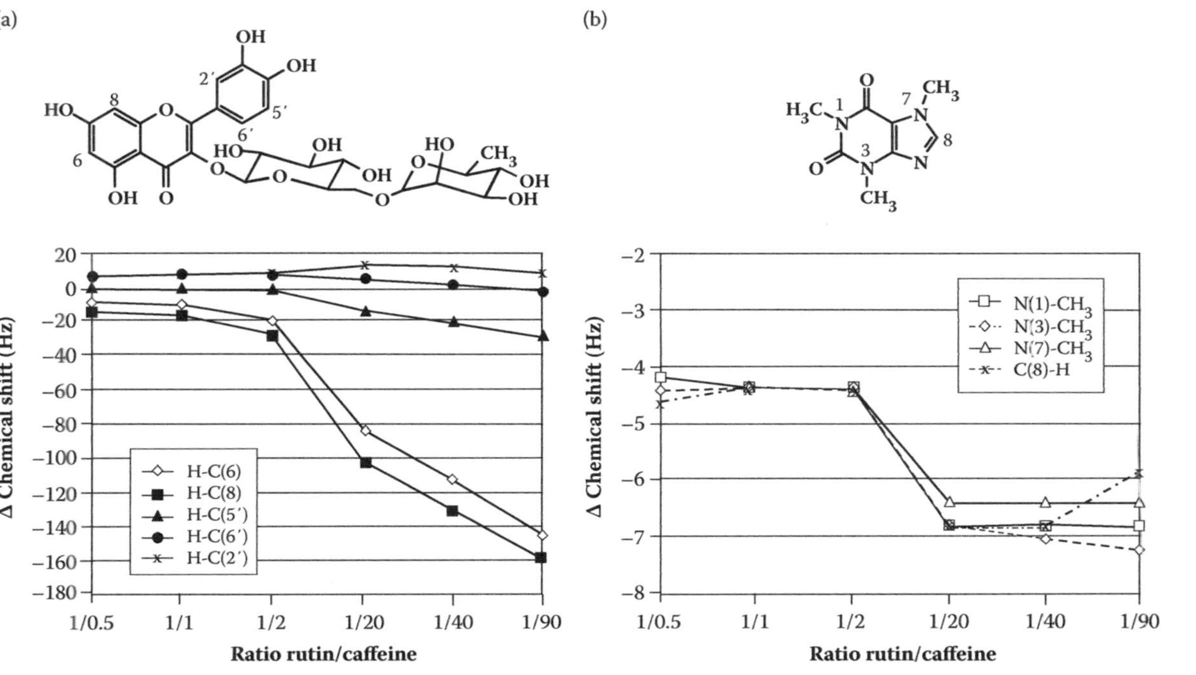

FIGURE 18.15 ^{1}H NMR titration studies (400 MHz) on solutions of rutin with increasing concentrations of caffeine in NaD_2PO_4/Na_2DPO_4 (0.1 M, pD 6.0) at 25°C. Influence of rutin/caffeine ratio on chemical shifts of selected protons is given for rutin (a) and caffeine (b), respectively.

REFERENCES

Acree, T., R.M. Butts, R.R. Nelson, and C.Y. Lee. 1976. Sniffer to determine the odor of gas chromatographic effluents. *Anal. Chem.* 48(12): 1821–1922.

Bautista, D.M., J. Siemens, J.M. Glazer, P.R. Tsuruda, A.I. Basbaum, Ch.L. Stucky, S.-E. Jordt, and D. Julius. 2007. The menthol receptor TRPM8 is the principal detector of environmental cold. *Nature* 448: 204–208.

Bautista, D.M., Y.M. Sigal, A.D. Milstein, J.L. Garrison, J.A. Zorn, P.R. Tsuruda, R.A. Nicoll, and D. Julius. 2008. Pungent agents from Szechuan peppers excite sensory neurons by inhibiting two-pore potassium channels. *Nat. Neurosci.* 11: 772–779.

Buettner, A., A. Beer, C. Hannig, M. Settles, and P. Schieberle. 2002. Physiological and analytical studies on flavor perception dynamics as induced by the eating and swallowing process. *Food Qual. Pref.* 13: 497–504.

Buettner, A., and P. Schieberle. 1999. Characterization of the most odor-active volatiles in fresh, hand-squeezed juice of grapefruit (*Citrus paradisi* Macfayden). *J. Agric. Food Chem.* 47: 5189–5192.

Buettner, A., and P. Schieberle. 2000. Exhaled odorant measurement (EXOM)—A new approach to quantify the degree of in-month release of food aroma compounds. *Lebensm.-Wiss. Technol.* 33: 553–559.

Buettner, A., and P. Schieberle. 2001. Evaluation of key aroma compounds in hand-squeezed grapefruit juice (*Citrus paradisi* Macfayden) by quantitation and flavor reconstitution experiments. *J. Agric. Food Chem.* 49: 1358–1363.

Buhr, K., C. Pammer, and P. Schieberle. 2010. Influence of water on the generation of Strecker aldehydes from dry processed foods. *Eur. Food Res. Technol.* 230: 375–381.

Cartoni, C., K. Yasumatsu, T. Ohkuri, N. Shigemura, R. Yoshida, N. Godinot, J. le Coutre, Y. Ninomiya, and S. Damak. 2010. Taste preference for fatty acids is mediated by GPR40 and GPR120. *J. Neurosci.* 30: 8376–8382.

Czepa, A., and T. Hofmann. 2003. Structural and sensory characterization of compounds contributing to the bitter off-taste of carrots (*Daucus carota* L.) and carrot puree. *J. Agric. Food Chem.* 51: 3865–3873.

Dunkel, A., J. Köster, and T. Hofmann. 2007. Molecular and sensory characterization of γ-glutamyl peptides as key contributors to the kokumi taste of edible beans (*Phaseolus vulgaris* L.). *J. Agric. Food Chem.* 55: 6712–6719.

Engel, W., W. Bahr, and P. Schieberle. 1999. Solvent assisted flavour evaporation—A new and versatile technique for the careful and direct isolation of aroma compounds from complex food matrices. *Eur. Food Res. Technol.* 209: 237–241.

Chandrashekar, J., K.L. Mueller, M.A. Hoon, E. Adler, L. Feng, W. Guo, C.S. Zuker, and N.J. Ryba. 2000. T2Rs function as bitter taste receptors. *Cell* 100: 703–711.

Czerny, M., M. Christlbauer, A. Fischer, M. Granvogl, M. Hammer, C. Hartl, N. Moran Hernandez, and P. Schieberle. 2008. Re-investigation on odour thresholds of key food aroma compounds and development of an aroma language based on odour qualities of defined aqueous odourant solutions. *Eur. Food Res. Technol.* 228: 265–273.

Ding, Z., S. Kuhr, and U.H. Engelhardt. 1992. Influence of catechins and theaflavins on the astringent taste of black tea brews. *Z. Lebensm.-Unters.-Forsch.* 195: 108–111.

Ekborg-Ott, K.H., A. Taylor, and D.W. Armstrong. 1997. Varietal differences in the total and enantiomeric composition of theanine in tea. *J. Sci. Food Agric.* 45: 353–363.

Frank, O., H. Ottinger, and T. Hofmann. 2001. Characterization of an intense bitter-tasting 1*H*,4*H*-quinolizinium-7-olate by application of the taste dilution analysis, a novel bioassay for the screening and identification of taste-active compounds in foods. *J. Agric. Food Chem.* 49: 231–238.

Greger, V., and P. Schieberle. 2007. Characterization of the key aroma compounds in apricots (*Prunus armeniaca*) by application of the molecular sensory science concept. *J. Agric. Food Chem.* 55: 5221–5228.

Hilton, P.J., and R.Z. Ellis. 1972. Estimation of the market value of central African tea by theaflavin analysis. *J. Sci. Food Agric.* 23: 227–232.

Hinman, H., H. Chuang, D.M. Bautista, and D. Julius. 2006. TRP channel activation by reversible covalent modification. *Proc. Natl. Acad. Sci. U.S.A.* 103: 19564–19568.

Laing, D.G., and G.W. Francis. 1989. The capacity of humans to identify odors in mixtures. *Physiol. Behav.* 46: 809–814.

Li, X., L. Staszewski, H. Xu, K. Durick, M. Zoller, and F. Adler. 2002. Human receptors for sweet and umami taste. *Proc. Natl. Acad. Sci. U.S.A.* 99: 4692–4696.

Liu, L., and S.A. Simon. 2000. Capsaicin, acid and heat-evoked currents in rat trigeminal ganglion neurons: Relationship to functional VR1 receptors. *Physiol. Behav.* 69: 363–378.

Millin, D.J., D.J. Crispin, and D. Swain. 1969. Nonvolatile components of black tea and their contribution to the character of the beverage. *J. Agric. Food Chem.* 17: 717–721.

Ohsu, T., Y. Amino, H. Nagasaki, T. Yamanaka, S. Takeshita, T. Hatanaka, Y. Maruyama, N. Miyamura, and Y. Eto. 2010. Involvement of the calcium-sensing receptor in human taste perception. *J. Biol. Chem.* 285: 1016–1022.

Poisson, L., and P. Schieberle. 2008a. Characterization of the most odor-active compounds in an American bourbon whisky by application of the aroma extract dilution analysis. *J. Agric. Food Chem.* 56: 5813–5819.

Poisson, L., and P. Schieberle. 2008b. Characterization of the key aroma compounds in an American bourbon whisky by quantitative measurements, aroma recombination, and omission studies. *J. Agric. Food Chem.* 56: 5820–5826.

Roberts, E.A.H., and R.F. Smith. 1961. Spectrophotometric measurements of theaflavins and thearubigins in black tea liquors in assessments of quality in teas. *Analyst* 86: 94–98.

Rothe, M. 1975. In *Aroma Research*, Proceedings of the International Symposium on Aroma Research, ed. H. Maarse and P.J. Groenen, 111–121. Berlin: Akademie-Verlag.

Scharbert, S., and T. Hofmann. 2005. Molecular definition of black tea taste by means of quantitative studies, taste reconstitution, and omission experiments. *J. Agric. Food Chem.* 53: 5377–5384.

Scharbert, S., N. Holzmann, and T. Hofmann. 2004. Identification of the astringent taste compounds in black tea infusions by combining instrumental analysis and human bioresponse. *J. Agric. Food Chem.* 52: 3498–3508.

Scharbert, S., M. Jezussek, and T. Hofmann. 2004. Evaluation of the taste contribution of theaflavins in black tea infusions using the taste activity concept. *Eur. Food Res. Technol.* 218: 442–447.

Schieberle, P. 1995. New developments in methods for analysis of volatile compounds and their precursors. In *Characterization of Food: Emerging Methods*, ed. A.G. Goankar, 403–431. Amsterdam, The Netherlands: Elsevier Science BV.

Schuh, S., and P. Schieberle. 2005. Characterization of (*E,E,Z*)-2,4,6-nonatrienal as a character impact aroma compound of oat flakes. *J. Agric. Food Chem.* 53: 8699–8705.

Schuh, S., and P. Schieberle. 2006. Characterization of the key aroma compounds in the beverage prepared from Darjeeling black tea: Quantitative differences between tea leaves and infusion. *J. Agric. Food Chem.* 54: 916–924.

Schwarz, S., and. T. Hofmann. 2007. Isolation, structure determination, and sensory activity of mouth-drying and astringent nitrogen-containing phytochemicals isolated from red currants (*Ribes rubrum*). *J. Agric. Food Chem.* 55: 1405–1410.

Soldo T., I. Blank, and T. Hofmann. 2003. (+)-(S)-Alapyridaine—A general taste enhancer? *Chem. Sens.* 28: 371–379.

Soldo, T., O. Frank, H. Ottinger, and T. Hofmann. 2004. Systematic studies of structure and physiological activity of alapyridaine. A novel food-borne taste enhancer. *Mol. Nutr. Food Res.* 48: 270–281.

Söllner, K., and P. Schieberle. 2009. Decoding the key aroma compounds of a Hungarian-type salami by molecular sensory science approaches. *J. Agric. Food Chem.* 57: 4319–4327.

Teranishi, R., R.G. Buttery, and D.G. Guadagni. 1974. Odor quality and chemical structure in fruit and vegetable flavors. *Ann. N. Y. Acad. Sci.* 237: 209–216.

Toelstede, S., A. Dunkel, and T. Hofmann. 2009. A series of kokumi peptides impart the long-lasting mouthfulness of matured Gouda cheese. *J. Agric. Food Chem.* 57: 1440–48.

Ueda, Y., T. Tsubuku, and R. Miyajima. 1994. Composition of sulfur-containing components in onion and their flavor characters. *Biosci. Biotechnol. Biochem.* 58: 108–110.

Ueda, Y., M. Yonemitsu, T. Tsubuku, M. Sakaguchi, and R. Miyajima. 1997. Flavor characteristics of glutathione in raw and cooked foodstuffs. *Biosci. Biotechnol. Biochem.* 61: 1977–1980.

Wagner, R., R.M. Czerny, J. Bielohradsky, and W. Grosch. 1999. Structure–odour-relationships of alkylpyrazines. *Z. Lebensm.-Unters.-Forsch.* 208: 308–316.

19 Methods for Sensory Analysis

Renata Zawirska-Wojtasiak

CONTENTS

19.1 INTRODUCTION

Flavor cannot be measured directly by instruments; it is an interaction between food and the consumer (von Sydow 1971, cited in Piggott 1995). Therefore, sensory methods must be used to measure flavor. According to Wikipedia (http://en.wikipedia.org/wiki/Sensory_analysis), "sensory analysis (or sensory evaluation) is a scientific discipline that applies principles of experimental design and statistical analysis to the use of human senses (sight, smell, taste, touch, and hearing) for the purposes of evaluating consumer products." In literature, there are other similar definitions. Sensory analysis of food relies on evaluation through the use of our senses (odor, taste, taction, temperature, pain, etc.) (Jellinek 1985), or it refers to the technique that highlights and describes organoleptic properties of a product by the sense organs (Majou 2001). Sensory analysis and sensory evaluation are the terms often used interchangeably, but according to Majou (2001), there is some difference between them. The discipline requires panels of human assessors, on whom the products are tested, and recording the responses made by them. Statistical techniques, which are applied to the results, may bring about important information on product quality or consumers' attitude toward them. Sensory analysis or sensory evaluation was formerly called organoleptic testing. However, this name is outdated, because sensations are received by the senses, not by the organs and it attempts to measure stimuli rather than just detect them. The sensory action is difficult to analyze because these molecules interact with saliva and with various other compounds present in food. Accordingly, the flavor profile of a particular food cannot be reduced to its chemical composition. Humans have used their senses to evaluate food for several thousands of years. According to Lawless (1991), the human sense of smell is the ultimate discriminator of food aroma and flavor quality: instrumental analysis is a poor substitute. Even in cases in which chemical components of food flavors have been identified, these must be cross-referenced against human sensitivities in order to estimate their sensory impact. Sensory evaluation describes product characteristics that are of great importance to consumer acceptance and, therefore, is significant to the development process of product quality. Flavor (smell and taste), particularly smell, dominates our life, in contrast to the sense of vision, as it is commonly believed. It affects our emotions and behavior, and makes life pleasant or disgusting. Smell reaches our memory and emotions more directly than the other senses.

Thus, sensory analysis introduced the human element to the monitoring of food quality; after all, it is people who consume the food (McEwan 1991). Optimization of sensory quality of a food product is very complicated and expensive, and for this reason, the development of methods leading to a rapid and cost-effective solution is very important. A combination of assessments by expert and trained assessors with consumer tests is the most effective way to product optimization (Pokorny 1991). This suggests the division of all sensory methods into two main general categories: the analytical and the consumer (or hedonic) approaches.

Hedonic approach describes consumer acceptance of a specific product, consumer preference for one product over the others, and sometimes the degree of pleasure caused by the product.

Analytical approach, as performed by a trained panel, describes the characteristics of a product, which explain the consumer's choice.

There is simultaneously an important link and an important difference between these two approaches, which together with particular methods used in both approaches, will be discussed later on.

Sensory assessment has been an integral part of the food industry for many years and there is a growing need for more scientific methods of assessment, which are well-replicated, objective, unbiased, and which can be applied routinely across a wide range of foods. Precise sensory analysis of food was initiated around the 1940s in the Scandinavian countries with the development of the triangle test, a difference test method. At the same time, independent and analogous studies were being carried out in the United States. European countries started to use modern sensory analysis as early as the 1950s. The first book on sensory analysis was written in 1957 by Tilgner, in Polish language and translated into Eastern European languages (Czech, Hungarian, and Russian). The second book (Masuyama and Miura 1962) was written in Japanese. The third, the standard work by Amerine, Pangborn, and Roessler, was published in 1965 (Jellinek 1985). Since then, numerous sensory methods have been developed. The classification of various sensory methods may somehow be a

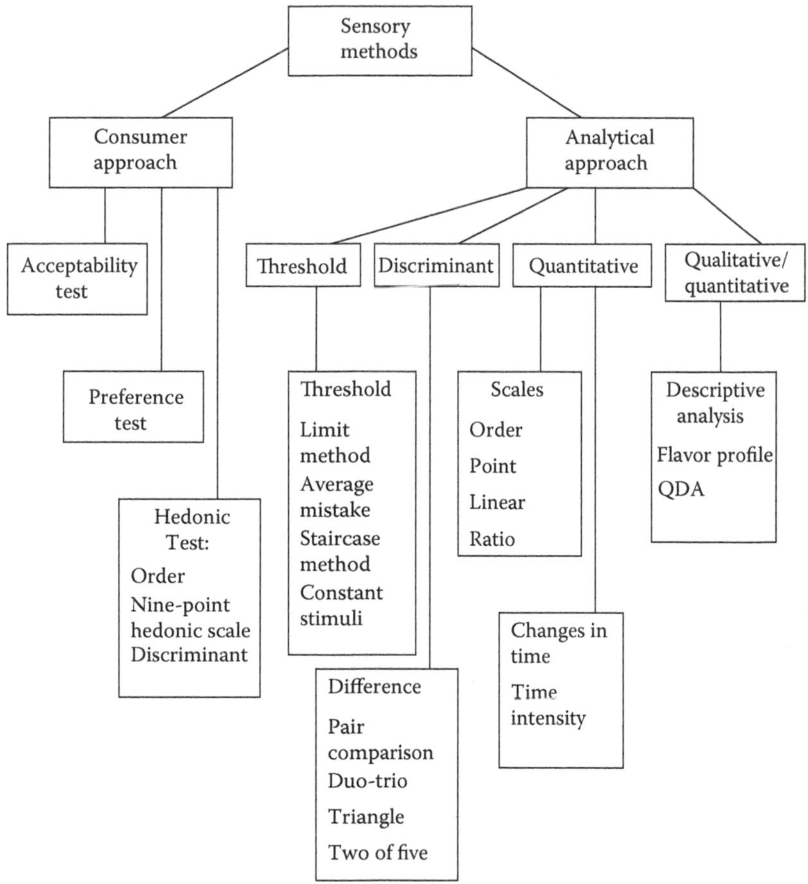

FIGURE 19.1 Possible division of the sensory evaluation methods. (Source: Author's own study, unpublished.)

matter of discussion, but the possible division of the sensory evaluation methods is presented in Figure 19.1.

19.2 CONSUMER APPROACH

The ultimate evaluation of any food product is the evaluation by the consumer during consumption (McEwan 1991). For a consumer evaluation, we need a consumer panel, which constitutes a representative group of the target population or of a market segment. It is a fixed sample of consumers, persons who used a product, episodically participating in hedonic tests, but have never taken part in discriminatory or descriptive tests (Majou 2001). Knowledge on consumer responses to sensory quality of food products is of great importance for understanding food choice. The results after the data collection have to be in some way related to sensory attributes of the product evaluated by the trained panel. That is the major achievement of sensory testing.

For the consumer testing—affective sensory evaluation—selected test methods can be applied, which are generally categorized into two groups: preference and acceptability (hedonic) evaluation. They must be simple, readily available, well-understandable, and not requiring specialist knowledge on sensory testing.

A paired comparison test can be used to estimate consumer preference. Two samples are presented and the consumer has to point out which of the two he or she likes more (prefers).

In the ranking test, the consumer ranks several samples—more than two, according to the degree of liking.

The hedonic test required estimation of the degree of liking for the product using a nine-point scale, but the method does not require training. The scale extends from "dislike extremely" (1), through "dislike very much" (2), "dislike moderately" (3), "dislike a little" (4), "neither not like nor dislike" (5), "like a little" (6), "like moderately" (7), "like very much" (8) to "like extremely" (9).

Consumer testing might be performed outside the laboratory, for example, in shopping centers, markets, restaurants, cafeterias, or at the consumer's home. The latter option sometimes might be of great value for extracting information. Knowledge on consumer food habits, food choice, and consumer attitudes to products are of great importance in determining relation between market information and sensory quality, which can be expressed in product specification. The links between consumer and sensory quality evaluation brings about the product development and optimization.

Consumer acceptance or preference depends on various reasons related both to the products and to consumers—physiological, psychological, and economical conditions. An important condition is the number of respondents as well as their characteristics: age, sex, income, place of living, family structure, also nationality, religion, local preferences, etc.

According to Lawless and Claassen (1993), there are three classes of questions asked by product developers: (1) Is the product liked by consumers? (2) Is the new product perceivably different from the standard? (3) What are its important points of difference? To answer such questions, appropriate analytical sensory methods have to be selected.

19.3 ANALYTICAL APPROACH

19.3.1 Threshold Tests

The sensory threshold of smell and taste refers to the minimum intensity of a stimulus (flavor substance), at which it can be detected by these senses. It is characteristic of the substance and may vary from person to person.

Three or even more different types of threshold values have to be considered: the detection threshold and/or recognition threshold, difference threshold, and the terminal threshold. Detection and recognition threshold are sometimes considered together as a combined meaning.

Detection threshold means the minimum intensity of stimuli at which it can be detected.

Recognition threshold means the minimum intensity of stimuli at which it can be recognized.

Difference threshold means just noticeable difference of the intensity between two stimuli of the same type.

Terminal threshold means a certain concentration above which an increasing concentration may no longer be perceived as different.

The threshold value for odorants is not easy to establish precisely. There are several methods used to perform this task, and the variation in threshold value may be related to the method used. According to Wysocki and Wise (2004), in virtually all available literature sources on odor sensitivity, the term "threshold" has no necessary relationship to the internal state of an organism; it rather refers to the concentration that produces some criteria for the level of performance on a given task. Accordingly, one must consider thresholds in light of the methods used to obtain them.

Four most frequently used methods could be distinguished (Figure 19.1) for the threshold-establishing process: limit method, average mistake, staircase method, and constant stimuli (Baryłko-Pikielna and Matuszewska 2009). However, tasks that subjects have to attempt are rather only of two types: "Yes–No" and "Forced-Choice" (Wysocki and Wise 2004). In the "Yes–No" model, the subject has to decide "yes" if he/she perceives the stimuli, or "no" if not. Such a model is used in the "limit method" or "staircase method." "Forced-Choice" experiments extract the identification of the sample with odorants against two or more blanks, whether the subject can distinguish the sample with odorants from the blanks. The subjects are forced to do it. According to Lawless and Heymann (1998), ascending forced-choice procedures are widely used techniques for threshold measurements for taste and smell. Classically, in the sensory research, a flavor threshold is estimated as the geometric mean of all panelists' thresholds. Very often "Forced-Choice" is performed as a triangle test (two blanks, one sample with an odorant) to establish the odor threshold. The triangle test seems to be the most powerful tool in threshold estimation of components in water solution. Santos et al. (2003) detected in this way the threshold of off-flavors in milk caused by proteolysis and lipolysis. The triangle test was also useful in a study conducted by Lűntzel et al. (2000), who established odor thresholds of 3-mercapto-2-methylpentanol diastereomers and enantiomers. Two beakers of water and one with the compound in the order of decreasing concentrations were presented to 15 judges. In another paper, odor threshold testing was established using the 1-butanol ascending staircase, two-bottle, and force-choice method in a study of odor memory over human lifespan (Lehrner et al. 1999). When thresholds in water are estimated, special containers have to be applied, instead of beakers or flasks. Teflon squeeze bottles with short-length Teflon tubing can be proposed in such a case (Buttery 1999).

Threshold values for different odorants vary significantly; they may be as low as 0.00000002 ppm for *p*-ment-1-ene-8-tiol (grapefruit flavor), or as high as 35 ppm for maltol (caramel-like aroma) (Rothe et al. 1987). The importance of a particular component to food aroma may be determined by the ratio of the component concentration in the food to its threshold in water. It can be expressed in aroma values, odor units, or odor activity values (OAVs) (Rothe and Thomas 1963; Teranishi et al. 1991; Teranishi 1997).

According to Pyysalo et al. (1977), OAV eliminates the synergistic and antagonistic influence of volatiles in the composition, as well as the influence of a medium on aroma. However, the medium sometimes has to be taken into consideration and threshold values are elaborated for the components in the medium, particularly in terms of the accepted or rejected threshold. The sensory threshold concept was useful for the dairy industry in determining the minimum concentration of an off-flavor compound that would be detected in pasteurized milk by consumers (Lawless and Heymann 1998). In the experiment performed by Saliba et al. (2009), the consumer rejection threshold (CRT) was measured for 1,8-cineol in red wine. Replicate paired comparison tests were conducted for each concentration level, using samples spiked with an increasing concentration of cineol. The negative flavor of eucalyptus may be different even for individual red wines and the CRT will depend on other masking/antagonistic as well as complementary/synergistic flavors in a particular wine.

From the threshold determined in water, the threshold in air or in oil–water mixture can be calculated (Buttery 1999), when we know the air-to-water partition coefficient or air-to-oil partition coefficient, respectively, of the compound in the water solution. The threshold in air may also be measured with an olfactometer, but recently new methods using gas chromatography (GC) have been developed. These methods may be performed by sniffing the effluent from the end of the GC column with regard to all separated compounds, even those not identified (Acree et al. 1984; Grosch and Schieberle 1988). New aspects of these methods were developed later on (Grosch 1993; Grosch et al. 1994; Acree and Barnard 1994) to establish key aroma compounds of the particular food. The methods now known as aroma extract dilution analysis or static headspace analysis and olfactometry (Gutz and Grosch 1999) rely mostly on sniffing the GC effluent after separation of the sample injected at concentrations reduced stepwise, until no odorant is detectable at the sniffing port.

19.3.1.1 Discrimination Tests

Discrimination tests are the sensory analysis tools for determining the perception of a difference between two or more objects with respect to certain characteristics.

19.3.2 Difference Testing

Sensory difference tests are very popular methods in sensory analysis. They are used as a simple and comparative procedure in sensory discriminability of similar types of stimuli. They are used mostly to detect small differences in sensory quality of food products that might be an effect of changes in the technological processes, storage time, new additives, or packaging methods. The method may be performed as paired difference test, a triangle test and duo–trio test, a multiple comparison procedure, and an ordinal test. Difference tests are used as analytical methods, but also in consumer testing. The panel members or consumers analyze whether there is a difference between samples, often between two samples only. In paired comparison tests, the panelists are asked to indicate which of the two samples has more of the specified characteristic. It is rather simple and sensitive, but the probability is only $p = 50\%$. The triangle test is the most widely used of all different tests for this problem. In this test, three coded samples are presented to every panelist, two of which are alike, whereas the third is different. Assessors are asked to find two equal samples. A possible questionnaire for the triangle test may be similar to that in Figure 19.2. The number of corrected answers is counted. If the required number of assessors according to the statistical tables finds a difference, it can be concluded that the samples are different. Expanded statistical tables are compiled for use in determining significance in paired-difference, triangle, and paired-preference tests, which are convenient for a rapid estimation of significance in the interpretation of laboratory sensory data as well as consumer responses (Roessler et al. 1978). Usually a 5% significance level is adopted. The test requires about 30 individual notes (Baryłko-Pikielna and Matuszewska 2009). The correct use of statistical methodology in sensory analysis will greatly improve the interpretation of the obtained sensory data.

Despite the statistical significance, the sensitivity of some assessors to a particular product or property should be taken into account, as it might vary. It is evident when tests are conducted in replications. Sensitive assessors will identify the odd sample each time. A number of such assessors might be of much more importance than the data obtained by a large group of nonselected panelists. There is also a reason for an extended triangle test to define the degree or character of the difference, particularly in the research but in the industrial practice as well. These methods may be used in the "two of five" version, where two objects are compared, but one of them is presented twice (two samples), whereas the other is in three replicates. However, this form is much more complicated and tiring for panelists, and as such, it is rarely used (Baryłko-Pikielna and Matuszewska 2009).

The duo–trio test is a kind of the triangle test from the point of view of the number of samples presented to assessors, as there are three samples. The distinction is that one sample is represented by the standard (reference mode). The assessor has to point out which of the two others samples are equal to the standard. The test may

TRIANGLE TEST

Product........
Data............ Name.................................

There are three triangles.
Try and estimate (taste, aroma, . . .) of three presented samples in every triangle. Two of them are alike, one is different. Circle the odd or two equal samples. Mark by "+" the more desired – the odd one or the identical two.

NUMBER OF TRIANGLE	SAMPLE CODE
I	23+ (56) 31+
II	5 22 (4+)
III	(47+) (12+) 32

FIGURE 19.2 Possible questionnaire for triangle test with possible answers. (Source: Author's own study, unpublished.)

be performed in two versions: as "the constant reference mode" or as "the balance reference mode." According to Lyon et al. (1992) (cited in Baryłko-Pikielna and Matuszewska 2009), the minimum number of assessors has to be either 32 if they are untrained persons, or 20 trained panelists.

In the ordinal test, there are several samples evaluated (three or more) and the judges are asked to order them according to the intensity of the specified attribute (from the lowest to the highest or vice versa). The subjects have to be well-informed before the evaluation about the direction of ordering (growing or decreasing), as well as about the criterion of ordering (attribute). The evaluation is rather simple and rapid, but it does not consider the degree of the differences. The data obtained by these methods are relative, as they regard only a specified group of compared samples, but using the method, it is possible to distinguish between the samples characterized by a rather small sensory difference (Baryłko-Pikielna et al. 2004). The data may be analyzed by means of nonparametric statistics to establish the statistical significance of the differences between samples.

The multiple comparison procedure is a kind of the difference test, which makes it possible to estimate the degree of the difference. It might be sometimes of importance from the technological point of view, when monitoring food quality particularly during storage. A number of samples, including a coded standard, are compared with a designated standard for a specific attribute. The size of the difference is expressed on a nine-category scale with five verbal anchors: none, very slight, slight, moderate, and large. The method makes it possible to find a difference between analyzed samples, as well as to point out the direction of changes—desired or undesired—on the basis of interpreted statistics. The test is performed in several repetitions to obtain 20–50 individual notes for every sample compared with the control.

19.3.3 Quantitative Methods

There is a special problem to overcome when quantitative methods are used in sensory measurements. This is because an attempt has to be made to express sensory sensations in mathematical values, that is, it means numerically rating the intensity of a sensory characteristic. However, a sensory sensation is created on the borderline of physiology and psyche, so however well it is perceived, is not easy to describe or measure. The subjects have to score the intensity using specified scales, whereas the types of scales are different.

19.3.3.1 Ordinal (Ranking) Test

A ranking means a method of evaluation as well as a type of scaling. It requires all samples to be present at the same time. It gives little indication of the magnitude of the difference between compared samples. This method is described above with the difference test, so it can be classified in both the difference and quantitative methods, or it is somewhere between these two (Baryłko-Pikielna and Matuszewska 2009).

19.3.3.2 Categories

Category scales consist of several verbal or numerical descriptions ordered according to their importance (hierarchy) of the attribute. Category scales can be used to

TABLE 19.1
Category Scale Schema for Two Selected Attributes (Taste and Odor) of Kefir

	Notes				
Attribute	**5**	**4**	**3**	**2**	**1**
Odor	Very fresh, slightly acidic	Fresh, slightly acidic	Acidic	Strong acidic	Very strong acidic, rancid
Taste	Characteristic, slightly acidic–milky	Slightly acidic–milky	Acidic	Strong acidic, slightly bitter	Very strong acidic, bitter

Source: Author's own study, unpublished.

evaluate intensity of a selected attribute, to estimate the overall quality as well as acceptability. They are constructed in various ways and consist of five, nine (hedonic), ten, or more degrees or levels. However, category scales with very many degrees are rarely used and do not bring about a more precise estimation. In some cases, numerical values are provided with a detailed description of the estimated attribute, when the schema is elaborated for a specified product (Table 19.1), particularly in quality control system. In the table, there are presented schema for two selected attributes (taste and odor) of kefir, so the chapter is concentrated on flavor sensory measurements. The hedonic scale was mentioned before in *Consumer Approach*, so it is most often used to establish consumer acceptance of products. However, elaborated for nine degrees of acceptance, it may be used in the version of seven, five, or even only three degrees. The latter is used in form of a mimic scale for studies with children (Baryłko-Pikielna and Matuszewska 2009).

19.3.3.3 Linear Scale

Linear scales make it possible to estimate the intensity of an attribute without using particular descriptions (numbers) to every degree in category scales. Using numbers might also be inconvenient for some judges. The linear scale means a line of specified length (often 10 cm), anchored on both sides in the following manner: "not detected"–"very intensive," "weak"–"strong," and "not desired"–"very desired." The scales may be structuralized (divided into sections of the same length) or unstructuralized. The latter type of scale is often used in flavor estimation, particularly in aroma profile analysis like in the example presented in Figure 19.3. The subjects mark their notes on the scale, and then the notes are converted into numerical values by measuring the marked lengths. Linear category scales may also be used to generate ordinal data. Linear scales appear to be very useful in cooperative research conducted by various laboratories—even those located in different countries (international studies). In such cases, very similar interpretations of sensory experiments were obtained (Baryłko-Pikielna et al. 1992).

Profile Analysis of Mushroom Aroma

Sample code.............. Data .... Name.............

	Not detectable	Very intensive
1. Mushroom-like		
2. Forest-like		
3. Earthy		
4. Musty		
5. Putrid		
6. Fishy		
7. Meaty		

FIGURE 19.3 An example of unstructuralized scale used in aroma profile analysis (for mushroom aroma in this example). (Source: Author's own study, unpublished.)

19.3.3.4 Ratio Scale

To avoid the same limitations of category scales, which are caused by the limited construction of the scale, the concept of ratio scales was created by Stevens in 1957 and developed later by other authors (Baryłko-Pikielna et al. 1992). It is well known as "magnitude estimation scaling." The subject has to express how many times the sample is more or less intensive than the standard or the sample first presented to him. Panelists choose an arbitrary point on the mental scale and ascribe it a number for the first sample. The other samples are then evaluated with respect to this sample. However simple and rather reliable, some disadvantages of this method are mentioned by Lawless and Malone (1986). It is not more precise in sample differentiation as compared to the other methods.

19.3.3.5 Changes in Time—"Time Intensity"

Generally speaking, sensory sensation is a dynamic phenomenon, thus, it means that it changes in the course of time; a method has been elaborated to measure intensity of the sensation in relation to the time of its perception. The method is known as "time–intensity." It is more related to taste, which changes during mastication in the mouth when eating. The evaluation starts when putting the sample into the mouth. The final point means that the perceived intensity is none. Experiments are performed applying a special device coupled with the computer system. The results are produced in the form of a special curve. The curves of the intensities related to time vary between judges, so, consequently, an average curve is drawn (Van Buuren 1992).

In relation to odor measurement, the idea of a "time–intensity" method may somehow be used in the gas chromatography/olfactometry (GC/O) system, when it is performed with a similar device working together with a gas chromatograph to produce aromagrams, described by Acree et al. (1994) as "charm analysis" or by Lopez et al. (1992), who referred to them as "Osme." For every sniffed odor (a separated substance on the GC column), a particular curve (peak) is obtained: the area under this curve corresponds to the total sensation stimulated by the compound. On the basis of such obtained aromagram, attention is drawn away from sensorially inert components and focused on those that are odor-active.

19.4 QUALITATIVE/QUANTITATIVE DESCRIPTIVE METHODS

One attempt of odor quality measurements is odor categorization to the group of the same similarity. However, the background of this categorization is not clear, thus, different authors perform it according to their own perception ability. One suggested way to explore flavor categories was to use multidimensional scaling and cluster analysis (Lawless 1989). The universally accepted schema should be of great value (Wysocki and Wise 2004).

Descriptive sensory analysis, which is much more comprehensive and useful than straightforward difference testing, provides information on food products. By coupling this with a scoring system, it provides the researcher with a powerful tool for understanding sensory quality (Williams 1982). It represents one of the most sophisticated of the available sensory methods. The main assumption of these methods is that aroma and flavor, similarly to texture, may be described not by one but several different attributes (descriptors). All the possible attributes have to be selected, verbally described, arranged according to the order of their perception and also quantified for their intensity. The main known version of the method is quantitative descriptive analysis (QDA), but previously, the flavor profiling method (FPM) was used. In FPM, the overall intensity of sensation (amplitude) was estimated, together with the intensity of individual descriptors, the order of their perception, and eventually the aftertaste. Panelists estimate samples individually; then during a discussion session directed by the leader, they have to reach some agreement in understanding the descriptions of the flavor attributes of the evaluated product, the number of descriptors, as well as to agree on the order of perception of the different sensations.

In QDA, the main version of descriptive analysis, the estimation to be performed requires special preparation with the development of descriptors using standardized verbal descriptions for flavor sensations and a 100-mm linear scale to measure their intensity, a trained group of the panelists (from several to a dozen), and finally introducing a statistical interpretation. QDA is particularly useful in flavor studies, so its modification in this direction is known as quantitative flavor profiling (QFP). QFP, as a form developed from QDA, is focused on flavor sensations created by food aromas, food products, or even cosmetics such as perfume. Panelists are selected and trained for a particular experiment. Descriptions are almost inherent in profiling, so at the starting point of the procedure, the vocabulary of descriptors has to be developed. This is usually done arbitrarily by the group of several aroma specialists, when a wide range of different quality coded samples are presented. Before this, assessors during several sessions give their own descriptions of perceiving odor attributes and then the list of possible descriptions (odor attributes) is created. The list may be improved and extended if necessary in the next screening sessions. The reference samples are collected. In order to produce proper definitions, multiple standards might be suitable reference samples to define sensory characteristics for descriptive analysis (Ishi and O'Mahony 1991). During training of judges, a set of standards is presented to produce a single standard definition of one descriptor—judges have to reach an agreement in their choice.

In the case of flavor estimation, mostly nonnumerical line scales are applied. The line scales in profile analysis are particularly common when a computing system of collecting and transforming sensory data is used. The main job of assessors is to estimate on the scale the intensity of individual attributes, not their own acceptance, or even independently from their acceptance to these attributes.

Intensive and demanding training of panelists as the next step of the QFP method has to be performed. As compared to QDA, panelists in OFP do not participate in completing the vocabulary of descriptors, but they are trained in several sessions. First, they are presented with the previously elaborated descriptors, then training is focused on the estimation of particular products.

A possible questionnaire used in profile analysis is presented in Figure 19.3. This was used by the author when investigating odor of different varieties of the *Pleurotus* species. The data obtained from these estimations are presented in different possible ways (Figures 19.4, 19.5, and 19.6): bar chart, a spider diagram, and a principal component analysis (PCA) graph. To be presented, data were collected from all panelists and all sessions, which yielded about 30 repetitions for every attribute, then means, medians, standard deviations, standard errors, and ranges were calculated, producing simple descriptive statistics. Examination of mean data may suggest a difference

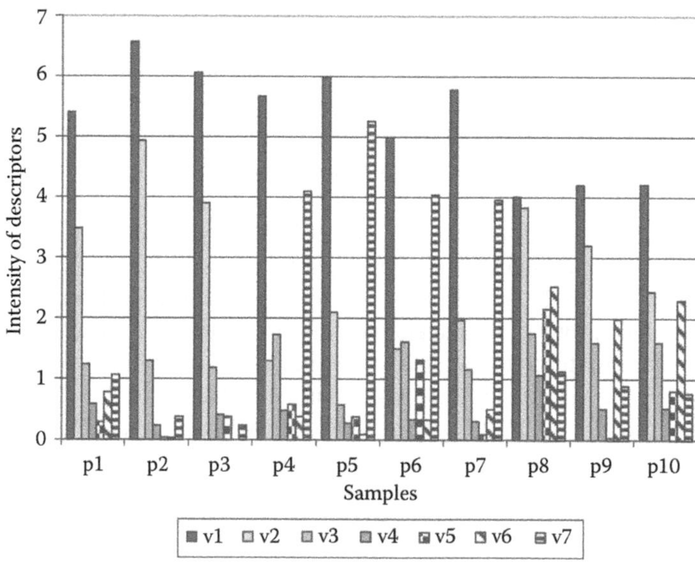

FIGURE 19.4 Bar chart from odor profile analysis of different varieties of the *Pleurotus* species. Samples: p1 P80T (*P. ostreatus*), p2 P80D (*P. ostreatus*), p3 P80D′ (*P. ostreatus*), p4 B62I (*P. djamor*), p5 B62II (*P. djamor*), p6 B123I (*P. djamor*), p7 B123II (*P. djamor*), p8 B70I (*P. citrinopileatus*), p9 B70II (*P. citrinopileatus*), and p10 B83I (*P. citrinopileatus*). Descriptors: v1 mushroom-like, v2 forest-like, v3 earthy, v4 musty, v5 putrid, v6 fishy, and v7 meaty. (Source: Author's own study, unpublished.)

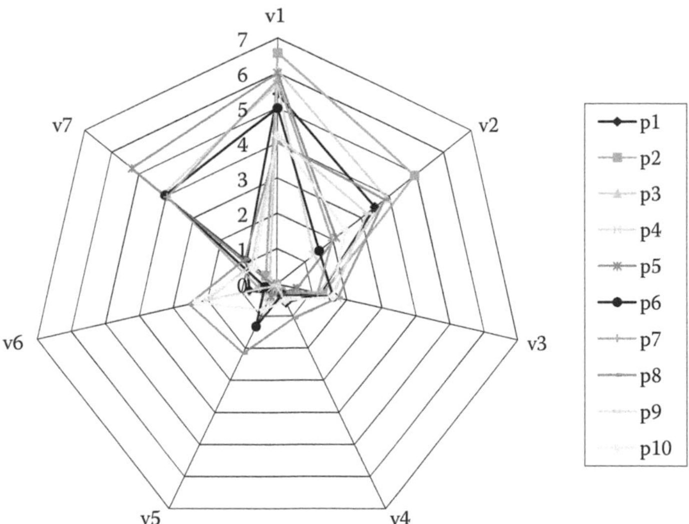

FIGURE 19.5 Spider diagram from odor profile analysis of different varieties of the *Pleurotus* species. Samples: p1 P80T (*P. ostreatus*), p2 P80D (*P. ostreatus*), p3 P80D′ (*P. ostreatus*), p4 B62I (*P. djamor*), p5 B62II (*P. djamor*), p6 B123I (*P. djamor*), p7 B123II (*P. djamor*), p8 B70I (*P. citrinopileatus*), p9 B70II (*P. citrinopileatus*), and p10 B83I (*P. citrinopileatus*). Descriptors: v1 mushroom-like, v2 forest-like, v3 earthy, v4 musty, v5 putrid, v6 fishy, and v7 meaty. (Source: Author's own study, unpublished.)

between samples on a particular attribute, which can be proved by the analysis of variance and particularly the last significance difference for every two samples. Bar charts or spider diagrams are the simplest visual presentation of the relation between samples and descriptors. Multidimensional data analysis in the form of PCA means improved data interpretation, so it is more informative. In Figure 19.6, the sensory differentiation between three strains of *Pleurotus*: *P. ostreatus*, *P. djamor*, and *P. citrinopileatus* is evident. The last one was the most different strain of the three. Recently, PCA is a widely used method for the statistical interpretation of data from various research projects; it is very useful in sensory profiling. It needs to be stressed here that it is also applied to interpreting data from electronic nose measurements, from a mass spectrometer working in the way similar to the electronic nose without separation of volatiles as well as GC analysis of volatiles. It is of some benefit when aroma of a product estimated by these various methods may be interpreted by the same statistical analysis, so it makes it possible to easily compare all these aroma measurements and identify actual sample differentiation. It is particularly interesting when comparing sensory analysis with the instrumental data (Mildner-Szkudlarz et al. 2003). Based on the developed sensory and headspace-solid phase microextraction (HS-SPME)–GC/MS methods, it was possible to distinguish between different oils and oils stored for various periods of time. PCA of chromatographic data was related to PCA sensory profile analysis and similarities in sample clustering were observed.

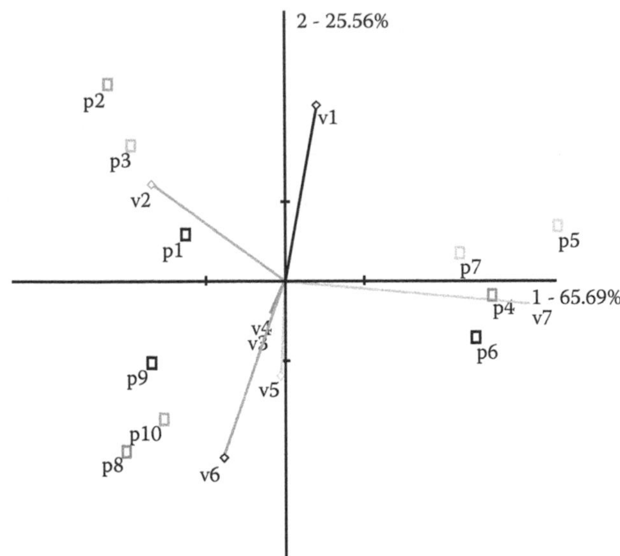

FIGURE 19.6 PCA plot from odor profile analysis of different varieties of the *Pleurotus* species. Samples: p1 P80T (*P. ostreatus*), p2 P80D (*P. ostreatus*), p3 P80D′ (*P. ostreatus*), p4 B62I (*P. djamor*), p5 B62II (*P. djamor*), p6 B123I (*P. djamor*), p7 B123II (*P. djamor*), p8 B70I (*P. citrinopileatus*), p9 B70II (*P. citrinopileatus*), and p10 B83I (*P. citrinopileatus*). Descriptors: v1 mushroom-like, v2 forest-like, v3 earthy, v4 musty, v5 putrid, v6 fishy, and v7 meaty. (Source: Author's own study, unpublished.)

In PCA interpretation of the data from descriptive analysis, samples are described by the mean scores given by the panelists to all sensory descriptors. All samples are represented in a space typically reduced to two dimensions. According to Husson et al. (2005), there might be a problem with stability of the PCA plane, because of low repeatability and reproducibility of sensory measurements. They proposed a confidence ellipse for sensory profiles obtained by PCA. The method adapts the resampling technique, leading to the construction of sensory evaluation of a virtual panel. The number of the assessors in the virtual panel is the same as in the actual panel, but assessors from the virtual panel are sampled by replacement. After statistical elaboration and determination of the PCA plane, finally, the confidence ellipses are built containing 95% of the sensory evaluation of virtual panels. The ellipses describe the variability of sensory evaluations and graphically present the significance of the difference between the products. The number of panelists, which would be necessary to differentiate the profile of two analyzed products, may be predicted using this technique.

Another possibility to use sensory descriptive profile analysis in aroma characteristics is free-choice profiling (FCP). In this variant developed by Williams et al. (cited in Beal and Mottram 1993), sensory descriptive analysis is used without previously elaborating list of attributes. In the FCP, the evaluation may be performed by an untrained panel and without establishing the consensus list of attributes. In several sessions conducted before the analysis, various samples of different sensory quality are presented to the panel. Assessors are asked to smell the samples and describe their odors using their own descriptions. Later on, assessors are asked to reassess the samples and finally to justify their descriptions. At this stage, the same terms might be excluded. In this method, however, a special statistical procedure is needed to elaborate the data obtained in such a way, which facilitates differentiation and description of the evaluated samples. However, there are also some disadvantages of the method; for example, a limited precision as well as some disadvantages in data interpretation and forms of their presentation (Baryłko-Pikielna and Matuszewska 2009).

QFP, or generally QDA, is widely used in sensory studies focused on the impact of various growing, technological, or storage conditions on food flavor and off-flavor (Baryłko-Pikielna et al. 1992; Zawirska-Wojtasiak et al. 1992, 1998, 2000, 2005, 2009; Mildner-Szkudlarz et al. 2003). ODA sensory studies may also concern processes of improving the sensory quality of a food product and its optimization (Stampanoni 1993; Beal and Mottram 1993). The above-mentioned papers more often compare the sensory flavor profile with instrumental data, whereas others do it with consumer studies (Baryłko-Pikielna 1995; Bolini et al. 2009; Cadena and Bolini 2009).

Sometimes "difference profiling" is applied as a version of QDA. In this type of descriptive analysis, the samples are analyzed in relation to the defined standard and this standard is placed at the "0" point of scale. In such a situation, particular odor descriptors are estimated as much or less intensive than the standard. It is the best when the standard is presented to assessors during the evaluation session. The data are presented in the form of a difference graph. Figure 19.7 shows such data for mushrooms of *P. citrinopileatus* and *P. djamor* in relation to *P. ostreatus*. The latter is the most popular species among the *Pleurotus* genus.

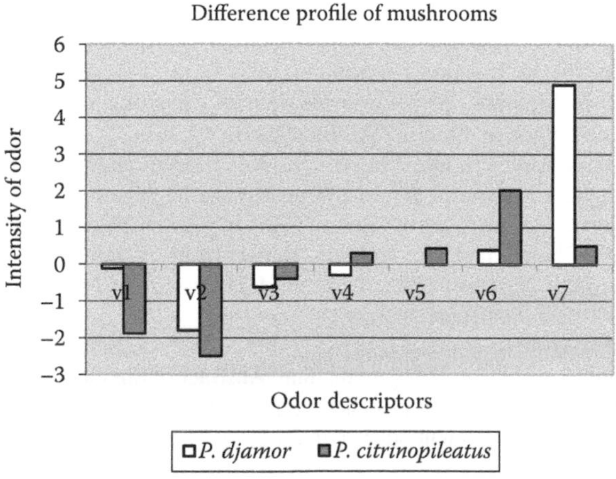

FIGURE 19.7 Difference profile of *P. djamor* and *P. citrinopileatus* in relation to *P. ostreatus*. Descriptors: v1 mushroom-like, v2 forest-like, v3 earthy, v4 musty, v5 putrid, v6 fishy, and v7 meaty. (Source: Author's own study, unpublished.)

19.5 CONCLUSION

This short review of sensory methods related mostly to flavor estimation does not cover all possible tools of sensory measurements. It presents the main sensory methods, which may be applied to estimate food flavor, but the choice of the methods depends on a particular case. Also pointed out in this chapter are the requirements for properly carrying out a sensory analysis: practical experience of the leader, often a trained panel, and the extended knowledge of the methods based on practice and literature.

REFERENCES

Acree, T.E., J. Barnard, and D.G. Cunningham. 1984. A procedure for the sensory analysis of gas chromatographic effluents. *Food Chem.* 14: 273–286.

Acree, T.E., and J. Barnard. 1994. Gas chromatography–olfactometry and charm analysis. In *Trends in Flavour Research*, ed. H. Maarse and D.G. van der Heij. Amsterdam: Elsevier.

Baryłko-Pikielna, N. 1995. Sensoryczna analiza profilowa i ocena konsumencka w opracowaniu nowych produktów żywnościowych (Sensory profile analysis and consumer evaluation in new food products elaboration). In *Food Product Development*, ed. J. Czapski, 207–220. Poznan, Poland: Wydawnictwo Akademii Rolniczej im. A. Cieszkowskiego w Poznaniu.

Baryłko-Pikielna, N., and I. Matuszewska. 2009. Sensoryczne badania żywności. Podstawy-Metody-Zastosowania (Sensory estimation of food. Background–Methods–Applications). Kraków: Wydawnictwo Naukowe PTTŻ.

Baryłko-Pikielna, N., I. Matuszewska, M. Peruszka, K. Kozłowska, A. Brzozowska, and W. Roszkowski. 2004. Discriminability and appropriateness of category scaling versus ranking methods to study sensory preferences in elderly. *Food Qual. Pref.* 15(2): 167–175.

Baryłko-Pikielna, N., R. Zawirska-Wojtasiak, C. Kornelson, and M. Rothe. 1992. Interlaboratory reproducibility of sensory profiling results of margarines. In *Aroma Production and Application*, ed. M. Rothe, and H.P. Cruse, 301–328. Proceedings of the 3rd Wartburg Aroma Symposium, Eisenach, 25–27 February, 1991. Potsdam, Germany: Eigenverlag Deutsches Institut für Ernährungsforschung.

Beal, A.D., and D.S. Mottram. 1993. An evaluation of the aroma characteristics of malted barley by free-choice profiling. *J. Sci. Food Agric.* 61: 17–22.

Bolini, H.M.A., D.B. Arellano, and J.M. Block. 2009. Relationship between sensory profile and acceptance of pecan nut oil during storage. Delegate manual. Abstract of 8th Pangborn Sensory Science Symposium, 26–30 July 2009, Florence, Italy. P1.1.13.

Buttery, R.G. 1999. Flavor chemistry and odor threshold. In *Flavor Chemistry. Thirty Years of Progress*, ed. R. Teranishi, E.L. Wick, and I. Hornstein, 353–356. New York: Kluwer Academic/Plenum Publishers.

Cadena, R.S., and H.M.A. Bolini. 2009. Sensory profile and consumer research of traditional and light vanilla ice cream. Delegate manual. Abstract of 8th Pangborn Sensory Science Symposium, 26–30 July 2009, Florence, Italy. P1.1.22.

Grosch, W. 1993. Detection of potent odorant in foods by aroma dilution analysis. *Trends Food Sci. Technol.* 4: 68–73.

Grosch, W., C. Milo, and S. Widder. 1994. Identification and quantification of odorants causing off-flavour. In *Trends in Flavour Research*, ed. H. Maarse and D.G. van der Heij. Amsterdam: Elsevier.

Grosch, W., and P. Scheberle. 1988. Bread flavour—Qualitative and quantitative analysis. In *Characterization, Production and Application of Food Flavours*, Proceedings of the 2nd Wartburg Aroma Symposium 1987, ed. M. Rothe, 139–152. Berlin: Academie-Verlag.

Gutz, H., and W. Grosch. 1999. Evaluation of important odorants in foods by dilution techniques. In *Flavor Chemistry—Thirty Years of Progress*, ed. R. Teranishi et al., 377–386. New York: Kluwer Academic/Plenum Publishers.

Husson, F., S. Lê, and J. Pagès. 2005. Confidence ellipse for sensory profiles obtained by principal component analysis. *Food Qual. Pref.* 16: 245–250.

Ishi, R., and M. O'Mahony. 1991. Use of multiple standards to define sensory characteristics for descriptive analysis: Aspects of concept formation. *J. Food Sci.* 56(3): 838–842.

Jellinek, G. 1985. *Sensory Evaluation of Food: Theory and Practice*. Chichester, England: Ellis Horwood Ltd.

Lawless, H.T. 1989. Exploration of fragrance categories and ambiguous odors using multidimensional scaling and cluster analysis. *Chem. Sens.* 14(3): 349–360.

Lawless, H.T. 1991. The sense of smell in food quality and sensory evaluation. *J. Food Qual.* 14: 33–60.

Lawless, H.T., and M.R. Claassen. 1993. Application of the central dogma in sensory evaluation. *Food Technol.* June: 139–146.

Lawless, H.T., and H. Haymann. 1998. *Sensory Evaluation of Food Principles and Practices*, New York: Chapman & Hall.

Lawless, H.T., and G.J. Malone. 1986. The discriminative efficiency of common scaling methods. *J. Sens. Stud.* 1(1): 85–98.

Lehrner, J.P., J. Gluck, and M. Laska. 1999. Odor identification, consistency of label use, olfactory threshold and their relationships to odor memory over the human lifespan. *Chem. Sens.* 24: 337–346.

Lopez, R.M., L.M. Libbey, B.T. Watson, and M.R. McDaniel. 1992. Odor analysis of pinot noir wines from grapes of different maturities by gas chromatography–olfactometry technique (Osme). *J. Food Sci.* 57: 985–994.

Lűntzel, Ch.S., S. Widder, T. Vőssing, and W. Pickenhagen. 2000. Enantioselective syntheses and sensory properties of 3-mercapto-methylpentanols. *J. Agric. Food Chem.* 48: 424–427.

Majou, D. 2001. *Sensory Evaluation. Guide of Good Practice*. Paris: Actia.

McEwan, J.A. 1991. Statistical analysis of sensory data. Paper presented on the conference organized by Polish Society of Food Technologists. Poland: Mogilany.

Mildner-Szkudlarz, S., H.H. Jeleń, R. Zawirska-Wojtasiak, and E. Wąsowicz. 2003. Application of headspace-solid phase microextraction and multivariate analysis for plant oils differentiation. *Food Chem.* 83: 515–522.

Piggott, J.R. 1995. Design questions in sensory and consumer science. *Food Qual. Pref.* 6: 217–220.

Pokorny, J. 1991. Sensory problems related to product optimisation. Paper presented at the conference organized by the Polish Society of Food Technologists. Poland: Mogilany.

Pyysalo, T., M. Suihko, and E. Honkanen. 1977. Odour thresholds of major volatiles identified in cloudberry (*Rubus chamaemorus* L.) and arctic bramble (*Rubus arcticus* L.). *Lebensm.-Wiss. Technol.* 10: 36–39.

Roessler, E.B., R.M. Pangborn, J.L. Sidel, and H. Stone. 1978. Expanded statistical tables for estimating significance in paired-preference, paired-difference, duo–trio and triangle tests. *J. Food Sci.* 43: 940–947.

Rothe, M., R. Ruttloff, R. Schrődter, and F.C. Strüber. 1987. Actual trends in food flavouring. In *Characterization, Production and Application of Food Flavours*, Proceedings of the 2nd Wartburg Aroma Symposium, ed. M. Rothe, 7–22. Berlin: Academie-Verlag.

Rothe, M., and B. Thomas. 1963. Aromastoffe in Brot. *Z. Lebensm.-Unters.-Forsch.* 109: 302–310.

Saliba, A.J., J. Bullock, and W.J. Hardie. 2009. Consumer rejection threshold for 1,8 cineole (eucalyptol) in Australian red wine. *Food Qual. Pref.* 20: 500–504.

Santos, M.V., Y. Ma, Z. Caplan, and D.M. Barbano. 2003. Sensory threshold of off-flavors caused by proteolysis and lipolysis in milk. *J. Dairy Sci.* 86: 1601–1607.

Stampanoni, Ch.R. 1993. Metoda ilościowego profilowania smakowitości (Method of Quantitative Flavour Profiling (QFP)). *Przemysł Spożywczy* 10: 277–280 (translated into Polish by Nina Baryłko-Pikielna).

Teranishi, R. 1997. Development and uses of odor threshold. In *Flavor Perception Aroma Evaluation*, Proceedings of the 5th Wartburg Aroma Symposium, Eisenach, ed. H.P. Kruse and M. Rothe, 171–177. Potsdam, Germany: Universitat Potsdam.

Teranishi, R., R.G. Buttery, D.J. Stern, and G. Takeoka. 1991. Use of odor threshold in aroma research. *Lebensm.-Wiss. Technol.* 24: 1–5.

Van Buuren, S. 1992. Analyzing time–intensity responses in sensory evaluation. *Food Technol.* February: 101–104.

Wikipedia. A Wikimedia project, powered by MediaWiki. Available online at http://en.wikipedia.org/wiki/Sensory_analysis.

Williams, A.A. 1982. Scoring methods used in sensory analysis of foods and beverages at Long Ashton Research Station. *J. Food Technol.* 17: 163–175.

Wysocki, Ch.J., and P. Wise. 2004. Methods, approaches and caveats for functionally evaluating olfaction and chemesthesis. In *Handbook of Flavor Characterization. Sensory Analysis, Chemistry, and Physiology*, ed. K.D. Deibler and J. Delwiche, 1–40. New York: Marcel Dekker, Inc.

Zawirska-Wojtasiak, R., M. Gośliński, M. Szwacka, J. Gajc-Wolska, and S. Mildner-Szkudlarz. 2009. Aroma evaluation of transgenic, thaumatin: II. Producing cucumber fruits. *J. Food Sci.* 74(3): 204–210.

Zawirska-Wojtasiak, R., E. Kamiński, and M. Rogalska. 1992. Application of sensory profile analysis for evaluation of the quality of cereal grain during storage. In *Aroma Production and Application*, ed. M. Rothe and H.P. Cruse, 351–367. Proceedings of the 3rd Wartburg Aroma Symposium, Eisenach, 25–27 February, 1991. Potsdam, Germany: Eigenverlag Deutsches Institut für Ernährungsforschung.

Zawirska-Wojtasiak, R., S. Mildner-Szkudlarz, H. Jeleń, and E. Wąsowicz. 2005. Estimation of rosemary aroma by sensory analysis, gas chromatography and electronic nose. In *State of the Art. Flavour Chemistry and Biology*, ed. T. Hofmann, M. Rothe, and P. Schieberle, 130–136. Garching, Germany: Deutsche Forschungsanstalt fur Lebensmittelchemie.

Zawirska-Wojtasiak, R., E. Wąsowicz, H. Jeleń, M. Rudzińska, E. Kamiński, and P. Błażczak. 1998. Aroma characteristics of dill seeds varieties grown in Poland. *Pol. J. Food Nutr. Sci.* 2: 181–191.

Zawirska-Wojtasiak, R., E. Wąsowicz, A. Komosa, W. Tyksiński, and A. Spychalski. 2000. Sensory quality of greenhouse tomato fruits grown in the traditional substrates and inert media. *Roczniki Akademii Rolniczej w Poznaniu*, CCCXXVII, *Ogrodnictwo* 32: 85–99.

20 Machine Olfaction: A Devices Approach to Measurement of Food Aroma

Corrado Di Natale

CONTENTS

20.1 INTRODUCTION

Artificial olfaction systems stem from the idea that an array of nonselective gas sensors can mimic the property of natural olfaction to identify and recognize odors. This basic structure has been investigated for more than two decades now using many different sensor technologies, and odor identification properties have been tested in several application fields.

Among them, food analysis has been perhaps the most considered. Olfaction plays an evident role in food acceptance, and there is a growing interest in developing a technological support for sensorial analysis. In spite of the great distance between artificial and natural olfaction, positive results have been obtained for many different foodstuffs.

Herewith, a review of the main features of electronic noses is given. In particular, this chapter is focused on sensor technologies and data analysis principles. Even if a detailed treatment of these arguments is beyond the scope of this paper, an introduction to the main physical and chemical features of sensors and the principles underlying the data treatments and presentation may provide increased awareness in the interpretation of electronic nose results.

20.2 ARTIFICIAL OLFACTION

Food analysis is a complex discipline aimed at connecting the general properties of foodstuffs with the physical and chemical properties.

For many kinds of applications, this connection is rather well established. For instance, the alteration of physical and chemical quantities concerned with the safety (e.g., the search for contaminants), the composition (to individuate the basic constituents), or the effects of food treatment and processing can be adequately studied. For these scopes, there are several methods to measure the relevant quantities; they span from classical analytical chemistry to the more advanced diagnostic imaging techniques such as nuclear magnetic resonance (Pare and Belenger 1997).

Besides the above-mentioned objectives, there is a strong increase in demand for certification of food quality. In an epoch of strong globalization of productions and markets, quality is becoming one of the latest trends in food marketing, for instance, in the protection of typical products that may be identified with certain animal or vegetable species, or with particular production methods.

Quality is a global character of food, and it is primarily concerned with the interaction between food and consumers. From this point of view, the instruments "par excellence" to determine quality are the human senses. In reality, trained panels of tasters are utilized to fix and label the criteria of quality, to assess the quality of food, and to help the development of new products.

Although sensory analysis has been assessed for many years, panels are in practice affected by imprecision, they are scarcely repeatable, and cannot be utilized for routine operation. One of the major difficulties with panels is the comparison of analysis done at different times. Given these limitations, the importance of panels is still growing. As an example, according to the European Communities regulation 2568/91, olive oil's market value is based on a sensorial panel evaluation.

The role of olfaction in sensorial analysis is of course of primary importance for food quality assessment, and since the advent of electronic noses, there have been a number of studies aimed at demonstrating that electronic noses can be trained as the olfaction in humans.

In order to replicate human senses, it is necessary to fulfill a twofold requirement. The first is the design of sensors that have as much as possible to share with human senses the same receptive field; the second is perhaps more complicated and is concerned with the human perception. Perception of sensations is often due not only to the physics and chemistry of the sensorial event but also to the interaction between the senses, between the senses and the memory, and between the senses and the emotional contest. In case of vision and hearing, technology is rather close to develop human-like sensorial systems, but in the case of chemical senses (olfaction and taste), the situation is completely different. In olfaction, indeed, current chemical sensors' sensitivity is still orders of magnitude far from the performance of olfactory receptors; furthermore, artificial sensors are quite sensitive to compounds for which human olfaction is insensitive (e.g., water vapor, carbon monoxide, and carbon dioxide). Concerning the perceptual aspects of odors, it is known that most olfactory evaluations are more related to the previous experience and the context than to the chemistry of the odorant molecule (Andersson et al. 2009), so that the same molecule (e.g., butanol) can be perceived as pleasant or unpleasant according to the additional information provided. Besides the strong connection with unconscious perceptions, olfaction is also characterized by an unusual scarcity of semantic expressions, limiting the communication of olfactive experiences.

In the development of artificial olfaction systems, it is necessary to consider the structure of natural olfaction. Since the last three decades, the physiology of olfaction has been largely understood. Models of receptors' mechanisms explaining the sensitivity to volatile compounds are now available, and the genetic repertoire expressing the olfactory receptors is known for many species (Buck and Axel 1991).

Recent studies have also begun to unveil the signal pathways starting from the generation of an olfactory neuron signal and leading to the cognitive identification of odors (Friedrich and Stopfer 2001). Nevertheless, as previously discussed, olfaction remains a mysterious sense, because with respect to other senses (vision, hearing, and touch)—for which technological correspondents have been available for more than a century now—the attempts to endow artificial systems with odor recognition features have been thwarted for a long time.

The primary element of an artificial olfaction system is the ensemble of sensors translating the chemical stimuli in an accessible signal, usually electric. For vision and hearing, physics provides a sufficient background to develop artificial receptors. In the case of odor, the interaction of solid-state devices with volatile molecules started to be systematically investigated only in the last few decades, making the development of gas sensors possible.

The analysis of odors, that is, the investigation of gaseous samples composed of several volatile compounds, is a typical subject of analytical chemistry where several methods of separation of mixtures in individual compounds are available (e.g., gas chromatography). The principle of sample separation is obviously very different from natural olfaction where the odor interacts at once with the totality of receptors. Nonetheless, recent studies evidenced that both the mucus and the turbinate's shape have a certain property of separating the airborne chemicals according to their chemical and physical properties (Stitzel et al. 2003).

There are attempts to introduce the concept of sample separation in artificial olfactory systems, but this practice can still be considered absent in the current electronic nose technology (Sánchez-Montañés et al. 2008; Che Harun et al. 2009; Dini et al. 2010).

To analyze a gaseous mixture without separation, it is then necessary to use a set of selective sensors because each sensor senses only one of the many molecular species. In reality, many solid-state sensors developed since the 1970s are intrinsically nonselective, making them unable to achieve the analytical goal of measuring the concentration of many different compounds at once.

The nonselectivity of chemical sensors was then considered one of the main problems limiting their diffusion for practical applications. Nonetheless, physiological investigations about olfaction receptors show that the nature strategy for odor recognition is completely different from the analytical approach. Receptors were found to be rather nonselective, that is, each receptor senses several kinds of molecules and each molecule is sensed by many receptors (Sicard and Holley 1984). After this discovery, it was proposed that arrays of nonselective chemical sensors might show properties similar to that of natural olfaction (Persaud and Dodds 1982). On the basis of this conjecture, the development of artificial olfaction became possible. Actually, the denomination of an electronic nose is improperly extended to any array of nonselective chemical sensor coupled with some multicomponent classifier even if they are not aimed at mimicking olfaction. Since the 1980s, almost all sensor technologies were used to assemble electronic noses. Odor classification properties of artificial systems were tested on several different fields proving that electronic noses could be in principle used to replace human olfaction in practical applications such as food quality and medical diagnosis (Röck et al. 2008).

The features of electronic noses are fundamentally dependent on the sensing properties of the artificial receptors. The possibility of having a versatile tool to tailor the sensitivity and selectivity of sensors is of primary importance to capture either large or narrow ranges of chemicals allowing for electronic nose application-oriented optimizations. To this end, organic synthetic receptors offer an unlimited number of possibilities to assemble molecules endowed with differentiated sensing features. Several brilliant examples of synthetic receptors have been provided where the key–lock principle is implemented (Anslyn 2007). More recently, supramolecular assemblies of molecular units, such as metalloporphyrins folded as nanotubes, were found to possess superior sensing characteristics with respect to the single molecular units (Di Natale et al. 2010).

In the following sections of this chapter, an overview of the existing chemical sensor technologies used for electronic noses and a general discussion on data analysis are provided. For each sensor technology, examples of electronic nose applications to food analysis are also discussed. The section dedicated to data analysis is based on the more popular methods used to process electronic nose data. These methods are largely derived from other disciplines and in particular from chemometrics. For this reason, this section could be of benefit even for readers not directly interested with sensors and electronic noses.

20.3 CHEMICAL SENSOR TECHNOLOGIES FOR ELECTRONIC NOSES

A sensor is an electronic device whose parameters depend on some external quantity of whatever nature (Fraden 2004). As an example, according to this definition, there are resistors whose resistance is a function of external temperature (thermistors) or diodes whose current–voltage relationship is strongly altered once they are illuminated by light (photodiodes). In the same way, there are devices that, from the electronic point of view, are resistors, capacitors, or even field effect transistors (FETs), whose electrical parameters may depend on the chemical composition of the environment at which they are in contact.

Electronic properties of materials may hardly be directly influenced by the ambiental chemistry, and then in order to achieve chemical sensors, a complex structure is necessary.

The device is composed of two parts. The first is a chemically interactive material, that is, a solid-state layer of molecules that can interact with the molecules in the environment. These interactions can be of different nature, and the more utilized are adsorption and reaction phenomena. The interaction with a target molecule (hereafter called analyte) and a solid-state layer is a chemical event that, as a consequence, can modify the physical properties of the sensing layer. Properties such as conductivity, work function, mass, or optical absorbance are among those that can be transduced into an electric signal by suitable transducers. These transducers are the second component of chemical sensors, and they are sometimes called "basic devices."

The matching between sensitive material and transducer is not univocal: a single sensitive material can be coupled with many different transducers and vice versa. In practice, there are many possibilities in assembling a chemical sensor. The optimal matching between sensitive layer and transducer is then fundamental to achieve a well-performing sensor.

In the following section, the basic principles of the most popular categories of chemical sensors are illustrated. Sensors are classified according to the measured physical intermediate quantity.

20.3.1 Sensors Based on Conductance Changes

20.3.1.1 Metal Oxide Semiconductors

Sensors based on conductance changes are among the most appreciated for practical use. Indeed, resistances are the simplest electronic components and there are infinite possibilities to translate a resistance value into an electric signal.

Conductance change can occur either for an alteration of the charge carriers' (electrons or holes) concentration or for an alteration of their mobility in the material. Both these effects are exploited in many popular physical sensors such as thermistors or strain gauges. Noteworthy, the most sensitive conductance-based sensors are semiconductors.

The most popular materials undergoing a conductance change upon the interaction with gases are the metal oxide semiconductors. These are oxides of transition metals, in which the most known and studied is SnO_2, a wide band gap n-type semiconductor (Madou and Morrison 1989). The main sensitivity mechanism is related to the role played by oxygen. At a sufficiently high temperature (above 200°C), dissociative adsorption sites of molecular oxygen are activated on the oxide surface. As a consequence, charge transfer occurs between the material and the adsorbed oxygen atom with the consequence that the conductance band in proximity of the surface becomes depleted and a surface potential barrier is formed. The amount of depletion and the barrier height are proportional to the number of adsorbed molecules. Because the material is a semiconductor, the number of conductance electrons is limited and it also limits the amount of oxygen molecules that can be adsorbed at the surface. As a consequence to the exposure to oxygen, the surface conductance is reduced. The exposure to any molecule interacting on the sensor surface with adsorbed oxygen atoms may result in a release of electrons back to the conductance band, a reduction of the surface conductance band depletion, and a lowering of the potential barrier. Paradigmatic, to this regard, is the case of carbon monoxide that reacts with the bounded oxygen to form carbon dioxide, releasing an electron back to the conductance band. This is only one of the many interactions taking place on the surface of metal oxides, and the sensitivity of these devices is extended to many different kinds of volatile compounds (Barsan et al. 2007). The sensitivity can be further modified by adding ultrathin amounts of noble catalytic metal atoms on the surface. It is important to remark that this kind of sensors requests to be operated at high temperatures, and as a consequence, an electrically actuated heater is integrated in the device.

Metal oxide semiconductor sensors can be prepared in many different ways. In any case, the general advice is to produce a nanocrystalline material in such a way that the modulation of the surface conductance band population becomes dominant in the whole sensor, thus providing maximum sensitivity. Recently, metal oxides' growth in regular shapes such as nanosized belts (Comini and Sberveglieri 2010) has shown peculiar properties. The characteristics of these structures, although interesting, have not yet resulted in practical improvements of performances.

Several electronic noses have been based on metal oxide semiconductor chemoresistors and many of them have been demonstrated in food quality applications. Examples of applications have been done for raw food such as meat (Musatov et al. 2010), eggs (Suman et al. 2007), and fruit juices (Mielle and Marquis 2001). Electronic noses are also requested to monitor food processes and processed food; to this regard, metal oxide semiconductor-based electronic noses have been used to discriminate wine vintages (Di Natale et al. 1996) or to characterize the geographical origin of a cultivar (Berna et al. 2009). Another interesting example is the monitoring of bread-baking process (Ponzoni et al. 2008) that introduces the possibility of monitoring food aroma changes during cooking.

20.3.1.2 Conducting Polymers and Molecular Aggregates

The conductance properties of organic materials based either on polymers or molecular aggregates have been studied for several years with broader scopes related to the possibility of developing a novel sort of electronics based on carbon chemistry (Heeger 2001). Chemical sensors based on conducting polymers may be considered as a lateral result of these studies. Indeed, aggregates of polypyrrole or polythiophene have a semiconducting character and their conductance can change after the exposure to volatile compounds. With respect to metal oxides, these sensors have two important advantages: they are operated at room temperature and, most importantly, their chemical sensitivity can be changed at the synthesis level, modifying the chemical structure of the monomer (Persaud 2005). Besides conducting polymers, molecular aggregates were also found with a measurable DC conductivity. To this regard, the most important molecular class comprises the phthalocyanines, whose aggregates have been used for chemical sensing in many different applications.

Another important feature of conducting polymer sensors is their relative facility of preparation and technological implementation. In recent years, a number of technological efforts have been devoted to the development of electronic systems based on organic molecules instead of the more traditional silicon. Several examples of printed electronics have been provided including all plastics electronic circuits. Even sensor preparation benefitted from these techniques and examples of ink-jet printed sensors have been provided (Skotadis et al. 2010).

The chemical versatility of conducting polymers and molecular aggregates allows the design of sensors for use in different applications. Examples of use in food analysis have been provided for olive oil for both geographical origin and variety of olives (Stella et al. 2000; Guadarrama et al. 2000).

20.3.1.3 Conductive Composite Polymers

Conductivity interests only a restricted subset of polymers, but there are a large number of possibilities to engineer polymers in order to orient the chemical selectivity toward selected kinds of chemicals. Gas chromatography columns and solid phase microextraction fibers are good examples of the use of the chemical sensitivity of polymers in chemical analysis. These polymers are usually nonconductive and, traditionally, they have been used only with mass transducers, as will be discussed later.

An ingenious technique to extract an electrical signal from a nonconducting polymer was introduced in the 1990s (Zheng et al. 2009). The basic idea consists in blending the conducting, but sensitive, polymer with conducting powders such as carbon black. If the relative concentration of the two constituents is optimal, a network of carbon black is formed inside the polymer matrix, and then the material is conductive. Upon the adsorption of molecules from the gas phase, the polymer matrix swells, and as a consequence, the carbon black grains tend to separate from each other. This effect is measured as an increase in the material resistance.

This technique was patented at the University of California and gave place to the Cyrano Nose. This instrument was one of the most diffused commercially available electronic noses, and it was used in several applications even for food quality, such as cereal quality (Lonergan et al. 1996; Balasubramanian et al. 2007) and apple ripeness (Pathange et al. 2006).

20.3.1.4 Amperometric Gas Sensors

Electrolytic cells based on either solid-state or liquid ionic conductors are used to detect several kinds of gases. The main mechanism is the catalytic reaction occurring on the surface of a noble metal electrode. Although designed for polluting gases, these sensors also demonstrated good sensitivities for volatile compounds relevant in food analysis such as alcohol and amines. These molecular families are actually important for fish freshness. For instance, sensors designed for CO are found to be sensitive toward alcohols, aldehydes, and esters; sensors for ammonia can detect amines; and sensors for SO_2 can detect volatile sulfides. These cross-selectivities enable the use of these sensors in electronic noses; nonetheless except for a few applications, these sensors found scarce attention. It has to be mentioned that an electronic nose entirely based on catalytic sensors was developed in Iceland by a company that is no longer in business (FreshSense). Noteworthy, the FreshSense instrument was demonstrated in applications aimed at detecting fish freshness (Olafsdottir et al. 1997). Amperometric gas sensors were used in a hybrid electronic nose combined with metal oxide semiconductors and quartz microbalances (Mitrovics et al. 1998).

20.3.1.5 Mass Transducers

The measurement of adsorbed mass is perhaps the most simple approach to chemical sensors. Indeed, the adsorption of molecules into a sorbent layer obviously produces a change in mass, and the measurement of the mass increase can allow the evaluation of the amount of adsorbed molecules. Of course, for this kind of application, it is necessary to measure very small amounts of mass. The measure of small mass variations is made possible by piezoelectric resonators. A piezoelectric resonator is a piece of piezoelectric crystal properly cut along a well-specified crystalline axis. Due to the piezoelectric effect, the mechanical resonance of the crystal is coupled with an electric resonance. Because crystal resonance is extremely efficient, the electric resonance is characterized by a very large quality factor (Q). This property is largely exploited in electronics to build stable oscillators as clock references. The same effect can be used for chemical sensing, adopting particularly shaped crystals such as in quartz microbalances (QMB). These are thin slabs of AT cut quartz oscillating at a frequency between 5 MHz and 50 MHz approximately (Ballantine et al. 1997). The frequency of the mechanical oscillation decreases almost linearly with the mass gravitating onto the quartz surface. If the quartz is connected to an oscillator circuit, the electric frequency decreases linearly with the mass. A typical QMB has a limit of detection around 1 ng, an amount that is sufficient in many practical applications.

Electronic noses based on QMB coated by sensitive layers were used for many applications since the very beginning of researches on electronic noses. Typically, these sensors were coated with gas chromatographic phases whose behavior was well known from the analytical chemistry practice. The necessity to develop proper materials for gas chromatography gave rise to a number of models describing the sensitivity of polymers with regard to basic interactions that can take place between the solid phase and an airborne molecule. Among these models, linear sorption energy relationship (LSER) is particularly efficient and it was used to optimize polymer-coated QMB arrays. A brilliant development of these sensors was done in the 1980s at the University of Tübingen where on the basis of LSER, polysiloxanes were properly engineered to cope with the basic interactions (Hierlemann et al. 1995).

Besides polymers, even molecular films were used. To this regard, the use of cavitands (Pinalli et al. 1999) and porphyrins (Di Natale et al. 2007) is noteworthy. In particular, porphyrins were instrumental to the development of an electronic nose that has been successfully used in different fields such as food quality (Di Natale et al. 1997) and medical diagnosis (Di Natale et al. 2003).

Specific examples on the use of QMB-based electronic noses in food analysis were given for tomatoes (Berna et al. 2004), olive oil (Escuderos et al. 2011), cheese ripening (Bargon et al. 2003), and to estimate optimal picking time of apples (Saevels et al. 2003).

Piezoelectric effect can also be exploited in other configurations such as those based on surface acoustic waves. More sophisticated mass transducers were proposed by using resonant cantilevers similar to those adopted in atomic force microscopies (Battiston et al. 2002). In spite of the claimed properties, these sensors were never demonstrated in practical applications.

20.3.1.6 Field Effect Transistors

FETs are among the most important devices in electronics where they are used for the most important analog and digital circuits.

Most of the properties of FETs depend on the difference between the work function of electrons in the metal gate and in the semiconductor. It was demonstrated at the end of the 1970s that this difference can be modulated by a layer of electric dipoles that can reach the metal oxide interface. This principle was adequately exploited with a palladium gate FET exposed to hydrogen gas (Lundström et al. 1975). H_2 molecules dissociate into atomic hydrogen at the palladium surface, and hydrogen atoms can diffuse through the palladium film into the oxide surface where they form an ordered dipole layer. As a result, although under constant bias, the current flowing in the FET changes, revealing the chemical interaction.

This basic structure was successively modified by changing the gate metal and thickness in order to extend the range of measured gases. In this way, sensitivity to ammonia, an important gas for fish freshness and quality, was also obtained (Winquist et al. 1983). FET structures were also modified to accommodate, as sensing part, organic molecular layers, such as metalloporphyrins (Andersson et al. 2001). Because of the necessity of a microelectronic fabrication, FETs were only seldom used in electronic noses, and in practice, their utilization was limited to the instrument produced by the NST, then Applied Sensors. These devices, both in research and commercial version, were used for many applications in food analysis such as for the detection of boar taint in pork meat (Vestergaard et al. 2006), the flavor released in cooked meat (Tikk et al. 2008), and microbial quality of cereals (Jonsson et al. 1997).

20.3.2 Color Indicators

Colorimetry is an important method in chemistry. There is indeed a strong relationship between molecular structure and color, and several molecules change their color in the presence of particular physical and chemical conditions. The best known example is the litmus paper used for pH evaluation.

Although known for several years (Tozawa et al. 1971), the colorimetric detection of food properties, such as fish freshness, received an upsurge of interest recently. In the case of fish freshness, the importance of amines as spoilage markers leads to consider their reducing role and then the possibility to detect them with functional layers sensitive to pH changes. The feasibility of this approach has been demonstrated using a film of sodium salt (bromocresol green) as sensitive layer (Pacquit et al. 2007). This salt exhibits a rather large change in color also appreciable by the naked eye. Nonetheless, the use of pH indicators is limited by the fact that mainly amines are considered (limiting the detection not to freshness but rather to spoilage). Furthermore, the visual determination limits the performance and may vary greatly between individuals.

Chemical sensing based on optical sensitive layers is a captivating strategy due to the strong influence of target chemicals on the absorption and fluorescence spectra of chosen indicators (Gauglitz 2006). Nonetheless, the relatively simple chemistry involved in this approach is badly balanced by the transducer counterpart. Indeed, standard optical instrumentations are usually expensive. On the other hand, in the past decade, a fast growth of performance in fields such as consumer electronics has taken place. It is giving rise to a number of low-cost advanced optical equipment such as digital scanners, cameras, and screens, actually embedded in other devices, but whose characteristics largely fit the requirements necessary to capture the change of optical properties of sensitive layers in many practical applications.

The first demonstration in this direction was given by Suslick and colleagues when they showed that a digital scanner has enough sensitivity to detect the color changes in chemical dyes due to the adsorption of volatile compounds (Rakow and Suslick 2000). The method demonstrated also the possibility to identify toxic chemicals and also food ingredients such as sugars.

Furthermore, Lundström and Filippini proved that it is possible to assemble a sort of spectrophotometer using the computer screen monitor as a programmable source and a web cam as detector (Filippini et al. 2003). This last technique, known as computer screen photo assisted technique (CSPT) is based on the fact that a computer screen can be easily programmed to display millions of colors, combining wavelengths in the optical range. Compared with the use of digital scanners, to probe the sample with a variable combination of wavelengths instead of using the white light of scanners, offers the possibility to perform an optical fingerprint measurement allowing a simultaneous evaluation of absorbance and fluorescence of samples. Because of the large diffusion of portable computers, PDAs, and cellular phones all endowed with color screen, camera, and an even more extended computation capabilities, the application of the CSPT concept may be foreseen as to greatly extend the analytical capacity worldwide. CSPT has demonstrated its utility in particular to classify airborne chemicals reading absorbance and fluorescence changes in chemical dyes such as metalloporphyrins (Filippini et al. 2006). Standard optochemical sensors are based either on absorbance or on fluorescence, whereas CSPT arrangement gives the possibility of evaluating both effects at the same time.

Digital scanner evaluation of color changes of chemical indicators was applied to specific food analysis such as the discrimination of coffees (Suslick et al. 2010); another noteworthy application of this kind of sensor array is the identification of natural and artificial sweeteners (Musto et al. 2009). CSPT evaluation of a set of porphyrins was shown to be able to monitor fish freshness decay (Alimelli et al. 2007).

20.4 DATA ANALYSIS FOR ELECTRONIC NOSES

Although the development of sensor arrays is oriented to exploit the analogies with natural receptors (e.g., cross-selectivity and redundancy), the data analysis still does not take advantage of natural data treatment paradigms. Given the differences between sensors and receptors, the principles of data processing of natural olfaction receptors signals do not provide simple methods to process sensors data. Nonetheless, the incorporation of olfactory processing paradigm into artificial systems is evolving and first attempts to derive analysis methodologies from natural paradigms have already been proposed (Pearce 1997). Except for these rare attempts to include biological paradigms, electronic nose data analysis is rather based on techniques borrowed from other disciplines and chosen among a vast class of algorithms known as pattern recognition.

Pattern recognition in chemistry appeared at the end of the 1960s. In gas sensing area, multivariate methods of pattern recognition are commonly required when sensor arrays, composed of real, nonselective, cross-sensitive sensors, are utilized. Pattern recognition, exploiting the cross-correlation, extracts information contained in the sensor outputs ensemble.

Formally, pattern recognition may be defined as "the mapping of a pattern from a given pattern space into a class-membership function" (Di Natale et al. 1995).

A pattern can be defined as any ordered set of real numbers. In analytical chemistry, the response of a multichannel instrument (such as gas chromatography or spectrophotometer) forms a pattern. For a sensor array, the responses of the sensors of the array take the name of sensors pattern.

Class-membership space represents either a quantitative or qualitative set of quantities. When quantification is considered, the class-membership function turns out to be a vectorial function defined in a metric space. In the case of class recognition, class-membership is an ensemble of abstract sets at which each measured sample is assumed to belong. In the first case, we prefer to talk of multicomponent analysis, whereas the term pattern recognition is usually referred to in the second case.

Mathematically, a pattern is represented by a vector belonging to a proper vector space. The space where sensor responses are represented is called sensor space. The simplest of these representations considers one scalar response for each sensor. This response is dubbed sensor feature and feature extraction is the operation of extracting from a sensor signal the synthetic descriptors that form the patterns. When a sensor (independently from its working principle) is exposed to a gas or a mixture of gases, it gives a response depending on the nature of the gas and its concentration. Almost the totality of the examples reported in literature considers the sensor response as either the absolute or the relative change in sensor signal measured in two steady-state conditions in the absence and in the presence of the gas, and when all the transients are ended. This definition leaves the dynamic behavior of the sensor signal unexploited. Several authors studied the optimization of feature extraction in order to maximize the array performance. These papers show that quantification error is minimized when descriptors considering the dynamic behavior are taken into account. On the other hand, it has to be mentioned that the dynamic behavior is very sensitive to the fluctuations of the sample delivery system.

Figure 20.1 shows the conceptual steps in a typical electronic nose experiment. It is worthy to note the importance of the cases extracted from the abstract classes and used in the experiment. Unbiased sampling is not always easy in practical applications where the definition of abstract classes may be vague, and the effective class

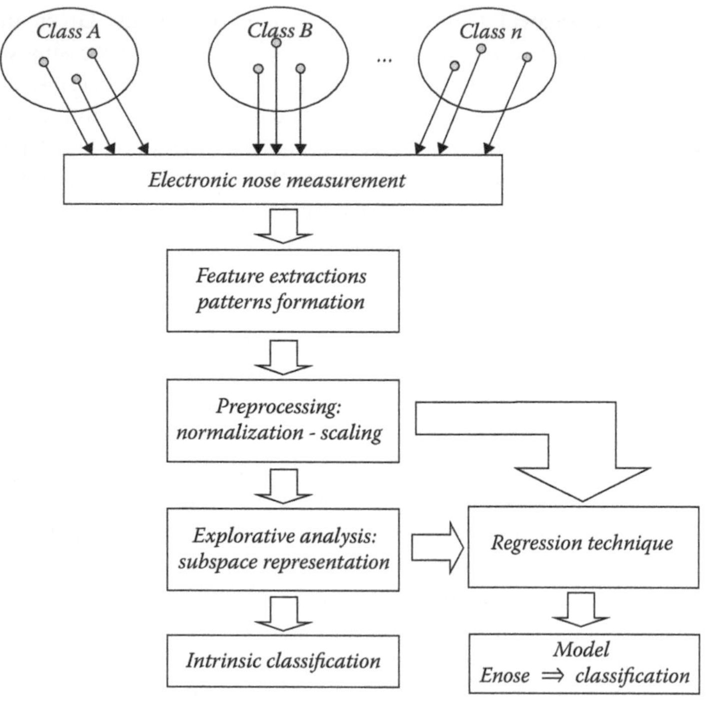

FIGURE 20.1 Sketch of a typical electronic nose experiment. Samples are extracted from abstract classes and measured by the instrument. Once features are extracted and a pattern has formed, preprocessing may be applied to standardize data and to remove influence of quantitative information. Hence, data may be explored, to discover intrinsic classification properties, or used to build a classifier through regression.

membership may not be fully certain. It is important to keep this problem in mind when the performances of an electronic nose have to be evaluated. In Figure 20.1, three main blocks are shown, which correspond to preprocessing, exploratory analysis, and classification. In the following, each of these items will be discussed, providing evidence on both theoretical and practical aspects.

20.4.1 Data Preprocessing: Scaling and Normalization

A usual procedure in pattern recognition is the scaling of the data. Instead of using raw data, two main scaling procedures are widely used: zero-centered and autoscaling.

Zero-centered data means that each sensor is shifted across the zero value, so that the mean of the responses is zero. Zero-centered scaling may be important when the assumption of a known statistic distribution of the data is used. For instance, in case of a normal distribution, zero-centered data are completely described only by the covariance matrix.

Autoscaling means to scale each sensor to zero-mean and unitary-variance. This operation equalizes the dynamics of the sensor responses, avoiding the fact that a sensor, with a larger response range, may hide the contribution of other sensors dynamically limited. Furthermore, autoscaling makes the sensor responses dimensionless. This feature becomes necessary when sensors whose signals are expressed in different units are joined in the same array. This is the case in hybrid arrays (different sensor technologies in the same array) and when electronic noses are fused with other instruments, for example, the fusion of electronic noses and electronic tongues (Di Natale et al. 2000).

Several years ago, it was shown that the application of a simple normalization of sensor data can greatly help in removing the quantitative information and putting in evidence the qualitative aspects of the data (Horner and Hierold 1990).

The normalization consists in dividing each sensor response by the sum of all the sensor responses to the same sample, so that the concentration information disappears.

The application of linear normalization to an array of linear sensors should produce, on the principal component analysis (PCA) score plot, one point for each compound, independent of its concentration, and achieving the highest possible recognition. Deviations from ideal behavior are due to nonlinear relationships between sensor response and concentration.

Normalization is of limited efficiency because the mentioned assumptions strictly hold for simple gases and they are fated to fail when mixtures of compounds are measured. Furthermore, in complex mixtures, temperature fluctuations do not result in a general concentration shift, but since individual compounds have different boiling temperatures, each component of a mixture changes differently so that both quantitative (concentration shift) and qualitative (pattern distortion) variations occur.

20.4.2 Principal Component Analysis

Given a set of data related to a number of measurements, exploratory techniques aim at studying the intrinsic characteristics of the data in order to discover eventual internal properties.

Exploratory analysis evidences the attitude of an electronic nose to be utilized for a given application, leaving to the supervised classification the task of building a model to be used to predict the class membership of unknown samples.

Two main groups of exploratory analysis may be identified: representation techniques and clustering techniques.

Representation techniques are a group of algorithms aimed at providing a representation of the data in a space of dimensions lower than the original sensor space. The most popular of these methods is the PCA (Jolliffe 1986).

The scope of PCA is a *reliable* portrayal of a data set in a subspace of reduced dimensionality. *Reliable* here means that data are reproduced, preserving their statistical properties.

PCA is part of a large set of linear transformations reassumed as: $\mathbf{S} = \mathbf{WX}$, where $\mathbf{X}$ is the original data set, $\mathbf{W}$ is the transformation matrix, and $\mathbf{S}$ are the data points in the representation space.

PCA was first introduced in the 1930s in psychology to describe the behavior of humans (Hotelling 1933). It has then been deeply developed in the chemometrics, where it has been introduced at the beginning to analyze spectroscopic and chromatographic data, which are characterized by a higher correlation among the spectra channels.

The possibility of a reliable representation of an electronic nose data set in a subspace of reduced dimensionality lies on the fact that the chemical sensors always exhibit a certain degree of correlation among them. The PCA consists in finding an orthogonal basis where the correlation among sensors disappears.

Correlation means that given a sensor space of dimension N, the effective dimension of the subspace occupied by the data is less than N.

PCA is calculated considering only the second momentum of the probability distribution of the data (covariance matrix). The covariance matrix ($\mathbf{X}^T\mathbf{X}$) completely describes zero-centered data only for normally distributed data. From a geometric point of view, any covariance matrix, because it is a symmetric matrix, is associated with a hyperellipsoid in the N-dimensional space. PCA corresponds to a coordinate rotation to represent the associated hyperellipsoid in its canonical form. In practice, the novel coordinate basis is coincident with the hyperellipsoid principal axis.

The reduction of an ellipsoid to its canonical form is a typical linear algebra operation; it is performed through the eigenvectors of the associated matrix. Thus, PCA can be calculated with the following rule. Let us consider a matrix $\mathbf{X}$ of data, let $\mathbf{C} = \mathbf{X}^T\mathbf{X}$ be the covariance matrix of $\mathbf{X}$. The ith principal component of $\mathbf{X}$ is $\mathbf{X}^T l(i)$, where $l(i)$ is the ith normalized eigenvector of $\mathbf{C}$ corresponding to the ith largest eigenvalue.

The eigenvalues happen to be directly proportional to the variance explained by their corresponding eigenvector, so that considering the relative values of the eigenvalues $l(i)$, it is possible to reduce the representation to only those components carrying most of the information.

Given a matrix of data, PCA provides two main information called scores and loadings. The scores are concerned with the measurements, and they are defined as the coordinates of each vector measurement (a row of matrix $\mathbf{X}$) in the principal components base.

The loadings are concerned with the sensors. They measure the contribution of each sensor to the PCA basis. A large loading, for a sensor, means that the principal component is aligned along the sensor direction.

It is important to note that the highest eigenvalues correspond to components defining the directions of highest correlation among the sensors, whereas the components characterized by smaller eigenvalues are related to uncorrelated directions. Because sensor noises are uncorrelated, the representation of the data using only the most meaningful components removes the noise of the sensors. In this way, PCA is also used to remove noise for multivariate data.

When applied to electronic nose data, the presence of various sources of correlated disturbances has to be considered. As an example, sample temperature fluctuations induce correlated disturbances, which are described by the principal components of highest order. When these disturbances are important, the first principal component has to be eliminated in order to emphasize the relevant data properties.

The hypothesis of normal distribution is a strong limitation that should always be kept in mind when PCA is used. In electronic nose experiments, samples are usually extracted from more than one class, and not always the totality of measurements results in a normally distributed data set. Nonetheless, PCA is frequently used to analyze electronic nose data. Because of the high correlation normally shown by electronic nose sensors, PCA allows a visual display of electronic nose data in either 2D or 3D plots.

Another limitation to the use of PCA comes from the fact that being a linear projection, it may introduce mistakes. Indeed, in the projection, data separated in the original space may result with a similar score, a phenomenon similar to that of producing constellation in the starred sky.

20.4.3 Supervised Classification

In supervised classification, the classes are *a priori* defined both as kind and number. This information has to be acquired from other considerations about the application under study.

Once classes are defined, supervised classification may be described as the search of a model of the following kind:

$$\vec{c} = f(\vec{s}) \tag{20.1}$$

where $\vec{c}$ is a vector describing the class assignment, $\vec{s}$ is the vector of features of the sensors in the array, and f is a generic function. To solve pattern recognition problems, class memberships must be encoded in a numerical form that allows treating the problem by numerical methods. The most common way to express class memberships is the so-called "one-of-many" code. In this codification, the dimension of $\vec{c}$ is equal to the number of classes. The component corresponding to the class at which a sample belongs is settled to 1, leaving 0 to the others.

In electronic nose applications, various sources of measurement errors may occur, and as a consequence, Equation 20.1 is written in a more realistic form as:

$$\vec{c} = f(\vec{s}) + \vec{e} \tag{20.2}$$

where the vector $\vec{e}$ contains everything that is not related to the classification scheme expressed by the vector $\vec{c}$.

Equation 20.2 is formally similar to the general problem of regression where the scope is the determination of the function f, in terms of functional form and parameters. Statistics provides the tools to estimate, from an experimental data set, the parameters of the function f, in order to approximate the measured experimental data. The classical approach is the least squares method. Many practical algorithms were proposed as practical solutions. Among them, chemometrics and the neural networks are those widely used.

It is important to reflect about the applicability of the least squares method in typical electronic nose experiments. Least squares method is based on the assumption that the variables are normally distributed and that the quantity $\vec{e}$ of Equation 20.2 is a random variable normally distributed with zero-mean. The assumption of zero-mean means that all the variables to which the sensor responses are sensitive, except those related to the classification of the samples, may fluctuate but not biasing the measurement. Usually, this does not hold true for electronic nose data where, except for the sensor's noise, the contributions to $\vec{e}$ (sensor's drift, sample temperature variations, sample dilution changes, etc.) are not zero-mean quantities.

Nevertheless, solutions based on the least squares method can be used to establish classification models, but it is important to be aware of the fundamental limitations of the methods.

Before we discuss practical solutions, it is necessary to detail the general frame. In pattern classification, it is very important to estimate the expected error rate after the classifier model has been assessed on a calibration data set. The expected error rate evaluates the efficiency of the model, given the probability of misclassifying future samples. The estimation of error rate on the same set used for calibration induces large optimistically biased estimate of the performances. This effect is often called "overfitting." The importance of over-fitting grows with the order of the regression function. In particular, it is important when highly nonlinear functions are used (e.g., in neural networks). In these cases, the model may be almost perfect to estimate the data on which the model is estimated failing completely in generalizing other data.

Overfitting may be more important for classification scopes, where the samples are extracted from sets that may realistically be not well defined, so that the assignment of samples to classes may be affected by errors due to vagueness of the classification scheme. In this situation, the possibility to generate models, which are not able to predict unknown samples at a sufficient accuracy, is high.

The straightforward solution for the estimation of error rate is to split the data set into two independent sets, and use one for calibration and the other to test the classifier and estimate the error rate. This method cannot be used when small sample sizes are available. In many practical cases, and electronic nose experiments are often among these, the data sets are small because of the difficulties to get a highly populated set. In these cases, all the data are necessary to estimate the classification model. Moreover, how to split samples is a nontrivial problem because the division should be done by keeping the distributions of the two sets as close as possible in order to avoid biasing the evaluation of performances.

A more reliable validation of a model is achieved using the "leave-one-out" technique (Lachenbruck and Mickey 1968). "Leave-one-out" repeats n times for n measures the model building, each time leaving one measure out for testing and the rest for training. The average test error rate over n trials is the estimated error rate.

In case of small data sets, the method of bootstrap has been proven to be more efficient than "leave-one-out." Both "leave-one-out" and bootstrap are kinds of resampling methods. Bootstrap method generates new samples (called "bootstrap samples") by drawing, with replacement, a number N of samples from the original samples (Hamamoto et al. 1997).

20.4.3.1 Linear Discrimination

The simplest way to estimate a supervised model is to consider that the descriptor of each class may be represented as a linear combination of the sensor responses, considering N sensors and M classes, the expressions can be written as:

$$\begin{cases} c_1 = \sum_{j=1}^{N} k_{1j} \cdot s_j + e \\ \quad \dots \\ c_M = \sum_{j=1}^{N} k_{Mj} \cdot s_j + e \end{cases} \tag{20.3}$$

Geometrically, this means to section the sensor space with straight lines each bisecting the space. The result is a partitioning of the space in volumes each defining one class.

Considering a set of P experimental data, the previous set of equations can be written in a compact matrix form as:

$$C_{M\times P} = K_{M\times N} \cdot s_{N\times P} + E_{N\times P} \tag{20.4}$$

The matrix $K_{M\times N}$ containing the model parameters can then be directly estimated considering the Gauss–Markov theorem to solve least squares solution of generic linear problems written in matrix form (Campbell and Meyer 1979):

$$K_{M\times N} = c_{M\times P} \cdot s^{+}_{P\times N} \tag{20.5}$$

where the matrix $s^{+}_{P\times N}$ is the generalized inverse, or pseudoinverse, of the matrix $s_{N\times P}$. The operation of pseudoinversion generalizes the inversion of square matrices to rectangular matrices. This solution is called in literature as multiple linear regression (MLR).

Once the model is assessed, it allows assigning any unknown samples to one class. Due to the presence of the above-mentioned error matrix ($E_{N\times P}$), the model provides a numerical estimation of the "one-of-many" encoding of class assignment. In practice, something different from 0 and 1 is obtained. The estimated class assignment vector is called "classification score" and the sample is assigned to the class represented by the component with the higher value. This gives the possibility to evaluate also a sort of goodness of the classification considering either the ratio between the first and the second value of the components of the estimated $\vec{c}$ or the difference between the highest value and 1 (target value).

The components of the matrix $K_{M\times N}$ define the importance of each sensor in the classification of each class. This information can also be used, as the loadings of PCA to optimize the sensor array composition.

The pseudoinversion, like the inversion of square matrices, is influenced by the partial correlation among the sensors. Chemometrics offers methods to solve problems with colinear sensors; a particularly useful method is the partial least squares (PLS) (Wolde et al. 2001). PLS is often used to solve classification problems. In this case, PLS offers not only a more robust solution of the classification problem, but plotting the latent variables is possible to graphically represent the class separation. It is worth mentioning that the PLS latent variables differ from the PCA factors because they are evaluated as the eigenvectors of the matrix $S^T \cdot C \cdot C^T \cdot S$. Geometrically, the PCA components are rotated in order to maximize their correlation with the components of the matrix C. Furthermore, PLS loadings can be used to study the contribution of each sensor to the solution of the classification problem.

In linear discrimination, only classes separable by straight lines are correctly classified. Classification improves if nonlinear boundaries between classes are used.

Increasing the order of the discriminant function may solve highly complex classes distribution. The extreme solution is to use a method where the choice of the function is not required. Neural networks offer the possibility to solve the classification problem disregarding the functional form. It is well known that optimized neural networks may reproduce any kind of nonlinear function. Neural networks derive all the knowledge from the experimental data, so that, increasing the size of the calibration data set increases the accuracy of the neural network-based classifier. Comparison studies among different algorithms do not indicate a clear indication (Shaffer et al. 1999).

20.5 CONCLUSIONS

Electronic noses have fascinated scientists for about 25 years. A standard model of electronic nose emerged since the very beginning of this research. This model is well described by the following sentence: "An electronic nose is an instrument, which comprises an array of electronic chemical sensors with partial specificity and an appropriate pattern-recognition system, capable of recognizing simple or complex odors" (Gardner and Bartlett 1994). In the two decades after the introduction of electronic noses, the progress in the comprehension of olfaction made the standard model depicted above rather primitive, but besides few examples of artificial olfactory systems where the distance between artificial and natural is shortened (Dickinson et al. 1996; Di Natale et al. 2008), the standard model of electronic noses is still the mainstream today. Although primitive with respect to natural olfaction, the standard model of electronic noses has been rather successfully applied in several contexts, and the quality of foods is certainly one of the most explored areas of applications of electronic noses. In many cases, the results are certainly interesting for the improvement of the field, but only rarely do they constitute a basis for immediate industrial exploitation. The field still requires more basic research. Most of the research reports have concentrated on the improvement of sensors, whereas other important areas, such as the reliability of the sampling systems, have not been completely investigated.

However, the results achieved so far are a promising basis for continuing toward more reliable and industrially applicable quality measurement systems. To progress in this direction, it is necessary to establish long-term cooperation between sensor developers, application designers, and food scientists to derive from a general-purpose technology, such as the electronic nose, the specific instruments and methods required by food industries.

REFERENCES

Alimelli, A., G. Pennazza, M. Santonico, R. Paolesse, D. Filippini, A. D'Amico, I. Lundström, and C. Di Natale. 2007. Fish freshness detection by a computer screen photoassisted based gas sensor array. *Anal. Chim. Acta* 582: 320–328.

Andersson, L., M. Bende, E. Millqvist, and S. Nordin. 2009. Attention bias and sensitization in chemical sensitivity. *J. Psychosom. Res.* 66: 407–416.

Andersson, M., M. Holmberg, I. Lundstrom, A. Lloyd-Spetz, P. Martensson, R. Paolesse, C. Falconi, E. Proietti, C. Di Natale, and A. D'Amico. 2001. Development of a ChemFET sensor with molecular films of porphyrins as sensitive layer. *Sens. Actuators B* 77: 567–571.

Anslyn, E. 2007. Supramolecular analytical chemistry. *J. Org. Chem.* 72: 687–699.

Balasubramanian, S., S. Panigrahi, B. Kottapalli, and C.E. Wolf-Hall. 2007. Evaluation of an artificial olfactory system for grain quality discrimination. *LWT—Food Sci. Technol.* 40: 1815–1825.

Ballantine, D.S. et al. 1997. *Acoustic Wave Sensors*. San Diego, CA: Academic Press.

Bargon, J., S. Braschoß, J. Flörke, U. Herrmann, L. Klein, J.W. Loergen, M. Lopez, et al. 2003. Determination of the ripening state of Emmental cheese via quartz microbalances. *Sens. Actuators B* 95: 6–19.

Barsan, N., D. Koziej, and U. Weimar. 2007. Metal oxide-based gas sensor research: How to? *Sens. Actuators B* 121: 18–35.

Battiston, F.M., J.-P. Ramseyer, H.P. Lang, M.K. Baller, Ch. Gerber, J.K. Gimzewski, E. Meyer, and H.-J. Guntherhodt. 2002. A chemical sensor based on a microfabricated cantilever array with simultaneous resonance frequency and bending readout. *Sens. Actuators B* 77: 122–123.

Berna, A., J. Lammertyn, S. Saevels, C. Di Natale, and B. Nicolaï. 2004. Electronic nose systems to study shelf life and cultivar effect on tomato aroma profile. *Sens. Actuators B* 97: 324–333.

Berna, A.Z., S. Trowell, D. Clifford, W. Cynkar, and D. Cozzolino. 2009. Geographical origin of Sauvignon Blanc wines predicted by mass spectrometry and metal oxide based electronic nose. *Anal. Chim. Acta* 648: 146–152.

Buck, L., and R. Axel. 1991. A novel multigene family may encode odorant receptors: A molecular basis for odor recognition. *Cell* 65: 175–187.

Campbell, S.L., and C.D. Meyer. 1979. *Generalized Inverses of Linear Transformations*. London, UK: Pitman.

Che Harun, F., J. Taylor, J. Covington, and J. Gardner. 2009. An electronic nose employing dual channel odor separation columns with large chemosensor arrays for odor discrimination. *Sens. Actuators B* 141: 134–140.

Comini, E., and G. Sberveglieri. 2010. Metal oxide nanowires as chemical sensors. *Mater. Today* 13: 36–44.

Dickinson, T.A., J. White, J.S. Kauer, and D.R. Walt. 1996. A chemical detecting system based on a cross-reactive optical sensor array. *Nature* 382: 697–700.

Di Natale, C., F. Davide, and A. D'Amico. 1995. Pattern recognition in gas sensing: Well-stated techniques and advances. *Sens. Actuators B* 23: 111.

Di Natale, C., F.A.M. Davide, A. D'Amico, P. Nelli, S. Groppelli, and G. Sberveglieri. 1996. An electronic nose for the recognition of the vineyard of a red wine. *Sens. Actuators B* 33: 83–88.

Di Natale, C., A. Macagnano, F. Davide, A. D'Amico, R. Paolesse, T. Boschi, M. Faccio, and G. Ferri. 1997. An electronic nose for food analysis. *Sens. Actuators B* 44: 521–526.

Di Natale, C., A. Macagnano, E. Martinelli, R. Paolesse, G. D'Arcangelo, C. Roscioni, A. Finazzi-Agrò, and A. D'Amico. 2003. Lung cancer identification by the analysis of breath by means of an array of non selective gas sensors. *Biosens. Bioelectron.* 18: 1209–1218.

Di Natale, C., E. Martinelli, R. Paolesse, A. D'Amico, D. Filippini, and I. Lundström. 2008. A biomimetic platform for artificial olfaction. *PLoS ONE* 3: e3139–e3150.

Di Natale, C., D. Monti, and R. Paolesse. 2010. Chemical sensitivity of porphyrin assemblies. *Mater. Today* 13: 46–52.

Di Natale, C., R. Paolesse, and A. D'Amico. 2007. Metalloporphyrins based artificial olfactory receptors. *Sens. Actuators B* 121: 238–246.

Di Natale, C., R. Paolesse, A. Macagnano, A. Mantini, A. D'Amico, A. Legin, L. Lvova, A. Rudnitskaya, and Y. Vlasov. 2000. Electronic nose and electronic tongue integration for improved classification of clinical and food samples. *Sens. Actuators B* 64: 15–21.

Dini, F., D. Filippini, R. Paolesse, A. D'Amico, I. Lundström, and C. Di Natale. 2010. Polymers with embedded chemical indicators as an artificial olfactory mucosa *Analyst* 135: 1245–1252.

Escuderos, M., S. Sánchez, and A. Jiménez. 2011. *Food Chem.* 124: 857–862.

Filippini, D., A. Alimelli, C. Di Natale, R. Paolesse, A. D'Amico, and I. Lundström. 2006. Chemical sensing with familiar devices. *Angew. Chem. Int. Ed.* 45: 3800–3803.

Filippini, D., S. Svensson, and I. Lundström. 2003. Computer screen as a programmable light source for visible absorption characterization of (bio)chemical assays. *Chem. Commun.* 240–241 (*Chem. Comm.* were not organized in volumes).

Fraden, J. 2004. *Handbook of Modern Sensors*. New York: AIP Press.

Friedrich, R.W., and M. Stopfer. 2001. Recent dynamics in olfactory population coding. *Curr. Opin. Neurobiol.* 11: 468–474.

Gardner, J., and P. Bartlett. 1994. A brief history of electronic nose. *Sens Actuators B* 18: 211–220.

Gauglitz, G. 2006. Optical sensing looks to new field. *Trends Anal. Chem.* 25: 748.

Guadarrama, A., M.L. Rodríguez-Méndez, J.A. de Saja, J.L. Ríos, and J.M. Olías. 2000. Array of sensors based on conducting polymers for the quality control of the aroma of the virgin olive oil. *Sens. Actuators B* 69: 276–282.

Hamamoto, Y., S. Uchimura, and S. Tomita. 1997. A bootstrap technique for nearest neighbor. *IEEE Trans. Pattern Anal. Mach. Intell.* 19: 73–79.

Heeger, A.J. 2001. Semiconducting and metallic polymers. *Angew. Chem., Int. Ed.* 40: 2591.

Hierlemann, A., U. Weimar, G. Kraus, M. Schweizer-Berberich, and W. Göpel. 1995. Polymer-based sensor arrays and multicomponent analysis for the detection of hazardous organic vapours in the environment. *Sens Actuators B* 26: 126–134.

Horner, G., and C. Hierold. 1990. Gas-analysis by partial model-building. *Sens. Actuators B* 2: 173–184.

Hotelling, H. 1933. Analysis of a complex of statistical variables into principal components. *J. Educ. Psychol.* 24: 498–520.

Jolliffe, T. 1986. *Principal Component Analysis*. New York: Springer Verlag.

Jonsson, A., F. Winquist, J. Schnurer, H. Sundgren, and I. Lundström. 1997. Electronic nose for microbial quality classification of grains. *Int. J. Food Microbiol.* 35: 187–193.

Lachenbruck, P.A., and R.M. Mickey. 1968. Estimation of error rates in discriminant analysis. *Technometrics* 10: 1–11.

Lonergan, M., E. Severin, B. Doleman, S. Beaber, R. Grubbs, and N. Lewis. 1996. Array-based vapor sensing using chemically sensitive, carbon black–polymer resistor. *Chem. Mater.* 8: 2298–2312.

Lundström, I. et al. 1975. A hydrogen sensitive MOS field effect transistor. *Appl. Phys. Lett.* 26: 55.

Madou, M., and S. Morrison. 1989. *Chemical Sensing with Solid State Devices*. San Diego, CA: Academic Press.

Mielle, P., and F. Marquis. 2001. One-sensor electronic olfactometer for rapid sorting of fresh fruit juices. *Sens. Actuators B* 76: 470–476.

Mitrovics, J., H. Ulmer, U. Weimar, and W. Göpel. 1998. Modular sensor systems for gas sensing and odor monitoring: The MOSES concept. *Acc. Chem. Res.* 31: 307–315.

Musatov, V.Yu., V.V. Sysoev, M. Sommer, and I. Kiselev. 2010. Assessment of meat freshness with metal oxide sensor microarray electronic nose: A practical approach. *Sens. Actuators B* 144: 99–103.

Musto, C., S. Lim, and K. Suslick. 2009. Colorimetric detection and identification of natural and artificial sweeteners. *Anal. Chem.* 81: 6526–6533.

Olafsdottir, G., E. Martinsdóttir, and E.H. Jónsson. 1997. Rapid gas sensor measurements to predict the freshness of capelin (*Mallotus villosus*). *J. Agric. Food Chem.* 45: 2654.

Pacquit, A., K. Lau, H. McLaughlin, J. Frisby, B. Quilty, and D. Diamond. 2007. Development of a smart packaging for the monitoring of fish spoilage. *Talanta* 102: 466.

Pare, J.R.J., and J.M.R. Belenger, eds. 1997. *Instrumental Techniques in Food Analysis*. Amsterdam, The Netherlands: Elsevier.

Pathange, L.P., P. Mallikarjunan, R.P. Marini, S. O'Keefe, and D. Vaughan. 2006. Non-destructive evaluation of apple maturity using an electronic nose system. *J. Food Eng.* 77: 1018–1023.

Pearce, T.C. 1997. Computational parallels between the biological olfactory pathway and its analogue "the electronic nose": Part I. Biological olfaction. *BioSystems* 41: 43–67.

Persaud, K. 2005. Polymers for chemical sensing. *Mater. Today* 8: 38–44.

Persaud, K., and G. Dodds. 1982. Analysis of discrimination mechanisms in the mammalian olfactory system using a model nose. *Nature* 299: 352–355.

Pinalli, R., F. Nachtingall, F. Ugozzoli, and E. Dalcanale. 1999. Supramolecular sensors for the detection of alcohols. *Angew. Chem. Int. Ed.* 38: 2377–2380.

Ponzoni, A., A. Depari, M. Falasconi, E. Comini, A. Flammini, D. Marioli, A. Taroni, and G. Sberveglieri. 2008. Bread baking aromas detection by low-cost electronic nose. *Sens. Actuators B* 130: 100–104.

Rakow, N., and K. Suslick. 2000. A colorimetric sensor array for odour visualization. *Nature* 406: 710.

Röck, F., N. Barsan, and U. Weimar. 2008. Electronic nose: Current status and future trends. *Chem. Rev.* 108: 705–725.

Saevels, S., J. Lammertyn, A. Berna, E. Veraverbeke, C. Di Natale, and B. Nicolaï. 2003. Electronic nose as a non-destructive tool to evaluate the optimal harvest date of apples. *Postharvest Biol. Technol.* 30: 3–14.

Sánchez-Montañés, M.A., J.W. Gardner, and T.C. Pearce. 2008. Spatiotemporal information in an artificial olfactory mucosa. *Proc. R. Soc. A* 464: 1057–1077.

Shaffer, R.E., S.L. Rose-Pehrsson, and R.A. McGill. 1999. A comparison study of chemical sensor array pattern recognition algorithms. *Anal. Chim. Acta* 384: 305–317.

Sicard, G., and A. Holley. 1984. A receptor cell responses to odorants: Similarities and differences among odorants. *Brain Res.* 292: 283–291.

Skotadis, E., J. Tang, V. Tsouti, and D. Tsoukalas. 2010. Chemiresistive sensor fabricated by the sequential ink-jet printing deposition of a gold nanoparticle and polymer layer. *Microelectron. Eng.* 87: 2258–2263.

Stella, R., J. Barisci, G. Serra, G. Wallace, and D. De Rossi. 2000. Characterization of olive oil by an electronic nose based on conducting polymers sensors. *Sens. Actuators B* 63: 1–9.

Stitzel, S.E., D.R. Stein, and D.R. Walt. 2003. Enhancing vapor sensor discrimination by mimicking a canine nasal cavity flow environment. *J. Am. Chem. Soc.* 125: 3684–3685.

Suman, M., G. Riani, and E. Dalcanale. 2007. MOS-based artificial olfactory system for the assessment of egg products freshness. *Sens. Actuators B* 125: 40–47.

Suslick, B., L. Feng, and K. Suslick. 2010. Discrimination of complex mixtures by a colorimetric sensor array: Coffee aromas. *Anal. Chem.* 82: 2067–2073.

Tikk, K., J.E. Haugen, H. Andersen, and M. Aaslyng. 2008. Monitoring of warmed-over flavour in pork using the electronic nose—Correlation to sensory attributes and secondary lipid oxidation products. *Meat Sci.* 80: 1254–1263.

Tozawa, H., K. Enokihara, and K. Amano. 1971. Proposed modification of Dyer's method for trimethylamine determination in cod fish. *Fish Inspection and Quality Control*, p. 187. London: Fishing News Books.

Vestergaard, J., J.E. Haugen, and D. Byrne. 2006. Application of an electronic nose for measurements of boar taint in entire male pigs. *Meat Sci.* 74: 564–577.

Winquist, F. et al. 1983. Modified palladium metal-oxide semiconductor structure with increased ammonia gas sensitivity. *Appl. Phys. Lett.* 43: 839.

Wolde, S., M. Sjostrom, and L. Eriksson. 2001. PLS-regression: A basic tool of chemometrics. *Chemom. Intell. Lab. Syst.* 58: 109–130.

Zheng, X., Y. Lan, J. Zhu, J. Westbrook, W.C. Hoffmann, and R.E. Lacey. 2009. Rapid identification of rice samples using an electronic nose. *J. Bionic Eng.* 6: 290–297.

Index

Page numbers followed by *f* and *t* indicate figures and tables, respectively.

F

G

H

I

M

N

O

Milton Keynes UK
Ingram Content Group UK Ltd.
UKHW031137141024
449569UK00024B/1256